Biofuels and Biorefineries

Volume 5

Aims and Scope of the Series

The Biofuels and Biorefineries Series aims at being a comprehensive and integrated reference for biomass, bioenergy, biofuels, and bioproducts. The Series provides leading global research advances and critical evaluations of methods for converting biomass into biofuels and chemicals. Scientific and engineering challenges in biomass production and conversion are covered that show technological advances and approaches for creating new bio-economies in a format that is suitable for both industrialists and environmental policy decision-makers.

The Biofuels and Biorefineries Series provides readers with clear and concisely-written chapters that are peer-reviewed on significant topics in biomass production, biofuels, bio-products, chemicals, catalysts, energy policy, economics and processing technologies. The text covers major fields in plant science, green chemistry, economics and economy, biotechnology, microbiology, chemical engineering, mechanical engineering and energy.

Series Description

Annual global biomass production is about 220 billion dry tons or 4,500 EJ, equivalent to 8.3 times the world's energy consumption in 2014 (543 EJ). On the other hand, world-proven oil reserves at the end of 2011 reached 1652.6 billion barrels, which can only meet 54.2 years of global production. Therefore, alternative resources are needed to both supplement and replace fossil oils as the raw material for transportation fuels, chemicals and materials in petroleum-based industries. Renewable biomass is a likely candidate, because it is prevalent over the Earth and is readily converted to other products. Compared with coal, some of the advantages of biomass are: (i) its carbon-neutral and sustainable nature when properly managed; (ii) its reactivity in biological conversion processes; (iii) its potential to produce bio-oil (ca. yields of 75%) by fast pyrolysis because of its high oxygen content; (iv) its low sulphur and lack of undesirable contaminants (e.g. metals, nitrogen content) (v) its wide geographical distribution and (vi) its potential for creating jobs and industries in energy crop productions and conversion plants. Many researchers, governments, research institutions and industries are developing projects for converting biomass including forest woody and herbaceous biomass into chemicals, biofuels and materials and the race is on for creating new "biorefinery" processes needed for future economies. The development of biorefineries will create remarkable opportunities for the forestry sector, biotechnology, materials, chemical processing industry, and stimulate advances in agriculture. It will help to create a sustainable society and industries that use renewable and carbon-neutral resources.

More information about this series at http://www.springer.com/series/11687

Zhen Fang • Richard L. Smith, Jr. • Xinhua Qi
Editors

Production of Hydrogen from Renewable Resources

Springer

Editors
Zhen Fang
Chinese Academy of Sciences
Xishuangbanna Tropical Botanical Garden
Kunming, China

Richard L. Smith, Jr.
Tohoku University
Sendai, Japan

Xinhua Qi
Nankai University
Tianjin, China

ISSN 2214-1537 ISSN 2214-1545 (electronic)
Biofuels and Biorefineries
ISBN 978-94-017-7329-4 ISBN 978-94-017-7330-0 (eBook)
DOI 10.1007/978-94-017-7330-0

Library of Congress Control Number: 2015955605

Springer Dordrecht Heidelberg New York London

Printed on acid-free paper

Springer Science+Business Media B.V. Dordrecht is part of Springer Science+Business Media (www.springer.com)

Preface

As a clean energy carrier with high energy capacity, hydrogen has the potential to supplement or replace traditional fossil fuels in the near future. The use of renewable biomass resources from many different substrates for hydrogen production is receiving much attention for innovative and robust processes that demonstrate hydrogen production. This text provides state-of-the-art reviews, current research, and prospects of producing hydrogen by fermentation, electrochemical, bioelectrochemical, gasification, pyrolysis, and solar techniques from many possible biomass resources. Hydrogen separation, storage, and applications are also covered.

This book is the fifth book of the series entitled *Biofuels and Biorefineries*, and it contains 12 chapters contributed by leading experts in the field. The text is arranged into four key areas:

Part I: Bioconversion (Chapters 1, 2, and 3)
Part II: Thermoconversion (Chapters 4, 5, 6, and 7)
Part III: Electrochemical and Solar Conversions (Chapters 8, 9, and 10)
Part IV: Separations and Applications with Fuel Cells (Chapters 11 and 12)

Chapter 1 focuses on the technological background of dark fermentative hydrogen production from lignocellulosic biomass. Chapter 2 gives an introduction to the feedstocks, primary technologies, feedstock pretreatment methods, microorganisms, fermenter, types and operational conditions for producing biohydrogen. Chapter 3 presents in vitro synthetic (enzymatic) pathways that can provide high-yield production of biohydrogen from carbohydrates and water. Chapter 4 focuses on the analysis of gasification processes and investigation on the possible options for hydrogen production from the product gas streams from a gasification process. Chapter 5 provides an overview of different catalytic routes for producing hydrogen from biomass via pyrolysis processes. Chapter 6 describes techniques for the production of hydrogen by hydrocarbon decarbonization. Chapter 7 focuses on innovative technology that generates hydrogen by supercritical water gasification of biomass-related compounds and various biomass sources such as glucose, cellulose, lignin, alcohols, industrial wastewaters and sewage sludge. Chapter 8 reviews the principles, status, and progress in the electrochemical hydrogen production process by water electrolysis such as low temperature alkaline electrolysis cells and polymer exchange membrane electrolysis cells. Chapter 9 deals with development of microbial electrolysis cell (MEC) technology and introduces critical factors affecting MEC performance, anodic biocatalysts and technical challenges, and provides perspectives and outlooks on hydrogen production from organic waste. Chapter 10 summarizes the current status of solar-aided hydrogen production technologies, with special emphasis on high temperature thermochemical concepts. Chapter 11 reviews the carbon dioxide and hydrogen sulfide-selective facilitated transport membranes for low-pressure and high-pressure applications. Chapter 12 covers recent findings on hydrogen generation from reforming processes of bioresources combined with membrane reactor technology.

The text provides current research and prospects of producing hydrogen by bio-, thermal, and electrochemical methods. Hydrogen separation, storage, and applications are also covered. The text should be of interest to students, researchers, academicians, and industrialists in the areas of energy, environmental and chemical sciences, engineering, resource development, biomass processing, sustainability, and the hydrogen economy.

Acknowledgments

First and foremost, we would like to cordially thank all the contributing authors for their great efforts in writing the chapters and insuring the reliability of the information given in their chapters. Their contributions have really made this project realizable.

Apart from the efforts of authors, we would also like to acknowledge the individuals listed below for carefully reading the book chapters and giving constructive comments that significantly improved the quality of many aspects of the chapters:

Prof Hidayet Argun, Pamukkale University, Turkey
Dr. Marcello De Falco, Università Campus Bio-Medico, Italy
Dr. Ángela Nuria Garcia, Universidad de Alicante, Spain
Prof. Ángel Pérez-Navarro Gómez, Polytechnic University of Valencia, Spain
Prof. John R. Grace, the University of British Columbia, Canada
Prof. Josephine M. Hill, University of Calgary, Canada
Prof. Jürgen Karl, Friedrich Alexander Universität Erlangen-Nürnberg, Germany
Dr. Jung Rae Kim, Pusan National University (PNU), Korea
Prof. Masaharu Komiyama, University of Yamanashi, Japan
Mrs. Efthymia-Ioanna Koytsoumpa, Mitsubishi Hitachi Power Systems Europe GmbH, Germany
Dr. Hai-Qing Lin, University at Buffalo, State University of New York, USA
Prof. G. Glenn Lipscomb, University of Toledo, Italy
Dr. Hong Liu, Oregon State University, USA
Prof. Tian-Gang Liu, Wuhan University, China
Dr. Youjun Lu, Xi'an Jiaotong University, China
Dr. Adélio Mendes, Universidade do Porto, Portugal
Mr. Theodoros Papadopoulos, Siemens AG, Corporate Technology Research and Technology Center, Germany
Dr. Alissara Reungsang, Khon Kaen University (KKU), Thailand
Prof. Theodore T. Tsotsis, University of Southern California, USA

Prof. Shigeyuki Uemiya, Gifu University, Japan
Prof. Krzysztof Urbaniec, Warsaw University of Technology, Poland
Dr. Francesca Varsano, Italian National Agency for New Technologies, Energy and Sustainable Economic Development (ENEA), Italy
Prof. Lionel Vayssieres, Xi'an Jiaotong University, China
Prof. Shao-Rong Wang, Shanghai Institute of Ceramics, Chinese Academy of Sciences, China
Prof. Zhi Wang, Tianjin University, China
Prof. Thomas Wetzel, Karlsruher Institut für Technologie (KIT), Germany
Dr. Walter Wukovits, Technische Universität Wien, Austria
Prof. Guo-Min Xiao, Southeast University, China
Dr. Yun-Fei Yan, Chongqing University, China
Prof. Shu-Zhong Zhan, South China University of Technology, China
Prof. Jian-Jiang Zhong, Shanghai Jiao Tong University, China

We are also grateful to Ms. Becky Zhao (senior editor) and Ms. Abbey Huang (editorial assistant) for their encouragement, assistance, and guidance during the preparation of the book.

Finally, we would like to express our deepest gratitude toward our families for their love, understanding, and encouragement, which helped us in the completion of this project.

June 19, 2015 in Kunming

(Zhen Fang)

June 19, 2015 in Sendai

(Richard L. Smith, Jr.)

June 19, 2015 in Tianjin

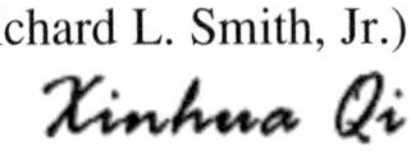

(Xinhua Qi)

Contents

Part I Bioconversion

1 **Dark Fermentative Hydrogen Production from Lignocellulosic Biomass** 3
Ioannis A. Panagiotopoulos

2 **Biohydrogen Production via Lignocellulose and Organic Waste Fermentation** 41
Chen-Yeon Chu and Bing-Shun Huang

3 **High-Yield Production of Biohydrogen from Carbohydrates and Water Based on In Vitro Synthetic (Enzymatic) Pathways** . . . 77
Jae-Eung Kim and Yi-Heng Percival Zhang

Part II Thermoconversion

4 **Hydrogen Production from Biomass Gasification** 97
Sotirios Karellas

5 **Hydrogen Production from Catalytic Biomass Pyrolysis** 119
Lucía García, Javier Ábrego, Fernando Bimbela, and José Luis Sánchez

6 **Low Carbon Production of Hydrogen by Methane Decarbonization** 149
Alberto Abánades

7 **Hydrogen Production by Supercritical Water Gasification of Biomass** 179
Ekin Kıpçak and Mesut Akgün

Part III Electrochemical and Solar Conversions

8 Hydrogen Production from Water and Air Through Solid Oxide Electrolysis 223
Kongfa Chen, Dehua Dong, and San Ping Jiang

9 Bioelectrochemical Production of Hydrogen from Organic Waste 249
In S. Kim, Euntae Yang, Mi-Jin Choi, and Kyu-Jung Chae

10 Solar Hydrogen Production 283
Athanasios G. Konstandopoulos, Chrysoula Pagkoura, Dimitrios A. Dimitrakis, Souzana Lorentzou, and George P. Karagiannakis

Part IV Separations and Applications with Fuel Cells

11 Separation and Purification of Hydrogen Using CO_2-Selective Facilitated Transport Membranes 315
Varun Vakharia and W.S. Winston Ho

12 Hydrogen Production for PEM Fuel Cells 339
Angelo Basile, Adolfo Iulianelli, Giuseppe Bagnato, and Francesco Dalena

Index 357

Contributors

Alberto Abánades Department of Energy Engineering, School of Industrial Engineering, Technical University of Madrid, Madrid, Spain

Javier Ábrego Thermochemical Processes Group (GPT), Aragón Institute for Engineering Research (I3A), Universidad de Zaragoza, Zaragoza, Spain

Mesut Akgün Chemical Engineering Department, SCFT-Supercritical Fluid Technologies Research Group, Yıldız Technical University, Davutpaşa Campus, Istanbul, Turkey

Giuseppe Bagnato Institute on Membrane Technology of the Italian National Research Council (ITM-CNR), University of Calabria, Rende, CS, Italy

Angelo Basile Institute on Membrane Technology of the Italian National Research Council (ITM-CNR), University of Calabria, Rende, CS, Italy

Fernando Bimbela Thermochemical Processes Group (GPT), Aragón Institute for Engineering Research (I3A), Universidad de Zaragoza, Zaragoza, Spain

Kyu-Jung Chae Department of Environmental Engineering, Korea Maritime and Ocean University, Busan, South Korea

Kongfa Chen Department of Chemical Engineering, Fuels and Energy Technology Institute, Curtin University, Perth, WA, Australia

Mi-Jin Choi Ulsan Development Institute, Ulsan, South Korea

Chen-Yeon Chu Green Energy Development Center, Feng Chia University, Taichung, Taiwan, Republic of China

Francesco Dalena Chemistry and Chemical Technologies Department, University of Calabria, Rende, CS, Italy

Dimitrios A. Dimitrakis Aerosol and Particle Technology Laboratory, Center for Research and Technology Hellas, Thessaloniki, Greece

Department of Chemical Engineering, Aristotle University, Thessaloniki, Greece

Dehua Dong Department of Chemical Engineering, Fuels and Energy Technology Institute, Curtin University, Perth, WA, Australia

Lucía García Thermochemical Processes Group (GPT), Aragón Institute for Engineering Research (I3A), Universidad de Zaragoza, Zaragoza, Spain

W.S.Winston Ho William G. Lowrie Department of Chemical and Biomolecular Engineering, The Ohio State University, Columbus, OH, USA

Department of Materials Science and Engineering, The Ohio State University, Columbus, OH, USA

Bing-Shun Huang Green Energy Development Center, Feng Chia University, Taichung, Taiwan, Republic of China

Adolfo Iulianelli Institute on Membrane Technology of the Italian National Research Council (ITM-CNR), University of Calabria, Rende, CS, Italy

San Ping Jiang Department of Chemical Engineering, Fuels and Energy Technology Institute, Curtin University, Perth, WA, Australia

George P. Karagiannakis Aerosol and Particle Technology Laboratory, Center for Research and Technology Hellas, Thessaloniki, Greece

Sotirios Karellas Laboratory of Steam Boilers and Thermal Plants, School of Mechanical Engineering, National Technical University of Athens, Athens, Greece

In S. Kim School of Environmental Science and Engineering, Gwangju Institute of Science and Technology, Gwangju, South Korea

Jae-Eung Kim Biological Systems Engineering Department, Virginia Tech, Blacksburg, VA, USA

Ekin Kıpçak Chemical Engineering Department, SCFT-Supercritical Fluid Technologies Research Group, Yildiz Technical University, Davutpaşa Campus, Istanbul, Turkey

Athanasios G. Konstandopoulos Aerosol and Particle Technology Laboratory, Center for Research and Technology Hellas, Thessaloniki, Greece

Department of Chemical Engineering, Aristotle University, Thessaloniki, Greece

Souzana Lorentzou Aerosol and Particle Technology Laboratory, Center for Research and Technology Hellas, Thessaloniki, Greece

Chrysoula Pagkoura Aerosol and Particle Technology Laboratory, Center for Research and Technology Hellas, Thessaloniki, Greece

Ioannis A. Panagiotopoulos School of Chemical Engineering, National Technical University of Athens, Zografou Campus, Athens, Greece

José Luis Sánchez Thermochemical Processes Group (GPT), Aragón Institute for Engineering Research (I3A), Universidad de Zaragoza, Zaragoza, Spain

Varun Vakharia William G. Lowrie Department of Chemical and Biomolecular Engineering, The Ohio State University, Columbus, OH, USA

Euntae Yang School of Environmental Science and Engineering, Gwangju Institute of Science and Technology, Gwangju, South Korea

Yi-Heng Percival Zhang Biological Systems Engineering Department, Virginia Tech, Blacksburg, VA, USA

Cell Free Bioinnovations Inc., Blacksburg, VA, USA

Tianjin Institute of Industrial Biotechnology, Chinese Academy of Sciences, Tianjin, China

About the Editors

Zhen Fang is leader and founder of biomass group, Xishuangbanna Tropical Botanical Garden, Chinese Academy of Sciences. He is also an adjunct full professor of Life Sciences, University of Science and Technology of China. He is the inventor of the "fast hydrolysis" process. He is listed in the "2014 Most Cited Chinese Researchers" in energy (Elsevier-Scopus). Professor Fang specializes in thermal/biochemical conversion of biomass, nanocatalyst synthesis and its applications, and pretreatment of biomass for biorefineries. He obtained his PhDs from China Agricultural University (Biological and Agricultural Engineering, 1991, Beijing) and McGill University (Materials Engineering, 2003, Montreal). Professor Fang is associate editor of Biotechnology for Biofuels and is serving on editorial boards of major international journals in energy.

Richard L. Smith, Jr. is professor of Chemical Engineering, Graduate School of Environmental Studies, Research Center of Supercritical Fluid Technology, Tohoku University, Japan. Professor Smith has a strong background in physical properties and separations and obtained his PhD in chemical engineering from the Georgia Institute of Technology (USA). His research focuses on developing green chemical processes especially those that use water and carbon dioxide as the solvents in their supercritical state. He has expertise in physical property measurements and in separation techniques with ionic liquids and published has more than 200 scientific papers, patents, and reports in the field of chemical engineering.

Professor Smith is the Asia regional editor for the *Journal of Supercritical Fluids* and has served on editorial boards of major international journals associated with properties and energy.

Xinhua Qi is professor of Environmental Science, Nankai University, China. Professor Qi obtained his PhD from the Department of Environmental Science, Nankai University, China. Professor Qi has a strong background in environmental treatment techniques in water and in chemical transformations in ionic liquids. His research focuses on the catalytic conversion of biomass into value-added chemicals and biofuels with green solvents. Professor Qi has published more than 60 scientific papers, books, and reports with a number of papers being in top-ranked international journals.

Part I
Bioconversion

Chapter 1
Dark Fermentative Hydrogen Production from Lignocellulosic Biomass

Ioannis A. Panagiotopoulos

Abstract This chapter focuses on the technological background of dark fermentative hydrogen production from lignocellulosic biomass. Firstly, the utilization of biomass with various conversion technologies for biofuel production is described, and specific biological hydrogen production processes are highlighted. Then, the basic principles of dark hydrogen fermentation and the current technology of the process are outlined. Among the different processes for biological hydrogen production, dark fermentation is advantageous because it is relatively inexpensive and it has low energy demands. The influence of the key process parameters on dark fermentative hydrogen production is summarized and, subsequently, the use of lignocellulosic raw materials for hydrogen production is examined in detail. Pretreatment of lignocellulosic biomass is typically required prior to fermentative hydrogen production because most hydrogen-producing microorganisms cannot directly utilize cellulose or hemicellulose as a carbon source to grow and produce hydrogen. In particular, the influence of biomass type and pretreatment method on hydrogen yield and productivity is discussed. Moreover, the potential of the coproducts to be further valorized toward the improvement of the economic profile of the process is explored. Finally, the challenges of dark fermentative hydrogen production from lignocellulosic biomass as well as the major conclusions drawn and the perspectives for further development are discussed.

Keywords Biological hydrogen production • Thermophilic fermentation • Agricultural residues • Energy crops • Pretreatment • Hydrolysis • Fermentable sugars • Inhibitors • Hydrogen yield

---

I.A. Panagiotopoulos (✉)
School of Chemical Engineering, National Technical University of Athens, Zografou Campus, Athens GR-15700, Greece
e-mail: iapan@chemeng.ntua.gr

Z. Fang et al. (eds.), *Production of Hydrogen from Renewable Resources*, Biofuels and Biorefineries 5, DOI 10.1007/978-94-017-7330-0_1

1.1 Introduction

Hydrogen is a gaseous biofuel that is a very important biofuel for the long-term future [1–3]. Today hydrogen is mainly produced from natural gas. However, the use of renewable resources such as biomass for hydrogen production [4–6] is expected to be developed [7]. Biomass is any organic material made from plants or animals. The use of biomass for hydrogen production has some general advantages such as the local availability of biomass, its renewability, the reduction in greenhouse gas emissions, and creation of new jobs [8]. In particular, lignocellulosic biomass, which includes agricultural and forestry residues, industrial wastes, and energy crops, has attracted high interest because of its abundance and its high content of sugars fermentable to hydrogen. Dark fermentation, commonly termed as "dark hydrogen fermentation" [9], is the process when the organic compounds, which break down toward hydrogen production, constitute the sole carbon and energy source providing metabolic energy.

This chapter focuses on the technological background of dark fermentative hydrogen (H_2) production from lignocellulosic biomass. Firstly, the utilization of biomass with various conversion technologies for biofuel production is described, and specific biological hydrogen production processes are discussed. Then, the basic principles of dark hydrogen fermentation and the current technology of the process are outlined (Sect. 1.2). Among the different processes for biological hydrogen production, dark fermentation is advantageous mainly because it is relatively inexpensive and it has low energy demands (Sect. 1.3). The influence of the key process parameters on dark fermentative H_2 production is summarized (Sect. 1.4). Subsequently, the use of lignocellulosic raw materials for hydrogen production is extensively discussed (Sects. 1.5, 1.6, and 1.7). Pretreatment of lignocellulosic biomass is typically required prior to fermentative hydrogen production because most hydrogen-producing microorganisms cannot directly utilize cellulose or hemicellulose as a carbon source to grow and produce hydrogen (Sect. 1.6). The type of pretreatment depends on the type and composition of the raw material. The role of pretreatment in the efficiency of hydrogen fermentation is reflected by the term "fermentability." The fermentability test is a rapid test that is applied to evaluate the potential for hydrogen production from a pretreated raw material. In particular, the influence of biomass type and pretreatment method on hydrogen yield and productivity is discussed (Sect. 1.7). Moreover, the potential of the coproducts to be further valorized toward the improvement of the economic profile of the process is explored (Sect. 1.8). Finally, the challenges of dark fermentative H_2 production from lignocellulosic biomass (Sect. 1.9) as well as the major conclusions drawn and the perspectives for further development (Sect. 1.10) are discussed.

1.2 Fundamentals of Dark Hydrogen Fermentations

When bacteria grow on organic substrates, these substrates are degraded by oxidation to provide building blocks and metabolic energy for growth. This oxidation leads to the production of electrons which need to be disposed of to maintain electrical neutrality. In aerobic environments, oxygen is reduced and water is produced. In anaerobic environments, other compounds act as electron acceptor and are reduced to molecular hydrogen. Although different organic compounds, such as carbohydrates, sugars, proteins, and lipids, enable the production of hydrogen during dark fermentation, estimations of potential yields are mostly based on hexose conversions. According to Eq. 1.1 the maximum theoretical yield of biohydrogen from glucose fermentation is 4 mol H_2 per mol of consumed glucose [10].

$$\begin{aligned} &C_6H_{12}O_6 + 4H_2O \rightarrow 2CH_3COO^- + 2HCO_3^- + 4H^+ + 4H_2 \\ &\Delta G'_0 = -206\ \mathrm{kJ/mol} \end{aligned} \tag{1.1}$$

Concurrent production of energy (206 kJ per mol of glucose) takes place and is sufficient to support microbial growth. The remainder of the hydrogen production from hexose is the byproduct acetate and, under non ideal circumstances, more reduced products like ethanol or lactate. The complete oxidation of glucose to H_2 and CO_2 yields a stoichiometry of 12 mol H_2 per mol of glucose, but in this case, no metabolic energy is obtained. In general, glucose fermentation can be accompanied by the generation of acetate, butyrate, and formate, as well as other, undesired byproducts such as lactate, propionate, ethanol, and butanol. These byproducts are typically produced in fermentations with mixed cultures. Their distribution mainly depends on the culture conditions. For example, it has been reported for *Escherichia coli* that for pH values below 7 the production of lactate is favored [11], thus resulting in reduced hydrogen yields.

Fermentative hydrogen can be produced either through mixed acidogenic microbial cultures, derived from natural environments such as soil or wastewater sludge, or through defined, pure cultures of hydrogen-producing microorganisms. Such microorganisms can be mesophilic (25–40 °C), thermophilic (40–65 °C), extreme thermophilic (65–80 °C), or hyperthermophilic (>80 °C), and they have quite diverse requirements not only in temperature but also in substrate preference, pH, and other parameters [12]. The main advantage of the use of mixed cultures is that no medium sterilization is required for the control and operation of the process, so the overall cost is reduced. The main advantages of the pure cultures are the easier manipulation of the metabolism with modification of the growth conditions and the higher hydrogen yields as a result of the reduced production of undesired coproducts.

There is a wide range of hydrogen-producing microorganisms, which include strict anaerobes (*Clostridia*, rumen bacteria, thermophiles, methanogens), facultative anaerobes (*Enterobacter, Escherichia coli, Citrobacter*), aerobes (*Alcaligenes,*

Bacillus), and co- and mixed cultures. *Clostridium butyricum* [13–15], *Clostridium beijerinckii* [16–18], *Clostridium tyrobutyricum* [19, 20], *Clostridium thermocellum* [21, 22], and *Clostridium paraputrificum* [23, 24] are examples of strict anaerobic microorganisms which produce hydrogen gas during the exponential growth phase. *Enterobacter aerogenes* [25–27] and *Enterobacter cloacae* [28, 29] have also been investigated for hydrogen production. The highest hydrogen yields have been obtained by hydrogen-producing extreme thermophilic anaerobic bacteria, with the strict anaerobic *Clostridia* producing hydrogen with higher yields than facultative anaerobes. The highest hydrogen production rates have been obtained by *Clostridia*, *Enterobacter,* and co- and mixed cultures. For the aforementioned reasons, among the hydrogen-producing bacteria, *Clostridia, Enterobacter,* and co- and mixed cultures are the most widely studied.

Thermophiles and in particular extreme thermophiles and hyperthermophiles are preferred for the production of hydrogen from lignocellulosic biomass because the increase of temperature in principle improves the reaction kinetics [30]. Another advantage of fermentations at elevated temperatures is that the process is less sensitive to contaminations by undesirable intruders, so a specific environment enabling maximum production of hydrogen is created. The main thermophiles that have been studied include *Caldicellulosiruptor saccharolyticus* [31–38], *Thermoanaerobacterium thermosaccharolyticum* [39–42], *Thermotoga neapolitana* [31, 33, 43–46, 90], and *Thermotoga maritima* [47, 48].

The growth of hydrogen-producing microorganisms can be inhibited with increased organic loadings [49, 50]; therefore, dilution of the raw waste is typically required in order to decrease the organic loading and thus prevent inhibition of the process. To extend the discussion, other parameters such as pH [51–53], temperature [54], hydraulic retention time (HRT) of the reactor [50, 55, 56], and gas stripping to avoid high partial hydrogen pressures [57–59] can influence the growth of the microorganisms, the hydrogen yield and productivity, and the economic profile of the process. The effects of the process parameters on various aspects of dark hydrogen fermentations are discussed in more detail in Sect. 1.4.

1.3 Advantages of Dark Hydrogen Fermentations

Fermentative processes for hydrogen production are advantageous against thermal processes such as gasification and supercritical water gasification. This happens because fermentative processes convert more efficiently biomass with high moisture content to hydrogen. Moreover, fermentative processes do not require large installations for economy of scale, so they can be small scale and cost effective. In particular, dark fermentative hydrogen production has many advantages against other options of biological hydrogen production. Dark fermentation can produce hydrogen all day long without the need of light. Moreover, dark fermentation systems have relatively simple construction, and operation presents low energy demands, mainly for mixing [9]. High hydrogen production rates are observed with dark fermentation, and under certain circumstances, the production can be maintained at non-aseptic conditions [60]. Another major advantage of dark fermentation is that the fermenting bacteria can utilize a wide range of organic substrates from hexoses, pentoses, and sucrose to starch, cellulose and pectin, and food wastes and municipal wastes [6, 61–63]. On the other hand, the greatest challenge for fermentative hydrogen production is that the hydrogen yield is low, due to thermodynamic barriers [64]. With increased yield, the hydrogen fermentation becomes thermodynamically unfavorable and the product gas mixture has to be separated from the contained CO_2 [65]. The challenges of dark hydrogen fermentations are discussed in detail in Sect. 1.9.

1.4 Effect of Process Parameters on Dark Hydrogen Fermentation

Temperature is one of the most important parameters affecting dark fermentative hydrogen production. Most studies have been based on mesophilic temperatures. However, increased H_2 yields and rates are observed with change in operational conditions from mesophilic to thermophilic temperatures. It has been suggested that the optimal temperature for fermentative hydrogen production seems to be either close to 37 °C or close to 55 °C [12]. However, the investigation of the effect of temperature on fermentative hydrogen production with real biomass substrates is recommended, given that so far mainly pure glucose and sucrose have been used. Some examples of raw materials that have been investigated in terms of the effect of temperature on fermentation are sugarcane bagasse [66] and wheat starch [67]. It should be noted that cultivation at high temperatures can be a viable strategy to inhibit hydrogen consumers and maximize hydrogen yield. However, the required energy for the operation of the reactor needs to be considerably less than the energy recovered as hydrogen in order to make the process economically sustainable.

One of the most studied parameters that influence dark fermentative hydrogen production is pH. The concentration of H^+ in the extracellular environment is used not only to control the directions of the metabolic pathways toward oxidized products but also to suppress the activity of methanogenic bacteria. Although there are large variations among different studies, it seems that the optimal pH for hydrogen production from carbohydrates is in the range of 5.2–7.0. The differences among various studies are mainly related to different raw materials, microbial populations, and operational conditions. In particular, an optimal pH of 5.5 has been found by studies which have used pure sugars [51, 68, 69]. An aciduric *E. aerogenes* strain HO-39 has been isolated and has been found able to grow and produce hydrogen at low pH of 4.5. Generally though, low H_2 yields are observed when the initial pH is below 5 [70, 71]. The aforementioned variations in optimal pH also exist when biomass hydrolyzates are used: Optimal initial pH for H_2 production from hydrolyzates has been found 6.5–7.0 with enrichment cultures from cow dung compost [72], 5.5 with *C. butyricum* [15], and 8.0 with dairy manure bacteria [70].

Mixed reactors for fermentative hydrogen production can operate in either batch or continuous mode. Batch-mode fermentative hydrogen production is typically performed at laboratory scale for research purposes. Batch-mode reactors offer ease of operation and flexibility. However, continuous bioprocesses are recommended for industrial-scale fermentative hydrogen production. In particular, a typical reactor for these purposes is the continuously stirred tank reactor (CSTR). The CSTR offers simple construction and simple control of pH and temperature. The most important operating parameter in a CSTR is the hydraulic retention time (HRT) of the reactor. Typically an HRT of 12–36 h provides complete conversion of carbohydrates and high hydrogen yields, avoiding the production of methane during dark hydrogen fermentation. A HRT of less than 6 h has been recommended to selectively washout the methanogens in continuous reactors [73]. At a high HRT product inhibition occurs due to accumulation of volatile fatty acids (VFAs).

The partial pressure of hydrogen (pH_2) is a very important parameter because high partial pressures of hydrogen inhibit hydrogen production. It is well known that thermophilic bacteria shift their end-product formation upon changes in partial pressure of hydrogen. This has been reported for *Caldicellulosiruptor saccharolyticus* and *Caldicellulosiruptor owensis* where increased pH_2 led to increased lactate formation [74–76]. One way to avoid high partial pressures is gas stripping [57, 77], which typically takes place with an inert gas such as N_2 [57]. CO_2 could also be an interesting stripping gas given that it is relatively easy to separate from hydrogen [58]. Regardless of the type of gas, the main disadvantage of the gas-stripping techniques is that the stripping gas dilutes the hydrogen content in the reactor and thus creates a further decrease in the separation efficiency. Membrane absorption techniques can be used for hydrogen removal from a gas mixture. However, the use of membrane-absorption techniques can result in the development of a biofilm which may favor the emergence of methanogenic bacteria.

Finally, hydrogen-producing microorganisms can have limited tolerance toward increased substrate loadings. The relation between substrate loading and hydrogen yield and productivity is discussed in Sect. 1.7.

1.5 Lignocellulosic Biomass Sources

Lignocellulosic biomass, which includes agricultural, forestry, agroindustrial residues, and energy crops, can be utilized for biological hydrogen production [78–80]. It is considered as a sustainable source of fermentable substrates due to its abundance and low cost. The collection, transport, and processing of lignocellulosic biomass pose significant challenges to its use in hydrogen production, as well as to their use in other types of energy production. Concerning the technical suitability of a lignocellulosic raw material for hydrogen production, the carbohydrate content and hydrogen fermentability of the raw material are the critical factors. Moreover,

Fig. 1.1 Sweet sorghum (**a**) on the field, (**b**) after removal of grains and leaves, (**c**) bagasse. Sweet sorghum bagasse before (**d**) and after (**e**) pretreatment with sodium hydroxide. The pretreatment with 10 % NaOH (w/w dry matter) resulted in a delignification of 46 % (Photo **a** by A. Glynos, Peloponnesos, Greece; photos **b**, **c** by I. Panagiotopoulos, National Technical University of Athens, Greece; photos **d**, **e** by I. Panagiotopoulos, Wageningen UR Food & Biobased Research, The Netherlands)

the characterization of the yield and the value of the coproduct obtained from the pretreatment step of the process is critical for the selection of the raw material (see Sect. 1.8). The type of pretreatment typically depends on the type and composition of the raw material. Raw materials such as energy crops, agricultural residues, and agroindustrial residues have attracted the highest attention among other raw material types for dark hydrogen fermentation. In particular, some representative examples of pretreated energy crops and energy crops residues that have been examined for dark hydrogen fermentation are *Miscanthus* [31], sweet sorghum bagasse (Fig. 1.1c) [32, 81], sugarcane bagasse [15], and reed canary grass [82]. Representative examples of pretreated agricultural residues are corn stover [40, 83, 84], corn cob [70, 85], wheat straw [37, 86, 87], barley straw [35, 85, 88], and rice straw [89–91]. It should be noted that although energy crops and agricultural residues have been tested for use as a substrate in fermentative hydrogen production, the exploration of the use of energy crop residues has only recently started. Finally, carrot pulp [45], potato steam peels [33], sugar beet molasses [92], and cotton-seed cake [93] are examples of agroindustrial residues with mixed chemical composition that have been investigated for dark hydrogen fermentation.

1.6 Methods of Lignocellulosic Biomass Pretreatment for Dark Hydrogen Fermentations

Biomass pretreatment is needed to make the available carbohydrates accessible for hydrogen fermentation. The degree of carbohydrate accessibility depends on the type of the raw material. In the case of sugary biomass, most of the sugars are readily fermentable, so the pretreatment is relatively simple. In case of raw materials with "mixed" chemical composition, such as wheat bran or sugar beet pulp, the pretreatment is usually more complex, often requiring heat and chemicals. Cellulose is by far more difficult to pretreat compared to starch [94], so in the case of typical cellulose-rich materials, such as agricultural or forestry residues, pretreatment is necessary and various pretreatment methods have been extensively described in order to promote the accessibility of polysaccharides in complex biomass. Typically, the pretreatment step is the first step of the biomass-to-hydrogen process, followed by enzymatic hydrolysis and one or more (see Sect. 1.9) fermentation steps (Fig. 1.2). The step of hydrolysis can be performed with concentrated acids, but currently enzymatic hydrolysis is typically performed with cellulases and hemicellulases.

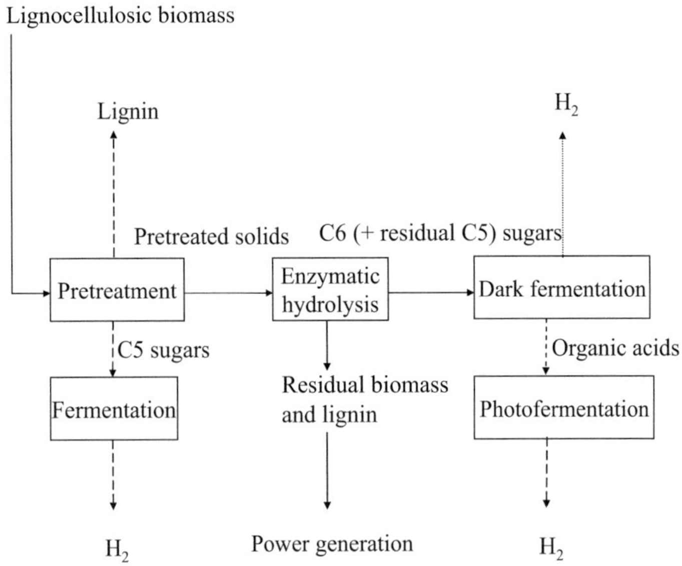

Fig. 1.2 A schematic outline of a biohydrogen production-based pretreatment process. The *dashed arrows* on lignin and C5 sugars represent the potential of the pretreatment for the fractionation of biomass into lignin, to be used as a substitute for polymeric materials, and C5 sugars, to be used for hydrogen fermentation. This potential depends on the pretreatment method. The *dashed arrows* on photofermentation represent the potential to couple dark fermentation with photofermentation in order to achieve increased hydrogen yields (see Sect. 1.9)

The pretreatments can be classified into four main groups: mechanical (including milling and grinding), thermal (including steam explosion and liquid hot water), chemical (including acids, alkali, ionic liquids, oxidizing compounds, etc.), and/or their combinations [95, 96]. Below we present the main characteristics of some selected pretreatment methods, and we discuss the key work that has been reported so far and the potential of each method toward hydrogen production.

Mechanical pretreatment leads to a reduction in particle size, which results in an increase of the available specific surface and a reduction in the degree of polymerization [97]. It is a method of high energy costs due to high power consumption, and it has not been used so far for biohydrogen production. Considering the need to decrease the cost of biomass pretreatment for the economically viable production of biohydrogen [98], it is likely that mechanical pretreatments are not economically feasible. It should be noted that pretreatments that combine mechanical pretreatment with chemical pretreatment (see text below) are also common.

Steam pretreatment is one of the most widely used methods for the pretreatment of lignocellulosic biomass mainly toward ethanol production [99–101]. It is a method which is expected to be incorporated into many of the first commercial biorefineries that are based on the hydrolysis procedure. Although the method is sometimes referred to as "steam explosion" describing the disruption of the biomass with the rapid decrease of pressure, it has been shown that explosion is not required to achieve good pretreatment and that the main mechanism is rather similar to dilute-acid hydrolysis [102]. In steam pretreatment, chips of biomass are conveyed into large vessels and high-pressure steam is applied (at 200–240 °C) for several minutes. At a set time, some steam is rapidly vented from the reactor to reduce the pressure, and the contents are discharged into a large vessel to flash cool the biomass [103, 104]. Iogen (Ottawa, Canada) has developed their own steam pretreatment process and along with Inbicon (Kalundborg, Denmark) and Abengoa (Salamanca, Spain) have used the process in a demonstration-scale ethanol plant, indicating that steam pretreatment could be close to commercialization.

Steam pretreatment has been shown to be particularly effective in the processing of agricultural residues and hardwoods such as corn stover and poplar, respectively. Under certain conditions and when sugar yields are considered, steam pretreatment can be a useful pretreatment method for softwood as well [105, 106]. However, high severities of steam pretreatment, usually reflected by high temperatures of 200 °C or more, are typically required in order to solubilize the hemicellulose of the material and achieve high (>75 %) cellulose to glucose conversions after subsequent enzymatic hydrolysis. These high temperatures result in the undesirable phenomenon of lignin condensation [107] which does not allow complete cellulose conversion mainly due to sorption of the cellulase enzymes [108]. To solve this problem, the application of various post-treatments has been suggested [109–111], but obviously, this additional step increases the chemical requirements of the overall pretreatment process without contributing to the efficient fractionation of the biomass into its valuable components of hemicellulose, cellulose, and lignin.

Steam pretreatment has been applied for biohydrogen production with corn stover [83], corn straw [71], and cornstalks [112], giving good results. Given that with steam pretreatment the hemicelluloses are solubilized thus making the cellulose better accessible for enzymatic hydrolysis, high C5 and C6 sugar yields are achieved with the method. This is in favor of applying steam pretreatment for biohydrogen production because hydrogen-producing microorganisms have been shown to be able to consume both C5 and C6 sugars. In particular, hydrogen-producing microorganisms, such as the extreme thermophilic anaerobic bacterium *Caldicellulosiruptor saccharolyticus* (see Sects. 1.6 and 1.7), consume simultaneously glucose and xylose [31, 32, 61], when these sugars are present in ratios which can be expected in hydrolysates from lignocellulosic biomass. Other microorganisms, such as *T. neapolitana*, seem to prefer glucose over xylose at the aforementioned sugar ratios [31]. Notable is that *C. saccharolyticus* seems to have a preference for xylose if both sugars are present at equal concentration [113]. Moreover, one of the concerns about using steam pretreatment for biofuel production through fermentative pathways is the generation of fermentation inhibitory compounds (see details further below) [114, 115]. However, early research work shows that hydrogen-producing microorganisms, such as *C. saccharolyticus* [31, 33, 74] and *C. beijerinckii* [116], are more tolerant to fermentation inhibitors compared to typical ethanol-producing microorganisms [114, 117, 118], so steam pretreatment could be more efficiently used for cellulosic biohydrogen production rather than cellulosic ethanol production. It should be noted that the inhibition phenomena in dark hydrogen fermentations have not been elucidated yet.

Pretreatment with liquid hot water (LHW) is similar to steam pretreatment, but in this case, liquid hot water is used instead of steam. LHW offers the potential for high xylose yields and reduction in cellulose recalcitrance to enzymatic hydrolysis, while maintaining the levels of inhibitory compounds at low levels. However, the high energy demands of this method, which are likely not met with the possible increases in hydrogen yields [119], make LHW pretreatment rather problematic. The use of such a method for hydrogen fermentation purposes has not been reported so far.

Unlike steam pretreatment which is primarily focused on solubilization of the hemicellulose component, organosolv pretreatment acts by extracting the lignin component by scission reactions that facilitate its solubilization [120]. The relatively high purity and reactivity of the organosolv lignin indicate its potential as a feedstock for the production of higher value products such as adhesives, antioxidants, and carbon fibers [121]. The effective solubilization of much of the lignin and hemicellulose during organosolv pretreatment results in a highly accessible cellulosic fraction which is highly amenable to subsequent enzymatic hydrolysis, at low enzyme loadings, regardless of the type of biomass that has been treated [122, 123]. However, organosolv pretreatment has mainly been considered so far as a typical method for woody biomass (hardwood, softwood) rather than agricultural residues. This is reasonable because of the relatively low lignin content of most of the agricultural residues and of the cellulase-inhibitory effect [124] of the increased free phenolic groups in the substrate [125] due to the delignification. Organosolv pretreatment for hydrogen production purposes has not been studied yet and is expected to attract more attention with the consideration of woody biomass as raw material for dark fermentative hydrogen production.

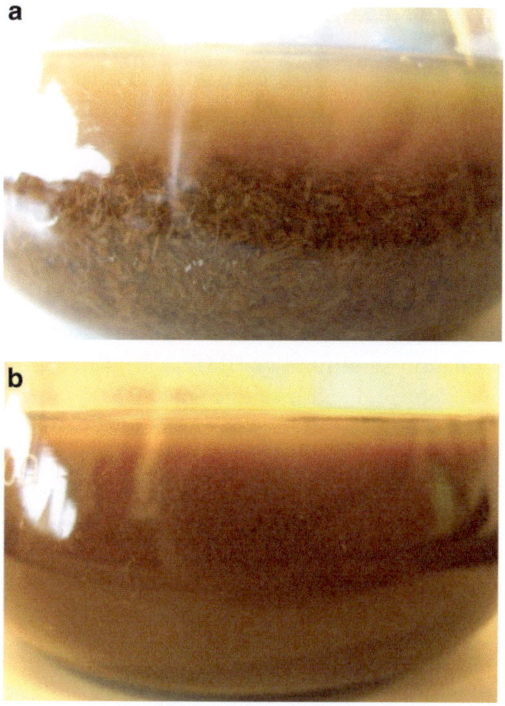

Fig. 1.3 Barley straw (**a**) before dilute-acid pretreatment and (**b**) after dilute-acid pretreatment (Photos by Ioannis Panagiotopoulos)

Acid pretreatment of lignocellulosic biomass has been studied extensively over the years [126, 127]. Concentrated acids have been traditionally used to treat lignocellulosic materials, but are toxic and corrosive and their use disturbs the economic feasibility of the biofuel production [128]. Although the use of concentrated acids has been mainly studied for ethanol production, the pretreatment of a hydrolysate of rice straw with concentrated sulfuric acid, resulting in low hydrogen yields, has been reported [129]. In general, dilute-acid pretreatment [130–132] has been more successful and can significantly improve the hydrolysis of polysaccharides, mainly hemicelluloses. The acid is mixed with the biomass at an acid concentration of 0.1–2.0 %, and the mixture is held at temperatures of 150–200 °C for periods ranging from seconds to minutes (Fig. 1.3). The hemicelluloses are consequently hydrolyzed. However, in addition to hydrolyzing the hemicelluloses, dilute-acid pretreatment releases substances that negatively affect the quality of the hydrolysates by decreasing their fermentability [31, 131, 133, 134]. These substances, known as fermentation inhibitors, include weak acids (mainly acetic acid), furans (mainly 5-hydroxymethylfurfural (HMF), a hexose degradation product and furfural, a pentose degradation product), and phenolic compounds such as vanillin, syringaldehyde, and catechol which are all formed by partial breakdown of lignin [135].

In general, fermentation inhibitors have toxic effects on the fermenting organisms, thus reducing the biofuel yield and productivity. The level of toxicity depends in part on fermentation variables including cell physiological conditions and pH of the medium. In addition, the fermenting organisms may, to some extent, be resistant to inhibitors or may become gradually adapted to their presence. Compounds inhibiting microbial activity can be either originally present in the biomass or produced due to too severe pretreatment conditions. The inhibitors that are present in lignocellulosic biomass are liberated relatively easy during the pretreatment step. For instance, under relatively mild conditions, acetylated hemicellulose is hydrolyzed, and beside monomeric sugars, the organic acid acetic acid is liberated. Inhibitors that are produced under too extreme pretreatment conditions generally consist of degradation compounds of sugars. For example, acidic conditions combined with high temperatures during lignocellulose pretreatment often lead to formation of furfural or HMF [136]. Both furfural and HMF are further degraded to formic acid and levulinic acid.

Acetic acid is derived from the acetyl groups in hemicellulose. At low pH in the fermentation medium, the acetic acid is in the undissociated form and may diffuse into the cells. Once in the cell, the acid dissociates causing a lowering of cell pH that inhibits cell activity. The toxicity varies according to the fermentation conditions. Since the formation of acetic acid is inherent to hemicellulose hydrolysis, its formation cannot be prevented. However, a higher fermentation pH can reduce this effect or the acid can be neutralized before fermentation.

Furfural and HMF affect cell growth and respiration. While HMF is considered less toxic than furfural and its concentration in (hemi)cellulose hydrolysates is usually low, it is generally considered that extensive degradation of (hemi)cellulose is responsible for the formation of the latter inhibitor compounds. Kinetic studies have shown that the production of furfural strongly increases with temperature and reaction time [137]. In contrast to undissociated weak organic acids which penetrate microbial cells and decrease the intracellular pH, furfural derivatives interfere with fermentative enzymes and disturb the membrane integrity of diverse microorganisms. Nevertheless, both furfural and HMF (at 1 g/L) have been reported to stimulate the growth of *C. beijerinckii* BA101 as well as the production of acetone-butanol-ethanol through non-H_2-producing pathways [138]. A higher decrease in H_2 yield of *T. thermosaccharolyticum* W16 with HMF (50.2 %) than with furfural (17.9 %) at an initial concentration of 1 g/L has been reported [139]. The study of the relationship between the concentration of HMF and furfural in various lignocellulosic hydrolysates and hydrogen fermentability [85] indicated that the inhibition of fermentation might be caused not only by typical fermentation inhibitors, such as HMF and furfural, but also by inhibiting compounds either originally present in the biomass or by other, undetermined potential inhibitors, such as phenolic compounds from lignin (Fig. 1.4).

Phenolic compounds have a considerable inhibitory effect on fermentation and are generally more toxic than furfural and HMF, even at low concentrations. In particular, phenolic compounds are toxic to microbial cells because they cause partition and loss of integrity of cell membranes of the fermenting organisms reducing cell growth and sugar assimilation. At very low concentrations, some compounds may improve fermentation kinetics, whereas at higher concentrations

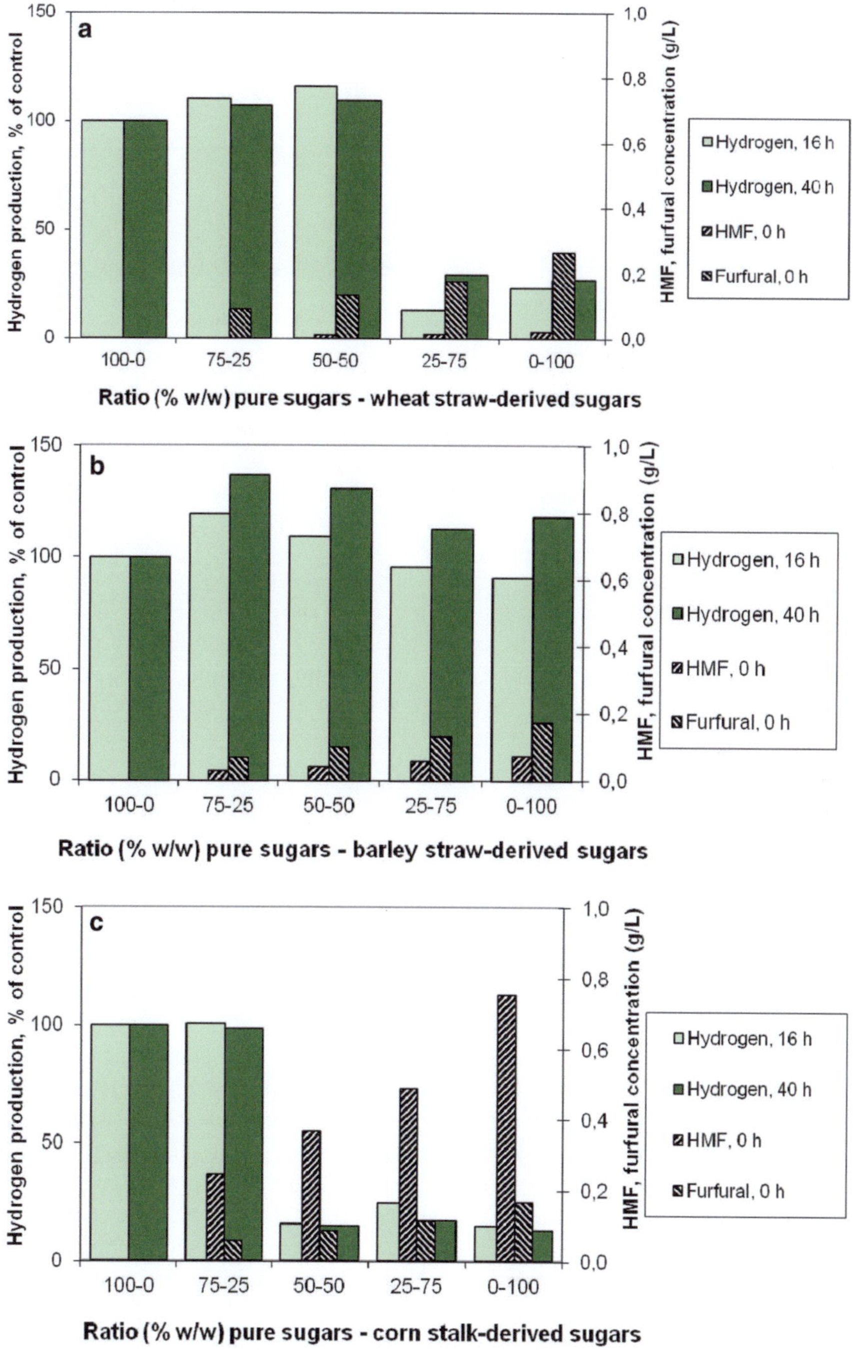

Fig. 1.4 Effect of the concentration of HMF and furfural on the fermentability of (**a**) wheat straw, (**b**) barley straw, (**c**) corn stalk, and (**d**) corn cob for hydrogen production. *Note*: The data on hydrogen are based on measurements after 16 and 40 h of fermentation. The data on HMF and furfural are based on measurements at the start of the fermentation (Reprinted with permission from Panagiotopoulos et al. [85], Copyright © 2011 The Japan Institute of Energy)

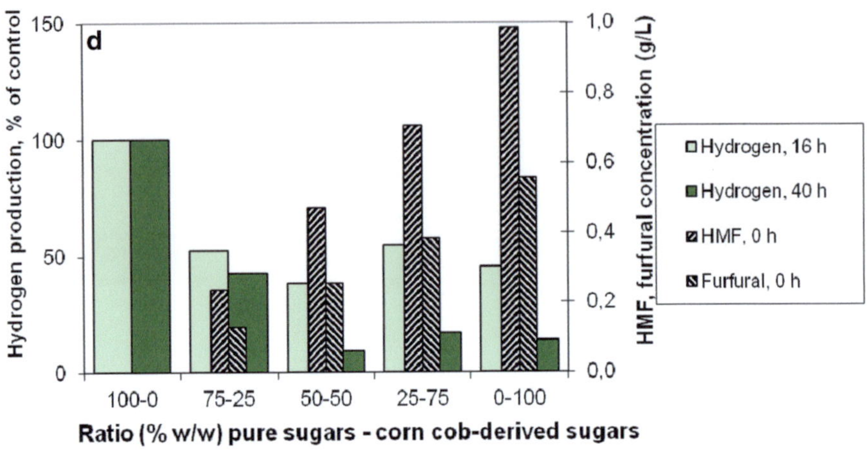

Fig. 1.4 (continued)

they lead to inhibition. According to Quéméneur et al. [87] who used mixed cultures, the addition of phenolic compounds (phenol, syringaldehyde, and vanillin) has less impact on the H_2 yields than the furan derivatives. In particular, the addition of syringaldehyde at the concentration of 1 g/L resulted in only 23 % inhibition in hydrogen production. On the other hand, Cao et al. [139], who used *T. thermosaccharolyticum* W16, showed a reduction in H_2 production of 54 % with syringaldehyde at 1 g/L. Ezeji et al. [138] reported a higher inhibition of *C. beijerinckii* cell growth by phenol than by syringaldehyde (at a concentration of 1 g/L).

The use of the ratio of Σsugars/Σinhibitors (Σs/Σi) has been proposed for the evaluation of the suitability of barley straw hydrolysates for hydrogen fermentation purposes [131]. The term Σsugars includes glucose, xylose, and arabinose, but it depends on the type of lignocellulosic material. For example, if wood is pretreated, then mannose or galactose may also need to be considered. However, it should be noted that the method of dilute-acid pretreatment seems to be advantageous for agricultural residues. The term Σinhibitors includes acetic acid, HMF, furfural, levulinic, and formic acid. Given that the formation of acetic acid occurs, to some extent, naturally from the hydrolysis of the acetyl groups in the hemicelluloses of the straw and thus its formation cannot be prevented, the investigation of the Σs/Σi can be performed either with or without consideration of acetic acid. The formation of levulinic acid and formic acid takes place with further degradation of HMF. Moreover, formic acid is formed when furfural breaks down [140]. With use of the tool of Σs/Σi, the dilute-acid pretreatment of barley straw has been optimized at the conditions of 170 °C and 60 min [131]. Under these conditions the production of undesired inhibitors is 2.1 g acetic acid/L, 0.1 g HMF/L, 0.4 g furfural/L, 0.0 g levulinic acid/L, and 0.0 g formic acid/L, which should not be problematic for hydrogen production.

Various types of agricultural residues have been dilute-acid pretreated toward the production of biohydrogen. Nguyen et al. [90] used dilute sulfuric acid to pretreat rice straw for biohydrogen production with *T. neapolitana*. Cao et al. [40] used dilute sulfuric acid to pretreat corn stover for biohydrogen production with *T. thermosaccharolyticum* W16. In this study the production of furanic compounds (0.9 g/L), phenolic compounds (0.1 g/L), and acetate (1.9 g/L) was reported. Furfural (0.6 g/L) and HMF (0.3 g/L) were the main furanic compounds. The main phenols in the hydrolysate were vanillin and syringaldehyde, which were present at very low concentrations. Sodium acetate at concentrations up to 6 g/L showed stimulatory effect on H_2 production by *T. thermosaccharolyticum* W16. On the other hand, furfural and HMF and potentially vanillin and syringaldehyde were inhibitors of growth and hydrogen production by *T. thermosaccharolyticum* W16, depending on their concentrations. Synergistic effects were shown in combined inhibitors. This observation is in agreement with the results obtained from *Escherichia coli* fermentations carried out in the combinations of furfural with other aldehydes [141].

The fermentability test is a rapid test that is applied to evaluate the potential for hydrogen production from a raw material pretreated with dilute acid or other chemicals [6]. With the fermentability test, the pretreatment method can be rapidly adapted to improve the fermentability of the hydrolysate. Inhibition of fermentation is quantified by determining the concentration of the hydrolysate sugars at which 25 % inhibition occurs, compared to the control samples with sugars of analytical grade. The fermentability of four dilute-acid pretreated and hydrolyzed lignocellulosic raw materials (barley straw, wheat straw, corn stalk, corn cob) has been investigated with the fermentability test, indicating the increased fermentability of barley straw with hydrolysate sugar concentrations up to 20 g/L [85]. Similar experimental work was performed by Monlau et al. [132] who studied inhibition of hydrogen production with increasing volumes of dilute acid hydrolyzate from sunflower stalks. A sharp decrease of the hydrogen yield was observed from 2.0 to 0.0 mol H_2/mol hexose (C6) for volumes higher than 15 % of added hydrolyzate. However, it should be noted that the dilute acid hydrolysis of that work used an acid load of 4 % (w/w dry matter) at 170 °C for 1 h, and the degree of inhibition may depend on the specific chemical and thermal conditions of the pretreatment method.

Given that sulfuric acid seems to be effective as a pretreatment agent in producing substrates fermentable to hydrogen (Fig. 1.5), Panagiotopoulos et al. [85] investigated to which extent the dilute sulfuric acid pretreatment of agricultural residues for hydrogen production is raw material specific. In this study, barley straw, wheat straw, corn stalk, and corn cob were dilute-acid pretreated and fermented to hydrogen production with *Caldicellulosiruptor saccharolyticus*. With the use of typical pretreatment conditions, it was shown that dilute sulfuric acid pretreatment is more suitable for hydrogen fermentation with the straws, particularly with barley straw, compared with the corn residues (Fig. 1.4). It should be noted that the HMF and furfural concentrations of 1–2 g/L have been reported to result in 50 % inhibition of hydrogen production [31]. The release of acetic acid during dilute-acid pretreatment does not seem to be a barrier in hydrogen

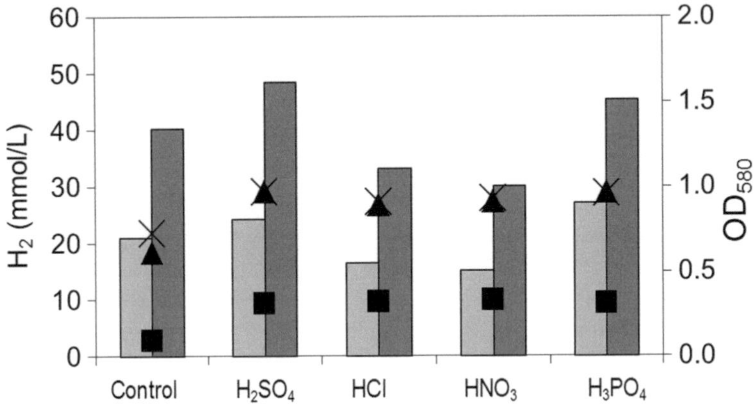

Fig. 1.5 Hydrogen production by culture of *C. saccharolyticus* grown on control medium with pure sugars (control) and on media with barley straw, pretreated with acid and enzymatically hydrolyzed. Measurements were done after 16 (*light gray bars*) and 40 (*dark gray bars*) h after the start of the fermentation. The growth of *C. saccharolyticus* (optical density measured at 580 nm, OD_{580}) after 0 (■), 16 (▲), and 40 (×) h of fermentation is also indicated (Figure slightly modified and reprinted with permission from Panagiotopoulos et al. [35], Copyright © 2012 International Association of Hydrogen Energy)

fermentations given that the acetic acid concentrations typically observed with dilute-acid pretreatment should be tolerable by hydrogen-producing microorganisms. For instance, the acetic acid concentration of 9 g/L was reported to lead to complete inhibition of the growth of *C. saccharolyticus* [33], being in agreement with a previous study of van Niel et al. [74]. *T. neapolitana* is even more tolerant, being able to grow at acetic acid concentrations as high as 18 g/L [33]. *C. saccharolyticus* has been found to tolerate acetic acid concentrations lower than 3 g/L in fermentations with real biomass substrates [6]. In conclusion, there is currently a lack of a clear relation between the concentration of sugar degradation products in the hydrolysates of dilute-acid pretreated biomass and hydrogen fermentability.

Although the research work on the comparison of the various pretreatment methods for biological hydrogen production is currently at an early stage, it should be noted that higher H_2 yields are more likely to be obtained by acid than other such as alkaline pretreatments [142]. However, this comparison needs to be raw material dependent, so the aforementioned statement should be considered with great care. Another aspect of dilute-acid pretreatment that has not been elucidated yet, though important, is the possible presence of sulfate remaining in the hydrolyzates. Sulfate may support the growth of sulfate-reducing bacteria that compete with hydrogen producers and consume the produced hydrogen [143]. Similarly, although the higher HCl concentration has been shown to be in favor of the conversion of lignocellulosic biomass to fermentable sugars, the high Cl^- anion concentration inhibits hydrogen production [72].

Alkaline pretreatment of lignocellulosic biomass for biofuel production, mostly bioethanol, has received significant attention. Alkaline pretreatments use sodium [32, 144, 145], calcium [146–148], potassium, and ammonium hydroxide as reactants. The extensive use of sodium hydroxide over calcium hydroxide is mostly due to the higher solubility of sodium hydroxide and the higher biomass digestibility achieved. In particular, treatment with NaOH causes swelling, leading to an increase in internal surface area, a decrease in the degree of polymerization, a decrease in crystallinity, separation of structural linkages between lignin and polysaccharides, and disruption of the lignin structure. De Vrije et al. [145] achieved a 77 % delignification of *Miscanthus* with addition of 12 % NaOH (w/w dry matter) to the biomass during extrusion at 70 °C, whereas Panagiotopoulos et al. [32] observed a 46 % delignification of sweet sorghum bagasse with 10 % NaOH (w/w dry matter) and milder pretreatment conditions (Fig. 1.1d, e). Maas [148] reported a 47 % delignification of wheat straw with 12 % NaOH while the delignification determined with up to 15 % lime was insignificant. When severe experimental conditions are applied, lime pretreatment can result in significant but still low delignification [146].

Alkaline pretreatment is more effective on agricultural residues than on wood materials. Lignin has a role on this because of its relatively low content in agricultural residues and high content in woody raw materials. Given that the efficient recovery of both lignin and hemicelluloses is a key feature of an ideal pretreatment [149, 150], it is important that alkaline pretreatment recovers lignin in a high-purity form and with a minimum use of water. The loss of hemicelluloses which are present in the non-fermentable lignin-rich liquor, known as dark liquor, is a challenge not only of alkaline pretreatment but also of organosolv pretreatment (see text above). The loss of hemicelluloses mainly depends on the type of alkali, the alkali level, and the type of biomass. For instance, the loss of hemicelluloses with a NaOH load of 12 % (w/w dry matter) has been reported to be 11–13 % [32, 148].

De Vrije et al. [31] investigated the fermentability of alkaline-pretreated *Miscanthus* hydrolysates for thermophilic hydrogen fermentation. They reported poor fermentability of lime-pretreated *Miscanthus* whereas fermentation of sodium hydroxide pretreatment hydrolysate yielded 3.2–3.3 mol H_2/mol hexose. These hydrogen yields are among the highest yields obtained in the fermentation of sugars in lignocellulosic hydrolysates reported to date (Table 1.1). These results are comparable with the results of Panagiotopoulos et al. [32] who used the fermentability test to adapt the severity of the alkaline pretreatment toward maximum hydrogen production and observed good fermentability of NaOH-pretreated sweet sorghum bagasse (Fig. 1.6). The aforementioned results suggest that the application of alkaline pretreatment has a high potential for enhanced hydrogen production in the future. One of the explanations of the good fermentability of the NaOH-pretreated biomass is that with mild pretreatment conditions at relatively low temperatures (70–75 °C), the generation of fermentation inhibitors is avoided, thus ameliorating the hydrogen fermentation.

Table 1.1 Hydrogen yields and production rates from lignocellulosic biomass

Biomass	Substrate	Microorganism	g sugar/L	Reactor operation mode	Y_{H2} mol/mol C6	Max. Q_{H2} mmol/(L • h)	Reference
Miscanthus	Glucose, xylose	*T. neapolitana*	14	Batch	3.2	12.3	[31]
Miscanthus	Glucose, xylose	*C. saccharolyticus*	14	Batch	3.3	10.4	[31]
Sweet sorghum bagasse	Glucose, xylose, sucrose	*C. saccharolyticus*	10	Batch	2.6	10.6	[32]
Sweet sorghum bagasse	Glucose, xylose, sucrose	*C. saccharolyticus*	20	Batch	1.3	10.2	[32]
Sweet sorghum bagasse	Glucose, xylose, arabinose	*R. albus*	3	Batch	2.6	–	[81]
Sugarcane bagasse	Glucose, xylose	*C. butyricum*	20[a]	Batch	1.7	1.6[b]	[15]
Wheat straw	Glucose, xylose, arabinose	*T. neapolitana*	5	Batch	2.6	1.2	[44]
Wheat straw	Glucose, xylose, arabinose	Mixed culture	3	Continuous	1.4	0.2[b]	[86]
Wheat straw	Glucose, xylose, arabinose	Mixed culture	3	Continuous	1.6	0.8[b]	[167]
Rice straw	Xylose	*C. butyricum*	9.2	Batch	0.8[c]	0.6[b]	[89]

Rice straw	Total sugar	Heat-treated sludge	20	Continuous	0.7	10.0[b]	[129]
Corn stover	Xylose	*T. thermosaccharolyticum*	9	Batch	2.2[c]	–	[40]
Corn stover	Glucose	Activated sludge	5	Batch	3.0	10.6	[83]
Corn stover	–	Activated sludge	1[a]	Batch	1.5	–	[168]
Carrot pulp	Glucose, fructose	*C. saccharolyticus*	10	Batch	2.8	15.7	[45]
Carrot pulp	Glucose, fructose	*T. neapolitana*	10	Batch	2.7	12.5	[45]
Wood fibers	Cellulose	*C. thermocellum*	1	Batch	1.5	–	[21]
Paper sludge	Glucose, xylose	*C. saccharolyticus*	8	Batch	2.0–2.1	5.3–6.0	[61]
Artificial medium	Glucose, xylose	*C. saccharolyticus*	10	Batch	3.0	17.0	[113]
Artificial medium	Cellulose	*T. neapolitana*	5	Batch	2.2	–	[169]
Artificial medium	Cellulose	*C. thermocellum*	4	Continuous	1.3	5.1	[170]

Data is based on extreme thermophilic fermentations except for Refs. [15, 83, 85, 91, 131] and Refs. [21, 142, 170, 172], which are based on mesophilic and thermophilic fermentations, respectively

[a] g COD/L

[b] L/L • d

[c] mol/mol C5

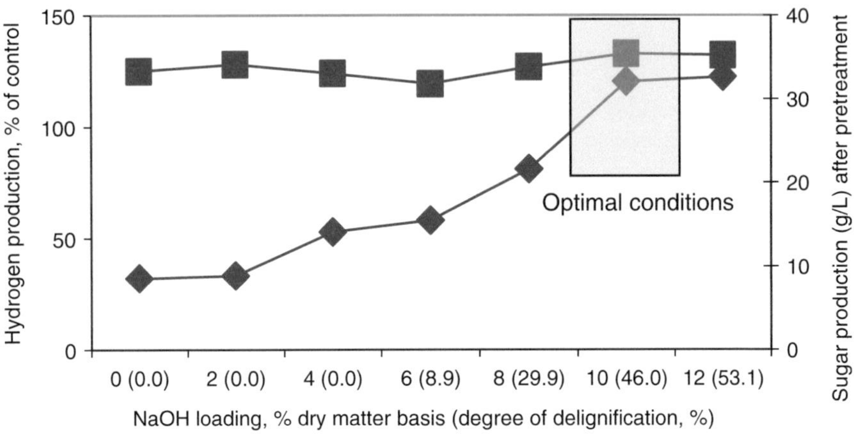

Fig. 1.6 Optimization of the pretreatment of sweet sorghum bagasse based on the chemical (NaOH) loading (and concomitant degree of delignification), the production of fermentable sugars, and the fermentability to hydrogen. In control fermentations pure sugars were used. Hydrogen production during fermentation, % of control (*square*). Sugar production (g/L) after pretreatment (*diamond*) (Unpublished material by Ioannis Panagiotopoulos)

The method of ionic liquids is a new concept in the fractionation of lignocellulosic biomass and has been applied so far for energy crops such as switchgrass [151, 152], agricultural residues such as sugarcane bagasse [153], and also woody biomass [154]. One of the challenges of ionic liquids, apart from their very high cost, is the stability and activity of cellulases in the presence of small amounts of ionic liquids co-precipitated with the recovered cellulose [155]. Recently the effect of the ionic liquids on the growth of *Saccharomyces cerevisiae* and bioethanol production was studied [156], but so far, no work has been reported on hydrogen-producing microorganisms. Therefore, no detailed discussion on ionic liquids is provided in the current chapter.

Wet oxidation is an effective method in fractionating cellulose from lignin and hemicelluloses. The main challenge of it is the loss of sugars which takes place due to non-selective oxidation. In case of wheat straw, the loss of hemicelluloses can be 40–45 % or higher, depending on the process conditions [157, 158]. This is high compared to the hemicelluloses loss occurred in, for example, alkaline pretreatment (see text above). Research efforts have focused on the optimization of the process conditions toward increased hemicelluloses recovery. For example, it has been shown that an increase in oxygen pressure results in an increase in hemicelluloses recovery [159, 160]. The non-selective oxidation also results in the oxidation of lignin which typically results in the formation of undesired aromatic compounds [158]. This phenomenon can result in inhibition in hydrogen fermentations and can also decrease the potential income from lignin at industrial scale for biohydrogen production in the biorefinery. In order to decrease the production of fermentation inhibitors, Martin et al. [161] used alkaline pH, which yet leads to reduced hemicelluloses recovery. Considering all the aforementioned characteristics of wet oxidation and the present status of its development, it is rather unlikely that wet oxidation will be suitable for biohydrogen production any time soon. Although wet oxidized wheat straw has been reported in the literature to produce hydrogen with *T. neapolitana* [44], the aforementioned technical characteristics of wet oxidation along with the high cost of oxygen render its application rather problematic.

1.7 Hydrogen Yields and Productivities from Lignocellulosic Hydrolysates

Two key parameters used to characterize the hydrogen production efficiency in controlled experiments with use of bioreactors are hydrogen yield, which is expressed as mol hydrogen per mol C6 sugar, and hydrogen productivity, which is typically used in its volumetric version and expressed as mmol hydrogen per L culture medium and h of fermentation. Typically, the maximum volumetric hydrogen productivity is calculated from the time interval with the highest percentage of hydrogen in the off gas. Given that the reproducibility of research works which employ undefined, mixed microflora is difficult, Table 1.1 mainly summarizes fermentation experiments which employed well-defined species. Moreover, Table 1.1 mainly includes data of representative studies where the two basic parameters of hydrogen production, yield and rate, are presented in mol H_2/mol hexose and mmol/(L • h), respectively. Only exceptionally Table 1.1 includes data from studies where hydrogen production was normalized as per, for example, chemical oxygen demand. It should be noted that for a fair comparison of the results, the hydrogen yields should always be reported as H_2 on hexose (mol/mol). Moreover, consideration of results should be performed with great care because hydrogen productivity and yield depend on the operating conditions as well as the raw material and the microorganism used.

At the present stage of development of hydrogen production from lignocellulosic biomass, pretreatment of biomass (and subsequent enzymatic hydrolysis) is required to bring the sugars in a soluble, fermentable form and thus result in efficient conversion of the sugars to hydrogen. Several studies on thermophilic bacteria growing on untreated raw materials have shown low yields compared with the yields observed with pretreated raw materials. According to Eq. 1.2, the maximum theoretical yield of biohydrogen is 4 mol H_2 per mol of glucose. This yield decreases with the generation of undesired byproducts which is not accompanied by hydrogen production. Barley hulls have been used without any pretreatment to produce hydrogen at a yield of 0.2 mol H_2/mol glucose [162]. Several studies on thermophilic bacteria growing on untreated wastewater cellulose have shown yields between 0.8 and 1.2 mol H_2/mol glucose [162, 163]. Co-culture studies of *C. thermocellum* and *T. saccharolyticum* on hydrogen production from microcrystalline cellulose resulted in 1.8 mol H_2/mol glucose [22]. The aforementioned hydrogen yields compare low against the hydrogen yields of 2–3 mol/mol glucose most typically observed with pretreated hydrolysates from lignocellulosic biomass. Mixed culture studies (35 and 50 °C) on corn stover pretreated with steam explosion and dilute sulfuric acid resulted in 2.8 mol H_2/mol glucose [83]. Acid hydrolysis of corn stover and subsequent fermentation of the hydrolysate by *T. saccharolyticum* resulted in 2.2 mol H_2/mol glucose [40]. The maximal hydrogen yield observed in batch experiments with *C. saccharolyticus* on sweet sorghum bagasse hydrolysates under controlled conditions was 2.6 mol/mol C6 sugar and the maximal volumetric hydrogen production rate ranged from 10.2 to 10.6 mmol/(L • h) [32]. At higher substrate concentrations, the production of lactic acid increased at the expense of hydrogen production (Fig. 1.7). This was also observed by Levin et al. [21], who used the thermophilic bacterium *C. thermocellum* in order to produce hydrogen from delignified wood fiber, and by de Vrije et al. [45] who used *C. saccharolyticus* in order to produce hydrogen from carrot pulp. The aforementioned trend was also observed in case of potato steam peels with *C. saccharolyticus* [33], as well as in case of waste ground wheat with use of anaerobic sludge [164]. However, fermentations with high concentrations of *Miscanthus* sugars did not result in high lactate production [31], so the phenomenon seems to be raw material specific [45].

As it can be seen in Table 1.1, thermophiles and in particular extreme thermophiles (65–80 °C) and hyperthermophiles (>80 °C) are preferred for the production of hydrogen from lignocellulosic biomass. This occurs because the increase of temperature in principle improves the reaction kinetics [30]. The main thermophiles that have been studied include *C. saccharolyticus* [31–36], *T. thermosaccharolyticum* [39, 141, 143], and *T. neapolitana* [31, 33, 44, 45, 88]. De Vrije et al. [31] used *C. saccharolyticus* and *T. neapolitana* in order to investigate the fermentation of alkaline-pretreated *Miscanthus* hydrolysates. They reported poor fermentability of lime-pretreated *Miscanthus* whereas fermentation of sodium hydroxide pretreatment hydrolysate yielded 3.2–3.3 mol H_2/mol hexose. These hydrogen yields are among the highest yields obtained in the fermentation of sugars in lignocellulosic hydrolysates reported to date (Table 1.1). With regard to hydrogen production rates, it should be noted that thermophiles usually grow at low densities and therefore typically show relatively low production rates. High cell densities, achieved through the development of reactor systems with physical retention of microbial biomass, are needed to maximize their hydrogen production rates.

Biological hydrogen production by a sequential operation of dark and photofermentation has attracted high research interest in the last 10 years [4, 165, 166]. In such a system, the anaerobic fermentation of the organic material produces organic intermediates, such as acetic, butyric, lactic, and propionic acids, Eq. 1.2, which can be readily fermented in the second step by photoheterotrophic bacteria, Eq. 1.3. The overall reactions of the process can be represented as:

$$C_6H_{12}O_6 + 2H_2O \rightarrow 2CH_3COOH + 2CO_2 + 4H_2 \qquad (1.2)$$

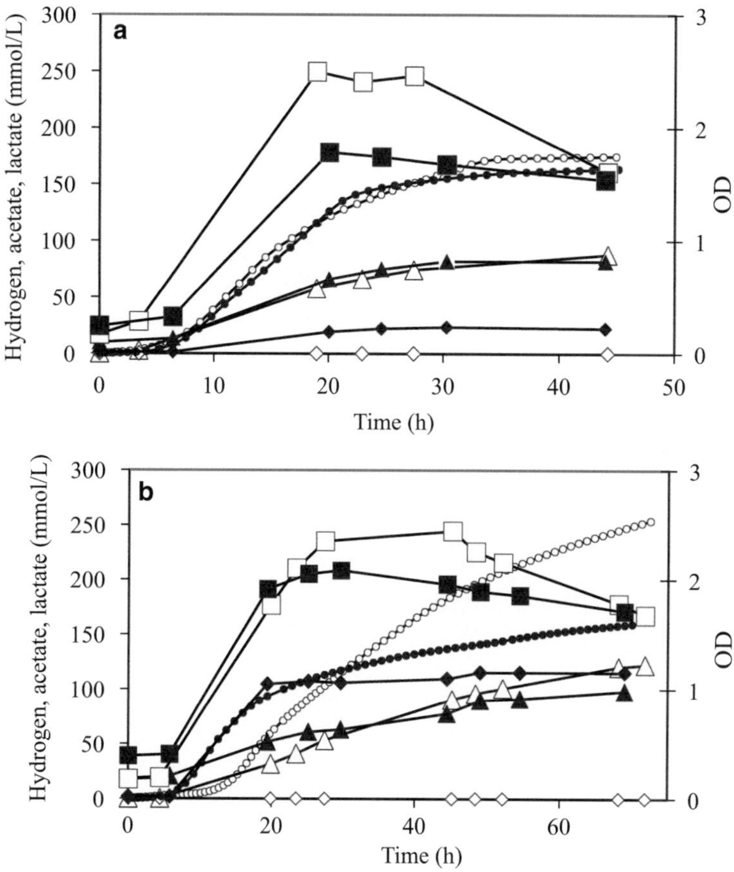

Fig. 1.7 Growth (*squares*), production of hydrogen (*circles*), acetate (*triangles*), and lactate (*diamonds*) in cultures of *C. saccharolyticus* grown on pure sugars (*open symbols*) and hydrolysate of sweet sorghum bagasse (*filled symbols*). The sum of glucose, xylose, and sucrose concentrations is (**a**) 10 g/L and (**b**) 20 g/L. The growth of *C. saccharolyticus* (optical density, OD) is also indicated (Reprinted with permission from Panagiotopoulos et al. [32], Copyright © 2010 International Association of Hydrogen Energy)

$$2CH_3COOH + 4H_2O \rightarrow 8H_2 + 4CO_2 \quad (1.3)$$

With use of glucose as the sole substrate in dark anaerobic fermentation, where acetic acid is the predominant metabolite product, a total of 12 mol hydrogen could be expected in a combined process from 1 mol of glucose. The detailed description of this combined process for biological hydrogen production is out of the scope of the present chapter.

1.8 Coproduct Valorization

Coproduct valorization is an important parameter of the technical and economic suitability of a lignocellulosic raw material for dark fermentative hydrogen production. Other parameters such as the water content of the raw material, hydrogen yield potential, sugar mobilization efficiency, and hydrogen fermentability also affect the technical suitability of a raw material for hydrogen production. Hydrogen yield potential is the maximum hydrogen yield based on stoichiometric hydrogen fermentation. Sugar mobilization efficiency is the percentage of all carbohydrates in the raw material that can be converted to fermentable sugars. Fermentability describes the tendency of a pretreated raw material to improve or inhibit fermentation. The part of the raw material which cannot be utilized for hydrogen production is characterized as coproduct. Coproduct valorization considers coproduct yield and value. In terms of value, some potential application fields of the non fermentable coproducts for each raw material are soil enhancement potential, energy value potential, animal feed potential, and higher value products potential. The coproduct yield is calculated based on the dry coproduct weight (in kg) per 100 kg dry raw material. Sugar-rich raw materials such as sugar beet juice and sweet sorghum juice show low coproduct yield mainly due to their low dry matter content. On the other hand, lignocellulosic raw materials such as wheat straw, barley straw, and *Miscanthus* show a high coproduct yield and value because their hydrolysis residues are rich in dry matter and lignin, respectively. It should be noted that the feasibility to purify the pretreatment/hydrolysis residue to a form suitable for the production of lignin-based biomaterials typically depends on the pretreatment type of a raw material (see Sect. 1.6). For example, the organosolv lignin has a relatively high extractability, purity, and reactivity, compared to lignins derived from acidic or alkaline pretreatments. This suggests that it has higher potential as a raw material for the production of higher value products [121, 171–174].

Apart from the valorization of pretreatment coproducts that cannot be utilized for hydrogen production, the valorization of agroindustrial coproducts, such as sugar beet pulp and wheat bran, is also interesting. Sugar beet pulp is a representative example of a coproduct which can be valorized toward the technoeconomic sustainability of dark fermentative hydrogen production. Sugar beet pulp is a coproduct from the sugar refining industry and contains 20–34 % hemicellulose, 19–27 % cellulose, 18–30 % pectin, and 1–2 % lignin [175–177]. The pulp is typically utilized for ruminant nutrition, but it has a relatively low protein content compared to the requirements of most ruminants, so it has to be supplemented with an extra protein source. Although no alternative industrial utilization of beet pulp in bulk quantities has taken place so far, its valorization has become an important field of research [178]. The anaerobic treatment of the pulp for biogas production has been investigated indicating that the energy from the produced biogas can have a significant contribution to the energy balance of a sugar beet factory [179]. In particular, this energy can cover up to 30 % of the daily energy consumption of the factory. Another example of sugar beet pulp utilization is its chemical treatment for the removal of copper from aqueous solutions [180]. Moreover, pectin gels from the pulp could be used for the biosorption of copper, cadmium, and lead [181].

Wheat bran is a coproduct of the industrial milling of wheat. During various milling operations, several side streams occur, which contain different sections of the whole wheat kernel and include the starchy endosperm (flour), the wheat germ, and the outer bran layers. The various side streams are combined into one, industrial-grade coproduct that is commonly referred to as wheat bran. Depending on the wheat variety and the milling process, wheat bran typically contains 12–26 % starch, 41–67 % non-starch polysaccharides, and up to 10 % lignin [182, 183]. Various biochemicals can be produced from wheat bran. The alkaline hydrogen peroxide extraction of the non-starch polysaccharides of wheat bran, mainly glucuronoarabinoxylans, has been studied [183]. The non-starch polysaccharides of wheat bran have applications as viscosity enhancers, emulsion or foam stabilizers, water absorbents, thickeners, gelling or filling agents, and fat replacers [183, 184]. Fermentative succinic acid production from wheat bran has also been studied [185]. Another use of wheat bran could be the production of biofuels. The production of hydrogen by mixed microflora from an untreated coproduct of white wheat flour milling has been investigated [186]. Moreover, the production of butanol from wheat bran with ABE (acetone, butanol, and ethanol) fermentation has been studied with use of *Clostridium beijerinckii* [187].

1.9 Challenges

The development of an effective biohydrogen production process at industrial scale is currently a challenge. The yields and rates of fermentative hydrogen production are not high enough to make the process economically viable. Based on known fermentation reactions, the theoretical maximum hydrogen yield is 4 mol H_2/mol glucose. Given that hydrogen production is more feasible at elevated temperatures [64, 188], H_2 yields are in general higher for (hyper)thermophiles, reaching easier the theoretical limit of 4 mol H_2/mol of hexose, compared to the mesophilic hydrogen producers [9, 64]. Regardless of the type of the microorganism, this theoretical maximum hydrogen yield can be achieved when the only electron sinks are H_2 and acetic acid, Eq. 1.2. With glucose as a substrate and acetic acid as the predominant metabolite product, 100 % conversion of the e^- equivalents of glucose to H_2 can increase H_2 yield to 12 mol H_2/mol glucose. One methodology to increase the conversion to biohydrogen is methanogenesis, which is a mature technology that can be coupled with fermentative hydrogen for the production of methane from acetate [189–191]. This coupling usually takes place with fermentations with mesophilic bacteria because these bacteria characteristically produce, besides hydrogen, other reduced intermediates such as acetate. Another option which has attracted high research interest is anoxygenic photosynthesis with purple non-sulfur bacteria. In such a system of sequential operation of dark and

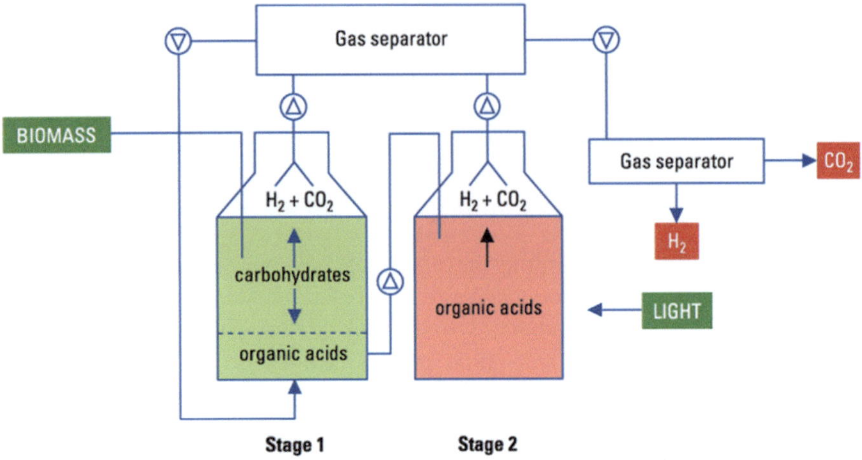

Fig. 1.8 Scheme of two-step fermentative hydrogen production from biomass (Drawing by courtesy of Pieternel Claassen and Truus de Vrije (Wageningen UR, The Netherlands))

photofermentation (Fig. 1.8), the bacteria can utilize the soluble organic fermentation products, such as acetic, Eq. 1.3, butyric, lactic, and propionic acids, as their electron donor for photosynthetic H_2 production [192–194].

The research on the sequential operation of dark and photofermentation has mainly focused so far on the use of pure sugars as the substrate. The observed yields with the two-step fermentations have been generally found to be higher than single step fermentations, ranging from 3 to 7 mol H_2/mol hexose. Some initial efforts to produce hydrogen in two steps from sugars from biomass, such as potato steam peels [194] and cassava starch [195, 196], have been recently reported. The photofermentative hydrogen production on effluents of thermophilic dark fermentations on a potato steam peels hydrolysate and molasses in indoor; batch fermentations has been investigated [194]. *Caldicellulosiruptor saccharolyticus* was used in the dark fermentation step, and *Rhodobacter capsulatus* was used in the photofermentation step. The overall hydrogen yield of the two-step fermentations was higher than the yield of single-step dark fermentations. Addition of buffer and nutrients to dark fermentor effluents was found to improve the overall efficiency of hydrogen production.

One of the key challenges of the sequential operation of dark and photofermentation is the low photosynthetic efficiency of the second step of the process, because at even moderate light intensities the main part of captured light is dissipated as heat [197]. This implies that large surface areas are required resulting in increasing the total cost of hydrogen production. One main disadvantage of the light-dependent processes is the more complex design of reactors. Due to the need to maintain a suitable proportion between reactor surface area and volume when scaling up, research focus is expected to target toward the design of reactors, which ideally allow high light availability, homogeneous distribution of light, and high hydrogen production efficiency.

The hydrogen production efficiency and the energy efficiency of this coupled dark and photofermentation can be compared to steam reforming of biogas. The production efficiency of the coupled dark and photofermentation seems to be comparable to this of the reforming of biogas but its energy efficiency is lower [198]. To make the discussion even broader, a techno-economic comparison between the two-step biological production of hydrogen from barley straw and biological production of ethanol showed that the production cost of the biohydrogen process is 20 times higher than the ethanol process, mainly due to low hydrogen productivity [199]. Given the existing market competition of biohydrogen with other biofuels, it is expected that in the near future research efforts will focus on the sequential dark and photofermentation. This will allow biohydrogen production to further approach practical application in the long-term future.

Another major challenge in the development of dark fermentative hydrogen production from biomass is the minimization of the cost of biomass pretreatment which is required to bring the carbohydrates into a soluble, fermentable form. This is particularly important in the case of lignocellulosic biomass because its pretreatment typically requires heat and chemicals to an extent that makes the process economically unsustainable. Moreover, pretreatment of lignocellulosic biomass can result in the release of undesired, inhibitory compounds in the hydrolysate. Fermentation inhibitors typically include aliphatic acids, furan derivatives, and phenolic compounds from lignin (see Sects. 1.6 and 1.7 for details). Although hydrogen-producing microorganisms [31, 85] may be more tolerant to sugar degradation products compared to ethanol-producing microorganisms [117, 118, 200], the amounts of fermentation inhibitors in the hydrolysate will have to be kept at low levels. This requires the development of tailor-made pretreatment techniques that result in easily fermentable hydrolysates. It should be noted that at the present stage of development of dark hydrogen fermentations, the inhibition phenomena have not been elucidated.

1.10 Conclusions and Future Outlook

Dark fermentative hydrogen production from biomass is an attractive and promising method for renewable hydrogen production. Depending on the type of biomass, pretreatment is needed to make the available carbohydrates accessible for fermentation toward hydrogen production. Sugar-rich raw materials are highly suitable for dark fermentative hydrogen production. However, sustainable hydrogen production needs to be based on the use of lignocellulosic biomass, which needs relatively complicated pretreatment procedures to become amenable to hydrogen fermentation. Some pretreatment methods, such as steam pretreatment and dilute-acid pretreatment, seem to be advantageous against others, but it is unlikely that one method will be suitable for all lignocellulosic raw materials. This becomes even more important if it is realized that the chemical composition of a lignocellulosic raw material is not enough to predict and characterize its suitability for hydrogen production. The pretreatment methods that have been developed so far target primarily toward high sugar yield and secondarily toward high hydrogen yield. The future research on specific pretreatment for biohydrogen production will focus on achieving high hydrogen fermentability of the produced sugars. Moreover, the ideal pretreatment should limit the release of inhibitory compounds in the hydrolysate because some hydrogen-producing microorganisms are not very tolerant to these compounds.

In addition to the pretreatment step, dark fermentative hydrogen production from lignocellulosic hydrolyzates needs to be optimized. At present, most of the studies on hydrogen fermentation from lignocellulosic hydrolyzates have been conducted in batch mode. However, continuous reactor studies on hydrogen production from hydrolyzates are required. In general, future technological progress in dark fermentative hydrogen production will be based on research efforts in the development of optimal reactor configurations and operating strategies, as well as in the isolation and development of microorganisms with high hydrogen yields. Moreover, an alternative method to enable dark fermentative hydrogen production to meet the needs of large-scale production is the combination of dark fermentation with a second-step energy recovery from the effluent organics such as acetate. The second-step process can be either methanogenesis, which is technologically ready today, or photofermentation with purple non-sulfur bacteria. The sequential operation of dark and photofermentation has attracted high research interest and is expected to approach practical application in the long-term future.

Commercial production of hydrogen from lignocellulosic biomass is expected to take place in 2020–2040. The cost reduction which is needed for this commercialization will be achieved by reducing costs in biomass pretreatment, by optimizing the efficiency and rate of the fermentations in the thermo- and more importantly in the photo-bioreactors, and by developing low-cost gas upgrading procedures.

Acknowledgements The author wishes to thank all partners in HYVOLUTION Integrated Project (Commission of the European Communities, 6th Framework Programme, Sustainable Energy Systems, Contract Nr. 019825) for their fruitful collaboration during the project. In particular, I thank Emmanuel Koukios (NTUA, Greece) for the helpful discussions on biomass pretreatment and Rob Bakker, Truus de Vrije, and Pieternel Claassen (Wageningen UR, The Netherlands) for their close experimental collaboration and helpful discussions on hydrogen fermentability. Moreover, Sofia Papadaki (NTUA) and Lazaros Karaoglanoglou (NTUA) are acknowledged for their collaboration on the coproduct valorization work during the project.

References

1. Veziroglu TN, Barbir F. Hydrogen: the wonder fuel. Int J Hydrog Energy. 1992;17 (6):391–404.
2. Wald ML. Questions about a hydrogen economy. Sci Am. 2004;42:41–7.
3. Stolten D, editor. Hydrogen and fuel cells – fundamentals, technologies and applications. Weinheim: Wiley–VCH; 2010.

4. Claassen PAM, de Vrije T. Non-thermal production of pure hydrogen from biomass: HYVOLUTION. Int J Hydrog Energy. 2006;31(11):1416–23.
5. Martínez-Pérez N, Cherryman SJ, Premier GC, Dinsdale RM, Hawkes DL, Hawkes FR, Kyazze G, Guwy AJ. The potential for hydrogen-enriched biogas production from crops: scenarios in the UK. Biomass Bioenergy. 2007;31(2–3):95–104.
6. Panagiotopoulos IA, Bakker RR, Budde MAW, de Vrije T, Claassen PAM, Koukios EG. Fermentative hydrogen production from pretreated biomass: a comparative study. Bioresour Technol. 2009;100(24):6331–8.
7. Muradov NZ, Veziroglu TN. "Green" path from fossil-based to hydrogen economy: an overview of carbon-neutral technologies. Int J Hydrog Energy. 2008;33(23):6804–39.
8. Hoogwijk M, Faaij A, van den Broek R, Berndes G, Gielen D, Turkenburg W. Exploration of the ranges of the global potential of biomass for energy. Biomass Bioenergy. 2003;25 (2):119–23.
9. de Vrije T, Claassen PAM. Dark hydrogen fermentations. In: Reith JH, Wijffels RH, Barten H, editors. Bio-methane & bio-hydrogen: status and perspectives of biological methane and hydrogen production. The Hague: Smiet Offset, Dutch Biological Hydrogen Foundation; 2003. p. 103–23.
10. Thauer RK, Jungermann K, Decker K. Energy conservation in chemotrophic anaerobic bacteria. Bacteriol Rev. 1977;41(3):100–80.
11. Kirkpatrick C, Maurer ML, Oyelakin NE, Yoncheva YN, Mauer R, Slonczewski JL. Acetate and formate stress: opposite responses in the proteome of *Escherichia coli*. J Bacteriol. 2001;183(21):6466–77.
12. Wang J, Wan W. Factors influencing fermentative hydrogen production: a review. Int J Hydrog Energy. 2009;34(2):799–811.
13. Heyndrickx M, Vansteenbeeck A, de Vos P, de Ley L. Hydrogen gas production from continuous fermentation of glucose in a minimal medium with *Clostridium butyricum* LMG 1213tl. Syst Appl Microbiol. 1986;8(3):239–44.
14. Chong ML, Raha AR, Shirai Y, Hassan MA. Biohydrogen production by *Clostridium butyricum* EB6 from palm oil mill effluent. Int J Hydrog Energy. 2009;34(2):764–71.
15. Pattra S, Sangyoka S, Boonmee M, Reungsang A. Bio-hydrogen production from the fermentation of sugarcane bagasse hydrolysate by *Clostridium butyricum*. Int J Hydrog Energy. 2008;33(19):5256–65.
16. Jeong T-Y, Cha G-C, Yeom SH, Choi SS. Comparison of hydrogen production by four representative hydrogen-producing bacteria. J Ind Eng Chem. 2008;14(3):333–7.
17. Lin P-Y, Whang L-M, Wu Y-R, Ren W-J, Hsiao C-J, Li S-L, Chang J-S. Biological hydrogen production of the genus *Clostridium*: metabolic study and mathematical model simulation. Int J Hydrog Energy. 2007;32(12):1728–35.
18. Skonieczny MT, Yargeau V. Biohydrogen production from wastewater by *Clostridium beijerinckii*: effect of pH and substrate concentration. Int J Hydrog Energy. 2009;34 (8):3288–94.
19. Jo HJ, Lee DS, Park D, Park JM. Biological hydrogen production by immobilized cells of *Clostridium tyrobutyricum* JM1 isolated from food waste treatment process. Bioresour Technol. 2008;99(14):6666–72.
20. Mitchell RJ, Kim J-S, Jeon B-S, Sang B-I. Continuous hydrogen and butyric acid fermentation by immobilized *Clostridium tyrobutyricum* ATCC 25755: effects of the glucose concentration and hydraulic retention time. Bioresour Technol. 2009;100(21):5352–5.
21. Levin DB, Islam R, Cicek N, Sparling R. Hydrogen production by *Clostridium thermocellum* 27405 from cellulosic biomass substrates. Int J Hydrog Energy. 2006;31(11):1496–503.
22. Liu Y, Yu P, Song X, Qu Y. Hydrogen production from cellulose by co-culture of *Clostridium thermocellum* JN4 and *Thermoanaerobacterium thermosaccharolyticum* GD17. Int J Hydrog Energy. 2008;33(12):2927–33.

23. Evvyernie D, Yamazaki S, Morimoto K, Karita S, Kimura T, Sakka K, Ohmiya K. Identification and characterization of *Clostridium paraputrificum* M-21, a chitinolytic, mesophilic and hydrogen producing bacterium. J Biosci Bioeng. 2000;89(6):596–601.
24. Evvyernie D, Morimoto K, Karita S, Kimura T, Sakka K, Ohmiya K. Conversion of chitinouswastes to hydrogen gas by *Clostridium paraputrificum* M-21. J Biosci Bioeng. 2001;91(4):339–43.
25. Tanisho S, Ishiwata W. Continuous hydrogen production from molasses by the bacterium *Enterobacter aerogenes*. Int J Hydrog Energy. 1994;19(10):807–12.
26. Yokoi H, Ohkawara T, Hirose J, Hayashi S, Takasaki Y. Characteristics of hydrogen production by aciduric *Enterobacter aerogenes* strain HO-39. J Ferment Bioeng. 1995;80 (6):571–4.
27. Fabiano B, Perego P. Thermodynamic study and optimization of hydrogen production by *Enterobacter aerogenes*. Int J Hydrog Energy. 2002;27(2):149–56.
28. Kumar N, Das D. Enhancement of hydrogen production by *Enterobacter cloacae* IIT-BT 08. Process Biochem. 2000;35(6):589–93.
29. Kumar N, Ghosh A, Das D. Redirection of biochemical pathways for the enhancement of H_2 production by *Enterobacter cloacae*. Biotechnol Lett. 2001;23(7):537–41.
30. Levin DB, Carere CR, Cicek N, Sparling R. Challenges for biohydrogen production via direct lignocellulose fermentation. Int J Hydrog Energy. 2009;34(17):7390–403.
31. de Vrije T, Bakker RR, Budde MAW, Lai MH, Mars AE, Claassen PAM. Efficient hydrogen production from the lignocellulosic energy crop *Miscanthus* by the extreme thermophilic bacteria *Caldicellulosiruptor saccharolyticus* and *Thermotoga neapolitana*. Biotechnol Biofuels. 2009;2:12.
32. Panagiotopoulos IA, Bakker RR, de Vrije T, Koukios EG, Claassen PAM. Pretreatment of sweet sorghum bagasse for hydrogen production by *Caldicellulosiruptor saccharolyticus*. Int J Hydrog Energy. 2010;35(15):7738–47.
33. Mars AE, Veuskens T, Budde MAW, van Doeveren PFNM, Lips SJ, Bakker RR, de Vrije T, Claassen PAM. Biohydrogen production from untreated and hydrolyzed potato steam peels by the extreme thermophiles *Caldicellulosiruptor saccharolyticus* and *Thermotoga neapolitana*. Int J Hydrog Energy. 2010;35(15):7730–7.
34. VanFossen AL, Verhaart MRA, Kengen SMW, Kelly RM. Carbohydrate utilization patterns for the extremely thermophilic bacterium *Caldicellulosiruptor saccharolyticus* reveal broad growth substrate preferences. Appl Environ Microbiol. 2009;75(24):7718–24.
35. Panagiotopoulos IA, Bakker RR, de Vrije T, Claassen PAM, Koukios EG. Dilute-acid pretreatment of barley straw for biological hydrogen production using *Caldicellulosiruptor saccharolyticus*. Int J Hydrog Energy. 2012;37(16):11727–34.
36. Willquist K, van Niel EWJ. Growth and hydrogen production characteristics of *Caldicellulosiruptor saccharolyticus* on chemically defined minimal media. Int J Hydrog Energy. 2012;37(6):4925–9.
37. Panagiotopoulos IA, Bakker RR, de Vrije T, Claassen PAM, Koukios EG. Integration of first and second generation biofuels: fermentative hydrogen production from wheat grain and straw. Bioresour Technol. 2013;128:345–50.
38. Pawar SS, Nkemka VN, Zeidan AA, Murto M, van Niel EWJ. Biohydrogen production from wheat straw hydrolysate using *Caldicellulosiruptor saccharolyticus* followed by biogas production in a two-step uncoupled process. Int J Hydrog Energy. 2013;38(22):9121–30.
39. O-Thong S, Prasertsan P, Karakashev D, Angelidaki I. Thermophilic fermentative hydrogen production by the newly isolated *Thermoanaerobacterium thermosaccharolyticum* PSU-2. Int J Hydrog Energy. 2008;33(4):1204–14.
40. Cao G, Ren N, Wang A, Lee D-J, Guo W, Liu B, Feng Y, Zhao Q. Acid hydrolysis of corn stover for biohydrogen production using *Thermoanaerobacterium thermosaccharolyticum* W16. Int J Hydrog Energy. 2009;34(17):7182–8.

41. Cao G-L, Ren N-Q, Wang A-J, Guo W-Q, Yao J, Feng Y-J, Zhao Q-L. Statistical optimization of culture condition for enhanced hydrogen production by *Thermoanaerobacterium thermosaccharolyticum* W16. Bioresour Technol. 2010;101(6):2053–8.
42. Cao G-L, Zhao L, Wang A-J, Wang Z-Y, Ren N-Q. Single-step bioconversion of lignocellulose to hydrogen using novel moderately thermophilic bacteria. Biotechnol Biofuels. 2014;7:82.
43. Ngo TA, Nguyen TH, Bui HTV. Thermophilic fermentative hydrogen production from xylose by *Thermotoga neapolitana* DSM 4359. Renew Energy. 2012;37(1):174–9.
44. Eriksen NT, Riis ML, Holm NK, Iversen N. H_2 synthesis from pentoses and biomass in *Thermotoga* spp. Biotechnol Lett. 2011;33(2):293–300.
45. de Vrije T, Budde MAW, Lips SJ, Bakker RR, Mars AE, Claassen PAM. Hydrogen production from carrot pulp by the extreme thermophiles *Caldicellulosiruptor saccharolyticus* and *Thermotoga neapolitana*. Int J Hydrog Energy. 2010;35(15):13206–13.
46. Frascari D, Cappelletti M, Mendes JDS, Alberini A, Scimonelli F, Manfreda C, Longanesi L, Zannoni D, Pinelli D, Fedi S. A kinetic study of biohydrogen production from glucose, molasses and cheese whey by suspended and attached cells of *Thermotoga neapolitana*. Bioresour Technol. 2013;147:553–61.
47. Nguyen TAD, Kim JP, Kim MS, Oh YK, Sim SJ. Optimization of hydrogen production by hyperthermophilic eubacteria, *Thermotoga maritima* and *Thermotoga neapolitana* in batch fermentation. Int J Hydrog Energy. 2008;33(5):1483–8.
48. Nogales J, Gudmundsson S, Thiele I. An *in silico* re-design of the metabolism in *Thermotoga maritima* for increased biohydrogen production. Int J Hydrog Energy. 2012;37(17):12205–18.
49. Davila-Vazquez G, Alatriste-Mondragon F, de Leon-Rodriguez A, Razo-Flores E. Fermentative hydrogen production in batch experiments using lactose, cheese whey and glucose: influence of initial substrate concentration and pH. Int J Hydrog Energy. 2008;33(19):4989–97.
50. Ntaikou I, Kourmentza C, Koutrouli E, Stamatelatou K, Zampraka A, Kornaros M, Lyberatos G. Exploitation of olive oil mill wastewater for combined biohydrogen and biopolymers production. Bioresour Technol. 2009;100(15):3724–30.
51. van Ginkel S, Sung S, Lay JJ. Biohydrogen production as a function of pH and substrate concentration. Environ Sci Technol. 2001;35(24):4726–30.
52. Lee YJ, Miyahara T, Noike T. Effect of pH on microbial hydrogen fermentation. J Chem Technol Biotechnol. 2002;77(6):694–8.
53. Antonopoulou G, Gavala HN, Skiadas IV, Lyberatos G. Influence of pH on fermentative hydrogen production from sweet sorghum extract. Int J Hydrog Energy. 2010;35(5):1921–8.
54. Wang JL, Wan W. Effect of temperature on fermentative hydrogen production by mixed cultures. Int J Hydrog Energy. 2008;33(20):5392–7.
55. Arooj M-F, Han S-K, Kim S-H, Kim D-H, Shin H-S. Effect of HRT on ASBR converting starch into biological hydrogen. Int J Hydrog Energy. 2008;33(22):6509–14.
56. Wu SY, Hung CH, Lin CY, Lin PJ, Lee KS, Lin CN, Chang FY, Chang J-S. HRT-dependent hydrogen production and bacterial community structure of mixed anaerobic microflora in suspended, granular and immobilized sludge systems using glucose as the carbon substrate. Int J Hydrog Energy. 2008;33(5):1542–9.
57. Mizuno O, Dinsdale R, Hawkes FR, Hawkes DL, Noike T. Enhancement of hydrogen production from glucose by nitrogen gas sparging. Bioresour Technol. 2000;73(1):59–65.
58. Willquist K, Claassen PAM, van Niel EWJ. Evaluation of the influence of CO_2 on hydrogen production by *Caldicellulosiruptor saccharolyticus*. Int J Hydrog Energy. 2009;34(11):4718–26.
59. Ljunggren M, Willquist K, Zacchi G, van Niel EWJ. A kinetic model for quantitative evaluation of the effect of hydrogen and osmolarity on hydrogen production by *Caldicellulosiruptor saccharolyticus*. Biotechnol Biofuels. 2011;4:31.

60. Hallenbeck PC, Abo-Hashesh M, Ghosh D. Strategies for improving biological hydrogen production. Bioresour Technol. 2012;110:1–9.
61. Kádár Z, de Vrije T, van Noorden GE, Budde MAW, Szengyel Z, Réczey K. Yields from glucose, xylose, and paper sludge hydrolysate during hydrogen production by the extreme thermophile *Caldicellulosiruptor saccharolyticus*. Appl Biochem Biotechnol. 2004;114 (1–3):497–508.
62. Li CL, Fang HHP. Fermentative hydrogen production from wastewater and solid wastes by mixed cultures. Crit Rev Environ Sci Technol. 2007;37(1):1–39.
63. Guo XM, Trably E, Latrille E, Carrère H, Steyer J-P. Hydrogen production from agricultural waste by dark fermentation: a review. Int J Hydrog Energy. 2010;35(19):10660–73.
64. Kengen SWM, Goorissen HP, Verhaart MRA, Stams AJM, van Niel EWJ, Claassen PAM. Biological hydrogen production by anaerobic microorganisms. In: Soetaert W, Verdamme EJ, editors. Biofuels Soetaert W. Chichester: Wiley; 2009. p. 197–221.
65. Das D, Veziroglu TN. Advances in biological hydrogen production processes. Int J Hydrog Energy. 2008;33(21):6046–57.
66. Chairattanamanokorn P, Penthamkeerat P, Reungsang A, Lo YC, Lu WB, Chang JS. Production of biohydrogen from hydrolyzed bagasse with thermally preheated sludge. Int J Hydrog Energy. 2009;34(18):7612–7.
67. Cakir A, Ozmichi S, Kargi F. Comparison of bio-hydrogen production from hydrolyzed wheat starch by mesophilic and thermophilic dark fermentation. Int J Hydrog Energy. 2010;35(24):13214–128.
68. Fang H, Liu H. Effect of pH on hydrogen production from glucose by a mixed culture. Bioresour Technol. 2002;82(1):87–93.
69. Khanal SK, Chen WH, Li L, Sung S. Biological hydrogen production: effects of pH and intermediate products. Int J Hydrog Energy. 2004;29(11):1123–31.
70. Pan C, Zhang S, Fan Y, Hou H. Bioconversion of corncob to hydrogen using anaerobic mixed microflora. Int J Hydrog Energy. 2010;35(7):2663–9.
71. Li D, Chen H. Biological hydrogen production from steam exploded straw by simultaneous saccharification and fermentation. Int J Hydrog Energy. 2007;32(12):1742–8.
72. Fan YT, Zhang GS, Guo XY, Xing Y, Fan MH. Biohydrogen production from beer lees biomass by cow dung compost. Biomass Bioenergy. 2006;30(5):493–6.
73. Chen W, Chen S, Kumar Khanal S, Sung S. Kinetic study of biological hydrogen production by anaerobic fermentation. Int J Hydrog Energy. 2006;31(15):2170–8.
74. van Niel EWJ, Claassen PAM, Stams AJM. Substrate and product inhibition of hydrogen production by the extreme thermophile, *Caldicellulosiruptor saccharolyticus*. Biotechnol Bioeng. 2003;81(3):255–62.
75. Zeidan AA, van Niel EWJ. A quantitative analysis of hydrogen production efficiency of the extreme thermophile *Caldicellulosiruptor owensensis* OL^T. Int J Hydrog Energy. 2010;35 (3):1128–37.
76. Bielen AAM, Verhaart MRA, VanFossen AL, Blumer-Schuette SE, Stams AJM, van der Oost J, Kelly RM, Kengen SWM. A thermophile under pressure: transcriptional analysis of the response of *Caldicellulosiruptor saccharolyticus* to different H_2 partial pressures. Int J Hydrog Energy. 2013;38(4):1837–49.
77. Kraemer JT, Bagley DM. Improving the yield from fermentative hydrogen production. Biotechnol Lett. 2007;29(5):685–95.
78. Claassen PAM, van Lier JB, Contreras AML, van Niel EWJ, Sijtsma L, Stams AJM, de Vries SS, Weusthuis RA. Utilisation of biomass for the supply of energy carriers. Appl Microbiol Biotechnol. 1999;52(6):741–55.
79. Cheng CL, Lo YC, Lee KS, Lee DJ, Lin CY, Chang JS. Biohydrogen production from lignocellulosic feedstock. Bioresour Technol. 2011;102(18):8514–23.
80. Urbaniec K, Bakker RR. Biomass residues as raw material for dark hydrogen fermentation – a review. Int J Hydrog Energy. 2015;40(9):3648–58.

81. Ntaikou I, Gavala HN, Kornaros M, Lyberatos G. Hydrogen production from sugars and sweet sorghum biomass using *Ruminococcus albus*. Int J Hydrog Energy. 2008;33(4):1153–63.
82. Lakaniemi AM, Koskinen PEP, Nevatalo L, Kaksonen AH, Puhakka JA. Biogenic hydrogen and methane production from reed canary grass. Biomass Bioenergy. 2011;35(2):773–80.
83. Datar R, Huang J, Maness P-C, Mohagheghi A, Czernik S, Chornet E. Hydrogen production from the fermentation of corn stover biomass pretreated with a steam-explosion process. Int J Hydrog Energy. 2007;32(8):932–9.
84. Ren N, Wang A, Gao L, Xin L, Lee DJ, Su A. Bioaugmented hydrogen production from carboxymethyl cellulose and partially delignified corn stalks using isolated cultures. Int J Hydrog Energy. 2008;33(19):5250–5.
85. Panagiotopoulos IA, Bakker RR, de Vrije T, van Niel EWJ, Koukios EG, Claassen PAM. Exploring critical factors for fermentative hydrogen production from various types of lignocellulosic biomass. J Jpn I Energy. 2011;90:363–8.
86. Kongjan P, O-Thong S, Kotay M, Min B, Angelidaki I. Biohydrogen production from wheat straw hydrolysate by dark fermentation using extreme thermophilic mixed culture. Biotechnol Bioeng. 2010;105(5):899–908.
87. Quéméneur M, Bittel M, Trably E, Dumas C, Fourage L, Ravot G, Steyer J-P, Carrère H. Effect of enzyme addition on fermentative hydrogen production from wheat straw. Int J Hydrog Energy. 2012;37(14):10639–47.
88. Özgür E, Peksel B. Biohydrogen production from barley straw hydrolysate through sequential dark and photofermentation. J Clean Prod. 2013;52:14–20.
89. Lo Y-C, Lu W-C, Chen C-Y, Chang J-S. Dark fermentative hydrogen production from enzymatic hydrolysate of xylan and pretreated rice straw by *Clostridium butyricum* CGS5. Bioresour Technol. 2010;101(15):5885–91.
90. Nguyen T-AD, Kim K-R, Kim MS, Sim SJ. Thermophilic hydrogen fermentation from Korean rice straw by *Thermotoga neapolitana*. Int J Hydrog Energy. 2010;35(24):13392–8.
91. Chen C-C, Chuang Y-S, Lin C-Y, Lay C-H, Sen B. Thermophilic dark fermentation of untreated rice straw using mixed cultures for hydrogen production. Int J Hydrog Energy. 2012;37(20):15540–6.
92. Özgür E, Mars AE, Peksel B, Louwerse A, Yücel M, Gündüz U, Claassen PAM, Eroğlu I. Biohydrogen production from beet molasses by sequential dark and photofermentation. Int J Hydrog Energy. 2010;35(2):511–7.
93. Panagiotopoulos IA, Pasias S, Bakker RR, de Vrije T, Papayannakos N, Claassen PAM, Koukios EG. Biodiesel and biohydrogen production from cotton-seed cake in a biorefinery concept. Bioresour Technol. 2013;136:78–86.
94. Klass DL. Biomass as a nonfossil fuel source. Washington, DC: American Chemical Society; 1981.
95. Hendriks ATWM, Zeeman G. Pretreatments to enhance the digestibility of lignocellulosic biomass. Bioresour Technol. 2009;100(1):10–8.
96. Kumar P, Barrett DM, Delwiche MJ, Stroeve P. Methods for pretreatment of lignocellulosic biomass for efficient hydrolysis biofuel production. Ind Eng Chem Res. 2009;48(8):3713–29.
97. Palmowski L, Muller J (1999) Influence of the size reduction of organic waste on their anaerobic digestion. In: Proceedings of II international symposium on anaerobic digestion of solid waste, Barcelona. p. 137–44.
98. Hyvolution (2010) Final cost estimation of fully integrated process routes. Commission of the European Communities, HYVOLUTION Integrated Project, Deliverable 5.30 (unpublished).
99. Saddler JN, Gregg DJ (1997) Steam explosion and bioconversion of softwoods. In: Proceedings of the 1997 ethanol research and development workshop, Ottawa. p. 135–8.
100. Söderström J, Pilcher L, Galbe M, Zacchi G. Two-step steam pretreatment of softwood with SO_2 impregnation for ethanol production. Appl Biochem Biotechnol. 2002;98–100(1–9):5–21.

101. Linde M, Galbe M, Zacchi G. Steam pretreatment of acid-sprayed and acid-soaked barley straw for production of ethanol. Appl Biochem Biotechnol. 2006;130(1–3):546–62.
102. Brownell HH, Yu EKC, Saddler JN. Steam-explosion pretreatment of wood: effect of chip size, acid, moisture content and pressure drop. Biotechnol Bioeng. 1986;28(6):792–801.
103. Mosier N, Wyman C, Dale B, Elander R, Lee YY, Holtzapple M, Ladisch M. Features of promising technologies for pretreatment of lignocellulosic biomass. Bioresour Technol. 2005;96(6):673–86.
104. Hayes DJ. An examination of biorefining processes, catalysts and challenges. Catal Today. 2009;145(1–2):138–51.
105. Stenberg K, Tengborg C, Galbe M, Zacchi G. Optimisation of steam pretreatment of SO_2-impregnated mixed softwoods for ethanol production. J Chem Technol Biotechnol. 1998;71 (4):299–308.
106. Ewanick SM, Bura R, Saddler JN. Acid-catalyzed steam pretreatment of lodgepole pine and subsequent enzymatic hydrolysis and fermentation to ethanol. Biotechnol Bioeng. 2007;98 (4):737–46.
107. Li J, Henriksson G, Gellerstedt G. Lignin depolymerization/ repolymerization and its critical role for delignification of aspen wood by steam explosion. Bioresour Technol. 2007;98 (16):3061–8.
108. Shevchenko SM, Beatson RP, Saddler JN. The nature of lignin from steam explosion/ enzymatic hydrolysis of softwood. Structural features and possible uses. Appl Biochem Biotechnol. 1999;77–79(1–3):867–76.
109. Yang B, Boussaid A, Mansfield SD, Gregg DJ, Saddler JN. Fast and efficient alkaline peroxide treatment to enhance the enzymatic digestibility of steam-exploded softwood substrates. Biotechnol Bioeng. 2002;77(6):678–84.
110. Kumar L, Chandra R, Saddler J. Influence of steam pretreatment severity on post-treatments used to enhance the enzymatic hydrolysis of pretreated softwoods at low enzyme loadings. Biotechnol Bioeng. 2011;108(10):2300–11.
111. Panagiotopoulos IA, Chandra RP, Saddler JN. A two-stage pretreatment approach to maximise sugar yield and enhance reactive lignin recovery from poplar wood chips. Bioresour Technol. 2013;130:570–7.
112. Lu Y, Lai Q, Zhang C, Zhao H, Ma K, Zhao X, Chen H, Liu D, Xing X-H. Characteristics of hydrogen and methane production from cornstalks by an augmented two- or three-stage anaerobic fermentation process. Bioresour Technol. 2009;100(12):2889–95.
113. Zeidan AA, van Niel EWJ. Developing a thermophilic hydrogen producing co-culture for efficient utilization of mixed sugars. Int J Hydrog Energy. 2009;34(10):4524–8.
114. Oliva JM, Sáez F, Ballesteros I, González A, Negro MJ, Manzanares P. Effect of lignocellulosic degradation compounds from steam explosion pretreatment on ethanol fermentation by thermotolerant yeast *Kluyveromyces marxianus*. Appl Biochem Biotechnol. 2003;105 (1–3):141–53.
115. Cantarella M, Cantarella L, Gallifuoco A, Spera A, Alfani F. Effect of inhibitors released during steam-explosion treatment of poplar wood on subsequent enzymatic hydrolysis and SSF. Biotechnol Prog. 2004;20(1):200–6.
116. Quéméneur M, Hamelin J, Barakat A, Steyer J-P, Carrère H, Trably E. Inhibition of fermentative hydrogen production by lignocellulose-derived compounds in mixed cultures. Int J Hydrog Energy. 2012;37(4):3150–9.
117. Nigam JN. Ethanol production from wheat straw hemicellulose hydrolysate by *Pichia stipitis*. J Biotechnol. 2001;87(1):17–27.
118. Franden MA, Pienkos PT, Zhang M. Development of a high-throughput method to evaluate the impact of inhibitory compounds from lignocellulosic hydrolysates on the growth of *Zymomonas mobilis*. J Biotechnol. 2009;144(4):259–67.
119. Jung KW, Kim DH, Shin HS. Fermentative hydrogen production from *Laminaria japonica* and optimization of thermal pretreatment conditions. Bioresour Technol. 2011;102 (3):2745–50.

120. McDonough TJ. The chemistry of organosolv delignification. Tappi J. 1993;76:186–93.
121. Ragauskas AJ, Beckham GT, Biddy MJ, Chandra R, Chen F, Davis MF, Davison BH, Dixon RA, Gilna P, Keller M, Langan P, Naskar AK, Saddler JN, Tschaplinski TJ, Tuskan GA, Wyman CE. Lignin valorization: improving lignin processing in the biorefinery. Science. 2014;344(6185):1246843.
122. Pan X, Gilkes N, Kadla J, Pye K, Saka S, Gregg D, Ehara K, Xie D, Lam D, Saddler J. Bioconversion of hybrid poplar to ethanol and co-products using an organosolv fractionation process: optimization of process yields. Biotechnol Bioeng. 2006;94(5):851–61.
123. Huijgen WJJ, Smit AT, Reith JH, den Uil H. Catalytic organosolv fractionation of willow wood and wheat straw as pretreatment for enzymatic cellulose hydrolysis. J Chem Technol Biotechnol. 2011;86(11):1428–38.
124. Berlin A, Balakshin M, Gilkes N, Kadla J, Maximenko V, Kubo S, Saddler JN. Inhibition of cellulase, xylanase and β-glucosidase activities by softwood lignin preparations. J Biotechnol. 2006;125(2):198–209.
125. Sannigrahi P, Ragauskas AJ, Miller SJ. Lignin structural modifications resulting from ethanol organosolv treatment of loblolly pine. Energy Fuels. 2010;24(1):683–9.
126. McMillan JD. Pretreatment of lignocellulosic biomass. In: Himmel ME, Baker JO, Overend RP, editors. Enzymatic conversion of biomass for fuels production, vol. 566. Washington, DC: American Chemical Society; 1994. p. 292–324.
127. Sun Y, Cheng JJ. Dilute acid pretreatment of rye straw and bermudagrass for ethanol production. Bioresour Technol. 2005;96(14):1599–606.
128. Sivers MV, Zacchi G. A techno-economical comparison of three processes for the production of ethanol from pine. Bioresour Technol. 1995;51(1):43–52.
129. Liu CM, Chu CY, Lee WY, Li YC, Wu SY, Chou YP. Biohydrogen production evaluation from rice straw hydrolysate by concentrated acid pre-treatment in both batch and continuous systems. Int J Hydrog Energy. 2013;38(35):15823–9.
130. Saha BC, Iten LB, Cotta MA, Wu YV. Dilute acid pretreatment, enzymatic saccharification and fermentation of wheat straw to ethanol. Process Biochem. 2005;40(12):3693–700.
131. Panagiotopoulos IA, Bakker RR, de Vrije T, Koukios EG. Effect of pretreatment severity on the conversion of barley straw to fermentable substrates and the release of inhibitory compounds. Bioresour Technol. 2011;102(24):11204–11.
132. Monlau F, Aemig Q, Trably E, Hamelin J, Steyer J-P, Carrere H. Specific inhibition of biohydrogen-producing *Clostridium* sp. after dilute-acid pretreatment of sunflower stalks. Int J Hydrog Energy. 2013;38(28):12273–82.
133. Larsson S, Palmqvist E, Hahn-Hägerdal B, Tengborg C, Stenberg K, Zacchi G, Nilvebrant N-O. The generation of fermentation inhibitors during dilute acid hydrolysis of softwood. Enzyme Microb Technol. 1999;24(3–4):151–9.
134. Panagiotopoulos IA, Lignos G, Bakker RR, Koukios EG. Effect of low severity dilute-acid pretreatment of barley straw and decreased enzyme loading hydrolysis on the production of fermentable substrates and the release of inhibitory compounds. J Clean Prod. 2012;32:45–51.
135. Palmqvist E, Hahn-Hagerdal B. Fermentation of lignocellulosic hydrolysates. II: inhibitors and mechanisms of inhibition. Bioresour Technol. 2000;74(1):25–33.
136. Mussatto SI, Roberto IC. Alternatives for detoxification of diluted-acid lignocellulosic hydrolyzates for use in fermentative processes: a review. Bioresour Technol. 2004;93(1):1–10.
137. McKillip WJ, Collin G. Ullmann's encyclopedia of industrial chemistry. 6th ed. Weinheim: Wiley-VCH; 2002.
138. Ezeji T, Qureshi N, Blaschek HP. Butanol production from agricultural residues: impact of degradation products on *Clostridium beijerinckii* growth and butanol fermentation. Biotechnol Bioeng. 2007;97(6):1460–9.
139. Cao GL, Ren NQ, Wang AJ, Guo WQ, Xu JF, Liu BF. Effect of lignocellulose-derived inhibitors on growth and hydrogen production by *Thermoanaerobacterium thermosaccharolyticum* W16. Int J Hydrog Energy. 2010;35(24):13475–80.

140. Almeida JRM, Modig T, Petersson A, Hahn-Hägerdal B, Liden G, Gorwa-Grauslund MF. Increased tolerance and conversion of inhibitors in lignocellulosic hydrolysates by *Saccharomyces cerevisiae*. J Chem Technol Biotechnol. 2007;82(4):340–9.
141. Zaldivar J, Ingram LO. Effect of organic acids on the growth and fermentation of ethanologenic *Escherichia coli* LY01. Biotechnol Bioeng. 1999;66(4):203–10.
142. Cui M, Shen J. Effects of acid and alkaline pretreatments on the biohydrogen production from grass by anaerobic dark fermentation. Int J Hydrog Energy. 2012;37(1):1120–4.
143. Colleran E, Pender S, Philpott U, O'Flaherty V, Leahy B. Fullscale and laboratory-scale anaerobic treatment of citric acid production wastewater. Biodegradation. 1998;9 (3–4):233–45.
144. Koullas DP, Christakopoulos PF, Kekos D, Koukios EG, Macris BJ. Effect of alkali delignification on wheat straw saccharification by *fusarium oxysporum* cellulases. Biomass Bioenergy. 1993;4(1):9–13.
145. de Vrije T, de Haas GG, Tan GB, Keijsers ERP, Claassen PAM. Pretreatment of *Miscanthus* for hydrogen production by *Thermotoga elfii*. Int J Hydrog Energy. 2002;27 (11–12):1381–90.
146. Chang VS, Nagwani M, Holtzapple MT. Lime pretreatment of crop residues bagasse and wheat straw. Appl Biochem Biotechnol. 1998;74(3):135–59.
147. Kim S, Holtzapple MT. Lime pretreatment and enzymatic hydrolysis of corn stover. Bioresour Technol. 2005;96(18):1994–2006.
148. Maas RHW (2008) Microbial conversion of lignocellulose-derived carbohydrates into bioethanol and lactic acid. Ph.D Thesis, Wageningen University.
149. Agbor VB, Cicek N, Sparling R, Berlin A, Levin DB. Biomass pretreatment: fundamentals toward application. Biotechnol Adv. 2011;29(6):675–85.
150. Galbe M, Zacchi G. Pretreatment: the key to efficient utilization of lignocellulosic materials. Biomass Bioenergy. 2012;46:70–8.
151. Singh S, Simmons BA, Vogel KP. Visualization of biomass solubilization and cellulose regeneration during ionic liquid pretreatment of switchgrass. Biotechnol Bioeng. 2009;104 (1):68–75.
152. Li C, Knierim B, Manisseri C, Arora R, Scheller HV, Auer M, Vogel KP, Simmons BA, Singh S. Comparison of dilute acid and ionic liquid pretreatment of switchgrass: biomass recalcitrance, delignification and enzymatic saccharification. Bioresour Technol. 2010;101 (13):4900–6.
153. Tan SSY, MacFarlane DR, Upfal J, Edye LA, Doherty WOS, Patti AF, Pringle JM, Scott JL. Extraction of lignin from lignocellulose at atmospheric pressure using alkylbenzene-sulfonate ionic liquid. Green Chem. 2009;11:339–45.
154. Sun N, Rahman M, Qin Y, Maxim ML, Rodriguez H, Rogers RD. Complete dissolution and partial delignification of wood in the ionic liquid 1-ethyl-3-methylimidazolium acetate. Green Chem. 2009;11:646–55.
155. Datta S, Holmes B, Park JI, Chen Z, Dibble DC, Hadi M, Blanch HW, Simmons BA, Sapra R. Ionic liquid tolerant hyperthermophilic cellulases for biomass pretreatment and hydrolysis. Green Chem. 2010;12:338–45.
156. Ouellet M, Datta S, Dibble DC, Tamrakar PR, Benke PI, Li C, Singh S, Sale KL, Adams PD, Keasling JD, Simmons BA, Holmes BM, Mukhopadhyay A. Impact of ionic liquid pretreated plant biomass on *Saccharomyces cerevisiae* growth and biofuel production. Green Chem. 2011;13:2743–9.
157. Schmidt AS, Thomsen AB. Optimization of wet oxidation pretreatment of wheat straw. Bioresour Technol. 1998;64(2):139–51.
158. Klinke HB, Ahring BK, Schmidt AS, Thomsen AB. Characterization of degradation products from alkaline wet oxidation of wheat straw. Bioresour Technol. 2002;82(1):15–26.
159. Ahring BK, Licht D, Schmidt AS, Sommer P, Thomsen AB. Production of ethanol from wet oxidized wheat straw by *Thermoanaerobacter mathranii*. Bioresour Technol. 1999;68 (1):3–9.

160. Arvaniti E, Bjerre AB, Schmidt JE. Wet oxidation pretreatment of rape straw for ethanol production. Biomass Bioenergy. 2012;39:94–105.
161. Martin C, Klinke HB, Thomsen AB. Wet oxidation as a pretreatment method for enhancing the enzymatic convertibility of sugarcane bagasse. Enzyme Microb Technol. 2007;40(3):426–32.
162. Magnusson L, Islam R, Sparling R, Levin D, Cicek N. Direct hydrogen production from cellulosic waste materials with a single-step dark fermentation process. Int J Hydrog Energy. 2008;33(20):5398–403.
163. Liu H, Zhang T, Fang HHP. Thermophilic H_2 production from cellulose containing wastewater. Biotechnol Lett. 2003;25(4):365–9.
164. Sagnak R, Kargi F, Kapdan IK. Bio-hydrogen production from acid hydrolyzed waste ground wheat by dark fermentation. Int J Hydrog Energy. 2011;36(20):12803–9.
165. Uyar B, Schumacher M, Gebicki J, Modigell M. Photoproduction of hydrogen by *Rhodobacter capsulatus* from thermophilic fermentation effluent. Bioprocess Biosyst Eng. 2009;32(5):603–6.
166. Afsar N, Özgür E, Gürgan M, Akköse S, Yücel M, Gündüz U, Eroglu I. Hydrogen productivity of photosynthetic bacteria on dark fermenter effluent of potato steam peels hydrolysate. Int J Hydrog Energy. 2011;36(1):432–8.
167. Kongjan P, Angelidaki I. Extreme thermophilic biohydrogen production from wheat straw hydrolysate using mixed culture fermentation: effect of reactor configuration. Bioresour Technol. 2010;101(20):7789–96.
168. Liu CZ, Cheng XY. Improved hydrogen production via thermophilic fermentation of corn stover by microwave-assisted acid pretreatment. Int J Hydrog Energy. 2010;35(17):8945–52.
169. Nguyen TAD, Han SJ, Kim JP, Kim MS, Oh YK, Sim SJ. Hydrogen production by the hyperthermophilic eubacterium, *Thermotoga neapolitana*, using cellulose pretreated by ionic liquid. Int J Hydrog Energy. 2008;33(19):5161–8.
170. Magnusson L, Cicek N, Sparling R, Levin D. Continuous hydrogen production during fermentation of α-cellulose by the thermophillic bacterium *Clostridium thermocellum*. Biotechnol Bioeng. 2009;102(3):759–66.
171. Kelley SS, Glasser WG, Rials TG, Ward TC. Engineering plastics from lignin XVII. Effect of molecular weight on polyurethane film properties. J Appl Polym Sci. 1989;37(10):2961–71.
172. Thring RW, Vanderlaan MN, Griffin SL. Polyurethanes from alcell® lignin. Biomass Bioenergy. 1997;13(3):125–32.
173. Kadla JF, Kubo S, Venditti RA, Gilbert RD, Compere AL, Griffith W. Lignin-based carbon fibers for composite fiber applications. Carbon. 2002;40(15):2913–20.
174. Lora JH, Glasser WG. Recent industrial applications of lignin: a sustainable alternative to nonrenewable materials. J Polym Environ. 2002;10(1–2):39–48.
175. Cheilas T, Stoupis T, Christakopoulos P, Katapodis P, Mamma D, Hatzinikolaou DG, Kekos D, Macris BJ. Hemicellulolytic activity of *Fusarium oxysporum* grown on sugar beet pulp. Production of extracellular arabinanase. Process Biochem. 2000;35(6):557–61.
176. Energy Research Centre of The Netherlands (ECN) Phyllis. The composition of biomass and waste. Available from: http://www.ecn.nl/phyllis/info.asp. Accessed date 28 Aug 2014.
177. Zheng Y, Cheng Y-S, Yu C, Zhang R, Jenkins BM, VanderGheynst JS. Improving the efficiency of enzyme utilization for sugar beet pulp hydrolysis. Bioproc Biosyst Eng. 2012;35(9):1531–9.
178. Voragen AGJ, Bergmans MEF, Oosteveld A, Schols HA, Beldman G. Utilization of cell wall polysaccharides from cereal by products and beet pulp. In: van Bekkum H, Roeper H, Voragen F, editors. Carbohydrates as organic raw materials III. Weinheim: VCH; 1996. p. 1–16.
179. Hutnan M, Drtil M, Mrafkova L. Anaerobic biodegradation of sugar beet pulp. Biodegradation. 2000;11(4):203–11.
180. Altundogan HS, Arslan NE, Tumen F. Copper removal from aqueous solutions by sugar beet pulp treated by NaOH and citric acid. J Hazard Mater. 2007;149(2):432–9.

181. Mata YN, Blázquez ML, Ballester A, González F, Muñoz JA. Sugar-beet pulp pectin gels as biosorbent for heavy metals: Preparation and determination of biosorption and desorption characteristics. Chem Eng J. 2009;150(2–3):289–301.
182. Bergmans MEF, Beldman G, Gruppen H, Voragen AGJ. Optimisation of the selective extraction of (glucurono)arabinoxylans from wheat bran: use of barium and calcium hydroxide solution at elevated temperatures. J Cereal Sci. 1996;23(3):235–45.
183. Maes C, Delcour JA. Alkaline hydrogen peroxide extraction of wheat bran non-starch polysaccharides. J Cereal Sci. 2001;34(1):29–35.
184. Du C, Lin SKC, Koutinas A, Wang R, Dorado P, Webb C. A wheat biorefining strategy based on solid-state fermentation for fermentative production of succinic acid. Bioresour Technol. 2008;99(17):8310–5.
185. Lin CSK, Luque R, Clark JH, Webb C, Du C. Wheat-based biorefining strategy for fermentative production and chemical transformations of succinic acid. Biofuel Bioprod Bioref. 2012;6(1):88–104.
186. Hawkes F, Forsey H, Premier G, Dinsdale R, Hawkes D, Guwy A, Maddy J, Cherryman S, Shine J, Auty D. Fermentative production of hydrogen from a wheat flour industry co-product. Bioresour Technol. 2007;99(11):110–9.
187. Liu Z, Ying Y, Li F, Ma C, Xu P. Butanol production by *Clostridium beijerinckii* ATCC 55025 from wheat bran. J Ind Microbiol Biotechnol. 2010;37(5):495–501.
188. Verhaart MRA, Bielen AAM, van der Oost J, Stams AJM, Kengen SWM. Hydrogen production by hyperthermophilic and extremely thermophilic bacteria and archaea: mechanisms for reductant disposal. Environ Technol. 2010;31(8–9):993–1003.
189. Gavala HN, Skiadas IV, Ahring BK, Lyberatos G. Potential for biohydrogen and methane production from olive pulp. Water Sci Technol. 2005;52(1–2):209–15.
190. Ueno Y, Fukui H, Goto M. Operation of a two-stage fermentation process producing hydrogen and methane from organic waste. Environ Sci Technol. 2007;41(4):1413–9.
191. Antonopoulou G, Stamatelatou K, Venetsaneas N, Kornaros M, Lyberatos G. Biohydrogen and methane production from cheese whey in a two-stage anaerobic process. Ind Eng Chem Res. 2008;47(15):5227–33.
192. Chen CY, Yeh KL, Lo YC, Wang HM, Chang JS. Engineering strategies for the enhanced photo-H_2 production using effluents of dark fermentation processes as substrate. Int J Hydrog Energy. 2010;35(24):13356–64.
193. Claassen PAM, de Vrije T, Koukios E, van Niel EWJ, Eroglu I, Modigell M, Friedl A, Wukovits W, Ahrer W. Non-thermal production of pure hydrogen from biomass: HYVOLUTION. J Clean Prod. 2010;18(1):S4–8.
194. Özgür E, Afsar N, de Vrije T, Yücel M, Gündüz U, Claassen PAM, Eroglu I. Potential use of thermophilic dark fermentation effluents in photofermentative hydrogen production by *Rhodobacter capsulatus*. J Clean Prod. 2010;18(1):S23–8.
195. Su H, Cheng J, Zhou J, Song W, Cen K. Improving hydrogen production from cassava starch by combination of dark and photo fermentation. Int J Hydrog Energy. 2009;34(4):1780–6.
196. Zong W, Yu R, Zhang P, Fan M, Zhou Z. Efficient hydrogen gas production from cassava and food waste by a two-step process of dark fermentation and photo-fermentation. Biomass Bioenergy. 2009;33(10):1458–63.
197. Hoekema S, Douma RD, Janssen M, Tramper J, Wijffels RH. Controlling light-use by *Rhodobacter capsulatus* continuous cultures in a flat-panel photobioreactor. Biotechnol Bioeng. 2006;95(4):613–26.
198. Miltner A, Wukovits W, Pröll T, Friedl A. Renewable hydrogen production: a technical evaluation based on process simulation. J Clean Prod. 2010;18(1):S51–62.
199. Ljunggren M, Wallberg O, Zacchi G. Techno-economic comparison of a biological hydrogen process and a 2nd generation ethanol process using barley straw as feedstock. Bioresour Technol. 2011;102(20):9524–31.
200. Oliva JM, Saez F, Ballesteros I, Gonzalez A, Negro MJ, Manzanares P. Effect of lignocellulosic degradation compounds from steam explosion pretreatment on ethanol fermentation by thermotolerant yeast *Kluyveromyces marxianus*. Appl Biochem Biotechnol. 2003;105–108:141–54.

Chapter 2
Biohydrogen Production via Lignocellulose and Organic Waste Fermentation

Chen-Yeon Chu and Bing-Shun Huang

Abstract Hydrogen is a promising energy carrier and a replacement for fossil fuels, since it is clean and has high energy and its application does not contribute to the greenhouse effect. Renewable resources, such as lignocellulosic materials and organic wastes, in particular, dark fermentative hydrogen methods, as the feedstock for hydrogen production have great potential for supplying hydrogen.

The development of novel and effective cellulase enzymes, the optimization and improvement of cellulase systems, and engineering approaches for cellulose pretreatment and saccharification to produce biohydrogen have high interest in the scientific community. This chapter gives an introduction to the feedstocks (lignocellulosic materials and organic wastes), primary technologies (physical, chemical, physicochemical, and biological), process of feedstock pretreatments, microorganisms, fermenter types (continuous stirred tank reactors, upflow anaerobic sludge blanket, anaerobic biofilm and granule reactor, membrane bioreactor, etc.), and operational conditions (substrate concentrations, nutrients, pH, temperature, hydraulic retention time, etc.) for producing biohydrogen.

Pretreatment and saccharification are at the heart of producing biological hydrogen from lignocellulosic feedstocks. Efficient production of biohydrogen via lignocellulosis and organic waste depends largely on the fermenter type. Selection of the pretreatment system and fermenter type is the key for economic success of the biohydrogen production plant. This chapter also aims to develop a fundamental understanding of key technologies and variables during biohydrogen production from lignocellulosic raw materials and organic feedstock.

Keywords Biohydrogen • Lignocellulosic materials • Organic wastes • Pretreatment • Fermenter type

C.-Y. Chu (✉) • B.-S. Huang
Green Energy Development Center, Feng Chia University, Taichung 40724, Taiwan, Republic of China
e-mail: cychu@fcu.edu.tw

Z. Fang et al. (eds.), *Production of Hydrogen from Renewable Resources*,
Biofuels and Biorefineries 5, DOI 10.1007/978-94-017-7330-0_2

2.1 Introduction to Feedstocks

Biofuel sources can be classified into first-, second-, and third-generation feedstocks. The first-generation biofuel feedstocks are traditionally food related such as corn for ethanol, vegetable oil, and animal fats. Second-generation biofuel feedstocks are of the nonfood crop type, wastewaters, lignocellulosic agriculture, and forest residues. However, one issue is that the energy crops required for the biofuel source can compete with land use for agriculture [1]. The key challenge for developing the next generation of biofuels is acquiring an economically viable feedstock. Feedstock costs contribute 80–90 % to the final fuel price for many processes and therefore they are important to the economics of biofuels [2].

The feedstock is made up of organic compounds such as carbohydrates that can be converted to biohydrogen by biological metabolism [3]. Biological hydrogen is produced via dark fermentation of organic wastes [4]. Fermentative hydrogen production is carried out by anaerobic bacteria that ferment organic compounds to volatile fatty acids (VFAs), alcohols, carbon dioxide, and hydrogen (Fig. 2.1). Many different substrates can be fermented to produce hydrogen [5]. Dark fermentation has advantages, for example, high rate of bacterial growth, low energy requirement, no oxygen limitation problems, low cost [6, 7], and biohydrogen production without light, and various carbon sources can be utilized as the substrate [8].

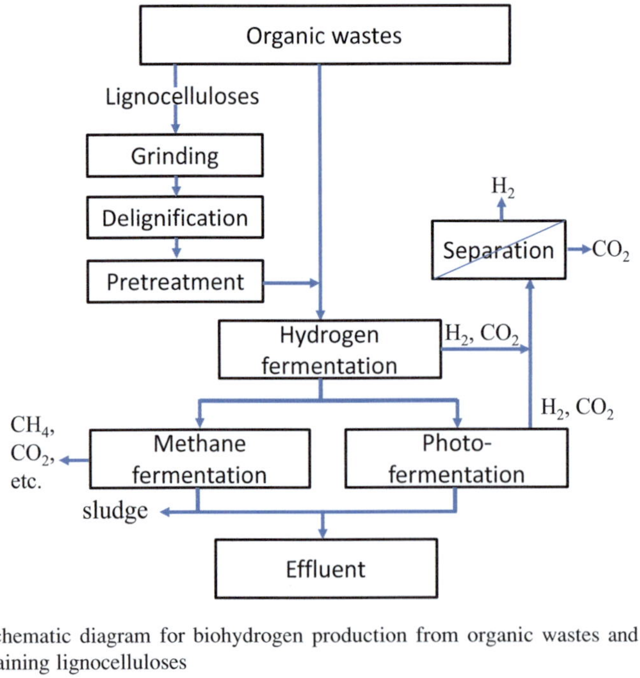

Fig. 2.1 Schematic diagram for biohydrogen production from organic wastes and agricultural wastes containing lignocelluloses

Waste materials that have been used as substrate for biohydrogen include palm oil mill effluent (POME) [6, 9–17], starch-based materials [18–28], food waste [29–33], and condensed molasses fermentation soluble (CMS) [34–38]. Sugary wastewater [39–42] is a more efficient source of carbohydrates than raw materials for biohydrogen production. Simple sugars, such as sucrose and glucose, can be converted at high temperatures into hydrogen at high conversion efficiencies [43]. Glucose, which is an easily biodegradable carbon source, is present in many industrial effluents and can be obtained abundantly from agricultural wastes [44]. Therefore, sugary wastewater can be considered to be most useful for industrial hydrogen production.

2.1.1 *Organic Wastes*

Various organic solid wastes or wastewaters have attracted considerable attention for biohydrogen production due to the advantages of high organic loading possibilities, low nutrient requirements, concurrent wastewater treatment, and positive net energy gain. Table 2.1 shows complex organic wastes that are considered as feedstock, such as kitchen, food processing, mixed, and municipal wastes for biohydrogen production. These organic wastes usually have high concentrations of protein and fat, making their hydrogen conversion efficiencies lower than that of the carbohydrate-based wastewaters. As a matter of fact, previous studies show that the hydrogen production prospective of carbohydrate-based wastes was higher than that of fat- and protein-based wastes by about 20 times [57]. This is partially due to the protein degradation that produces nitrogen that consumes the free hydrogen. In many kinds of wastes, the organic fraction of municipal solid waste (OFMSW) is considered to be quite favorable as a potential raw material for biohydrogen production, because it is able to represent up to 70 % of the total MSW produced, consisting of paper (up to 40 %), food wastes, garden residues, and wood [56]. It should be noted that the process will lead to an extra cost, because an initial selection/separation process would be necessary in order to obtain the suitable substrates. Starchy- and sugary-based biomass and wastes are readily fermented by microorganisms for hydrogen generation. Although lignocellulosic biomass is abundant in agricultural residues, it needs pretreatment to be useable.

2.1.2 *Lignocelluloses*

Lignocellulose is the most abundant renewable biomass in the world with an annual quantity of about 220 billion tons (dry weight) per year [58]. To use this feedstock, pretreatment is necessary to decompose the lignin structure and loosen the crystalline cellulose structure to promote enzyme accessibility. Pretreated lignocellulosic materials (e.g., sugarcane bagasse, corncob, wheat straw, cornstalks, grass, energy

Table 2.1 Fermentative hydrogen production from different types of waste and wastewaters

Type	Microorganism	Operation mode	H_2 production rate	Maximum H_2 yield	Ref.
Rice winery wastewater	Mixed culture	Continuous	380 mmol/g VSS/day 155.5 mmol/l·d	2.14 mol/mol hexoses	[42]
Food waste	Mixed thermo-philic culture	Batch	11.8 mmol/g VSS/day	1.8 mol/mol hexoses	[31]
Food waste sewage sludge	Mixed mesophilic culture	Batch	109 mmol/g VSS/day	122.9 ml/g COD carbohydrate	[30]
Dairy wastewater	Mixed mesophilic culture	Continuous	1.59 mmol/l·d	–	[45]
Molasses	Mixed mesophilic culture	Continuous	8.1 mmol/l·d	–	[46]
Cheese whey	Mixed mesophilic culture	Batch	194 mmol/l·d	5.9 mol/mol lactose	[47]
Dairy wastewater	Mixed mesophilic culture	Batch	47.67 mmol/g VSS/day	17.2 mmol/g COD	[48]
Olive pulp	Mixed mesophilic culture	Continuous	10.6 mmol/l·d	0.19 mol/kg TS	[49]
Olive oil mill wastewater	Mixed mesophilic culture	Continuous	8.2 mmol/l·d	196.2 ml/g hexose	[50]
Coffee drink manufacturing wastewater	Mixed thermo-philic culture	Continuous	4,153 mmol/l·d	2.57 mol H_2/mol hexoses	[51]
Sugar factory wastewater	Mixed mesophilic culture	Continuous	4.1 mmol·gML VSS/l·d	–	[52]
Starch wastewater	Mixed mesophilic culture	Sequencing batch	140.8 mmol/l·d	5.79 ± 0.41 mmol H_2/g COD_{added}	[53]
Textile wastewater	Mixed mesophilic culture	Intermittent-flow, stirred tank reactor (IFSTR)	408 mmol/l·d	0.97 mol H_2/mol hexoses	[54]
Enzymatic hydrolyzed food waste	Mixed mesophilic culture	Continuous mixed immobilized sludge reactor (CMISR)	346.7 mmol/l·d	–	[55]

Adapted from Ntaikou et al. [56] with kind permission from Springer Science and Business Media, Copyright © 2010

crops, oil palm trunk, and beer lees [59]) can be used for fermentative hydrogen production. Some studies have focused on biohydrogen and biomethane production from raw or pretreated solid wastes such as olive pulp, household solid waste, and potato waste [60–68]. The feedstock costs are high due to processing (shredding, densifying, pulverizing, and handling), collection, and transportation. Despite the challenges, these second-generation feedstock options are plentiful and produce great amounts of fuel. Lignocellulosic material basically consists of three different types of polymers, namely, cellulose, hemicellulose, and lignin. Agricultural and forestry residues are rich in carbohydrates and the cost of obtaining these residues is negligible. However, these do not contain readily accessible free sugars necessary for efficient fermentation. To convert cellulose, it is necessary to transform the carbohydrate polymers into fermentable sugars through the use of enzymes and change the structure of the cellulosic biomass. Therefore, biotransforming it into hydrogen is a difficult task in most cases. In Table 2.2 different types of residues used as feedstock for hydrogen production are presented, along with the achieved hydrogen yields and rates.

2.2 Pretreatment of Lignocellulosic Feedstock

Pretreatment is widely accepted to be an essential step for making lignocellulosic biomass accessible to enzymatic attack by breaking the lignin seal, removing hemicellulose, or disrupting the crystalline structure of cellulose [56]. An effective and economical pretreatment should meet the following requirements: (a) delignify feedstock for enzymatic attack, (b) avoid destruction of hemicelluloses and cellulose, (c) avoid formation of possible inhibitors, (d) minimize energy demand, (e) reduce cost for size reduction of feedstock, (f) reduce reactor costs, (g) produce low residues, and (h) decrease chemical costs [82]. Several methods have been introduced for pretreating lignocellulosic materials, namely, prior enzymatic hydrolysis or digestion. These methods are classified into physical, chemical, physicochemical, and biological pretreatments. The main principles of each pretreatment method are illustrated below.

2.2.1 Physical

The objective of the physical pretreatment is to reduce particle size, pore size, crystallinity of lignocellulosic material, and the degree of polymerization and increase the specific surface [83, 84]. Different types of physical processes such as milling (e.g., ball milling, two-roll milling, hammer milling, colloid milling, and vibro energy milling) and irradiation (e.g., by gamma rays, electron beam, or microwaves) can be used to improve the enzymatic hydrolysis or biodegradability of lignocellulosic waste materials [82].

Table 2.2 Fermentative hydrogen production from lignocellulosic residues

Lignocellulosic residue	Pretreatment	Microorganism	Operation mode	H_2 production rate (mmol /l·d)	Maximum H_2 yield (mol/mol consumption hexose)	Ref.
Rice straw hydrolysate	Concentrated acid 55 % + calcium hydroxide to remove the SO^{4-}	Mixed mesophilic cultures	Continuous	666.1	1.02	[4]
Rice straw hydrolysate	Concentrated acid 55 % + calcium hydroxide to remove the SO_4^{2-}	Mixed mesophilic cultures	Continuous	408.2	0.69	[69]
Wood fibers	Mechanical	*Clostridium thermocellum*	Batch	–	1.47	[70]
Corn stover	Steam explosion (90–220 °C, 3–5 min)	Mixed mesophilic cultures	Continuous	10.56	3	[71]
Sugarcane bagasse hydrolysate	Acid–thermal hydrolysis H_2SO_4 0.27–7(v/v), +121 °C, 60 min	*Clostridium butyricum*	Batch	–	1.73 mol/mol total sugar	[72]
Fodder maize juice	Mechanical	Mixed mesophilic cultures	Continuous	–	69.4 ml H_2/g dry mass	[73]
Sweet sorghum residues	Mechanical	*Ruminococcus albus*	Batch	–	2.59	[74]
Wheat straw	Mechanical	*Caldicellulosiruptor saccharolyticus*	Batch	–	3.8 (44.7 l/kg dry biomass)	[75]
Maize leaves	Mechanical	*Caldicellulosiruptor saccharolyticus*	Batch	–	3.6 (81.5 l/kg dry biomass)	[75]

Wheat straw	Pretreated with 7.4 % (w/w) $Ca(OH)_2$	Mixed mesophilic culture	Batch	–	58.78 ml/g-VS	[76]
Wheat straw	Pretreatment via steam explosion	*Clostridium* sp.	Batch	–	2.54 ± 0.2 mol H_2/mol reducing sugar	[77]
Bagasse	Alkali–thermal 0.2–4 g/l NaOH, 100 °C, 2 h	Mixed thermophilic cultures	Batch	0.28 mmol/h.g TVS	13.39 mmol H_2/g TVS	[78]
Corn stover	Acid–thermal hydrolysis H_2SO_4 0.25–4(v/v), +121 °C, 30–180 min	*Thermoanaerobacterium thermosaccharolyticum*	Batch	–	2.24	[79]
Disposable wooden chopsticks (DWC) waste	Alkaline pretreatment and enzymatic hydrolysis	Enriched hot spring culture	Batch	–	195 ml H_2/g total sugars consumed	[80]
Rice husk	Enzyme treated	*Clostridium beijerinckii*	Batch	9.7 mmol/l	2.93 mmol H_2/g of reducing sugar	[81]

Adapted from Ntaikou et al. [56] with kind permission from Springer Science and Business Media, Copyright © 2010

Milling can be employed to alter the inherent ultrastructure of lignocellulosic material and the degree of crystallinity and consequently make it more amenable to cellulase [85]. Milling and size reduction have been applied prior to enzymatic hydrolysis or even other pretreatment processes such as dilute acid, steam, or ammonia [85, 86]. Among the milling processes, the colloid mill, fibrillator, and dissolver are the only ones suitable for wet materials, e.g., wet paper from domestic waste separation or paper pulps. However, the extruder, roller mill, cryogenic mill, and hammer mill are usually used for dry materials.

Irradiation by gamma rays, electron beam, and microwaves can improve enzymatic hydrolysis of lignocellulosic materials as well. The combination of radiation and other methods such as acid treatment can further accelerate enzymatic hydrolysis [87]. Irradiation has enhanced enzymatic degradation of cellulose into glucose. However, pre-irradiation was found to be more effective in air than in an acid solution [88].

Ultrasound can be used for disintegration of waste-activated sludge and aquacultural effluents [89, 90] due to its advantage on the mechanical properties of sludge hydrolysis. In this method, the sludge is disintegrated and the bacterial cell walls are disrupted [91]. Several factors such as ultrasonic density and intensity, sludge pH, and sludge concentration have an impact on disintegration [92].

2.2.2 Chemical

Chemical pretreatment for lignocellulosic feedstocks employs different chemicals such as acids, alkalis, and oxidizing agents, e.g., peroxide and ozone [93]. Among these methods, dilute acid pretreatment using H_2SO_4 is the most widely used method. Depending on the type of chemical used, pretreatment can have different effects on lignocellulose structural components. Alkaline pretreatment, ozonolysis, and peroxide and wet oxidation pretreatments are more effective in lignin removal, whereas dilute acid pretreatment is more efficient in hemicellulose solubilization [94–96].

Alkali pretreatment refers to the application of alkaline solutions such as NaOH, $Ca(OH)_2$ (lime), or ammonia to remove lignin and a part of the hemicellulose. The purpose of alkali pretreatment is to (1) induce swelling of the biomass and lead to an increase of internal surface area, (2) separate cellulose from hemicellulose and lignin, (3) reduce crystallinity of cellulose, (4) disrupt lignin structure, (5) eliminate both hydrolysis and fermentation inhibitors, and (6) improve accessibility of cellulose and hemicellulose toward enzymatic hydrolysis [97, 98]. Alkaline peroxide is an effective method for the pretreatment of biomass. In this method, the lignocellulose is soaked in pH-adjusted water (e.g., to pH 11–12 using NaOH) containing H_2O_2 at room temperature for a period of time (e.g., 6–24 h). The process can improve the enzymatic hydrolysis after delignification.

Acid pretreatment has received considerable research attention over the years [99]. Dilute sulfuric acid has been added to cellulosic materials for some years to commercially manufacture furfural [100]. Dilute sulfuric acid is mixed with a biomass to hydrolyze hemicellulose to xylose and other sugars and then continue to break xylose down to form furfural. The most widely used and tested approaches are based on dilute sulfuric acid. However, nitric acid, hydrochloric acid, and phosphoric acid have also been studied.

Processing of lignocellulosic biomass with ionic liquids (IL) and other solvents has gained importance in the last decade due to the tunability of the solvent chemistry and hence the ability to dissolve a wide variety of biomass types. Ionic liquids are salts, typically composed of a small anion and a large organic cation, which exist as liquids at room temperature and have very low vapor pressure [101].

2.2.3 Physicochemical

Pretreatments that combine both chemical and physical methods are referred to as physicochemical processes. Physicochemical pretreatment for lignocellulosic feedstock employs different methods such as steam explosion, steam explosion with addition of SO_2, ammonia fiber explosion (AFEX), liquid hot-water pretreatment, and microwave–chemical pretreatment [71, 102–105]. Steam-explosion pretreatment is one of the most commonly used pretreatment options, as it uses both chemical and physical techniques to break the structure of the lignocellulosic material. This hydrothermal pretreatment method subjects the material to high pressures and temperatures for a short duration of time after which it rapidly depressurizes the system, disrupting the structure of the fibrils. The disruption of the fibrils increases the accessibility of the cellulose to the enzymes during hydrolysis. Particle size is a major contributing factor on the effectiveness of the process, and it has been observed that relatively large particle sizes have been able to yield maximum sugar concentrations [106]. The steam-explosion pretreatment process is a proven technique for the pretreatment of different biomass feedstocks as it is able to generate complete sugar recovery while utilizing a low capital investment. Steam-explosion pretreatment also has a low environmental impact in regard to the chemicals being used and the way the process is implemented. Steam-explosion pretreatment method is highly efficient [106].

Acid catalysts have been used within the steam-explosion processes in dilute quantities to improve hemicellulose hydrolysis during the pretreatment stage and cellulose digestibility in later stages of the process. Dilute acids have the ability to decrease retention times and temperatures of the current operating systems and allow for the use of softwoods in this pretreatment technique, where it was originally thought to be uneconomical. By decreasing the retention time and temperature with the addition of this acid catalyst, a reduction of inhibitory compounds formed is observed, nearly all the hemicellulose is removed, and there is an increase rate of hydrolysis later on in the production [107].

Table 2.3 Effect of pretreatment methods on the chemical composition and chemical/physical structure of lignocellulosic biomass

Methods	Effects
Uncatalyzed steam explosion, liquid hot water, pH-controlled hot water, flow-through liquid hot water	Increases accessible surface area and removes hemicellulose, low effect on altering lignin structure
Dilute acid, flow-through acid, biological pretreatment	Increases accessible surface area, removes hemicellulose, and alters lignin structure
Thermal alkaline	Increases accessible surface area, removes hemicellulose, and alters lignin structure. Low effect on removing hemicellulose
AFEX (ammonia fiber expansion), ARP (ammonia recycled percolation)	Increases accessible surface area, decreases cellulose crystallinity, removes hemicellulose, and alters lignin structure. Low effect on removing hemicellulose
Lime	Increases accessible surface area, removes hemicellulose, and alters lignin structure. Low effect on removing hemicellulose
Ozonolysis	Removes lignin
Organosolv	Removes hemicellulose and removes lignin
PEF (pulsed electric field)	Increases accessible surface area and alters lignin structure
Ionic liquids	Increases accessible surface area, removes hemicellulose, lowers cellulose crystallinity, and removes hemicellulose

2.2.4 *Biological*

Biological hydrolysis of cellulose is carried out by cellulolytic microorganisms or by the cellulose enzyme complex [108–111]. In nature, cellulosic materials are degraded by microorganisms, of which brown-white and soft-rot fungi have proven to readily degrade lignin and hemicellulose in waste materials and are used in biological pretreatment processes [112]. A mixed culture [113, 114] comprising of cellulolytic bacterium and a noncellulolytic bacterium could degrade natural cellulosic materials aerobically or anaerobically without sterilization, thereby having a high degree of stability to degrade cellulosic material for long periods of time. The advantages of biological pretreatment include minimal cost, low energy requirement, and mild impact on the environment. However, utilizing these microorganisms and enzymes to process natural cellulosic materials without pretreatment and/or sterilization is difficult and the rate of hydrolysis is also low.

The main aim of pretreatment is to increase accessible surface area, to decrystallize cellulose, and to remove hemicellulose and lignin. The effects of different pretreatments are listed in Table 2.3. Several factors are mentioned to have a positive effect on the overall economy of the process. It is, for example, favorable to avoid the production of inhibitors [115], because the detoxification of the liquid fractions showed to be costly and/or ineffective [116, 117], leaving the lignin with the substrate and removing it after the hydrolysis of the (hemi)cellulose will minimize the overall cost of the process [118], and the use of low concentrations of water, energy, and alkali/acid during pretreatment can be attractive for industrial applications [106].

2.2.5 *Organosolv Pretreatment*

It is known that organosolvent (organosolv) pretreatment can be applied with a large number of organic or aqueous–organic solvent systems with or without added catalysts in the temperature range of 100–250 °C [119], while organic acid pretreatment can be applied under mild conditions, even at room temperature [120, 121]. For most organosolv processes, there is no need for acid addition if the pretreatment is conducted at high temperatures (185–210 °C), as it is believed that organic acids act as catalysts for breaking the lignin–carbohydrate complex [122]. However, when acid catalysts are added, the rate of delignification is increased and high yields of xylose are obtained. Mineral acids (hydrochloric acid, sulfuric acid, and phosphoric acid) are good catalysts to accelerate delignification and xylan degradation, while some organic acids such as formic, oxalic, acetylsalicylic, and salicylic acid also can be used as catalysts [123, 124]. Most of the hemicellulose and lignin are solubilized, but the cellulose remains as solid. The organic solvents used in the process need to be recycled to reduce the cost. On the other hand, removal of solvents from the system is necessary because the solvent may be inhibitory to the growth of organisms, enzymatic hydrolysis, and fermentation. Organosolv pretreatment yields three separate fractions: dry lignin, an aqueous hemicellulose stream, and a relatively pure cellulose fraction [122].

Heterogeneous catalysis for lignocellulosic biomass conversion is gaining attention in the literature [125–130]. This type of acid catalyst is a good alternative to concentrated sulfuric acid for hydrolysis reaction. It has numerous advantages over sulfuric acid in terms of activity, selectivity, catalyst lifetime, and reusability. Moreover, the use of solid acid reduces liquid pollutants and cost of wastewater treatment and thus reduces the costs [131–135].

2.3 Fermentative Hydrogen Production

Biohydrogen production performance is directly determined by operational strategy and key process parameters such as microorganism, fermenter type, substrate concentration, pH, temperature, and hydraulic retention time (HRT).

2.3.1 Microorganisms

In a previous review paper, the major hydrogen-producing bacteria identified are related strictly to facultative anaerobic genera (*Escherichia coli*, *Enterobacter*, *Citrobacter*), to anaerobic genera (*clostridia*, *methylotrophs*, *rumen bacteria*, *methanogenic bacteria*, *archaea*), and to aerobic genera (*Alcaligenes*, *Bacillus*). Table 2.4 shows hydrogen-producing bacteria and their characteristics. Dark

Table 2.4 Hydrogen-producing bacteria and their characteristics

Organisms	Functions	Characteristics	References
Clostridium spp.	H_2 production	Obligate and mesophilic anaerobes	[136–139]
		The most popular H_2 producer	
		Ferment a wide range of carbohydrates and produce H_2	
		E.g., *Clostridium butyricum*, *C. acetobutylicum*, *C. tyrobutyricum*, *C. saccharolyticum*	
Thermoanaerobacterium spp.	H_2 production	Obligate and thermophilic anaerobes	[140]
		E.g., *Thermoanaerobacterium thermosaccharolyticum*	
Ethanoligenens spp.	H_2 production	Facultative anaerobes	[141]
		May possess important features such as salt tolerance	
		E.g., *Bacillus megaterium*	
Bacillus spp.	H_2 production	Facultative anaerobes	[142]
		May possess important features such as salt tolerance	
		E.g., *Bacillus megaterium*	
Enterobacter spp.	H_2 production	Facultative anaerobes	[141]
		Have better tolerance against oxidative stress	
		E.g., *Enterobacter aerogenes*	
Klebsiella spp.	H_2 production	Facultative anaerobes	[143]
		Have better tolerance against oxidative stress	
		E.g., *Klebsiella pneumonia*	
Methanogens	H_2 consumption	Obligate anaerobes	[144]
		Utilize H_2 for methane production	
		E.g., *Methanobacterium* spp., *Methanococcus* spp., etc.	
Other H_2-consuming bacteria	H_2 consumption	Obligate/facultative anaerobes	[145, 146]
		Utilize H_2 as electron donor and precursors for metabolic compounds	
		E.g., *Lactobacillus* spp. and *Bifidobacterium* spp.	

Adapted from Wong et al. [147] with permission, Copyright © 2014 Elsevier

fermentative hydrogen production is the most practical to be applied among the various biological hydrogen production methods [148, 149] due to its efficient processes to convert organic substrates to energy and electrons. Three types of metabolism to dark fermentative hydrogen production are as follows. The first type is for *Escherichia coli* and *Enterobacteriaceae* [150, 151], which has two major enzymes: (1) pyruvate formate lyase (PFL) and (2) formate hydrogen lyase (FHL). Pyruvate formed via the Embden–Meyerhof–Parnas (EMP) pathway is split into acetyl-CoA and formate by PFL under anaerobic conditions. H_2 and CO_2 are then generated from formate by FHL [152]. The second type typical for *Clostridium* species [153] includes pyruvate:ferredoxin oxidoreductase (PFOR) and Fd-dependent hydrogenase (HydA) [152]. Pyruvate:ferredoxin oxidoreductase catalyzes the oxidative decarboxylation of pyruvate to form acetyl-CoA and CO_2 under anaerobic conditions. The electrons are first transferred to Fd_{ox} with a highly negative potential (−420 mV) [154]. The electrons in Fd_{rd} are then transferred to protons to generate hydrogen by Hyd_A. The third type is reported to exist in many thermophilic bacteria and some *Clostridium* species for utilizing NAD(P)H to form hydrogen. This biochemical reaction is catalyzed by two major enzymes, NAD(P) H:ferredoxin oxidoreductase (NFOR) and Hyd_A [155]. NAD(P)H formed during carbon metabolism by Fd_{ox} reaction. This hydrogen-producing reaction is then processed by Fd_{rd} and Hyd_A.

2.3.2 *Fermenter Types*

2.3.2.1 CSTR

Continuous stirred tank reactors (CSTR) are commonly used for continuous biohydrogen production [7, 56, 156–158]. In a CSTR, hydrogen-producing microbes are completely mixed and suspended in the reactor liquor by the mixing pattern. Biomass is well suspended in the mixed liquor, which has the same biomass concentration in the effluent [159]. Under such hydrodynamics, good substrate–microbe contact and mass transfer can be accomplished. On the other hand, the CSTR is unable to maintain high levels of fermentative biomass because of the rapidly mixed operating pattern. Biomass washout may occur at short hydraulic retention times (HRTs) [160]; thus, the hydrogen production rates are considerably restricted [7]. To retain high biomass concentrations in reactors, various techniques have been developed for hydrogen fermentation, including sludge immobilization [35, 157, 158], utilization of the upflow reactor [42], and immobilization on a porous support such as loofah sponges, expanded clays, activated carbons [161], and membrane reactors [162, 163].

2.3.2.2 UASB

An upflow anaerobic sludge blanket (UASB) process is a widely applied anaerobic treatment system that has high treatment efficiency and a short hydraulic retention time (HRT). UASB hydrogen production systems have been used in granulation enhancement and granule microstructure [164–166]. Numerous works have dealt with hydrogen-producing UASBr, since hydrogen production granule (HPG) formation was first reported by Fang et al. [167], and as mentioned above, this reactor generally demonstrates a high and stable performance. However, for most studies in this field, synthetic wastewater is generally applied as a substrate. Chang and Lin [168] produced hydrogen from sucrose using a UASBr seeded with heat-pretreated sewage sludge. The highest HY (hydrogen yield) and HPR (hydrogen production rate) values were 0.75 mol H_2/mol hexose and 0.25 l H_2/l·h, respectively, at a HRT of 8 h. In an effort to decrease the start-up period in the UASBr, Jung et al. [169, 170] inoculated heat-treated sludge to a CSTR, and then the mixed liquor in the CSTR was transferred to the UASBr as a seeding source. As a result, hydrogen production granule with an average size of 1.9 mm was successfully formed in the UASBr after 45 days of operation using coffee drink manufacturing wastewater (CDMW), which was the first report on the formation of HPG from actual wastewater.

2.3.2.3 Anaerobic Biofilm and Granule Reactor

To overcome biomass washout problems, an addition of immobilized cells into the conventional CSTR for increasing the biomass retention in biohydrogen-producing fermenters has been previously attempted. Biohydrogen-producing fermenters, such as a carrier-induced granular sludge bed reactor (CIGSB) [160] and continuously stirred anaerobic bioreactor (CSABR) used with the silicone immobilized cells and agitated granular sludge bed reactor (AGSBR), have been investigated [35, 171–173]. Attempts to enhance biomass retention by immobilized cells exhibited a better hydrogen production performance than that of conventional CSTR, with HPR ranging from 6 to 360 l/l·d [160, 172]. Consequently, immobilized cells created by natural or synthetic matrices [174] were often used to allow better retention of hydrogen-producing bacterial cells for stable operations at high feeding rates. Cell immobilization by surface attachment [175] or self-flocculation [35, 173], [168, 176, 177] may have higher feasibility in practical environmental applications. Wu et al. [178] studied the hydrogen production from a sucrose-rich wastewater in a fluidized bed reactor by immobilized cell. Results showed that a stable yield of 182 ml H_2/g hexose and a hydrogen production rate of 22.3 l/l·d were obtained in the fluidized bed reactor.

2.3.2.4 Membrane Bioreactor

The membrane bioreactor (MBR) has emerged as an effective means of attaining performance improvement in wastewater treatment and has been applied to anaerobic processes due to its capability of increasing biomass retention via membrane separation [179, 180]. Attempts have been made to apply the MBR process to hydrogen production, but relatively little research has been carried out so far. Oh et al. [181] demonstrated that HPR increased by 25 %–0.32 l/l.h due to a 164 % increase in biomass concentration from 3.53 to 5.8 g/l with an increase of the slurry retention time (SRT) from 3.3 to 12 h using an external cross-flow membrane. Membrane fouling is a key process limitation and remains one of the most challenging issues with future MBR development [182].

2.3.3 Environmental Operational Conditions

2.3.3.1 Substrate Concentration

Many review reports [61, 65, 183, 184] have summarized the optimum values of substrate concentration, although most studies focus on lab-scale systems. The indexes for identifying high biogas production efficiency are biohydrogen or biomethane production yield (HY or MY, defined as the biohydrogen or biomethane production per unit weight of consumed substrate, mol H_2/g COD or mol CH_4/g COD) and biohydrogen/biomethane production rate (HPR or MPR, defined as the biohydrogen or biomethane production per unit working volume per day, l/l·d).

Finding the optimal substrate concentration in continuous operation mode is more meaningful and practical, since the batch mode does not take into consideration the hydrodynamic effect, steady state of the substrate concentration, and pH condition for bacterial growth. The best performance is found to be at 30 g sucrose COD/l with HY of 1.09 mol H_2/mol hexose using a CSTR [185]. At inlet substrate concentrations below 20 g COD/l, the HY decreases along with a significant decrease in the n-butyrate/acetate ratio. The appearance of hydrogen-consuming bacteria and decrease of substrate removal efficiency was observed at over 35 g COD/l [185].

High substrate concentration allows more energy-efficient operation but product inhibition is likely to set the upper limit. Certain levels of metabolic products in the dark fermentative hydrogen production reactor may inhibit the hydrogen-producing pathway as well as microbial activity. It is known that butyrate has the highest inhibiting effect on *Clostridium sp.*, among various acids; thus many attempts have been made to alleviate butyrate inhibition, mostly by chemical extraction [186].

Table 2.5 shows the biohydrogen production performance data in continuous operation mode in 2013–2015. Most investigators try to use agriculture waste or the residue from biofuel production processes or food waste to extend the feedstock

Table 2.5 The biohydrogen production performance in continuous operation mode in 2013–2015

Substrate	Reactor type	Microorganism	Substrate concentration	pH	Temperature (°C)	HRT (h)	H_2 production rate (mmol/l·d)	Maximum H_2 yield	References
Synthetic substrate based on sucrose	UASB	Mixed Culture	2 g COD/l	–	55	2	80.7	1.73 mol H_2/mol sucrose	[187]
Palm oil mill effluent wastewater	UASB	PEG-immobilized *Clostridium* sp. LS2	60 g COD/l	5.5	37	12	360	0.35 l H_2/g $COD_{removed}$	[188]
Rice straw hydrolysate	Continuously stirred anaerobic bioreactor (CSABR)	Mixed culture	20 g total sugar/l	5.5	37	4	446	0.69 mol H_2/mol total sugar	[69]
Rice straw hydrolyzate	Continuously external circulating bioreactor (CECBR)	Mixed culture	20 g total sugar/l	5.5	37	4	728	1.02 mol H_2/mol hexoses	[4]
Crude glycerol from biodiesel industry	CSTR	*C. pasteurianum* CH_4	2 g glycerol/l	–	35	12	177.8	0.77 mol H_2/mol glycerol	[189]

Molasses with liquid swine manure	Anaerobic sequencing batch reactor (ASBR)	Mixed culture	10 g sugar/l	5.32	37	15.62	99.6	1.57 mol H_2/mol sugar	[190]
De-oiled jatropha waste (DJW)	ASBR	Mixed culture	200 g DJW/l	6.5	55	48	66	8.7 ml H_2/g volatile $solid_{added}$	[191]
Food waste	CSTR	Mixed culture	29.17 g COD/l	5.0	35	20	16.9	261 ml H_2/g VS_{added}	[32]
Food waste	Continuous mixed immobilized sludge reactor (CMISR)	*A. awamori* and *A. oryzae*	10 g food waste hydrolysate/l	4	55	6	379	85.6 ml H_2/g food waste	[55]
Textile wastewater (TW)	Intermittent-flow, stirred tank reactor (IFSTR)	Mixed culture	33.1 g hexoses/l	5.5	35	24	44.6	0.97 mol H_2/mol hexoses	[54]

resources for the biohydrogen production. The study trend for enhancement of hydrogen production rate seems to change the hydrodynamic properties by changing the fermenter type.

To recover more energy from the biomass substrate, researchers have begun to explore two-stage (biohydrogen+biomethane) [192–195] production technology system recently. Two-stage biohydrogen and biomethane production systems can really increase the energy gain from biomass resource by around 8–43 % energy compared with one-stage anaerobic digestion systems [192, 196].

2.3.3.2 Nutrients and Metals

Excluding the main substrate, carbohydrate materials, dark fermentative hydrogen production (DFHP) requires nutrients for bacterial activity like all biological treatment processes. The nutrients include nitrogen (N), phosphorous (P), ferrous (Fe), and some trace metals. Among the many kinds of nutrients, N is the most essential one for bacterial growth. Optimal C/N ratio is 47 according to Lin and Lay [197]. P and Fe concentrations affect the metabolic pathway of *Clostridium* sp., and hydrogen production potential decreases when their concentrations are limited.

The effect of iron has been investigated many times in DFHP, since it is an essential component of hydrogenase. Lin and Lay [198] studied the requirement of 11 trace metals in hydrogen fermentation. Magnesium, sodium, zinc, and iron were found to be the important trace metals with magnesium being the most significant one. Hydrogen production is enhanced by 30 % at optimal combined concentrations, 4.8 mg Mg^{2+}/l, 393 mg Na^{+}/l, 0.25 mg Zn^{2+}/l, and 1 mg Fe^{2+}/l.

The effect of metal ions on the fermentative hydrogen production has been widely studied such as Ni [199–201], Fe [199, 202, 203], Cu [201, 204–206], Cr [201, 204], Zn [201, 204, 206], Cd [201], and Pb [201] ions. Hydrogenase enzymes catalyze the reduction of proton to H_2. Hydrogenase enzymes are classified into [Ni–Fe] and [Fe–Fe] hydrogenases, according to the metal content at their active site [207]. [Ni–Fe] hydrogenases are extensively distributed among bacteria [208], and both nickel and iron have important effects on fermentative H_2 yields [199, 200, 202, 209].

In a biohydrogen production process, electrons are transported via an intramolecular electron transfer chain from the redox partner of the [Ni–Fe] hydrogenases to the active site, and then the protons are reduced by producing biohydrogen [210, 211]. Since nickel is a fundamental component making up the [Ni–Fe] hydrogenases, it plays an important role in fermentative hydrogen production.

Karadag and Puhakka [199] investigated the effect of Fe^{2+} and Ni^{2+} on continuous hydrogen production in anaerobic completely stirred tank reactor (ACSTR). They found that hydrogen production increased about by 71 % with the increasing of iron and nickel supplementation, and the highest yields were achieved at the concentrations of 50 mg Fe^{2+}/l and 25 mg Ni^{2+}/l. Wang and Wan [200] reviewed the effects of Fe^{2+} on anaerobic hydrogen production and reported some inconsistency on the optimal Fe^{2+} concentration. They also found that increasing Ni^{2+}

concentration up to 0.2 mg/l enhanced the hydrogen production by using batch experiments at 35 °C. Metabolic pathway shifted at different Ni^{2+} concentrations and higher Ni^{2+} concentration promoted the growth of hydrogen-producing bacteria. Lee et al. [202] investigated the effect of iron on the efficiency of continuous hydrogen production in a submerged membrane bioreactor system. They found $FeSO_4$ concentration is the key factor affecting the fermentation pathway for hydrogen production with the membrane bioreactor. Both increase in the hydrogen production rate and the hydrogen yield were obtained by adding $FeSO_4$. They indicated that iron sulfate increased hydrogenase activity and hydrogen production in a membrane bioreactor when $FeSO_4$ concentration closes to 10.9 mg/l.

Lin and Shei [204] investigated the effect of Cr, Cu, and Zn ions on biohydrogen production using anaerobic sewage sludge microflora. Cr, Cu, and Zn significantly affect hydrogen-producing microflora enriched from sewage sludge with Zn and Cr being the most and least toxic metals, respectively. The microflora's hydrogen production activity could be reduced by 50 % for a biomass in contact with 4.5 mg Zn/l, 6.5 mg Cu/l, and 60 mg Cr/l. However, low concentrations of 2 mg Cu/l and 15 mg Cr/l resulted in peak hydrogen production by 20 and 10 %, respectively. Zheng and Yu have reported that the specific hydrogen production rate was enhanced by the dosage of Cu at 50–100 mg/l, but was inhibited by Cu over 200 mg/l from glucose by enriched anaerobic culture [206]. Li and Fang found that Cu strongly inhibited the bioactivity of hydrogen-producing sludge [201]. Han et al. [205] investigated Cu^{2+} concentration effect in a sucrose-fed CSTR on fermentative hydrogen production by mixed cultures. The result shows that 6.4 mg/L Cu^{2+} is the optimal concentration for the CSTR at HRT 4 h. In addition, copper causes shift in the metabolic pathway.

Li and Fang [201] studied the inhibition of six heavy metals and found the bioactivity of hydrogen-producing sludge in the following order: Cu (most toxic) > Ni >Zn > Cr > Cd > Pb (least toxic). Hydrogen-producing sludge exhibited in general higher resistance to metal toxicity than methanogenic granular sludge. Furthermore, Han et al. [203] studied the effects of hematite nanoparticle concentration on hydrogen production in batch system. The optimum hematite nanoparticle concentration was 200 mg/l, with the maximum hydrogen yield of 3.21 mol H_2/mol sucrose which was 32.64 % higher than the blank test. The slow release of hematite nanoparticles had been verified by transmission electron microscopy (TEM). In addition, TEM analysis indicated that the hematite nanoparticles can increase the length and narrow the width of bacteria.

2.3.3.3 pH

The control of pH is crucial to fermentative hydrogen and methane production due to its effects on hydrogenase activity and metabolic pathways. When the pH of a fermentation medium is too low, hydrogenase activity and methanogens would be inhibited or there would be a switch in metabolic pathway resulting in cessation of hydrogen and methane generation. Anaerobic hydrogen production process is typical during the exponential growth phase of *clostridia* [212]. The reactions shift from a hydrogen/acid production phase to a solvent production phase when the population reaches the stationary growth phase. The accumulation of volatile fatty acids such as butyric and propionic acids and hydrogen during the exponential growth phase prompts this shift. Some researchers claimed that this shift occurred when the pH dropped to 4.5 or below [213, 214], while others found that the shift occurred at pH levels above 5.7 due to enzyme synthesis or enzyme activation, which led to solvent production [215]. Thus, it is important to remove excess hydrogen from the system and control the pH at an optimal range to maintain hydrogen production. Otherwise, the biohydrogen production will stop due to the microbial population shift caused by pH uncontrolled in the desired range.

Khanal et al. [216] investigated the effect of pH on biological hydrogen production using sucrose and starch as organic substrates in batch system. Based on the evaluation of maximum hydrogen production rate, the optimum operational pH range was about 5.5–5.7. This result could be applied in continuous-flow processes to maintain a high rate of hydrogen production. Lin and Lay [217] found that phosphate acted as a better buffer source for hydrogen production than carbonate. Its addition enhanced hydrogen production by 1.9 times and decreased the lag period. Cavinato et al. [218] found that recirculation of anaerobic digested sludge after a mild solid separation to control the pH in an optimal hydrogen production range of 5–6 resulted in a stable hydrogen production output. The importance of pH control for continuous hydrogen production has been investigated extensively. The rapid pH depletion could cause a metabolic change of the microorganisms in the hydrogen production process, resulting in the shift of intermediate production pathway and a decrease in hydrogen production or system upset [216].

Reported optimal pH values for different systems or substrates differed substantially from 4.0 to 6.5, but for each specific situation, the optimal pH range was quite narrow within 0.5 [219]. Chu et al. [220] also reported a pH-phased two-stage fermentation process (combining thermophilic hydrogen production and mesophilic methane production) with recirculating digested sludge. These results show that a recirculation of precipitated digester sludge to a hydrogen reactor can maintain the hydrogen reactor pH at an optimal range without adding any reagents. Recently, most researches operated the hydrogen production fermenter at pH 5–5.5 as an optimal condition at mesophilic temperature [54, 69, 221–223].

2.3.3.4 Temperature

Fermentative hydrogen production via mixed cultures is conducted mostly under mesophilic (20–40 °C) and thermophilic (50–60 °C) conditions with only few studies being carried out under hyperthermophilic (65–75 °C) conditions. Biohydrogen production temperatures within 23–60 °C show that hydrogen production yield and hydrogen production rate increase along with the temperature increment [183].

Vatsala et al. [224] improved hydrogen production from a sugarcane distillery effluent using co-cultures at 100 m^3 reactor and found that unsteady hydrogen production was presumably due to temperature variation during daytime (32–39 °C) and nighttime (26–32 °C). Zhang et al. [225] reported biogas production from brown grease at mesophilic temperature (34.3–37.9 °C) in a pilot-scale high-rate anaerobic digester with a methane yield of 0.40–0.77 m^3 CH_4/kg-VS (higher than a typical range of other food wastes, 0.11–0.42 m^3 CH_4/kg-VS), a mean methane content of 75 %, and <200 ppm of hydrogen sulfide.

Cheong and Hansen [226] indicate that thermophilic acidogenesis enhances hydrogen production consistent with the biochemical pathway of butyrate fermentation. Under thermophilic temperatures (55 °C), the maximum hydrogen production potential of 134 ml with a specific hydrogen production rate of 25 ml H_2/h.g cell can be achieved. Throughout the thermophilic batch experiments, the main intermediate metabolites were acetate, n-butyrate, and ethanol. Propionate formation was suppressed completely during fermentation. Yields of produced hydrogen were correlated with increasing concentrations of n-butyrate, and quantities of ethanol present were significant in the batches producing lower yields of hydrogen.

Zhang et al. [19] studied conversion of starchy wastewater into hydrogen at thermophilic condition (55 °C) with batch experiments. The mixed liquor was composed mostly of acetate (40.2–53.4 %) and butyrate (26.0–40.9 %). Luo et al. [227] evaluated the pretreatment methods on mixed inoculum for both batch and continuous thermophilic biohydrogen production from cassava stillage. They found that butyrate was predominant and accounted for more than 75 % of the total amount of VFA/ethanol except the case of loading-shock pretreated sludge (56 % in this study). Butyrate concentration was observed to be correlated with the hydrogen production. Chen et al. [228] also found highest biohydrogen production was obtained when butyrate was predominate (70–85 % of the total VFA/ethanol).

Lin et al. [229] studied biohydrogen production with mesophilic conditions with a mixed microflora on a pilot scale. They found that the primary soluble microbial products (SMP) were butyrate (iso- and n-butyrate) accounting for 44.4–53.2 % of SMP. Acetate was also produced and accounted for 21.3–26.4 % of SMP. Productions of propionate and ethanol ranged from 7.2–10.6 % to 14.3–22.3 % of SMP, respectively. However, propionate and ethanol are unfavorable metabolites for hydrogen production [158, 172, 230]. The ratios of ethanol/acetate and acetate/butyrate have been used to indicate the performance of hydrogen production [231, 232]. Wu et al. [172] indicated that there might be an optimal acetate/butyrate ratio for hydrogen production, but the ratio is highly dependent on the anaerobic culture or the carbon substrate used.

2.3.3.5 HRT

Shortening hydraulic retention times (HRTs) is a well-used and effective operation strategy to enhance hydrogen production from organic wastewater and solid wastes because of its ability to exclude methanogens which have longer generation time.

The proper HRTs for hydrogen and methane production from organic fractions of municipal solid wastes (OFMSW) are 1–2 days and 10–15 days, respectively. A thermophilic hydrogen production reactor operating at HRT 1.3 days and a mesophilic methane production reactor operating at HRT 5.0 day have been combined to convert OFMSW into a pilot-scale two-phase fermentation system [220]. A pilot-scale two-phase hydrogen/methane fermentation system for food waste was operated at HRT 21 h with a peak hydrogen yield of 1.82 H_2 mol/mol glucose. Over 80 % of the methane was produced in the methane fermentation tank with acetic acid as the dominant organic acid. An economic evaluation shows that two-phase hydrogen/methane fermentation has greater potential for recovering energy than that of methane fermentation [233]. Cavinato et al. [218] operated pilot-scale hydrogen and methane fermenters by using HRTs of 3.3 and 12.6 days resulting in a specific hydrogen yield of 66.7 l/kg total volatile solids (TVS) and a specific biogas yield of 0.72 m^3/kg TVS respectively.

For most studies on continuously dark fermentative hydrogen production (DFHP), continuous systems are expected to operate at a low HRT 36–12 h [222, 234], very low of HRT 12–2 h [69, 229, 235–239], for obtaining a high biohydrogen production that can be operated at extremely low of HRT 2–0.5 h [35, 157, 240–243] with immobilized cell in the biohydrogen production fermenters. A mixture of food industry wastewater with rice straw hydrolyzate as substrate was conducted in a continuously stirred anaerobic bioreactor (CSABR) at HRT 4 h and found the hydrogen production of 10 l/l·d [69]. The rice straw hydrolyzate as sole substrate in continuously external circulating bioreactor (CECBR) prevented biomass washout by using a high volumetric flow rate with HRT of 4–2 h. It was found that the value of hydrogen production rate of 16.32 l/l·d at HRT 4 h was three times more than that at HRT 8 h [4].

2.4 Conclusions and Future Outlook

In recent years, the goal of biohydrogen systems has been to design economically viable hydrogen production processes. In this chapter, the potential feedstock, pretreatment processes, microorganisms, fermenter types, and operational conditions have been introduced. From this segment, readers will have attained a basic knowledge on biohydrogen production technologies from biomass waste. However, commercial biohydrogen production plants are not yet established because abundant and suitable feedstocks are not easily accessible yet. The economic feasibility of the dark fermentative hydrogen production process is dependent on the availability of cellulosic materials and organic wastes. Enhancements in the hydrogen production rate and yield from these feedstocks are important subjects for metabolic engineering. From an engineering point of view, the easily converted and abundant feedstock should be the first option to supply the biohydrogen production plants. Carbohydrate-rich organic wastes are a promising feedstock for anaerobic biohydrogen production. The sugary wastewater has a big potential for hydrogen

Fig. 2.2 Two-stage biohydrogen and biomethane production pilot plant in Feng Chia University campus with hydrogen fermenter of 0.4 m^3 and methane fermenter of 2 m^3

fermentation for future industrial applications. The potential economic gains from the internal rate of return (IRR) evaluation result in the motivation to apply this technology. From a global economic perspective, sugary wastewater could have higher profits with biogas energy used on-site for replacing natural gas and also could satisfy the energy requirements of some local areas. The adoption of fermentative hydrogen production from organic wastes will potentially lead to great advancements in energy and the environment. Finally, there are many pilot plants still at work in Spain [244], Taiwan (see Fig. 2.2) [229, 245], Italy [246], and the UK [247]. Toyota already has the first commercial fuel cell car (Mirai) [248] selling in Europe and North America right now. We believe that profitable and sustainable hydrogen production from biomass waste will be achieved in the near future.

References

1. Thompson W, Meyer S. Second generation biofuels and food crops: co-products or competitors? Glob Food Secur. 2013;2:89–96.
2. De Souza ACC, Silveira JL. Hydrogen production utilizing glycerol from renewable feedstocks—the case of Brazil. Renew Sustain Energy Rev. 2011;15:1835–50.
3. Kraemer J, Bagley D. Improving the yield from fermentative hydrogen production. Biotechnol Lett. 2007;29:685–95.
4. Liu C-M, Wu S-Y, Chu C-Y, Chou Y-P. Biohydrogen production from rice straw hydrolyzate in a continuously external circulating bioreactor. Int J Hydrog Energy. 2014;39:19317–22.

5. Abo-Hashesh M, Hallenbeck P. Fermentative hydrogen production. In: Hallenbeck PC, editor. Microbial technologies in advanced biofuels production. New York: Springer; 2012. p. 77–92.
6. Siriporn Yossan SO-T, Prasertsan P. Effect of initial pH, nutrients and temperature on hydrogen production from palm oil mill effluent using thermotolerant consortia and corresponding microbial communities. Int J Hydrog Energy. 2012;37:13806–14.
7. Show K-Y, Lee D-J, Chang J-S. Bioreactor and process design for biohydrogen production. Bioresour Technol. 2011;102(102):8524–33.
8. Das D, Veziroglu TN. Advances in biological hydrogen production processes. Int J Hydrog Energy. 2008;33:6046–57.
9. Rasdi Z, Rahman NAA, Abd-Azi S, Lai-Yee P, Zulkhairi M, Yusoff M, et al. Statistical optimization of biohydrogen production from palm oil mill effluent by natural microflora. Open Biotechnol J. 2009;3:79–86.
10. Khaleb NA, Jahim JM, Kamal SA. Biohydrogen production using hydrolysates of Palm Oil Mill Effluent (Pome). J Asian Sci Res. 2011;2:705–10.
11. Kamal SA, Jahim JM, Anuar N, Hassan O, Daud WRW, Mansor MF, et al. Pre-treatment effect of Palm Oil Mill Effluent (POME) during hydrogen production by a local isolate clostridium butyricum. Int J Adv Sci Eng Inf Technol. 2012;2(4):54–60. ISSN: 2088–5334.
12. Chong M-L, Rahim RA, Shirai Y, Hassan MA. Biohydrogen production by Clostridium butyricum EB6 from palm oil mill effluent. Int J Hydrog Energy. 2009;34:764–71.
13. Prasertsan P, O-Thong S, N-Kr B. Optimization and microbial community analysis for production of biohydrogen from palm oil mill effluent by thermophilic fermentative process. Int J Hydrog Energy. 2009;34:7448–59.
14. O-Thong S, Mamimin C, Prasertsan P. Effect of temperature and initial pH on biohydrogen production from palm oil mill effluent: long-term evaluation and microbial community analysis. Microb Biotechnol Electron J Biotechnol. 2011;14(5).
15. Atif AAY, Fakhru'l-Razi A, Ngan MA, Morimoto M, Iyuke SE, Veziroglu NT. Fed batch production of hydrogen from palm oil mill effluent using anaerobic microflora. Int J Hydrog Energy. 2005;30:1393–7.
16. O-Thong S, Prasertsana P, Intrasungkha N, Dhamwichukorn S, N-Kr B. Optimization of simultaneous thermophilic fermentative hydrogen production and COD reduction from palm oil mill effluent by Thermoanaerobacterium-rich sludge. Int J Hydrog Energy. 2008;33:1221–31.
17. O-Thong S, Prasertsan P, Intrasungkha N. Improvement of biohydrogen production and treatment efficiency on palm oil mill effluent with nutrient supplementation at thermophilic condition using an anaerobic sequencing batch reactor. Enzym Microb Technol. 2007;41:583–90.
18. Akutsu Y, Li Y-Y, Harada H, Yu H-Q. Effects of temperature and substrate concentration on biological hydrogen production from starch. Int J Hydrog Energy. 2009;34:2558–66.
19. Zhang T, Liu H, Fang HHP. Biohydrogen production from starch in wastewater under thermophilic condition. J Environ Manag. 2003;69:149–56.
20. Masset J, Calusinska M, Hamilton C, Hiligsmann S, Joris B, Wilmotte A, et al. Fermentative hydrogen production from glucose and starch using pure strains and artificial co-cultures of Clostridium spp. Biotechnol Biofuels. 2012;5:3.
21. Chen S-D, Lee K-S, Lo Y-C, Chen W-M, Wu J-F, Lin C-Y, et al. Batch and continuous biohydrogen production from starch hydrolysate by Clostridium species. Int J Hydrog Energy. 2008;33:1803–12.
22. O-Thong S, Hniman A, Prasertsan P, Imai T. Biohydrogen production from cassava starch processing wastewater by thermophilic mixed cultures. Int J Hydrog Energy. 2011;36:3409–16.
23. Chen S-D, Sheu D-S, Chen W-M, Huang Y-CL-I, Lin C-Y, Chang J-S. Dark hydrogen fermentation from hydrolyzed starch treated with recombinant amylase originating from Caldimonas taiwanensis On1. Biotechnol Prog. 2007;23:1312–20.

24. Hussy I, Hawkes FR, Dinsdale R, Hawkes DL. Continuous fermentative hydrogen production from a wheat starch co-product by mixed microflora. Biotech Bioeng. 2003;84:619–26.
25. Liu G, Shen J. Effects of culture and medium conditions on hydrogen production from starch using anaerobic bacteria. J Biosci Bioeng. 2004;98(4):251–6.
26. Wang C-H, Lu W-B, Chang J-S. Feasibility study on fermentative conversion of raw and hydrolyzed starch to hydrogen using anaerobic mixed microflora. Int J Hydrog Energy. 2007;32:3849–59.
27. Lee K-S, Hsu Y-F, Lo Y-C, Lin P-J, Lin C-Y, Chang J-S. Exploring optimal environmental factors for fermentative hydrogen production from starch using mixed anaerobic microflora. Int J Hydrog Energy. 2008;33:1565–72.
28. Arooj MF, Han S-K, Kim S-H, Kim D-H, Shin H-S. Continuous biohydrogen production in a CSTR using starch as a substrate. Int J Hydrog Energy. 2008;33:3289–94.
29. Zong W, Yu R, Zhang P, Fan M, Zhou Z. Efficient hydrogen gas production from cassava and food waste by a two-step process of dark fermentation and photo-fermentation. Biomass Bioenergy. 2009;33:1458–63.
30. Kim S-H, Han S-K, Shin H-S. Feasibility of biohydrogen production by anaerobic co-digestion of food waste and sewage sludge. Int J Hydrog Energy. 2004;29:1607–16.
31. Shin H-S, Youn J-H, Kim S-H. Hydrogen production from food waste in anaerobic mesophilic and thermophilic acidogenesis. Int J Hydrog Energy. 2004;29:1355–63.
32. Reungsang A, Sreela-or C, Plangklang P. Non-sterile bio-hydrogen fermentation from food waste in a continuous stirred tank reactor (CSTR): performance and population analysis. Int J Hydrog Energy. 2013;38:15630–7.
33. Han S-K, Shin H-S. Biohydrogen production by anaerobic fermentation of food waste. Int J Hydrog Energy. 2004;29:569–77.
34. Hsiao C-L, Chang J-J, Wu J-H, Chin W-C, We F-S, Huang C-C, et al. Clostridium strain co-cultures for biohydrogen production enhancement from condensed molasses fermentation solubles. Int J Hydrog Energy. 2009;34:7173–81.
35. Chu C-Y, Wu S-Y, Hsieh C, Lin C-Y. Biohydrogen production from immobilized cells and suspended sludge systems with condensed molasses fermentation solubles. Int J Hydrog Energy. 2011;36:14078–85.
36. Lay C-H, Wu J-H, Hsiao C-L, Chang J-J, Chen C-C, Lin C-Y. Biohydrogen production from soluble condensed molasses fermentation using anaerobic fermentation. Int J Hydrog Energy. 2010;35:13445–51.
37. Chen C-C, Wu J-H, Lay C-H, Sen B, Chang J-S. Kinetics of hydrogen production from condensed molasses fermentation solubles using sewage sludge in a continuous stirred tank reactor. Sustain Environ Res. 2011;21(2):117–21.
38. Chang J-J, Wu J-H, Wen F-S, Hung K-Y, Chen Y-T, Hsiao C-L, et al. Molecular monitoring of microbes in a continuous hydrogen-producing system with different hydraulic retention time. Int J Hydrog Energy. 2008;33:1579–85.
39. Liu H, Fang HHP. Hydrogen production from wastewater by acidogenic granular sludge. Water Sci Technol. 2002;47(1):153–8.
40. Oh S, Logan BE. Hydrogen and electricity production from a food processing wastewater using fermentation and microbial fuel cell technologies. Water Res. 2005;39:4673–82.
41. Ueno Y, Otsuka S, Morimoto M. Hydrogen production from industrial wastewater by anaerobic microflora in chemostat culture. J Ferment Bioeng. 1996;82:194–7.
42. Yu H, Zhu Z, Hu W, Zhang H. Hydrogen production from rice winery wastewater in an upflow anaerobic reactor by using mixed anaerobic cultures. Int J Hydrog Energy. 2002;27:1359–65.
43. Ginkel SV, Oh S-E, Logan BE. Biohydrogen gas production from food processing and domestic wastewaters. Int J Hydrog Energy. 2005;30:1535–42.
44. Kapdan IK, Kargi F. Bio-hydrogen production from waste materials. Enzym Microb Technol. 2006;38:569–82.

45. Mohan SV, Babu VL, Sarma P. Anaerobic biohydrogen production from dairy wastewater treatment in sequencing batch reactor (AnSBR): effect of organic loading rate. Enzym Microb Technol. 2007;41:506–15.
46. Ren N, Chua H, Chan S, Tsang Y, Wang Y, Sin N. Assessing optimal fermentation type for bio-hydrogen production in continuous-flow acidogenic reactors. Bioresour Technol. 2007;98:1774–80.
47. Davila-Vazquez G, Alatriste-Mondragón F, de León-Rodríguez A, Razo-Flores E. Fermentative hydrogen production in batch experiments using lactose, cheese whey and glucose: influence of initial substrate concentration and pH. Int J Hydrog Energy. 2008;33:4989–97.
48. Gadhe A, Sonawane SS, Varma MN. Enhancement effect of hematite and nickel nanoparticles on biohydrogen production from dairy wastewater. Int J Hydrog Energy. 2015;40:4502–11.
49. Koutrouli EC, Kalfas H, Gavala HN, Skiadas IV, Stamatelatou K, Lyberatos G. Hydrogen and methane production through two-stage mesophilic anaerobic digestion of olive pulp. Bioresour Technol. 2009;100:3718–23.
50. Ntaikou I, Kourmentza C, Koutrouli E, Stamatelatou K, Zampraka A, Kornaros M, et al. Exploitation of olive oil mill wastewater for combined biohydrogen and biopolymers production. Bioresour Technol. 2009;100:3724–30.
51. Jung K-W, Kim D-H, Lee M-Y, Shin H-S. Two-stage UASB reactor converting coffee drink manufacturing wastewater to hydrogen and methane. Int J Hydrog Energy. 2012;37:7473–81.
52. Zhu G, Liu C, Li J, Ren N, Liu L, Huang X. Fermentative hydrogen production from beet sugar factory wastewater treatment in a continuous stirred tank reactor using anaerobic mixed consortia. Front Environ Sci Eng. 2013;7:143–50.
53. Xie L, Dong N, Wang L, Zhou Q. Thermophilic hydrogen production from starch wastewater using two-phase sequencing batch fermentation coupled with UASB methanogenic effluent recycling. Int J Hydrog Energy. 2014;39:20942–9.
54. Lay CH, Sen B, Kuo SY, Chen CC, Lin CY. Biohydrogen production from textile wastewater by mixed microflora in an intermittent-flow, stirred tank reactor: effect of feeding frequency. J Chin Chem Soc. 2014;61:791–6.
55. Han W, Liu DN, Shi YW, Tang JH, Li YF, Ren NQ. Biohydrogen production from food waste hydrolysate using continuous mixed immobilized sludge reactors. Bioresour Technol. 2015;180:54–8.
56. Ntaikou I, Antonopoulou G, Lyberatos G. Biohydrogen production from biomass and wastes via dark fermentation: a review. Waste Biomass Valor. 2010;1:21–39.
57. Lay J-J, Fan K-S. Influence of chemical nature of organic wastes on their conversion to hydrogen by heat-shock digested sludge. Int J Hydrog Energy. 2003;28:1361–7.
58. Yuan X, Wen B, Ma X, Zhu W, Wang X, Chen S, et al. Enhancing the anaerobic digestion of lignocellulose of municipal solid waste using a microbial pretreatment method. Bioresour Technol. 2014;154:1–9.
59. Ishola MM, Jahandideh A, Haidarian B, Brandberg T, Taherzadeh MJ. Simultaneous saccharification, filtration and fermentation (SSFF): a novel method for bioethanol production from lignocellulosic biomass. Bioresour Technol. 2013;133:68–73.
60. Quéméneur M, Bittel M, Trably E, Dumas C, Fourage L, Ravot G, et al. Effect of enzyme addition on fermentative hydrogen production from wheat straw. Int J Hydrog Energy. 2012;37:10639–47.
61. Zhang M-L, Fan Y-T, Xing Y, Pan C-M, Zhang G-S, Lay J-J. Enhanced biohydrogen production from cornstalk wastes with acidification pretreatment by mixed anaerobic cultures. Biomass Bioenergy. 2007;31:250–4.
62. Pan C, Fan Y, Hou H. Fermentative production of hydrogen from wheat bran by mixed anaerobic cultures. Ind Eng Chem Res. 2008;47:5812–8.

63. Zhu H, Parker W, Basnar R, Proracki A, Falletta P, Béland M, et al. Buffer requirements for enhanced hydrogen production in acidogenic digestion of food wastes. Bioresour Technol. 2009;100:5097–102.
64. Wang X, Y-c Z. A bench scale study of fermentative hydrogen and methane production from food waste in integrated two-stage process. Int J Hydrog Energy. 2009;34:245–54.
65. Guo XM, Trably E, Latrille E, Carrère H, Steyer J-P. Hydrogen production from agricultural waste by dark fermentation: a review. Int J Hydrog Energy. 2010;35:10660–73.
66. Yang Z, Guo R, Xu X, Fan X, Li X. Enhanced hydrogen production from lipid-extracted microalgal biomass residues through pretreatment. Int J Hydrog Energy. 2010;35:9618–23.
67. Carver SM, Hulatt CJ, Thomas DN, Tuovinen OH. Thermophilic, anaerobic co-digestion of microalgal biomass and cellulose for H2 production. Biodegradation. 2011;22:805–14.
68. Lakaniemi A-M, Hulatt CJ, Thomas DN, Tuovinen OH, Puhakka JA. Biogenic hydrogen and methane production from Chlorella vulgaris and Dunaliella tertiolecta biomass. Biotechnol Biofuels. 2011;4:34.
69. Liu C-M, Chu C-Y, Lee W-Y, Li Y-C, Wu S-Y, Chou Y-P. Biohydrogen production evaluation from rice straw hydrolysate by concentrated acid pre-treatment in both batch and continuous systems. Int J Hydrog Energy. 2013;38:15823–9.
70. Levin DB, Islam R, Cicek N, Sparling R. Hydrogen production by Clostridium thermocellum 27405 from cellulosic biomass substrates. Int J Hydrog Energy. 2006;31:1496–503.
71. Datar R, Huang J, Maness P-C, Mohagheghi A, Czernik S, Chornet E. Hydrogen production from the fermentation of corn stover biomass pretreated with a steam-explosion process. Int J Hydrog Energy. 2007;32:932–9.
72. Pattra S, Sangyoka S, Boonmee M, Reungsang A. Bio-hydrogen production from the fermentation of sugarcane bagasse hydrolysate by Clostridium butyricum. Int J Hydrog Energy. 2008;33:5256–65.
73. Kyazze G, Dinsdale R, Hawkes FR, Guwy AJ, Premier GC, Donnison I. Direct fermentation of fodder maize, chicory fructans and perennial ryegrass to hydrogen using mixed microflora. Bioresour Technol. 2008;99:8833–9.
74. Ntaikou I, Gavala H, Kornaros M, Lyberatos G. Hydrogen production from sugars and sweet sorghum biomass using Ruminococcus albus. Int J Hydrog Energy. 2008;33:1153–63.
75. Ivanova G, Rákhely G, Kovács KL. Thermophilic biohydrogen production from energy plants by Caldicellulosiruptor saccharolyticus and comparison with related studies. Int J Hydrog Energy. 2009;34:3659–70.
76. Reilly M, Dinsdale R, Guwy A. Mesophilic biohydrogen production from calcium hydroxide treated wheat straw. Int J Hydrog Energy. 2014;39:16891–901.
77. Patel AK, Debroy A, Sharma S, Saini R, Mathur A, Gupta R, et al. Biohydrogen production from a novel alkalophilic isolate Clostridium sp. IODB-O3. Bioresour Technol. 2015;175:291–7.
78. Chairattanamanokorn P, Penthamkeerati P, Reungsang A, Lo Y-C, Lu W-B, Chang J-S. Production of biohydrogen from hydrolyzed bagasse with thermally preheated sludge. Int J Hydrog Energy. 2009;34:7612–7.
79. Cao G, Ren N, Wang A, Lee D-J, Guo W, Liu B, et al. Acid hydrolysis of corn stover for biohydrogen production using Thermoanaerobacterium thermosaccharolyticum W16. Int J Hydrog Energy. 2009;34:7182–8.
80. Phummala K, Imai T, Reungsang A, Chairattanamanokorn P, Sekine M, Higuchi T, et al. Delignification of disposable wooden chopsticks waste for fermentative hydrogen production by an enriched culture from a hot spring. J Environ Sci. 2014;26:1361–8.
81. Saratale GD, Kshirsagar SD, Sampange VT, Saratale RG, Oh S-E, Govindwar SP, et al. Cellulolytic enzymes production by utilizing agricultural wastes under solid state fermentation and its application for biohydrogen production. Appl Biochem Biotechnol. 2014;174:2801–17.
82. Taherzadeh MJ, Karimi K. Pretreatment of lignocellulosic wastes to improve ethanol and biogas production: a review. Int J Mol Sci. 2008;9:1621–51.

83. Meng X, Ragauskas AJ. Recent advances in understanding the role of cellulose accessibility in enzymatic hydrolysis of lignocellulosic substrates. Curr Opin Biotechnol. 2014;27:150–8.
84. Zakaria MR, Fujimoto S, Hirata S, Hassan MA. Ball milling pretreatment of oil palm biomass for enhancing enzymatic hydrolysis. Appl Biochem Biotechnol. 2014;173:1778–89.
85. Mais U, Esteghlalian AR, Saddler JN, Mansfield SD. Enhancing the enzymatic hydrolysis of cellulosic materials using simultaneous ball milling. Springer; New York, 2002.
86. Muller CD, Abu-Orf M, Novak JT. Application of mechanical shear in an internal-recycle for the enhancement of mesophilic anaerobic digestion. Water Environ Res. 2007;79:297–304.
87. Yang C, Shen Z, Yu G, Wang J. Effect and aftereffect of γ radiation pretreatment on enzymatic hydrolysis of wheat straw. Bioresour Technol. 2008;99:6240–5.
88. Mamar S, Hadjadj A. Radiation pretreatments of cellulose materials for the enhancement of enzymatic hydrolysis. Int J Radiat Appl Instrum Part C Radiat Phys Chem. 1990;35:451–5.
89. Subhedar PB, Gogate PR. Intensification of enzymatic hydrolysis of lignocellulose using ultrasound for efficient bioethanol production: a review. Ind Eng Chem Res. 2013;52:11816–28.
90. Jeun J-P, Lee B-M, Lee J-Y, Kang P-H, Park J-K. An irradiation-alkaline pretreatment of kenaf core for improving the sugar yield. Renew Energy. 2015;79:51–5.
91. Chu CP, Lee DJ, Chang B-V, You CS, Tay JH. "Weak" ultrasonic pre-treatment on anaerobic digestion of flocculated activated biosolids. Water Res. 2002;36:2681–8.
92. Wang F, Wang Y, Ji M. Mechanisms and kinetics models for ultrasonic waste activated sludge disintegration. J Hazard Mater. 2005;123:145–50.
93. Talebnia F, Karakashev D, Angelidaki I. Production of bioethanol from wheat straw: an overview on pretreatment, hydrolysis and fermentation. Bioresour Technol. 2010;101:4744–53.
94. Galbe M, Zacchi G. A review of the production of ethanol from softwood. Appl Microbiol Biotechnol. 2002;59:618–28.
95. Sanchez OJ, Cardona CA. Trends in biotechnological production of fuel ethanol from different feedstocks. Bioresour Technol. 2008;99:5270–95.
96. Tomas-Pejo E, Oliva J, Ballesteros M. Realistic approach for full-scale bioethanol production from lignocellulose: a review. J Sci Ind Res. 2008;67:874.
97. Chen Y, Stevens MA, Zhu Y, Holmes J, Xu H. Understanding of alkaline pretreatment parameters for corn stover enzymatic saccharification. Biotechnol Biofuels. 2013;6:1–10.
98. Monlau F, Barakat A, Trably E, Dumas C, Steyer J-P, Carrère H. Lignocellulosic materials into biohydrogen and biomethane: impact of structural features and pretreatment. Crit Rev Environ Sci Technol. 2013;43:260–322.
99. Mosier N, Wyman C, Dale B, Elander R, Lee Y, Holtzapple M, et al. Features of promising technologies for pretreatment of lignocellulosic biomass. Bioresour Technol. 2005;96:673–86.
100. Zeitsch KJ. The chemistry and technology of furfural and its many by-products. Amsterdam: Elsevier; 2000.
101. Groff D, George A, Sun N, Sathitsuksanoh N, Bokinsky G, Simmons BA, et al. Acid enhanced ionic liquid pretreatment of biomass. Green Chem. 2013;15:1264–7.
102. Mackie K, Brownell H, West K, Saddler J. Effect of sulphur dioxide and sulphuric acid on steam explosion of aspenwood. J Wood Chem Technol. 1985;5:405–25.
103. Hu F, Ragauskas A. Pretreatment and lignocellulosic chemistry. Bioenergy Res. 2012;5:1043–66.
104. Jiang W, Chang S, Li H, Oleskowicz-Popiel P, Xu J. Liquid hot water pretreatment on different parts of cotton stalk to facilitate ethanol production. Bioresour Technol. 2015;176:175–80.
105. Cheng J, Su H, Zhou J, Song W, Cen K. Microwave-assisted alkali pretreatment of rice straw to promote enzymatic hydrolysis and hydrogen production in dark-and photo-fermentation. Int J Hydrog Energy. 2011;36:2093–101.

106. Brodeur G, Yau E, Badal K, Collier J, Ramachandran K, Ramakrishnan S. Chemical and physicochemical pretreatment of lignocellulosic biomass: a review. Enzym Res. 2011;2011:787532. p. 1–17.
107. Qiu W, Chen H. Enhanced the enzymatic hydrolysis efficiency of wheat straw after combined steam explosion and laccase pretreatment. Bioresour Technol. 2012;118:8–12.
108. Christy PM, Gopinath L, Divya D. A review on anaerobic decomposition and enhancement of biogas production through enzymes and microorganisms. Renew Sustain Energy Rev. 2014;34:167–73.
109. Dutta S, Wu KC-W. Enzymatic breakdown of biomass: enzyme active sites, immobilization, and biofuel production. Green Chem. 2014;16:4615–26.
110. Hyeon JE, You SK, Kang DH, Ryu S-H, Kim M, Lee S-S, et al. Enzymatic degradation of lignocellulosic biomass by continuous process using laccase and cellulases with the aid of scaffolding for ethanol production. Process Biochem. 2014;49:1266–73.
111. Gusakov AV, Salanovich TN, Antonov AI, Ustinov BB, Okunev ON, Burlingame RP, et al. Construction of highly efficient cellulase compositions for enzymatic hydrolysis of cellulose. 2014. US Patents 8,916,363 B2, p. 1–84.
112. Saratale GD, Chen S-D, Lo Y-C, Saratale RG, Chang J-S. Outlook of biohydrogen production from lignocellulosic feedstock using dark fermentation – a review. J Sci Ind Res. 2008;67:962.
113. Odom JM, Wall JD. Photoproduction of H_2 from cellulose by an anaerobic bacterial coculture. Appl Environ Microbiol. 1983;45:1300–5.
114. Lewis SM, Montgomery L, Garleb KA, Berger LL, Fahey GC. Effects of alkaline hydrogen peroxide treatment on in vitro degradation of cellulosic substrates by mixed ruminal microorganisms and Bacteroides succinogenes S85. Appl Environ Microbiol. 1988;54:1163–9.
115. Ramos LP. The chemistry involved in the steam treatment of lignocellulosic materials. Quim Nova. 2003;26:863–71.
116. Gregg D, Saddler JN. A techno-economic assessment of the pretreatment and fractionation steps of a biomass-to-ethanol process. Seventeenth symposium on biotechnology for fuels and chemicals. Springer; New York, 1996. p. 711–27.
117. Sivers MV, Zacchi G, Olsson L, Hahn-Hügerdal B. Cost analysis of ethanol production from willow using recombinant Escherichia coli. Biotechnol Prog. 1994;10:555–60.
118. Shevchenko SM, Beatson RP, Saddler JN. The nature of lignin from steam explosion/enzymatic hydrolysis of softwood. Appl Biochem Biotechnol. 1999;79:867–76.
119. Muurinen E. Organosolv pulping: a review and distillation study related to peroxyacid pulping. Finland: Oulun yliopisto; 2000.
120. Xb Z, Wang L, Liu D. Effect of several factors on peracetic acid pretreatment of sugarcane bagasse for enzymatic hydrolysis. J Chem Technol Biotechnol. 2007;82:1115–21.
121. Teixeira LC, Linden JC, Schroeder HA. Optimizing peracetic acid pretreatment conditions for improved simultaneous saccharification and co-fermentation (SSCF) of sugar cane bagasse to ethanol fuel. Renew Energy. 1999;16:1070–3.
122. Duff SJ, Murray WD. Bioconversion of forest products industry waste cellulosics to fuel ethanol: a review. Bioresour Technol. 1996;55:1–33.
123. Sun Y, Cheng J. Hydrolysis of lignocellulosic materials for ethanol production: a review. Bioresour Technol. 2002;83:1–11.
124. Ostovareh S, Karimi K, Zamani A. Efficient conversion of sweet sorghum stalks to biogas and ethanol using organosolv pretreatment. Ind Crop Prod. 2015;66:170–7.
125. Degirmenci V, Hensen EJ. Development of a heterogeneous catalyst for lignocellulosic biomass conversion: glucose dehydration by metal chlorides in a silica-supported ionic liquid layer. Environ Prog Sustain Energy. 2014;33:657–62.
126. Grilc M, Likozar B, Levec J. Hydrotreatment of solvolytically liquefied lignocellulosic biomass over NiMo/Al 2 O 3 catalyst: reaction mechanism, hydrodeoxygenation kinetics and mass transfer model based on FTIR. Biomass Bioenergy. 2014;63:300–12.

127. Larabi C, Al Maksoud W, Szeto KC, Garron A, Arquilliere PP, Walter JJ, et al. Multifunctional heterogeneous catalyst for one step transformation of lignocellulosic biomass into low oxygenated hydrocarbons. Appl Catal A Gen. 2015;495:162–72.
128. Kobayashi H, Ohta H, Fukuoka A. Conversion of lignocellulose into renewable chemicals by heterogeneous catalysis. Catal Sci Technol. 2012;2:869–83.
129. Koo B-W, Kim H-Y, Park N, Lee S-M, Yeo H, Choi I-G. Organosolv pretreatment of Liriodendron tulipifera and simultaneous saccharification and fermentation for bioethanol production. Biomass Bioenergy. 2011;35:1833–40.
130. Hu J, Shen D, Wu S, Zhang H, Xiao R. Composition analysis of organosolv lignin and its catalytic solvolysis in supercritical alcohol. Energy Fuel. 2014;28:4260–6.
131. Lin Y-C, Huber GW. The critical role of heterogeneous catalysis in lignocellulosic biomass conversion. Energy Environ Sci. 2009;2:68–80.
132. Lee J-W, Jeffries TW. Efficiencies of acid catalysts in the hydrolysis of lignocellulosic biomass over a range of combined severity factors. Bioresour Technol. 2011;102:5884–90.
133. Guo F, Fang Z, Xu CC, Smith RL. Solid acid mediated hydrolysis of biomass for producing biofuels. Prog Energy Combust Sci. 2012;38:672–90.
134. Lanzafame P, Temi D, Perathoner S, Spadaro A, Centi G. Direct conversion of cellulose to glucose and valuable intermediates in mild reaction conditions over solid acid catalysts. Catal Today. 2012;179:178–84.
135. Feng G, Fang Z. Solid-and nano-catalysts pretreatment and hydrolysis techniques. Pretreatment techniques for biofuels and biorefineries. Berlin: Springer; 2013. p. 339–66.
136. Kamalaskar LB, Dhakephalkar P, Meher K, Ranade D. High biohydrogen yielding< i> Clostridium sp.</i> DMHC-10 isolated from sludge of distillery waste treatment plant. Int J Hydrog Energy. 2010;35:10639–44.
137. Chen W-M, Tseng Z-J, Lee K-S, Chang J-S. Fermentative hydrogen production with< i> Clostridium butyricum</i> CGS5 isolated from anaerobic sewage sludge. Int J Hydrog Energy. 2005;30:1063–70.
138. Baghchehsaraee B, Nakhla G, Karamanev D, Margaritis A, Reid G. The effect of heat pretreatment temperature on fermentative hydrogen production using mixed cultures. Int J Hydrog Energy. 2008;33:4064–73.
139. Adav SS, Lee D-J, Wang A, Ren N. Functional consortium for hydrogen production from cellobiose: concentration-to-extinction approach. Bioresour Technol. 2009;100:2546–50.
140. O-Thong S, Prasertsan P, Birkeland N-K. Evaluation of methods for preparing hydrogen-producing seed inocula under thermophilic condition by process performance and microbial community analysis. Bioresour Technol. 2009;100:909–18.
141. Ren N-Q, Guo W-Q, Wang X-J, Xiang W-S, Liu B-F, Wang X-Z, et al. Effects of different pretreatment methods on fermentation types and dominant bacteria for hydrogen production. Int J Hydrog Energy. 2008;33:4318–24.
142. Liu H, Wang G. Hydrogen production of a salt tolerant strain Bacillus sp. B2 from marine intertidal sludge. World J Microbiol Biotechnol. 2012;28:31–7.
143. Wu K-J, Saratale GD, Lo Y-C, Chen W-M, Tseng Z-J, Chang M-C, et al. Simultaneous production of 2, 3-butanediol, ethanol and hydrogen with a< i> Klebsiella</i> sp. strain isolated from sewage sludge. Bioresour Technol. 2008;99:7966–70.
144. Whitman WB, Bowen TL, Boone DR. The methanogenic bacteria. The prokaryotes. Springer; New York, 2006. p. 165–207.
145. Liu H, Wang G, Zhu D, Pan G. Enrichment of the hydrogen-producing microbial community from marine intertidal sludge by different pretreatment methods. Int J Hydrog Energy. 2009;34:9696–701.
146. Baghchehsaraee B, Nakhla G, Karamanev D, Margaritis A. Fermentative hydrogen production by diverse microflora. Int J Hydrog Energy. 2010;35:5021–7.
147. Wong YM, Wu TY, Juan JC. A review of sustainable hydrogen production using seed sludge via dark fermentation. Renew Sustain Energy Rev. 2014;34:471–82.

148. Levin DB, Pitt L, Love M. Biohydrogen production: prospects and limitations to practical application. Int J Hydrog Energy. 2004;29:173–85.
149. Brentner LB, Peccia J, Zimmerman JB. Challenges in developing biohydrogen as a sustainable energy source: implications for a research agenda. Environ Sci Technol. 2010;44:2243–54.
150. Hallenbeck PC, Benemann JR. Biological hydrogen production; fundamentals and limiting processes. Int J Hydrog Energy. 2002;27:1185–93.
151. Kim S, Seol E, Oh Y-K, Wang G, Park S. Hydrogen production and metabolic flux analysis of metabolically engineered Escherichia coli strains. Int J Hydrog Energy. 2009;34:7417–27.
152. Hallenbeck P. Fundamentals of the fermentative production of hydrogen. Water Sci Technol. 2005;52:21–9.
153. Mathews J, Wang G. Metabolic pathway engineering for enhanced biohydrogen production. Int J Hydrog Energy. 2009;34:7404–16.
154. Chabrière E, Vernède X, Guigliarelli B, Charon M-H, Hatchikian EC, Fontecilla-Camps JC. Crystal structure of the free radical intermediate of pyruvate: ferredoxin oxidoreductase. Science. 2001;294:2559–63.
155. Wang S, Huang H, Moll J, Thauer RK. NADP+ reduction with reduced ferredoxin and NADP + reduction with NADH are coupled via an electron-bifurcating enzyme complex in Clostridium kluyveri. J Bacteriol. 2010;192:5115–23.
156. Ren N-Q, Tang J, Liu B-F, Guo W-Q. Biological hydrogen production in continuous stirred reactor systems with suspended and attached microbial growth. Int J Hydrog Energy. 2010;35:2807–13.
157. Wu S-Y, Chu C-Y, Yeh W-Z. Aspect ratio effect of bioreactor on fermentative hydrogen production with immobilized sludge. Int J Hydrog Energy. 2013;38:6154–60.
158. Hawkes FR, Hussy I, Kyazze G, Dinsdale R, Hawkes DL. Continuous dark fermentative hydrogen production by mesophilicmicroflora: principles and progress. Int J Hydrog Energy. 2007;32:172–84.
159. Wang J, Wan W. Factors influencing fermentative hydrogen production: a review. Int J Hydrog Energy. 2009;3:799–811.
160. Lee K-S, Lo Y-C, Lin P-J, Chang J-S. Improving biohydrogen production in a carrier-induced granular sludge bed by altering physical configuration and agitation pattern of the bioreactor. Int J Hydrog Energy. 2006;31:1648–57.
161. Chang J-S, Lee K-S, Lin P-J. Biohydrogen production with fixed-bed bioreactors. Int J Hydrog Energy. 2002;27:1167–74.
162. Lee K-S, Lin P-J, Fangchiang K, Chang J-S. Continuous hydrogen production by anaerobic mixed microflora using a hollow-fiber microfiltration membrane bioreactor. Int J Hydrog Energy. 2007;32:950–7.
163. Show KY, Zhang ZP, Lee DJ. Design of bioreactors for biohydrogen production. J Sci Ind Res. 2008;67:941–9.
164. Xu H, Gong S, Sun Y, Ma H, Zheng M, Wang K. High-rate hydrogenotrophic methanogenesis for biogas upgrading: the role of anaerobic granules. Environ Technol. 2015;36:529–37.
165. Ab Halim MH, Anuar AN, Azmi SI, Jamal NSA, Wahab NA, Ujang Z, et al. Aerobic sludge granulation at high temperatures for domestic wastewater treatment. Bioresour Technol. 2015;185:445–9.
166. Lu X, Zhen G, Estrada AL, Chen M, Nic J, Hojo T, et al. Operation performance and granule characterization of upflow anaerobic sludge blanket (UASB) reactor treating wastewater with starch as the sole carbon source. Bioresour Technol. 2015;180:264–73.
167. Fang HH, Liu H, Zhang T. Characterization of a hydrogen-producing granular sludge. Biotechnol Bioeng. 2002;78:44–52.
168. Chang F-Y, Lin C-Y. Biohydrogen production using an up-flow anaerobic sludge blanket reactor. Int J Hydrog Energy. 2004;29:33–9.

169. Jung K-W, Kim D-H, Shin H-S. A simple method to reduce the start-up period in a H 2-producing UASB reactor. Int J Hydrog Energy. 2011;36:1466–73.
170. Jung K-W, Kim D-H, Shin H-S. Fermentative hydrogen production from Laminaria japonica and optimization of thermal pretreatment conditions. Bioresour Technol. 2011;102:2745–50.
171. Wu S-Y, Lin C-N, Chang J-S, Chang J-S. Biohydrogen production with anaerobic sludge immobilized by ethylene-vinyl acetate copolymer. Int J Hydrog Energy. 2005;30:1375–81.
172. Wu SY, Hung CH, Lin CN, Chen HW, Lee AS, Chang JS. Fermentative hydrogen production and bacterial community structure in high-rate anaerobic bioreactors containing silicone-immobilized and self-flocculated sludge. Biotechnol Bioeng. 2006;93:934–46.
173. Chu C-Y, Wu S-Y, Wu Y-C, Sen B, Hung C-H, Cheng C-H, et al. Phase holdups and microbial community in high-rate fermentative hydrogen bioreactors. Int J Hydrog Energy. 2011;36:364–73.
174. Wu SY, Lin CN, Chang JS, Lee KS, Lin PJ. Microbial hydrogen production with immobilized sewage sludge. Biotechnol Prog. 2002;18:921–6.
175. Lee K-S, Lo Y-S, Lo Y-C, Lin P-J, Chang J-S. H2 production with anaerobic sludge using activated-carbon supported packed-bed bioreactors. Biotechnol Lett. 2003;25:133–8.
176. Liu H, Fang H. Hydrogen production from wastewater by acidogenic granular sludge. Water Sci Technol. 2003;47:153–8.
177. Chu CY, Wu SY, Wu YC, Lin CY. Hydrodynamic behaviors in fermentative hydrogen bioreactors by pressure fluctuation analysis. Bioresour Technol. 2011;102:8669–75.
178. Wu SY, Lin CN, Chang JS. Hydrogen production with immobilized sewage sludge in three-phase fluidized-bed bioreactors. Biotechnol Prog. 2003;19:828–32.
179. Lee D-Y, Li Y-Y, Noike T. Influence of solids retention time on continuous H 2 production using membrane bioreactor. Int J Hydrog Energy. 2010;35:52–60.
180. Vallero M, Lettinga G, Lens P. Assessment of compatible solutes to overcome salinity stress in thermophilic (55¡ C) methanol-fed sulfate reducing granular sludges. Water Sci Technol. 2003;48:195–202.
181. Oh SE, Iyer P, Bruns MA, Logan BE. Biological hydrogen production using a membrane bioreactor. Biotechnol Bioeng. 2004;87:119–27.
182. Yang H, Shao P, Lu T, Shen J, Wang D, Xu Z, et al. Continuous bio-hydrogen production from citric acid wastewater via facultative anaerobic bacteria. Int J Hydrog Energy. 2006;31:1306–13.
183. Lin C-Y, Lay C-H, Sen B, Chu C-Y, Kumar G, Chen C-C, et al. Fermentative hydrogen production from wastewaters: a review and prognosis. Int J Hydrog Energy. 2012;37:15632–42.
184. Nissilä ME, Lay C-H, Puhakka JA. Dark fermentative hydrogen production from lignocellulosic hydrolyzates – a review. Biomass Bioenergy. 2014;67:145–59.
185. Kim S-H, Han S-K, Shin H-S. Effect of substrate concentration on hydrogen production and 16S rDNA-based analysis of the microbial community in a continuous fermenter. Process Biochem. 2006;41:199–207.
186. Jung K-W, Kim D-H, Kim S-H, Shin H-S. Bioreactor design for continuous dark fermentative hydrogen production. Bioresour Technol. 2011;102:8612–20.
187. FM Braga A, Júnior NF, Antônio D, Zaiat M. Thermophilic biohydrogen production using a UASB reactor: performance during long-term operation. J Chem Technol Biotechnol. 2015. doi:10.1002/jctb.4665. Article first published online: 5 Mar 2015.
188. Singh L, Siddiqui MF, Ahmad A, Rahim MHA, Sakinah M, Wahid ZA. Application of polyethylene glycol immobilized Clostridium sp. LS2 for continuous hydrogen production from palm oil mill effluent in upflow anaerobic sludge blanket reactor. Biochem Eng J. 2013;70:158–65.
189. Lo Y-C, Chen X-J, Huang C-Y, Yuan Y-J, Chang J-S. Dark fermentative hydrogen production with crude glycerol from biodiesel industry using indigenous hydrogen-producing bacteria. Int J Hydrog Energy. 2013;38:15815–22.

190. Wu X, Lin H, Zhu J. Optimization of continuous hydrogen production from co-fermenting molasses with liquid swine manure in an anaerobic sequencing batch reactor. Bioresour Technol. 2013;136:351–9.
191. Kumar G, Lin C-Y. Biogenic hydrogen conversion of de-oiled jatropha waste via anaerobic sequencing batch reactor operation: process performance, microbial insights, and reduction efficiency. Sci World J. 2014;2014:946503. p. 1–9.
192. Massanet-Nicolau J, Dinsdale R, Guwy A, Shipley G. Utilising biohydrogen to increase methane production, energy yields and process efficiency via two stage anaerobic digestion of grass. Bioresour Technol. 2015;189:379–83.
193. Fradler KR, Kim JR, Shipley G, Massanet-Nicolau J, Dinsdale RM, Guwy AJ, et al. Operation of a bioelectrochemical system as a polishing stage for the effluent from a two-stage biohydrogen and biomethane production process. Biochem Eng J. 2014;85:125–31.
194. Buitrón G, Kumar G, Martinez-Arce A, Moreno G. Hydrogen and methane production via a two-stage processes (H2-SBR + CH4-UASB) using tequila vinasses. Int J Hydrog Energy. 2014;39:19249–55.
195. Hsu C-W, Li Y-C, Chu C-Y, Liu C-M, Wu S-Y. Feasibility evaluation of fermentative biomass-derived gas production from condensed molasses in a continuous two-stage system for commercialization. Int J Hydrog Energy. 2014;39:19389–93.
196. Schievano A, Tenca A, Lonati S, Manzini E, Adani F. Can two-stage instead of one-stage anaerobic digestion really increase energy recovery from biomass? Appl Energy. 2014;124:335–42.
197. Lin C, Lay C. Carbon/nitrogen-ratio effect on fermentative hydrogen production by mixed microflora. Int J Hydrog Energy. 2004;29:41–5.
198. Lin C, Lay C. A nutrient formulation for fermentative hydrogen production using anaerobic sewage sludge microflora. Int J Hydrog Energy. 2005;30:285–92.
199. Karadag D, Puhakka JA. Enhancement of anaerobic hydrogen production by iron and nickel. Int J Hydrog Energy. 2010;35:8554–60.
200. Wang J, Wan W. Influence of Ni 2+ concentration on biohydrogen production. Bioresour Technol. 2008;99:8864–8.
201. Li C, Fang HHP. Inhibition of heavy metals on fermentative hydrogen production by granular sludge. Chemosphere. 2007;67:668–73.
202. Lee D-Y, Li Y-Y, Oh Y-K, Kim M-S, Noike T. Effect of iron concentration on continuous H 2 production using membrane bioreactor. Int J Hydrog Energy. 2009;34:1244–52.
203. Han H, Cui M, Wei L, Yang H, Shen J. Enhancement effect of hematite nanoparticles on fermentative hydrogen production. Bioresour Technol. 2011;102:7903–9.
204. Lin C-Y, Shei S-H. Heavy metal effects on fermentative hydrogen production using natural mixed microflora. Int J Hydrog Energy. 2008;33:587–93.
205. Han H, Jia Q, Wei L, Shen J. Influence of Cu2+ concentration on the biohydrogen production of continuous stirred tank reactor. Int J Hydrog Energy. 2014;39:13437–42.
206. Zheng X-J, Yu H-Q. Biological hydrogen production by enriched anaerobic cultures in the presence of copper and zinc. J Environ Sci Health A. 2004;39:89–101.
207. Frey M. Hydrogenases: hydrogen-activating enzymes. ChemBioChem. 2002;3:153–60.
208. Casalot L, Rousset M. Maturation of the [NiFe] hydrogenases. Trends Microbiol. 2001;9:228–37.
209. Yang H, Shen J. Effect of ferrous iron concentration on anaerobic bio-hydrogen production from soluble starch. Int J Hydrog Energy. 2006;31:2137–46.
210. de Lacey AL, Fernández VM, Rousset M. Native and mutant nickel–iron hydrogenases: unravelling structure and function. Coord Chem Rev. 2005;249:1596–608.
211. Kim S, Seol E, Raj SM, Park S, Oh Y-K, Ryu DD. Various hydrogenases and formate-dependent hydrogen production in Citrobacter amalonaticus Y19. Int J Hydrog Energy. 2008;33:1509–15.
212. Minton NP, Clarke DJ. Clostridia. New York: Springer Science & Business Media; 1989.

213. KIM BH, Zeikus J. Importance of hydrogen metabolism in regulation of solventogenesis by Clostridium acetobutylicum. Dev Ind Microbiol. 1984;26:549–56.
214. Kim IS, Hwang MH, Jang NJ, Hyun SH, Lee ST. Effect of low pH on the activity of hydrogen utilizing methanogen in bio-hydrogen process. Int J Hydrog Energy. 2004;29:1133–40.
215. Gottwald M, Gottschalk G. The internal pH of clostridium acetobutylicum and its effect on the shift from acid to solvent formation. Arch Microbiol. 1985;143:42–6.
216. Khanal SK, Chen WH, Li L, Sung S. Biological hydrogen production: effects of pH and intermediate products. Int J Hydrog Energy. 2004;29:1123–31.
217. Lin C-Y, Lay C. Effects of carbonate and phosphate concentrations on hydrogen production using anaerobic sewage sludge microflora. Int J Hydrog Energy. 2004;29:275–81.
218. Cavinato C, Giuliano A, Bolzonella D, Pavan P, Cecchi F. Bio-hythane production from food waste by dark fermentation coupled with anaerobic digestion process: a long-term pilot scale experience. Int J Hydrog Energy. 2012;37:11549–55.
219. Wu X, Yao W, Zhu J. Effect of pH on continuous biohydrogen production from liquid swine manure with glucose supplement using an anaerobic sequencing batch reactor. Int J Hydrog Energy. 2010;35:6592–9.
220. Chu C-F, Li Y-Y, Xu K-Q, Ebie Y, Inamori Y, Kong H-N. A pH- and temperature-phased two-stage process for hydrogen and methane production from food waste. Int J Hydrog Energy. 2008;33:4739–46.
221. Lee C, Lee S, Han S-K, Hwang S. Effect of operational pH on biohydrogen production from food waste using anaerobic batch reactors. Water Sci Technol. 2014;69:1886–93.
222. Paul JS, Quraishi A, Thakur V, Jadhav S. Effect of ferrous and nitrate ions on biological hydrogen production from dairy effluent with anaerobic waste water treatment process. Asian J Biol Sci. 2014;7:165–71.
223. Chu C-Y, Tung L, Lin C-Y. Effect of substrate concentration and pH on biohydrogen production kinetics from food industry wastewater by mixed culture. Int J Hydrog Energy. 2013;38:15849–55.
224. Vatsala TM, Raj SM, Manimaran A. A pilot-scale study of biohydrogen production from distillery effluent using defined bacterial co-culture. Int J Hydrog Energy. 2008;33:5404–15.
225. Zhang P, Lin C-J, Liu J, Pongprueksa P, Evers SA, Hart P. Biogas production from brown grease using a pilot-scale high-rate anaerobic digester. Renew Energy. 2014;68:304–13.
226. Cheong D-Y, Hansen CL. Feasibility of hydrogen production in thermophilic mixed fermentation by natural anaerobes. Bioresour Technol. 2007;98:2229–39.
227. Luo G, Xie L, Zou Z, Wang W, Zhou Q. Evaluation of pretreatment methods on mixed inoculum for both batch and continuous thermophilic biohydrogen production from cassava stillage. Bioresour Technol. 2010;101:959–64.
228. Chen C-C, Lin C-Y, Lin M-C. Acid–base enrichment enhances anaerobic hydrogen production process. Appl Microbiol Biotechnol. 2002;58:224–8.
229. Lin C-Y, Wu S-Y, Lin P-J, Chang J-S, Hung C-H, Lee K-S, et al. A pilot-scale high-rate biohydrogen production system with mixed microflora. Int J Hydrog Energy. 2011;36:8758–64.
230. Li C, Fang HH. Fermentative hydrogen production from wastewater and solid wastes by mixed cultures. Crit Rev Environ Sci Technol. 2007;37:1–39.
231. Lin C-Y, Hung C-H, Chen C-H, Chung W-T, Cheng L-H. Effects of initial cultivation pH on fermentative hydrogen production from xylose using natural mixed cultures. Process Biochem. 2006;41:1383–90.
232. Koskinen PE, Beck SR, Örlygsson J, Puhakka JA. Ethanol and hydrogen production by two thermophilic, anaerobic bacteria isolated from Icelandic geothermal areas. Biotechnol Bioeng. 2008;101:679–90.
233. Lee Y-W, Chung J. Bioproduction of hydrogen from food waste by pilot-scale combined hydrogen/methane fermentation. Int J Hydrog Energy. 2010;35:11746–55.
234. Fountoulakis M, Dokianakis S, Daskalakis G, Manios T. Fermentative hydrogen production from carob pod: a typical mediterranean forest fruit. Waste Biomass Valor. 2014;5:799–805.

235. Wu SY, Chu CY, Shen YC. Effect of calcium ions on biohydrogen production performance in a fluidized bed bioreactor with activated carbon-immobilized cells. Int J Hydrog Energy. 2012;37:15496–502.
236. Ramírez-Morales JE, Torres Zúñiga I, Buitrón G. On-line heuristic optimization strategy to maximize the hydrogen production rate in a continuous stirred tank reactor. Proc Biochem. 2015;50:893–900.
237. Si B, Li J, Li B, Zhu Z, Shen R, Zhang Y, et al. The role of hydraulic retention time on controlling methanogenesis and homoacetogenesis in biohydrogen production using upflow anaerobic sludge blanket (UASB) reactor and packed bed reactor (PBR). Int J Hydrogen Energy. 2015; 40: 11414–21.
238. Chookaew T, Sompong O, Prasertsan P. Biohydrogen production from crude glycerol by immobilized Klebsiella sp. TR17 in a UASB reactor and bacterial quantification under non-sterile conditions. Int J Hydrog Energy. 2014;39:9580–7.
239. Park J-H, Kumar G, Park J-H, Park H-D, Kim S-H. Changes in performance and bacterial communities in response to various process disturbances in a high-rate biohydrogen reactor fed with galactose. Bioresour Technol. 2015;188:109–16.
240. Chu CY, Wu SY, Shen YC. Biohydrogen production performance in a draft tube bioreactor with immobilized cell. Int J Hydrog Energy. 2012;37:15658–65.
241. Rosa PRF, Santos SC, Sakamoto IK, Varesche MBA, Silva EL. Hydrogen production from cheese whey with ethanol-type fermentation: effect of hydraulic retention time on the microbial community composition. Bioresour Technol. 2014;161:10–9.
242. Rosa PRF, Santos SC, Sakamoto IK, Varesche MBA, Silva EL. The effects of seed sludge and hydraulic retention time on the production of hydrogen from a cassava processing wastewater and glucose mixture in an anaerobic fluidized bed reactor. Int J Hydrog Energy. 2014;39:13118–27.
243. Sivagurunathan P, Sen B, Lin C-Y. High-rate fermentative hydrogen production from beverage wastewater. Appl Energy. 2015;147:1–9.
244. Tenca A, Perazzolo F, Naldi E, Provolo G, Bodria L, Oberti R. Design and start-up of a two-stage farm-scale pilot plant for biohydrogen and biomethane production. Energy, biomass and biological residues international conference of Agricultural Engineering-CIGR-AgEng 2012: agriculture and engineering for a healthier life, Valencia, 8–12 Jul 2012: CIGR-EurAgEng; 2012. p. C-1068.
245. Cheng S-S, Chao Y-C, Yang K-H, Bai M-D. Process recovery of biohydrogenation in a pilot plant from methanogens invasion. Int J Hydrog Energy. 2011;36:8779–84.
246. Environment Park S.p.A. Pilot biomass pretreatment plant, Turin. http://www.envipark.com/en/green-chemistry/plants/. Accessed 1 June 2015.
247. Richard Dinsdale. Biohydrogen production from food production waste, University of Glamorgan, UK. http://www.ifr.ac.uk/waste/Reports/Biohydrogen_Rep_Glam_Univ.pdf. Accessed 1 June 2015.
248. Toyota Motor Corporation. Powering the future hydrogen fuel cell vehicles could change mobility forever. http://www.toyota-global.com/innovation/environmental_technology/fuelcell_vehicle/. Accessed 1 June 2015.

Chapter 3
High-Yield Production of Biohydrogen from Carbohydrates and Water Based on In Vitro Synthetic (Enzymatic) Pathways

Jae-Eung Kim and Yi-Heng Percival Zhang

Abstract Distributed production of green and low-cost hydrogen from renewable energy sources is necessary to develop the hydrogen economy. Carbohydrates, such as cellulose, hemicellulose, starch, sucrose, glucose, and xylose, are abundant renewable bioresources and can provide the source of hydrogen. In this chapter, in vitro synthetic (enzymatic) pathways that overcome the limiting yields of hydrogen-producing microorganisms are discussed. These in vitro synthetic pathways produce hydrogen with theoretical yields from polymeric and monomeric hexoses or xylose with water of 2 mol of hydrogen per carbon molecule of carbohydrate. In the past years, hydrogen production rate of in vitro synthetic enzymatic pathways has been improved to 150 mmol/L/h by 750-fold through systematic optimization. All of the thermostable enzymes used in the pathways have been recombinantly produced in *E. coli*, and some of them are immobilized for enhanced stability and simple recycling. Redox enzymes are being engineered to work on low-cost and highly stable biomimetic coenzymes. It is expected that low-cost green hydrogen can be produced at $2.00/kg hydrogen in small-sized atmospheric pressure bioreactors in the future.

Keywords Carbohydrate • Hydrogen production • Synthetic enzymatic pathway • In vitro synthetic biosystem • Innovative biomanufacturing • Systems biocatalysis

J.-E. Kim
Biological Systems Engineering Department, Virginia Tech, 304 Seitz Hall, Blacksburg, VA 24061, USA

Y.-H.P. Zhang (✉)
Biological Systems Engineering Department, Virginia Tech, 304 Seitz Hall, Blacksburg, VA 24061, USA

Cell Free Bioinnovations Inc., 2200 Kraft Drive, Suite 1200B, Blacksburg, VA 24060, USA

Tianjin Institute of Industrial Biotechnology, Chinese Academy of Sciences, 32 West 7th Avenue, Tianjin Airport Economic Area, Tianjin 300308, China
e-mail: ypzhang@vt.edu

Z. Fang et al. (eds.), *Production of Hydrogen from Renewable Resources*, Biofuels and Biorefineries 5, DOI 10.1007/978-94-017-7330-0_3

3.1 Introduction

3.1.1 Hydrogen

Dihydrogen gas is a colorless, odorless, but very flammable diatomic molecule. Later on throughout this chapter, hydrogen is used to mean dihydrogen gas for convenience. Hydrogen atoms widely exist in natural inorganic and organic matters, such as water, hydrocarbons, and carbohydrates. Hydrogen has been proposed as a future alternative fuel to reduce our demand on traditional fossil fuel-based energy consumption due to its clean energy property, enhanced energy conversion efficiency, and high specific energy density (J/kg) [7, 9, 54]. The combustion or electrochemical conversion of hydrogen produces only water as a final by-product. Vehicles equipped with hydrogen fuel cells are far more energy efficient than traditional internal combustion engine-based vehicles [53]. For example, Toyota will start producing a large number of affordable hydrogen fuel cell vehicles at selling prices of $~50,000 in 2016. Hydrogen also has a higher specific mass energy density than any other fuel sources including gasoline and diesel. However, the low volumetric energy density of hydrogen is currently the biggest issue for its practical use. Thus, the development of new types of green hydrogen production and storage technology remains as challenges.

3.1.2 Hydrogen Production Approaches

Currently most hydrogen is produced from natural gas or coals through a reforming process or gasification followed by water shifting, respectively. These processes are not environment-friendly, releasing CO_2, and the resources are not sustainable. Therefore, the use of hydrogen produced in such ways would have little impact on reducing our demand on traditional fossil fuel-based energy sources [43]. Alternatively, hydrogen can be generated by splitting water molecules with high-temperature thermal energy sources or electricity. Direct splitting of water molecules requires high-temperature thermal energy over 2,000 °C [43]. Sulfur-based thermochemical decomposition of water with heterogeneous metal oxide catalysts can take place at much lower temperatures than direct thermal decomposition, but it still requires at least 750 °C or higher [14]. The thermal energy required for water splitting can be derived from solar energy. Water splitting by solar energy to generate hydrogen, however, is a very slow and inefficient process due to photocatalysts using limited range of visible light and low insolation flux (e.g., ~200 W/m^2) [30, 58]. Most photocatalytic water-splitting processes have shown their production rate less than 1 mmol H_2/L/h [23]. Much higher rates of hydrogen production can be achieved with electricity. A stationary electrolyzer with current technology has been reported to be able to generate hydrogen from water at the rate of about 40 mol/L/h [22]. In spite of such high hydrogen production rates and purity of hydrogen generation by water electrolysis, its practical applicability is limited due to its high production cost (e.g., ~0.05 US dollars per kWh of electricity). Water electrolysis cannot be free from environmental issues as long as the electricity for electrolyzers is generated from coal or natural gas-powered electric generators [17].

Solar energy and electricity can also be applied to microorganisms to produce hydrogen through photo-fermentation and microbial electrolysis cells (MECs), respectively. Dark fermentation has a theoretical maximum hydrogen production yield of 4 mol of hydrogen per mole of glucose (~33 % efficiency) because of the Thauer limit [42]. Practical efficiency of dark fermentation would be lower than the Thauer limit [16, 21]. Microbial electrolysis cells can achieve much higher efficiencies of about 80 %, but high costs of apparatus and slow production rates (~5.4 mmol H_2/L/h) are the biggest challenges for large-scale hydrogen production from MECs [17, 28]. Overall, microbial fermentation is not an efficient way to produce hydrogen because of the microbial basal metabolism that competes with hydrogen production and eventually reducing the overall product yield.

3.1.3 *In Vitro (Cell-Free) Enzymatic Pathways for Water Splitting*

In vitro synthetic biosystems for water splitting can produce high-purity (i.e., zero CO production) hydrogen with high yields and rates [59]. In vitro synthetic biosystems are a new cell-free platform that assembles a number of (purified) enzymes and cofactors into different in vitro synthetic (enzymatic) pathways for implantation of various desired biochemical reactions [2, 10, 59, 63]. The optimal reaction condition for numerous enzymes can be found by examining different buffers, such as HEPES, Tris and PBS, a broad range of pH, and cofactor concentrations, and also multi-metal ions, such as Mg^{2+} and Mn^{2+}, to meet trade-off needs of different enzymes. If all enzymes are thermostable, high reaction temperature could be chosen as an optimal condition. These synthetic pathways can utilize different carbohydrates and ambient thermal energy as energy inputs to overcome thermodynamically unfavorable water-splitting reactions to produce hydrogen at mild conditions (below 100 °C). Near theoretical yields of 12 mol of hydrogen per mole of glucose unit consumed have been achieved [50, 60]. More important feature of these synthetic pathways is that they are able to produce hydrogen by absorbing low-temperature waste heat [63]. As the result, endothermic water-splitting reactions can generate more output of chemical energy in the form of hydrogen than input of chemical energy from carbohydrates and water (i.e., 122 % energy efficiency in terms of higher heating values). It is possible to achieve the energy efficiency over 100 % because the water-splitting reaction is a very unique entropy-driven chemical reaction [60]. The in vitro synthetic biosystem for water splitting is more advantageous for hydrogen production than microbial fermentation and photocatalytic water-splitting systems, because of the absence of cellular membranes and microbial complexity that lower mass transfer and increase biocatalyst density [37]. The highest hydrogen production rate achieved by an in vitro synthetic pathway is about 150 mmol/L/h [38], while microbial fermentation has been reported to produce hydrogen with the rate of 1.96 mol/L/h [52].

3.2 Design of In Vitro Synthetic Enzymatic Pathways

In vitro synthetic pathways for water splitting powered by carbohydrates are reconstituted nonnatural catabolic pathways consisting of more than ten enzymes in four modules: (1) generation of phosphorylated sugars from poly- or monosaccharides without ATP, (2) NADPH generation via the oxidative pentose phosphate pathway (PPP), (3) hydrogen generation, and (4) G6P regeneration via the non-oxidative PPP and gluconeogenesis (Fig. 3.1 and Table 3.1). From the first module, poly- or monosaccharides are converted to their phosphorylated sugar units by phosphorylases or kinases without the use of ATP. For example, the phosphorylation of starch to glucose 1-phosphate units is catalyzed by starch phosphorylase, and different sugar substrates are catalyzed by different enzymes, cellodextrins and cellobiose by cellodextrin and cellobiose phosphorylases, sucrose by sucrose phosphorylase, glucose by polyphosphate glucokinase, and xylulose by polyphosphate xylulokinase. All poly- and oligosaccharides are phosphorylated to be glucose 1-phosphate, which is converted to glucose 6-phosphate by phosphoglucomutase, and then enter the second module reactions to generate NADPH. Xylose takes a different pathway to that catalyzed by xylose isomerase and polyphosphate xylulokinase to produce xylulose 5-phosphate. During the NADPH generation, glucose 6-phosphate (six-carbon sugar) enters the oxidative pentose phosphate pathway and becomes ribulose 5-phosphate (five-carbon sugar) by generating 2 mol of NADPH, releasing 1 mol of CO_2, and absorbing 1 mol of water molecule. When 1 mol of six-carbon sugar is completely consumed, 6 mol of CO_2 are released with the generation of 12 mol of NADPH. Each mole of NADPH is equivalent to 1 mol of hydrogen production catalyzed by NADPH-dependent hydrogenase. Thus, in the hydrogen generation module, 12 mol of hydrogen can be produced when 1 mol of six-carbon sugar is consumed for water splitting. In the non-oxidative pentose phosphate pathway, ribulose 5-phosphates are converted to fructose 6-phosphates and glyceraldehyde 3-phosphates via a carbon skeleton rearrangement. During the G6P regeneration module, a pair of glyceraldehyde 3-phosphates and dihydroxyacetone phosphate is combined to form a fructose 6-phosphate by multiple enzymes via the gluconeogenesis pathway. At this step, one extra mole of water molecule is absorbed by fructose 1,6-bisphosphatase. Finally, all fructose 6-phosphates are catalyzed to regenerate equal moles of glucose 6-phosphate which enter back to the NADPH generation module by completing the cofactor-balanced synthetic pathway for water splitting. The overall

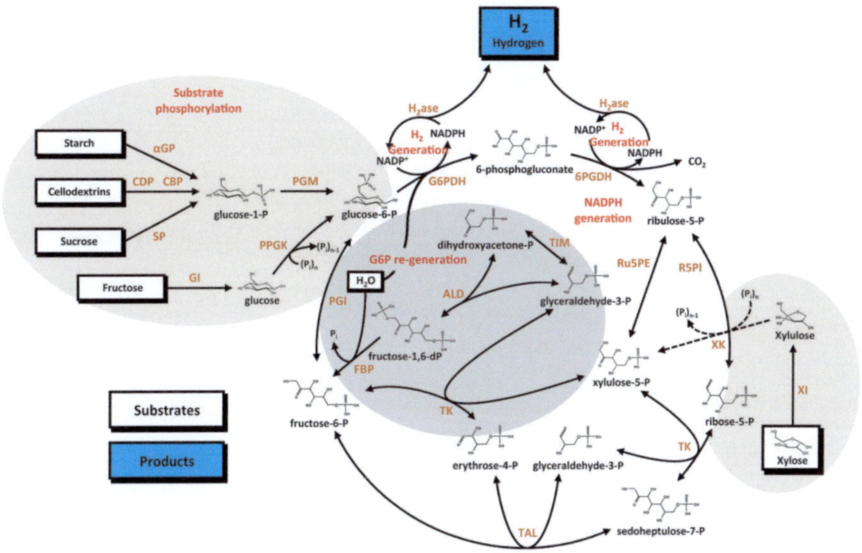

Fig. 3.1 ATP-free cofactor-balanced pathway for hydrogen generation from various types of carbohydrates with water

Table 3.1 Net reactions of each module and the overall reaction of in vitro synthetic biosystem for water splitting

Modules	Net reactions
Module #1	$C_6H_{10}O_5 + H_3PO_4\ (= P_i) \rightarrow$ glucose-6-P
Module #2 (oxidative phase) × 6	6 × (glucose-6-P + 2 $NADP^+$ + $\mathbf{H_2O}$ → ribulose-5-P + 2 NADPH + 2 H^+ + CO_2)
Module #2 (non-oxidative phase) × 2	2 × (3 ribulose-5-P → 2 fructose-6-P + glyceraldehyde-3-P)
Module #3 × 12	12 × (NADPH + H^+ → $NADP^+$ + H_2)
Module #4	2 glyceraldehyde-3-P + $\mathbf{H_2O}$ → fructose-6-P + P_i
Module #4 × 5	5 × (fructose-6-P → glucose-6-P)
Final reaction equation:	$C_6H_{10}O_5 + \mathbf{7H_2O} \rightarrow 12H_2 + 6CO_2$

reaction of this in vitro synthetic pathway for water splitting can be summarized by the equation (Eq. 3.1) as shown in Table 3.1. All of the enzymes used for this hydrogen-producing pathway are listed in Table 3.2.

$$CH_2O(aq) + H_2O(l) \rightarrow 2H_2(g) + CO_2(g) \tag{3.1}$$

Table 3.2 Enzymes used for in vitro synthetic biosystem for water splitting

Modules	E.C. #	Enzyme names (abbreviations)	Reactions
Substrate phosphorylation	2.4.1.1	Glycogen phosphorylase (αGP)	$Glycogen_{(n)} + P_i \rightarrow$ glucose-1-P + $glycogen_{(n-1)}$
	2.4.1.49	Cellodextrin phosphorylase (CDP)	$Cellodextrin_{(n)} + P_i \rightarrow$ glucose-1-P + $cellodextrin_{(n-1)}$
	2.4.1.20	Cellobiose phosphorylase (CBP)	Cellobiose + $P_i \rightarrow$ glucose-1-P + glucose
	2.4.1.7	Sucrose phosphorylase (SP)	Sucrose + $P_i \rightarrow$ glucose-1-P + fructose
	5.3.1.5	Glucose isomerase (GI)	Fructose → glucose
	2.7.1.63	Polyphosphate glucokinase (PPGK)	Glucose + $(P_i)_n \rightarrow$ glucose-1-P + $(P_i)_{n-1}$
	5.4.2.2	Phosphoglucomutase (PGM)	Glucose-1-P → glucose-6-P
	5.3.1.5	Xylose isomerase (XI)	Xylose → xylulose
	2.7.1.17	Polyphosphate xylulokinase (PPXK)	Xylulose + $(P_i)_n \rightarrow$ xylulose-5-P + $(P_i)_{n-1}$
NADPH generation	1.1.1.49	Glucose-6-phosphate dehydrogenase (G6PDH)	Glucose-6-P + $NADP^+$ + H_2O → 6-phosphogluconate + NAHPH + H^+
	1.1.1.44	6-phosphogluconic dehydrogenase (6PGDH)	6-phosphogluconate + $NADP^+$ → ribulose-5-P +NADPH + CO_2
	5.3.1.6	Ribose 5-phosphate isomerase (RPI)	Ribulose-5-P → ribose-5-P
	5.1.3.1	Ribulose-5-phosphate 3-epimerase (RPE)	Ribulose-5-P → xylulose-5-P
	2.2.1.1	Transketolase (TK)	Xylulose-5-P + ribose-5-P → sedoheptulose-7-P + glyceraldehyde-3-P
	2.2.1.1	Transketolase (TK)	Xylulose-5-P + erythrose-4-P → fructose-6-P + glyceraldehyde-3-P
	2.2.1.2	Transaldolase (TAL)	Sedoheptulose-7-P + glyceraldehyde-3-P → fructose-6-P + erythrose-4-P
Hydrogenation	1.12.1.3	Hydrogenase (H_2ase)	NADPH + H^+ → $NADP^+$ + H_2
G6P regeneration	5.3.1.1	Triose-phosphate isomerase (TIM)	Glyceraldehyde-3-P → dihydroxyacetone phosphate
	4.1.2.13	Aldolase (ALD)	Glyceraldehyde-3-P + dihydroxyacetone phosphate → fructose-1,6-bisphosphate
	3.1.3.11	Fructose-1,6-bisphosphatase (FBP)	Fructose-1,6-bisphosphate + H_2O → fructose-6-P + P_i
	5.3.1.9	Phosphoglucose isomerase	Fructose-6-P → glucose-6-P

3.3 Examples of Hydrogen Production from Carbohydrates

3.3.1 Hydrogen Production from Starch and Cellodextrins

In vitro enzymatic pathways to produce hydrogen from carbohydrates were first demonstrated by Woodward and colleagues using glucose, xylose, or sucrose with only a couple of enzymes [47, 48]. These simple enzymatic pathways could achieve less than 10 % of the theoretical yields of sugars, due to only one NADPH generation per hexose or pentose. To complete the oxidation, the oxidative pentose phosphate cycle was coupled with hydrogenase to produce hydrogen from glucose 6-phosphate, resulting in about 96 % of the theoretical yield [49]. High cost of glucose 6-phosphate prevents its practical application. These in vitro synthetic pathways were further improved by Zhang and collaborators to demonstrate hydrogen production from different types of carbohydrates: starch [60], cellulosic materials [50], xylose [31], and sucrose [32]. The first hydrogen production from starch by an in vitro synthetic enzymatic pathway proved its feasibility to produce low-cost hydrogen from inexpensive starch without ATP by achieving a high production yield surpassing the theoretical production yield of dark fermentation [60]. Most enzymes used for this proof-of-principle experiment producing hydrogen from starch were off-the-shelf enzymes, and the enzymatic pathway exhibited a production rate of 0.4 mmol/L/h with 43 % yield (5.2 mol H_2/mol glucose consumed) (Table 3.3). A couple of years later, the hydrogen production rates and yields were enhanced to 0.5 mmol/L/h with 93 % yield and 3.9 mmol/L/h with 68 % yield when cellobiose or cellodextrins were used as a substrate, respectively, and through minor optimizations including increased substrate concentration, reaction temperature, and more rate-limiting enzyme loadings [50] (Table 3.3). The hydrogen production rate using cellodextrins was greatly improved by increasing hydrogenase loading and substrate concentration to 8 mM. However, low-yield hydrogen production was observed due to the incomplete reaction. Cellobiose and cellodextrins were prepared as hydrolytic products of cellulose through incomplete enzymatic or mixed acid hydrolysis, respectively.

Table 3.3 Comparison of hydrogen production rates and yields from different carbohydrates

Substrate	Conc. (mM)	Reaction temperature (°C)	H_2 production rates (mmol H_2/L/h)	Yield (%)	Ref.
Starch	1	30	0.4	43	[60]
Cellobiose	2	32	0.5	93	[50]
Cellodextrins (cellopentaose)	8	32	3.9	68	[50]
Xylose	2	50	2.2	96	[31]
Sucrose	2	37	3.0	97	[32]
G6P	100	60	150	–	[38]
Biomass sugar	3.19	40	2.3	100	[36]
G6P	100	50	54	–	[36]

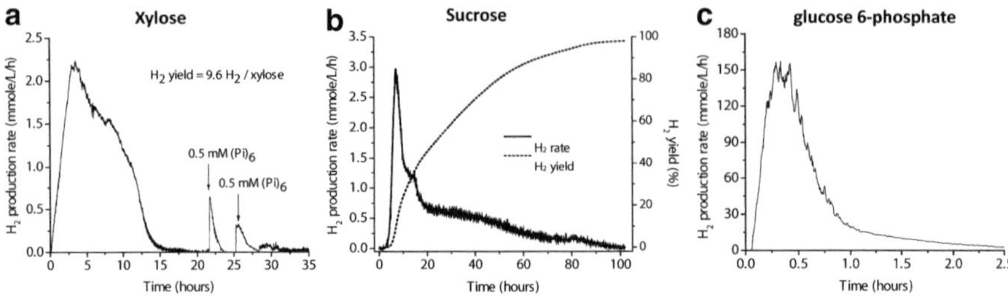

Fig. 3.2 Hydrogen production profiles from xylose (**a**), sucrose (**b**), and glucose 6-phosphate (**c**). (Reproduced with permission from [31, 32, 36])

3.3.2 Hydrogen Production from Xylose

Hemicellulose is another major component of lignocellulosic biomass besides cellulose. Xylose is the most abundant pentose and the major component of hemicellulose. Xylose composes about 20–30 % of lignocellulosic biomass by weight [31]. Thus, it is essential to use not only cellulosic materials but also the major portion of hemicellulosic materials to produce low-cost hydrogen from renewable carbohydrates. One mole of xylose can theoretically produce 10 mol of hydrogen by splitting 5 mol of water molecule when completely oxidized to carbon dioxide (Eq. 3.2). Since xylose is a five-carbon sugar, the in vitro synthetic pathway was modified by replacing hexose-phosphorylating enzymes with xylose isomerase and polyphosphate xylulokinase (Fig. 3.1). The hydrogen production from xylose exhibited the production rate of 2.2 mmol/L/h with 96 % yield (Table 3.3). Such high-yield hydrogen production was achieved with the addition of extra polyphosphate driving the reaction to completion (Fig. 3.2a), and the relatively high production rate was achieved with increased reaction temperature to 50 °C:

$$C_5H_{10}O_5(aq) + 5H_2O(l) \rightarrow 10H_2(g) + 5CO_2(g) \quad (3.2)$$

3.3.3 Hydrogen Production from Sucrose

Sucrose, also known as table sugar, is one of the cheapest carbohydrates because of its simple production process and abundance of cultivated sugar crops: sugarcane and sugar beets. Sucrose is a disaccharide composed of glucose and fructose. Therefore, two additional enzymes are required to convert fructose to equal moles of glucose 6-phosphate. An in vitro synthetic enzymatic pathway consisting of total 15 enzymes was designed to catalyze water-splitting process producing hydrogen powered by sucrose [32]. The hydrogen production from sucrose achieved the production rate of 3.0 mmol/L/h with 97 % yield (Fig. 3.2b). When increased sucrose concentration, the maximum hydrogen production rate was as high as 9.7 mmol/L/h [32].

3.3.4 Hydrogen Production from Biomass Sugars

The complete conversion of glucose and xylose from plant biomass to hydrogen and carbon dioxide has been achieved via an in vitro synthetic enzymatic pathway. Pretreated biomass was hydrolyzed to glucose and xylose by using a commercial cellulase. Glucose and xylose were simultaneously converted to hydrogen with the theoretical yield of 2 mol of hydrogen per each carbon molecule [36]. A genetic algorithm was used to find the best fitting parameters of a nonlinear kinetic model with experimental data. Global sensitivity analysis was used to identify the key enzymes that have the greatest impact on reaction rate and yield. After optimization of enzyme loadings using computational modeling and data analysis methods, the hydrogen production rate could be increased to 32 mmol/L/h. The production rate was further enhanced to 54 mmol/L/h by increasing reaction temperature, substrate, and enzyme concentrations. The production of hydrogen from locally produced biomass is a promising means to achieve global green hydrogen production.

3.3.5 High-Rate Hydrogen Production from Glucose 6-Phosphate

High-yield hydrogen production from these different carbohydrates has opened up a new way to produce low-cost hydrogen from renewable biomass. In the past years, the hydrogen production rate has been increased to 150 mmol/L/h by 750-fold when glucose 6-phosaphte is used as a substrate by an in vitro synthetic enzymatic pathway (Fig. 3.2c) [38, 49]. All enzymes used in the pathway have been replaced with recombinant thermostable enzymes produced in *E. coli*. Some of these enzymes are immobilized to enhance their stability. Hydrogen production rates and yields from different carbohydrates are summarized in Table 3.3.

3.4 Technical Obstacles to Low-Cost H_2 Production

To achieve low-cost hydrogen production from carbohydrates, a few obstacles have to be overcome. The ultimate hydrogen production costs are strongly related to the following factors: costs of substrate, enzyme cost, cofactor cost, and product-related downstream processing cost especially for product separation and purification [61]. It has been shown that the in vitro synthetic biosystem can produce hydrogen from various types of carbohydrates in near theoretical yields. Therefore, the use of inexpensive, abundant carbohydrate sources, such as cellulosic and hemicellulosic materials, to produce hydrogen can solve one of the major obstacles. Gaseous products can be easily separated and purified from the aqueous phase enzymatic reaction, leaving the other two obstacles unsolved. Hydrogen production rate is also an important factor, because it determines potential implementation of low-cost hydrogen production techniques from carbohydrates mainly related to capital investment.

3.4.1 Enzyme Cost and Stability

Enzyme costs are highly related to their production costs, turnover number (TTN), and stability. The current production costs for enzymes produced as recombinant proteins in *E. coli* at lab scales are high, for example, \$~1,000,000 per kg of dry protein, but it is expected that their production costs will decrease to the level of industrial bulk enzyme production costs of around \$5–100 per kg of dry protein [57]. Increasing TTNs can decrease enzyme costs exponentially through increasing enzyme efficiency [55]. It is estimated that TTNs over 10^8–10^9 are required to reduce enzyme costs low enough for industrial-scale use [11, 12, 51]. High TTNs can be achieved by using thermostable enzymes from thermophilic microbes, or through enzyme immobilization [33, 44, 54].

The use of thermostable enzymes can decrease their production costs by decreasing enzyme purification costs. Thermostable enzymes cloned from thermophilic microbes, such as *Thermotoga maritima*, can be stable at 60–70 °C, which is higher than the temperature where most of the other enzymes cloned from mesophilic hosts are stable. Therefore, purification procedures retaining only soluble thermostable enzymes after treating at high temperature for relatively short times (10–30 min) make it a simple and cost-effective way to purify target enzymes from cell lysates. Simple enzyme purification will eventually lower the overall production costs.

Enzyme immobilization is a technique in a relatively mature stage, and various immobilization techniques have been introduced and used to improve TTNs, enzyme stability, and catalytic efficiency [24, 33, 41]. Among many techniques, the cellulose-binding module (CBM) tagged protein immobilization combines enzyme purification and immobilization into one step [18]. This simple one-step enzyme purification and immobilization technique showed about an 80-fold enhanced half-life time of phosphoglucose isomerase (PGI) when the enzyme with CBM tag was immobilized on regenerated amorphous cellulose (RAC) (Fig. 3.3) [33]. Enzyme immobilization also enables enzymes to be recyclable. Enzymes immobilized on magnetic nanoparticles (MNPs) have shown to improve enzymatic reactions rates and also be able to recycle simply by using a magnetic force [34]. Green fluorescent protein (CBM-TGC) was used to demonstrate simply selective recycling of CBM-tagged GFP immobilized on Avicel-containing MNPs

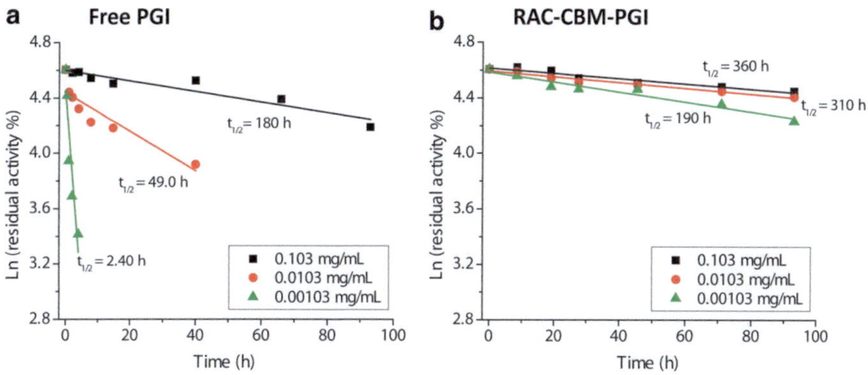

Fig. 3.3 Comparison of thermal stability between free PGI (**a**) and immobilized PGI (**b**) in different concentrations (Reproduced with permission from Myung et al. [33])

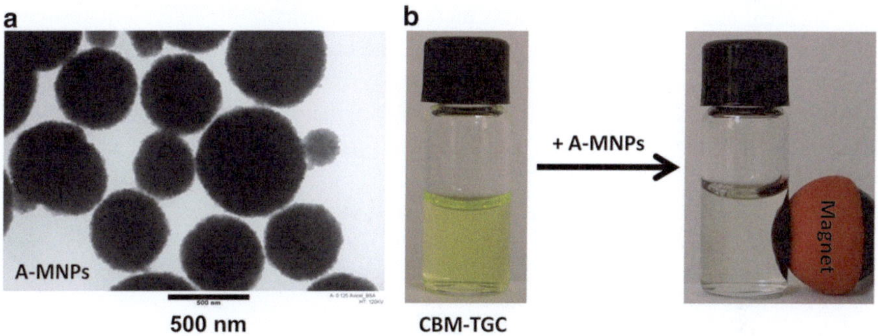

Fig. 3.4 Selective recycling of enzymes immobilized on Avicel-containing magnetic nanoparticles (A-MNPs) (**a**). The simple process of collecting CBM-tagged green fluorescent proteins (CBM-TGC) adsorbed on A-MNPs by a magnetic force (**b**) (Reproduced with permission from Myung et al. [34])

(A-MNPs) (Fig. 3.4). As the result of selective recycling, enzyme-related costs can be greatly reduced.

Protein protection additives, such as ligands and salts, have been studied as a simple approach to enhance enzyme storage and reaction stability. DMSO and glycerol are popular additives as cryoprotectants, stabilizing proteins during multi-freezing-and-thawing cycles. Various polyethylene glycols have been used as thermoprotectants. Thermostability of trypsin was increased from 49 to 93 °C without deteriorating its catalytic properties in the presence of glycol chitosan [13]. Amines, polyethylene glycol, and glycerol as additives improved catalase storage stability as well as its enzymatic performance in high temperature and alkaline pH [6]. These additives may, however, be potential inhibitors to the reaction system [19]. Therefore, additional dialysis or ultrafiltration may be required before the reactions.

3.4.2 Enzymatic Reaction Rates

Reaction rate is an important factor for determining potential applications of the enzymatic reaction and capital investment. Currently, the fastest hydrogen production rate achieved by an in vitro synthetic enzymatic pathway is about 150 mmol/L/h (=0.3 g H_2/L/h) [38]. This current rate is the fastest enzymatic hydrogen production from sugars and fast enough for distributed hydrogen generators, but it is slower than the need of on-demand hydrogen production for hydrogen fuel cell vehicles. It is anticipated that the production rate can be improved to 20 g H_2/L/h because microbial fermentation, which is usually slower than cell-free enzymatic reactions, has already achieved a production rate of over 20 g H_2/L/h [5, 54, 62]. Approaches made to increase enzymatic hydrogen production rate include elevated reaction temperature, optimized enzyme ratio, high substrate concentration, high enzyme loading, and substrate channeling among cascade enzymes. The substrate channeling was done by co-immobilizing multiple cascade enzymes, so that the local enzyme concentration is increased and susceptible intermediates have less time exposed to reaction solutions [20]. The cluster of multi-enzymes held by co-immobilizing scaffoldins results in increased reaction rates. Three important cascade enzymes, TIM, ALD, and FBP, in the enzymatic hydrogen production were expressed with dockerin domain which can self-assemble with mini-scaffoldin for co-immobilization on A-MNPs (Fig. 3.5a) [34]. In comparison with non-immobilized enzymes, the co-immobilized enzyme complex results in about 4.6 times increased reaction rate (Fig. 3.5b).

3.4.3 Cofactor Cost and Stability

Cofactors are chemical compounds required for enzymatic reactions, such as ATP and NAD(P)H. The issues from cofactor costs and stability have been addressed by various approaches including cofactor recycling systems [25, 46] and use of low-cost stable biomimetic cofactors [1, 29]. Although the regeneration of natural cofactors through recycling systems is economically beneficial to most current enzymatic reactions, the most farsighted solution would be to replace native cofactors with low-cost stable biomimetic cofactors. The structures of natural cofactors and biomimetic cofactors are shown in Fig. 3.6 with their estimated prices (USD, 2015). By sharing the nicotinamide moiety as a universal binding site for electron carriers, these natural and biomimetic cofactors vary in the chemical structures bound to the nitrogen atom in pyridine (Fig. 3.6). The alternative natural cofactors and biomimetic cofactors with simpler structures than NADP or NAD are estimated to have lower costs. However, most wild-type redox enzymes have no activities with such biomimetic cofactors. A number of studies have been done to change the cofactor specificity or preference through cofactor engineering. Cofactor engineering has three major types of approaches: rational design, directed

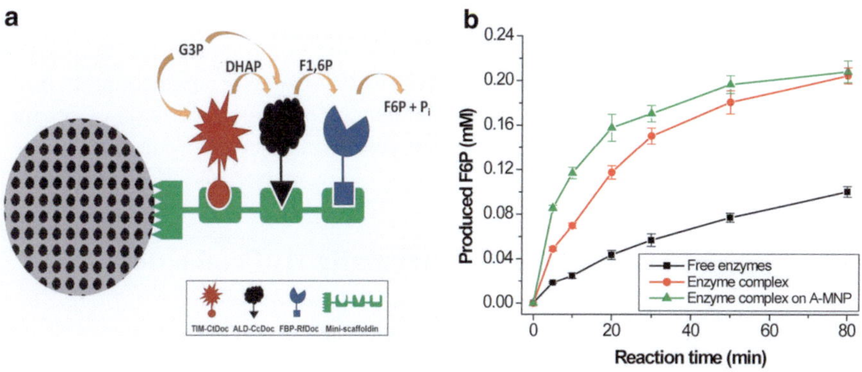

Fig. 3.5 The cascade enzymatic reaction among co-immobilized enzymes on MNPs for substrate channeling (**a**). Comparison of the reaction rates among free enzymes, non-immobilized enzyme complex, and the enzyme complex immobilized on A-MNP (**b**) (Reproduced with permission from Myung et al. [34])

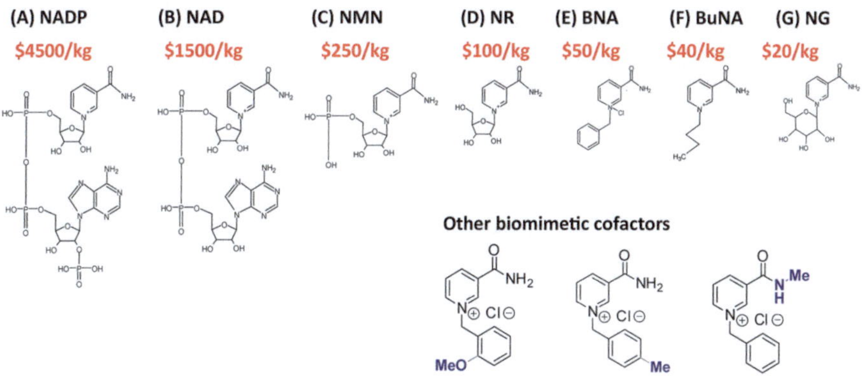

Fig. 3.6 Structures and estimated prices of natural cofactors (**A**, **B**), alternative natural cofactors (**C**, **D**), and biomimetic cofactors (**E–G**)

evolution, or swapping modules. Using these approaches, a number of studies have shown their redox enzymes with changed cofactor preferences from NADP to NAD [3, 8, 39], or from NAD to NADP [15, 45]. In 2012, Scott et al. discovered that their engineered alcohol dehydrogenase for broadened cofactor specificity and improved activity with NAD can utilize a minimal natural cofactor, NMN [4]. Fish et al. proposed the use of 1-benzyl-3-carbamoyl-pyridinium (BNA) as a biomimetic cofactor to replace NAD(P)H and discovered that two wild-type enzymes, horse liver alcohol dehydrogenase and monooxygenase, can actually utilize this cofactor [26, 27]. Clark collaborated with Fish and demonstrated that engineered P450 with two amino acid mutations can utilize BNA as a cofactor [40]. Recently, a large international collaborative group synthetized another biomimetic cofactor, 1-butyl-3-carbamoyl-pyridinium (BuNA), and demonstrated that wild-type enoate reductase can utilize it as a cofactor [35]. In most cases, such changes will decrease apparent activities greatly. The best example may be the engineered P450. The mutant P450 (W1064S/R966D) exhibited its activity on biomimetics up to seven times of that of wild-type enzyme on NADH [40]. Cofactor engineering is in its early stage, but its success will greatly influence enzymatic synthesis of organic chemicals and in vitro synthetic biosystems for biocommodity production.

3.5 Conceptual Obstacles to Enzymatic H_2 Production

In vitro synthetic biosystems consist of numerous enzymes as building bricks, enzyme complexes as building modules, and/or (biomimetic) coenzymes. These many components are assembled into in vitro synthetic pathways for implementing complicated bioreactions. They emerge as an alternative solution for accomplishing desired biotransformation without concerns of cell proliferation, complicated cellular regulation, and side-product formation. In addition to the capability of achieving high product yields as the most important advantage, in vitro synthetic biosystems feature several other biomanufacturing advantages, such as fast reaction rates, easy product separation, open process control, broad reaction conditions, and tolerance to toxic substrates or products.

The largest obstacle to the enzymatic hydrogen production is conceptual change. Microbe-based fermentation has been used by human beings for more than 10,000 years. As a result, most biotechnologists believe that living microbes are the best biocatalysts because they can duplicate themselves. Indeed, the primary goal of living microbes is their proliferation while bioconversions are side effects. Consequently, the success of several examples is needed to convince biotechnologists of accepting a new paradigm of in vitro synthetic biosystems.

3.6 Conclusions and Future Outlook

Hydrogen is an important commodity chemical with a global market size of approximately 100 billion US dollars. High-yield hydrogen has been produced from carbohydrates and water catalyzed by in vitro synthetic enzymatic pathways. The overall hydrogen production costs are mainly proportional to carbohydrate cost [59]. Assuming a substrate cost contributes a major portion of hydrogen product cost, it is expected that hydrogen can be produced at costs less than $2.00/kg hydrogen when the technology is well developed [59] (Fig. 3.7). Carbohydrates are the most abundant renewable natural resources on earth, and they are estimated to cost about $1.50/kg accounting for 75 % of the prospective hydrogen production cost [59] (Fig. 3.7). The next biggest portion of the hydrogen production cost is from the initial capital investment and operating expense accounting for about 17 %. The sum of capital and operating expenses would be about $0.35/kg hydrogen according to the similar expenses based on anaerobic digestion. Enzyme and

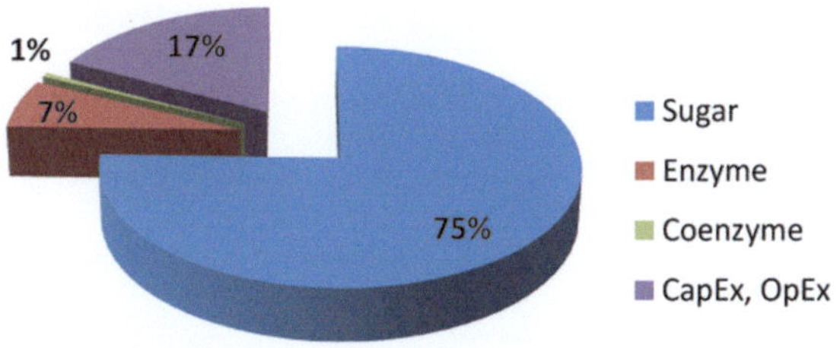

Sugar	$	1.50
Enzyme	$	0.13
Coenzyme	$	0.01
CapEx, OpEx	$	0.35
Total	$	1.99

Fig. 3.7 Prospective H_2 production cost and its contribution factors: substrate costs, enzyme costs, cofactor costs, and initial capital expense with operating expense

cofactor costs account for 7 % and 1 % of the enzymatic hydrogen production cost, respectively, when their TTN values are more than 10^9 and 10^7, respectively.

High-yield hydrogen production from carbohydrates will open up several potential applications for the hydrogen economy, where hydrogen is an alternative transportation fuel or a short-term electricity storage carrier. The potential applications include from small-sized distributed hydrogen refueling stations (e.g., 1–2 kg per day for a single house, or 50–200 kg for village) to the most ambitious application, such as a sugar fuel cell vehicles [53, 56].

The potential global market size for hydrogen as a future energy carrier replacing gasoline and diesel could be trillions of dollars. Such great potential, along with a bright future featuring enhanced energy conversion efficiency, nearly zero pollutants, and zero greenhouse gas emissions, will motivate the world to solve the remaining obstacles within next decades.

Acknowledgments YPZ was supported by the Virginia Tech Biological Systems Engineering Department, and subcontracts from NSF STTR I (IIP-1321528) and SBIR II (IIP-1353266) awards and Tianjin Institute of Industrial Biotechnology, Chinese Academy of Sciences. In addition, funding for this work was provided in part by the Virginia Agricultural Experiment Station and the Hatch Program of the National Institute of Food and Agriculture, US Department of Agriculture.

References

1. Ansell RJ, Lowe CR. Artificial redox coenzymes: biomimetic analogues of NAD(+). Appl Microbiol Biotechnol. 1999;51(6):703–10.
2. Ardao I, Hwang ET, Zeng AP. In vitro multienzymatic reaction systems for biosynthesis. Fundam Appl New Bioprod Syst. 2013;137:153–84.
3. Banta S, Swanson BA, Wu S, Jarnagin A, Anderson S. Optimizing an artificial metabolic pathway: engineering the cofactor specificity of Corynebacterium 2,5-diketo-D-gluconic acid reductase for use in vitamin C biosynthesis. Biochem-Us. 2002;41(20):6226–36.
4. Campbell E, Meredith M, Minteer SD, Banta S. Enzymatic biofuel cells utilizing a biomimetic cofactor. Chem Commun. 2012;48(13):1898–900.
5. Cooney MJ, Svoboda V, Lau C, Martin G, Minteer SD. Enzyme catalysed biofuel cells. Energy Environ Sci. 2008;1(3):320–37.
6. Costa SA, Tzanov T, Carneiro AF, Paar A, Gübitz GM, Cavaco-Paulo A. Studies of stabilization of native catalase using additives. Enzyme Microb Technol. 2002;30(3):387–91.

7. Crabtree GW, Dresselhaus MS, Buchanan MV. The hydrogen economy. Phys Today. 2004;57 (12):39–44.
8. Dohr O, Paine MJI, Friedberg T, Roberts GCK, Wolf CR. Engineering of a functional human NADH-dependent cytochrome P450 system. Proc Natl Acad Sci U S A. 2001;98(1):81–6.
9. Dresselhaus MS. Basic research needs for the hydrogen economy. Abstr Pap Am Chem S. 2004;227:U1084–5.
10. Dudley QM, Karim AS, Jewett MC. Cell-free metabolic engineering: biomanufacturing beyond the cell. Biotechnol J. 2015;10(1):69–82.
11. Eijsink VGH, Bjork A, Gaseidnes S, Sirevag R, Synstad B, van den Burg B, Vriend G. Rational engineering of enzyme stability. J Biotechnol. 2004;113(1–3):105–20.
12. Eijsink VGH, Gaseidnes S, Borchert TV, van den Burg B. Directed evolution of enzyme stability. Biomol Eng. 2005;22(1–3):21–30.
13. Fernández L, Gómez L, Ramírez HL, Villalonga ML, Villalonga R. Thermal stabilization of trypsin with glycol chitosan. J Mol Catal B: Enzym. 2005;34(1):14–7.
14. Ginosar DM, Petkovic LM, Glenn AW, Burch KC. Stability of supported platinum sulfuric acid decomposition catalysts for use in thermochemical water splitting cycles. Int J Hydrogen Energy. 2007;32(4):482–8.
15. Glykys DJ, Banta S. Metabolic control analysis of an enzymatic biofuel cell. Biotechnol Bioeng. 2009;102(6):1624–35.
16. Hallenbeck PC, Benemann JR. Biological hydrogen production; fundamentals and limiting processes. Int J Hydrogen Energy. 2002;27(11):1185–93.
17. Holladay JD, Hu J, King DL, Wang Y. An overview of hydrogen production technologies. Catal Today. 2009;139(4):244–60.
18. Hong J, Wang YR, Ye XH, Zhang Y-HP. Simple protein purification through affinity adsorption on regenerated amorphous cellulose followed by intein self-cleavage. J Chromatogr A. 2008;1194(2):150–4.
19. Iyer PV, Ananthanarayan L. Enzyme stability and stabilization—aqueous and non-aqueous environment. Process Biochem. 2008;43(10):1019–32.
20. Jia F, Narasimhan B, Mallapragada S. Materials-based strategies for multi-enzyme immobilization and co-localization: a review. Biotechnol Bioeng. 2014;111(2):209–22.
21. Kleerebezem R, van Loosdrecht M. Mixed culture biotechnology for bioenergy production. Curr Opin Biotechnol. 2007;18(3):207–12.
22. Levin DB, Pitt L, Love M. Biohydrogen production: prospects and limitations to practical application. Int J Hydrogen Energy. 2004;29(2):173–85.
23. Liao C-H, Huang C-W, Wu J. Hydrogen production from semiconductor-based photocatalysis via water splitting. Catalysts. 2012;2(4):490–516.
24. Liese A, Hilterhaus L. Evaluation of immobilized enzymes for industrial applications. Chem Soc Rev. 2013;42(15):6236–49.
25. Liu WF, Wang P. Cofactor regeneration for sustainable enzymatic biosynthesis. Biotechnol Adv. 2007;25(4):369–84.
26. Lo HC, Fish RH. Biomimetic NAD(+) models for tandem cofactor regeneration, horse liver alcohol dehydrogenase recognition of 1,4-NADH derivatives, and chiral synthesis. Angew Chem Int Edit. 2002;41(3):478–81.
27. Lo HC, Leiva C, Buriez O, Kerr JB, Olmstead MM, Fish RH. Bioorganometallic chemistry. 13. Regioselective reduction of NAD(+) models, 1-benzylnicotinamde triflate and beta-nicotinamide ribose-5′-methyl phosphate, with in situ generated [CP*Rh(Bpy)H](+): structure-activity relationships, kinetics, and mechanistic aspects in the formation of the 1,4-NADH derivatives. Inorg Chem. 2001;40(26):6705–16.
28. Logan BE, Call D, Cheng S, Hamelers HVM, Sleutels THJA, Jeremiasse AW, Rozendal RA. Microbial electrolysis cells for high yield hydrogen gas production from organic matter. Environ Sci Technol. 2008;42(23):8630–40.

29. Lutz J, Hollmann F, Ho TV, Schnyder A, Fish RH, Schmid A. Bioorganometallic chemistry: biocatalytic oxidation reactions with biomimetic NAD(+)/NADH co-factors and [Cp*Rh(bpy) H](+) for selective organic synthesis. J Organomet Chem. 2004;689(25):4783–90.
30. Maeda K, Domen K. Photocatalytic water splitting: recent progress and future challenges. J Phys Chem Lett. 2010;1(18):2655–61.
31. Martin del Campo JS, Rollin J, Myung S, Chun Y, Chandrayan S, Patino R, Adams MWW, Zhang Y-HP. High-yield production of dihydrogen from xylose by using a synthetic enzyme cascade in a cell-free system. Angew Chem Int Edit. 2013;52(17):4587–90.
32. Myung S, Rollin J, You C, Sun FF, Chandrayan S, Adams MWW, Zhang Y-HP. In vitro metabolic engineering of hydrogen production at theoretical yield from sucrose. Metab Eng. 2014;24:70–7.
33. Myung S, Zhang XZ, Zhang Y-HP. Ultra-stable phosphoglucose isomerase through immobilization of cellulose-binding module-tagged thermophilic enzyme on low-cost high-capacity cellulosic adsorbent. Biotechnol Prog. 2011;27(4):969–75.
34. Myung SW, You C, Zhang Y-HP. Recyclable cellulose-containing magnetic nanoparticles: immobilization of cellulose-binding module-tagged proteins and a synthetic metabolon featuring substrate channeling. J Mater Chem B. 2013;1(35):4419–27.
35. Paul CE, Gargiulo S, Opperman DJ, Lavandera I, Gotor-Fernández V, Gotor V, Taglieber A, Arends IW, Hollmann F. Mimicking nature: synthetic nicotinamide cofactors for C=C bioreduction using enoate reductases. Org Lett. 2012;15(1):180–3.
36. Rollin JA, Martin del Campo JS, Myung S, Sun FF, You C, Bakovic AE, Castro RL, Chandrayan S, Wu C-H, Adams MWW, Senger R, Zhang Y-HP. High-yield hydrogen production from biomass by in vitro metabolic engineering: mixed sugars coutilization and kinetic modelling. Proc Natl Acad Sci U S A. 2015;112:4964–9.
37. Rollin JA, Tam TK, Zhang Y-HP. New biotechnology paradigm: cell-free biosystems for biomanufacturing. Green Chem. 2013;15(7):1708–19.
38. Rollin JA, Ye X, del Campo JM, Adams MW, Zhang Y-HP. Novel hydrogen bioreactor and detection apparatus. Adv Biochem Eng Biotechnol. 2014;274:1–17.
39. Rosell A, Valencia E, Ochoa WF, Fita I, Pares X, Farres J. Complete reversal of coenzyme specificity by concerted mutation of three consecutive residues in alcohol dehydrogenase. J Biol Chem. 2003;278(42):40573–80.
40. Ryan JD, Fish RH, Clark DS. Engineering cytochrome P450 enzymes for improved activity towards biomimetic 1,4-NADH cofactors. Chembiochem. 2008;9(16):2579–82.
41. Sheldon RA, Sorgedrager M, Janssen MHA. Use of cross-linked enzyme aggregates (CLEAs) for performing biotransformations. Chim Oggi. 2007;25(1):62–7.
42. Thauer RK, Jungermann K, Decker K. Energy conservation in chemotrophic anaerobic bacteria. Bacteriol Rev. 1977;41(1):100.
43. Turner JA. Sustainable hydrogen production. Science. 2004;305(5686):972–4.
44. Wang YR, Zhang Y-HP. A highly active phosphoglucomutase from Clostridium thermocellum: cloning, purification, characterization and enhanced thermostability. J Appl Microbiol. 2010;108(1):39–46.
45. Watanabe S, Kodaki T, Makino K. Complete reversal of coenzyme specificity of xylitol dehydrogenase and increase of thermostability by the introduction of structural zinc. J Biol Chem. 2005;280(11):10340–9.
46. Weckbecker A, Groger H, Hummel W. Regeneration of nicotinamide coenzymes: principles and applications for the synthesis of chiral compounds. Adv Biochem Eng Biot. 2010;120:195–242.
47. Woodward J, Mattingly SM, Danson M, Hough D, Ward N, Adams M. In vitro hydrogen production by glucose dehydrogenase and hydrogenase. Nat Biotechnol. 1996;14(7):872–4.
48. Woodward J, Orr M. Enzymatic conversion of sucrose to hydrogen. Biotechnol Prog. 1998;14 (6):897–902.
49. Woodward J, Orr M, Cordray K, Greenbaum E. Biotechnology: enzymatic production of biohydrogen. Nature. 2000;405(6790):1014–5.

50. Ye XH, Wang YR, Hopkins RC, Adams MWW, Evans BR, Mielenz JR, Zhang Y-HP. Spontaneous high-yield production of hydrogen from cellulosic materials and water catalyzed by enzyme cocktails. Chemsuschem. 2009;2(2):149–52.
51. Ye XH, Zhang CM, Zhang Y-HP. Engineering a large protein by combined rational and random approaches: stabilizing the Clostridium thermocellum cellobiose phosphorylase. Mol Biosyst. 2012;8(6):1815–23.
52. Yoshida A, Nishimura T, Kawaguchi H, Inui M, Yukawa H. Enhanced hydrogen production from formic acid by formate hydrogen lyase-overexpressing Escherichia coli strains. Appl Environ Microbiol. 2005;71(11):6762–8.
53. Zhang Y-HP. A sweet out-of-the-box solution to the hydrogen economy: is the sugar-powered car science fiction? Energy Environ Sci. 2009;2(3):272–82.
54. Zhang Y-HP. Using extremophile enzymes to generate hydrogen for electricity. Microbe. 2009;4:560–5.
55. Zhang Y-HP. Production of biocommodities and bioelectricity by cell-free synthetic enzymatic pathway biotransformations: challenges and opportunities. Biotechnol Bioeng. 2010;105(4):663–77.
56. Zhang Y-HP. Renewable carbohydrates are a potential high-density hydrogen carrier. Int J Hydrogen Energy. 2010;35(19):10334–42.
57. Zhang Y-HP. Hydrogen production from carbohydrates: a mini-review. Acs Sym Ser. 2011;1067:203–16.
58. Zhang Y-HP. What is vital (and not vital) to advance economically-competitive biofuels production. Proc Biochem. 2011;46:2091–110.
59. Zhang Y-HP. Production of biofuels and biochemicals by in vitro synthetic biosystems: opportunities and challenges. Biotechnol Adv. 2014. doi:10.1016/j.biotechadv.2014.10.009.
60. Zhang Y-HP, Evans BR, Mielenz JR, Hopkins RC, Adams MWW. High-yield hydrogen production from starch and water by a synthetic enzymatic pathway. PLoS One. 2007;2(5): e456. doi:10.1371/journal.pone.0000456.
61. Zhang Y-HP, Mielenz JR. Renewable hydrogen carrier – carbohydrate: constructing the carbon-neutral carbohydrate economy. Energies. 2011;4(2):254–75.
62. Zhu ZG, Tam TK, Sun FF, You C, Zhang Y-HP. A high-energy-density sugar biobattery based on a synthetic enzymatic pathway. Nat Commun. 2014;5:3026.
63. Zhu ZG, Tam TK, Zhang Y-HP. Cell-free biosystems in the production of electricity and bioenergy. Fundam Appl New Bioprod Syst. 2013;137:125–52.

Part II
Thermoconversion

Chapter 4
Hydrogen Production from Biomass Gasification

Sotirios Karellas

Abstract Upgrading of gas streams formed from biomass gasification for the production of pure hydrogen or hydrogen-rich gases is facing many technical and technological challenges. Both gasification and hydrogen separation technologies play a significant role in the total efficiency of the production process. The aim of this chapter is to analyze gasification processes and examine the possible options for hydrogen production from the product gas streams from a gasification process. Gas conditioning and hydrogen purity will be taken into consideration as well as economic aspects of the integrated process.

Keywords Biomass gasification • Hydrogen production • Membrane separation • Hydrogen production economics

4.1 Introduction

One of the biggest advantages of the thermal gasification of solid fuels is that it gives the possibility not only to produce heat and power but also to provide valuable fuels. This is one of the main reasons why thermal gasification has been developed in the last decades and has been integrated to processes for the production of gaseous (substitute natural gas SNG, H_2) or liquid (bio-methanol, bio-ethanol, DME, etc.) fuels providing very important technological breakthroughs.

Hydrogen is an extensively investigated fuel, especially during the last few years, since it is a clean fuel. As environmental concerns increase, gasification technologies offer promising solutions for efficient application of green technologies in decentralized systems, since they allow the use of gas engines, microturbines, or fuel cells for energy production. The end use of the product gas is not limited to energy production, but it also includes the synthesis of other flexible products, since it is possible to upgrade the gas in order to produce chemicals, such as liquid fuels or gaseous fuels, hydrogen and methane being the

S. Karellas (✉)
Laboratory of Steam Boilers and Thermal Plants, School of Mechanical Engineering, National Technical University of Athens, 9, Heroon Polytechniou str., Athens 15780, Greece
e-mail: sotokar@mail.ntua.gr

Z. Fang et al. (eds.), *Production of Hydrogen from Renewable Resources*, Biofuels and Biorefineries 5, DOI 10.1007/978-94-017-7330-0_4

most representative, fertilizers, substitute natural gas, etc. Given the above considerations, hydrogen production from biomass seems to be a very attractive option, especially regarding decentralized applications [1–3].

The production of hydrogen from biomass can be divided in three steps. In the first step, woody biomass is thermally gasified producing a synthesis gas which mainly contains H_2 and CO as combustible gases. The second includes the treatment of the synthesis gas. In this step, fine particle filtering, desulfurization, and tar wash-cracking take place. Reforming and water-gas shift reactors can further increase the hydrogen content in the syngas. After this step, the treated gas is processed (e.g., by means of membranes) to produce pure hydrogen.

4.2 Biomass Gasification Technologies

The thermal gasification of biomass is a widely known process with a lot of applications in the field of energy production. The gasification is actually the third step in the thermochemical conversion process of solid biomass, as shown in Fig. 4.1, following the steps of drying and pyrolysis of the thermochemical conversion of solid biomass. In this process, biomass is gasified providing a product gas, which is combustible and can be further used in an apparatus for combined heat and power production (e.g., microturbine, fuel cell, internal combustion engines).

Gasification is the aggregated result of a great number of heterogeneous and homogeneous reactions. The difference between pyrolysis and gasification processes is that for the gasification, a gasification medium is needed (air, O_2, H_2O, or CO_2) as well as a way to provide the heat, for the endothermic reactions. The main reactions that take place in gasification are:

$$C + H_2O + \text{Heat} \leftrightarrow CO + H_2 \tag{4.1}$$

$$C + 2H_2O + \text{Heat} \leftrightarrow CO_2 + 2H_2 \tag{4.2}$$

$$C + CO_2 + \text{Heat} \leftrightarrow 2CO \tag{4.3}$$

$$C + 2H_2 \leftrightarrow CH_4 + \text{Heat} \tag{4.4}$$

$$CO + 3H_2 + \text{Heat} \leftrightarrow CH_4 + H_2O \tag{4.5}$$

$$CO + 3H_2O + \text{Heat} \leftrightarrow H_2 + CO_2 \tag{4.6}$$

There are a lot of gasification reactor designs, for example, cocurrent or countercurrent gasifiers, where the gasification medium and the biomass are contacted in a co- or counterflow, respectively. The five main types of gasifiers are presented by

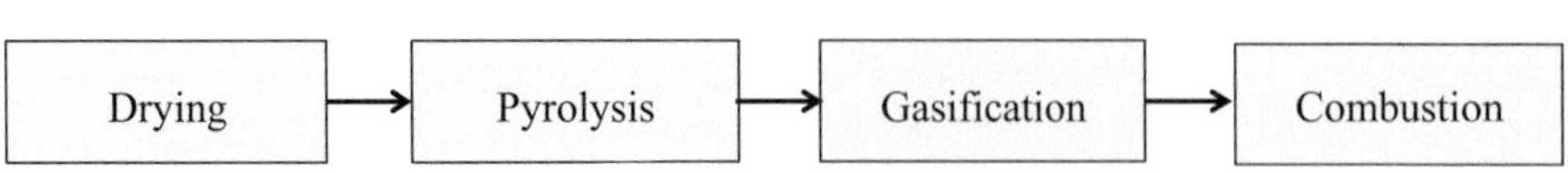

Fig. 4.1 Steps for the thermochemical conversion of solid biomass to gas

Knoef [4] and Karl [5] and are either fixed bed, fluidized bed, or entrained bed gasifiers. These gasification types can be seen in Table 4.1.

In Table 4.1, some of the main features of gasifiers are displayed. It can be seen that the counter-current gasifier has the highest tar content. Tars are higher carbon number hydrocarbons that are mainly produced in the pyrolysis zone of the gasifier. In a countercurrent gasifier, the product gas meets the pyrolysis zone before exiting the gasifier and the tars remain in the tar-rich product gas. On the other hand, in the case of the co-current gasifier, the product gas after the pyrolysis zone meets the combustion zone, where the tars are being combusted and, as a consequence, their content in the product gas is very limited. The fluidized bed gasifiers can be divided into bubbling (dense) and circulating fluidized bed types. They usually work at temperatures up to 900 °C, and their tar content is kept at low values. In entrained bed gasifiers, the gasification temperature is very high (up to 1500 °C), and as a consequence, it is a tar-free gasification process.

4.3 Autothermal and Allothermal Gasification

Gasifiers can be classified according to the gasification medium and the gasification process can be defined as autothermal, when heat us being generated from internal reactions, or allothermal, when heat is supplied from external sources.

In *autothermal gasification*, air (or oxygen) in low air ratio λ (about 0.2–0.3) is used as the gasification medium. For this case, the heat needed for the gasification is provided through the exothermic reaction of the partial oxidation of biomass.

In *allothermal or indirect gasification*, water, steam, or CO_2 is used as gasification medium. In this case, the required heat for the endothermic gasification reactions has to be provided to the reactor externally (e.g., from a combustion chamber through a heat exchanger).

Figure 4.2 shows a scheme for autothermal and allothermal gasification. In the autothermal gasification, about 60–85 % of the fuel input is chemically bounded energy, whereas 15–40 % corresponds to the sensible heat of the product gas due to its high temperature which results from the partial combustion of biomass. In the allothermal gasification process, as presented in Fig. 4.2, about 34 % of the energy input is introduced into the gasifier from an external combustion process. That means that the product gas contains approximately 134 % of energy, a part of which (about 64 %) is being released in the combustion chamber to produce the needed heat for the gasification. As a consequence, the product gas, which is further used, contains 70 % of the energy of the fuel input into the gasifier. As shown in Fig. 4.2, the heating value of the product gas from the allothermal gasifier is more than twice as high as the one produced in an autothermal gasifier.

Table 4.1 Gasifier types [4, 6]

	Fixed beds		Fluidized beds		Entrained beds
	Co-current	Counter-current	Dense	Circulating	
	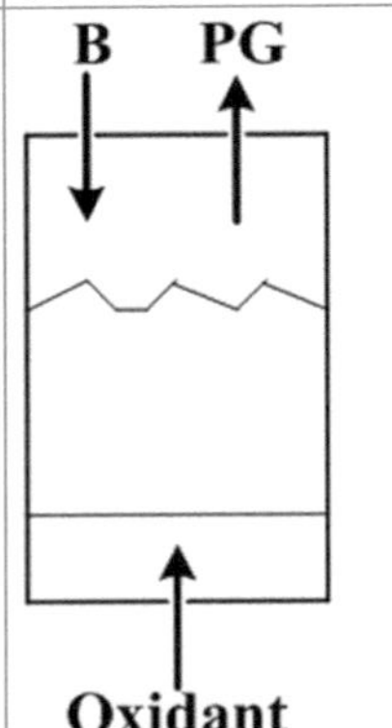	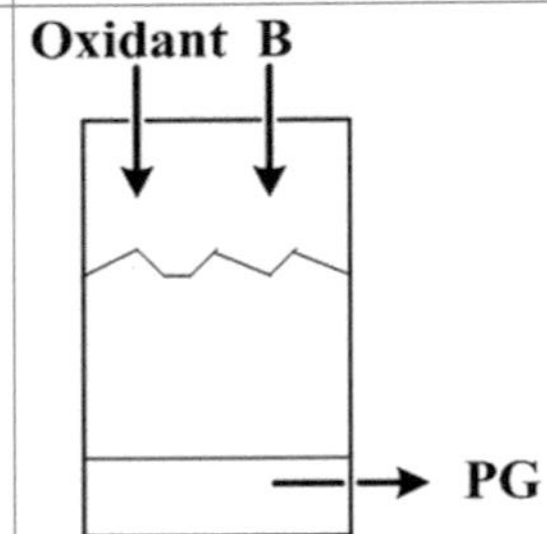	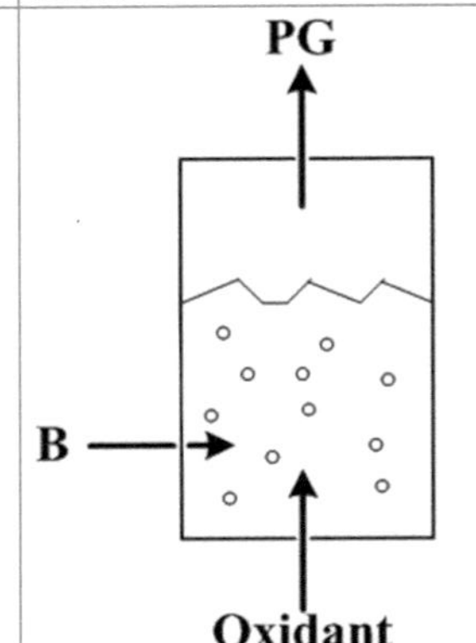	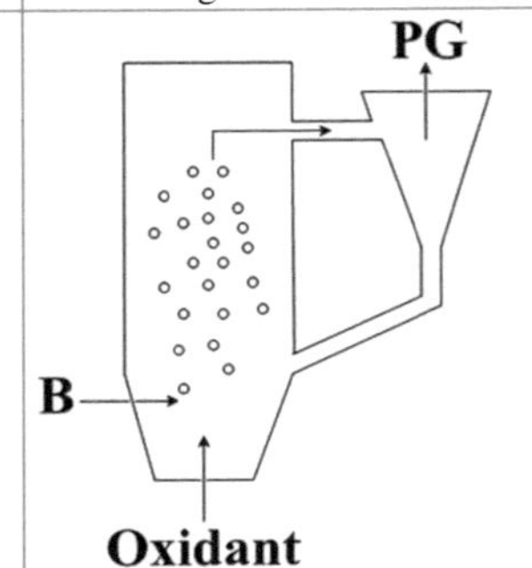	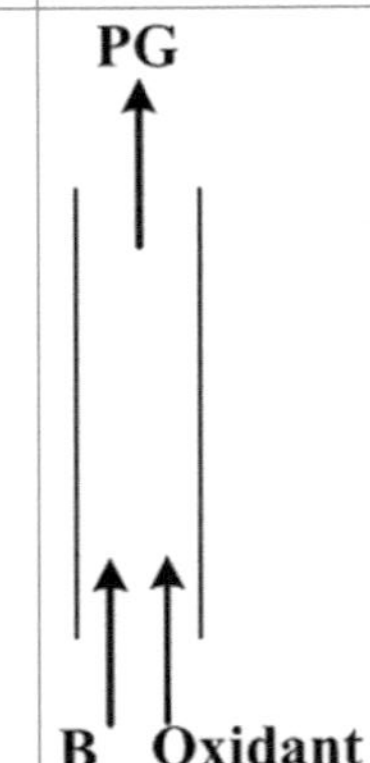
T (°C)	700–1,200	700–900	<900	<900	1500
Tars	Low	Very high	Intermediate	Intermediate	Absent
Control	Easy	Very easy	Intermediate	Intermediate	Very complex
Scale	$< 5MW_{\mathrm{th}}$	$< 20MW_{th}$	$10 < MW_{th} < 100$	$20 < MW_{th} < ?$	$> 100MW_{th}$
Feedstock	Very critical	Critical	Less critical	Less critical	Very fine particles

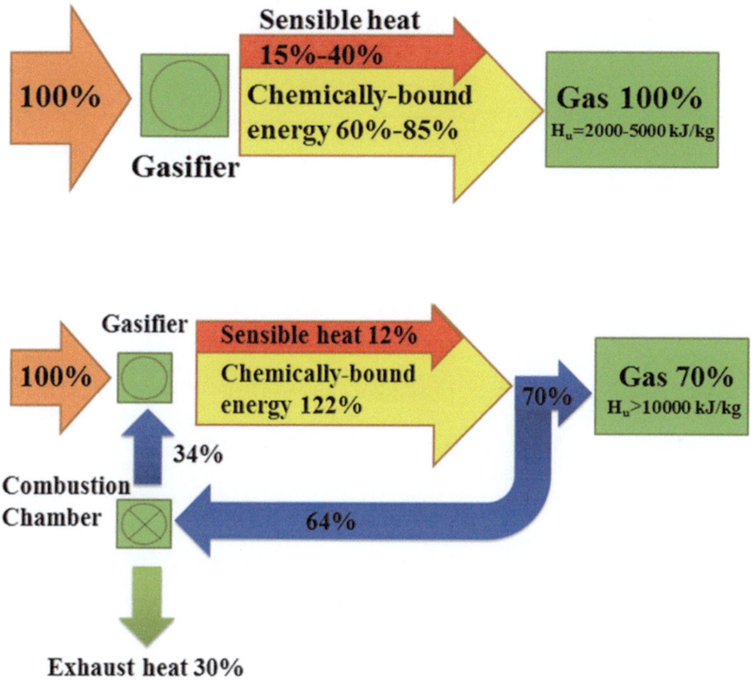

Fig. 4.2 Energy balance of autothermal and allothermal gasification (H_u stands for lower heating value)

4.4 Product Gas Quality

The composition of the product gas from biomass gasification depends on various factors, the most important of which are [7]:

- Type and shape of the fuel (e.g., dimensions, specific surface of the fuel particles, humidity, chemical composition)
- Type and quantity of the gasification medium (e.g., air, oxygen, water steam, carbon dioxide, mixtures of them)
- Design of the gasification reactor (mixture intensity of the fuel and the gasification medium, residence time of the fuel and the product gas in the reactor, etc.)
- Gasification temperature
- Use of catalysts
- Pressure conditions in the reactor

The product gas is a mixture of combustible gases like H_2, CH_4, CO, and higher hydrocarbons (C_2+) and noncombustible gases like H_2O, CO_2, and N_2.

The chemical reactions that take place during the gasification process can be divided into those that are heterogeneous and those that are homogeneous. In the heterogeneous reactions, gaseous molecules react with solid charcoal resulting in

Table 4.2 Heterogenous reactions that occur during the gasification of biomass [6]

Reaction	Enthalpy of reaction	Name	Type
$C + O_2 \rightarrow CO_2$	$\Delta H_R = -406$ kJ/mol	Oxidation of carbon	Exothermic
$2C + O_2 \rightarrow 2CO$	$\Delta H_R = -123$ kJ/mol	Partial oxidation of carbon	Exothermic
$C + H_2O \rightarrow CO + H_2$	$\Delta H_R = +119$ kJ/mol	Heterogeneous water – gas	Endothermic
$C + CO_2 \rightarrow 2CO$	$\Delta H_R = +162$ kJ/mol	Boudouard reaction	Endothermic
$C + 2H_2 \rightarrow CH_4$	$\Delta H_R = -87$ kJ/mol	Hydrogen gasification	Exothermic

Table 4.3 Homogenous reactions that occur during the gasification of biomass [6]

Reaction	Enthalpy of reaction	Name	Type
$CO + ½O_2 \rightarrow CO_2$	$\Delta H_R = -283$ kJ/mol	Oxidation of carbon monoxide	Exothermic
$H_2 + ½O_2 \rightarrow H_2O$	$\Delta H_R = -242$ kJ/mol	Oxidation of hydrogen	Exothermic
$CH_4 + 2O_2 \rightarrow CO_2 + 2H_2O$	$\Delta H_R = -802$ kJ/mol	Oxidation of methane	Exothermic
$CO + H_2O \rightarrow CO_2 + H_2$	$\Delta H_R = -42$ kJ/mol	Shift reaction	Exothermic
$CH_4 + H_2O \rightarrow CO + 3H_2$	$\Delta H_R = +206$ kJ/mol	Reforming	Endothermic

gaseous products. Table 4.2 presents the main heterogeneous reactions that take place during the gasification process.

The homogeneous reactions are the reactions between gaseous components. The most important reactions that occur during the biomass gasification process, as well as their endothermic or exothermic properties, are summarized in Table 4.3.

When the gases are in thermodynamic equilibrium, the composition of the product gas from the biomass gasification process can be calculated taking into account the above heterogeneous and homogeneous reactions. The gas composition, as a function of the temperature following the thermodynamic equilibrium, can be seen in Fig. 4.3. In Fig. 4.3, the composition of the product gas is presented on a wet and a dry basis.

Figure 4.3 depicts the theoretical thermodynamic equilibrium of the gases produced in the biomass steam gasification process. However, in reality, after gasification tests with various gasification technologies and gasification parameters, it can be concluded that there is a difference between the theoretical composition of the gas and the actual composition as it is measured from various gasification plants. The typical measured composition of the atmospheric gasification of biomass with air and water steam or oxygen is presented in Table 4.4. In Table 4.4, the average values are written in parenthesis.

As it can be seen in Table 4.4, because of the presence of N_2, the so-called poor gas or LCV (low calorific value) gas which is produced in an autothermal gasifier

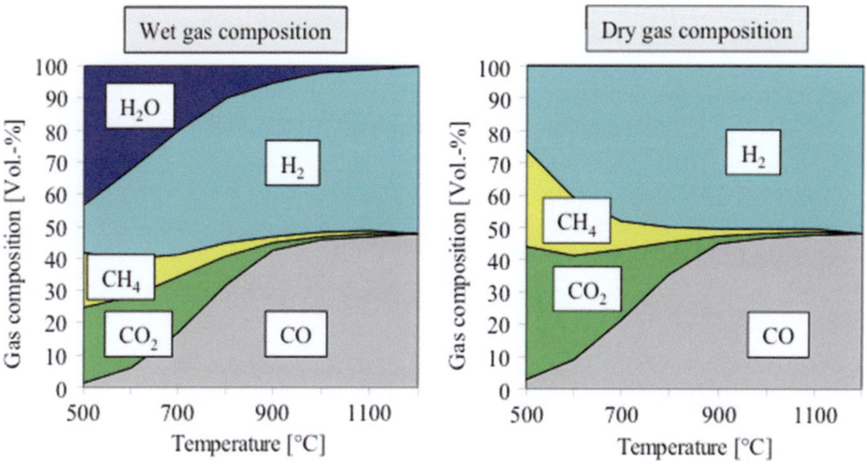

Fig. 4.3 Composition of the product gas for various gasification temperatures for stoichiometric atmospheric steam gasification (reforming)

Table 4.4 Typical composition of the product gas from the gasification of woody biomass [8]

	Gasification with	
Component	Air[a] (Vol.- %)	H_2O/O_2[a] (Vol.- %)
H_2	6.0–19 (12.5)	26–55 (38.1)
CO	9.0–21 (16.3)	20–40 (28.1)
CO_2	11–19 (13.5)	15–30 (21.2)
CH_4	3.0–7.0 (4.4)	4.0–14 (8.6)
C_2+	0.5–2.0 (1.2)	1.5–5.5 (3.0)
N_2	42–60 (52)	0
Heating value (MJ/Nm^3)	3.0–6.5 (5.1)	12–16 (13.2)

[a]Average values from data record: air 15, H_2O/O_2 9

with air has a lower heating value $H_u = 3.0–6.5\ MJ/Nm^3$. The main combustible component of this gas is CO. Its content can reach the value of 21 v/v %.

On the other hand, during allothermal gasification, or gasification with O_2, a hydrogen-rich gas is produced, and the lower heating value of the dry gas reaches values between 12 and 16 MJ/Nm^3.

Comparing the measured composition of the product gas from the gasification processes (Table 4.4) and the theoretical composition of the gas (Fig. 4.3), it can be concluded that the thermodynamic equilibrium between the gases in the gasifier is not always reached; thus, the gas composition can vary strongly from its theoretical value. As a consequence, the composition has to be continuously monitored, and the measured values have to be taken into consideration when simulating and planning gasification plants.

4.5 Supercritical Water Gasification Technology

Supercritical water gasification (SWG) is a process in which the gasifying agent is water in supercritical conditions of approximately 220 bar pressure (22 MPa) and a temperature over 37 °C. Via SWG, biomass can be gasified with a high conversion rate, and the product gas consists of H_2 and CO_2 [9]. The chemical conversion can be presented by the following reaction:

$$CH_xO_y + (2 - y)H_2O \rightarrow CO_2 + \frac{2 - y + x}{2} H_2 \quad (4.7)$$

This reaction can be divided into two steps, the steam reforming and the water-gas shift reactions, both described previously. The SWG gasification advantages over the conventional process are that it requires a single reactor. Furthermore, this technology is very efficient at low temperatures, below 700 °C, while it has no requirement for reducing the moisture content of the fuel allowing wet biomass to be used. Thus the energy-intensive drying processes, required for most solid fuels, can be avoided [10–12]. In addition, supercritical water has the ability to dissolve most organic substances, while it has excellent mass transfer performance [13, 14]. Another advantage is its high-pressure H_2, which reduces significantly the energy costs for its compression during storage [14]. On the contrary, there are several issues yet to be solved, concerning corrosion and plugging as well as economical issues, since this process requires external energy to preheat both the biomass and the reactor, increasing the overall costs.

The reaction performance is of great importance for SCW gasification systems. Matsumura [15] has formulated two different approaches for SCW type of gasification in terms of the temperature: the high temperature, between 773 and 1023 K, without the presence of any catalysts and the low temperature, approximately 623–873 K, catalytic SCW gasification.

4.6 Hydrogen Separation from Biomass Gasification

Biomass gasification technologies still have many challenges to face to become more competitive in the energy market. These challenges include the lack of consistent fuel availability, the low net calorific value (especially in the state-of-the-art autothermal gasification technologies) of the produced gas, as well as several transport and storing issues that make biomass more complex to handle compared to fossil fuels. Nevertheless, when biomass is used for the production of hydrogen, there are a wide variety of processes that can be applied.

The next chapters analyze the proposed modules for the separation of hydrogen from the product gas of biomass gasification.

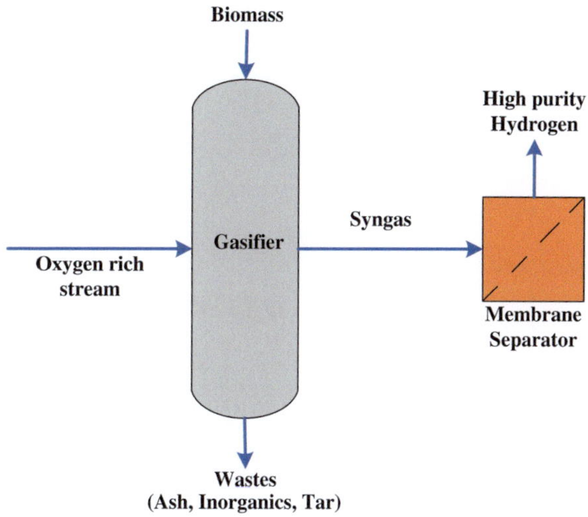

Fig. 4.4 Membrane separation of syngas

4.6.1 Membrane Separation

The preferentially selective permeation of hydrogen from a mixed gas stream through a polymer, metallic, or ceramic membrane is the base of the membrane separation technology [16, 17]. The mechanism beyond this technology varies based on the membrane material and its design (Fig. 4.4).

In microporous polymeric membranes, the separation of the molecules occurs by a molecular diffusion transport mechanism determined by the pore diameter and particle size [18]. In metallic membranes, the hydrogen molecule is dissociated into atoms that pass through the film and then recombine into hydrogen molecules on the other side of the membrane. On the other hand, in dense ceramic membranes, the separation of hydrogen is based on the transfer of hydrogen ions and electrons through the membrane. Dense ceramic membranes require higher operational temperatures to achieve comparable flux rates to those of other membrane technologies (Fig. 4.5).

Most suitable materials for hydrogen separation from biomass gasification products for pressures above 20 bars and temperatures over 700 °C are palladium-copper membranes [16]. Palladium membrane hydrogen purifiers operate via pressure-driven diffusion across palladium membranes [19].

One of the main advantages of this technology is the capability of producing extremely high purity hydrogen for fuel cell applications, i.e., PEM, which operate only with pure hydrogen. The technological challenges for this membrane technology include the high hydrogen flux at low pressure drop; the increase of the membrane tolerance to contaminants, especially sulfur and carbon monoxide; and the operation at system temperatures within the range of 250–500 °C [20]. Such advances would make it possible to achieve a reduction of the cost of membrane

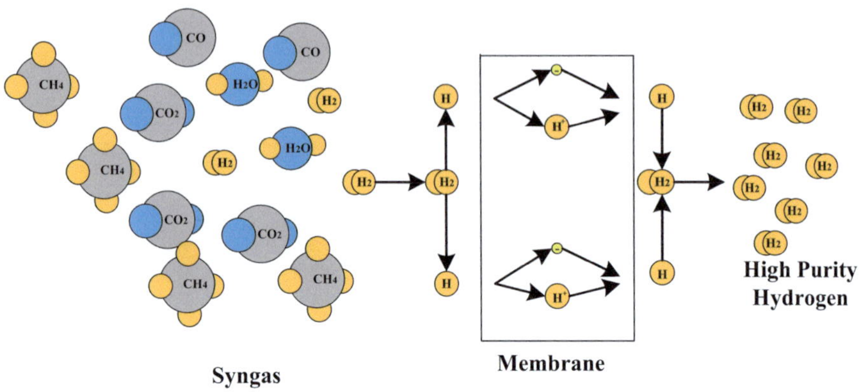

Fig. 4.5 Functionality of a gas separation membrane

separation technologies, an improvement in their efficiency, and simplify hydrogen separation and purification systems.

The main configurations in the use of selective membranes are described in the following chapters.

4.6.2 Membrane Integrated in the Gasification Reactor (Reformer)

In this concept, the selective membrane is assembled directly inside the gasification reaction environment to remove the produced hydrogen in situ. In the gasifier (reformer), catalyst can be used for higher hydrogen yield, and the membrane is in the form of a tube, allowing the transit of pure hydrogen, as shown in Fig. 4.6. Through the inner tube, steam is used to drag the permeated hydrogen. This configuration can also be accomplished by many smaller tubes, increasing the specific membranes surface per unit volume of the gasifier, and as a result the overall hydrogen flow rate. The types of membranes used in these reactors are mainly dense metallic membranes and ceramic membranes.

The main drawback of this technology is that a thermal threshold is needed for the reliable durability of the membrane as well as for stability problems due to stress [20].

4.6.3 Reformer and Membrane Modules

Contrary to the integrated system described before, in this configuration the selective membrane is assembled in proper units after the gasification reactor, to avoid the conflict of catalyst and membrane operating parameters, subdivided into a series

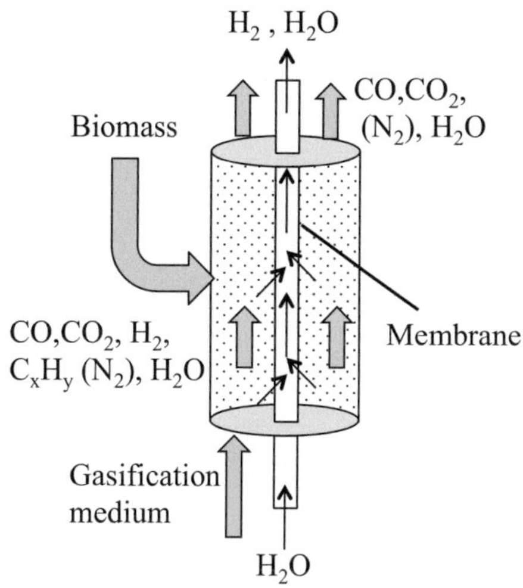

Fig. 4.6 Membrane reactor integrated in the biomass gasifier

of reaction-separation modules. The main benefits of this module are not only the ability to decouple, if needed, the reforming and separation operating conditions but also the simpler mechanical design of the membrane tubes compared with the ones in the membrane reactor concept. On the other hand, the membrane surface needed for the same methane conversion is greater compared with membrane reactors, leading to a higher capital cost [20].

Figure 4.7 presents the process in which the selective membrane is placed after the gasification reactor. Biomass is partially converted in hydrogen in an allothermal gasifier, and the produced hydrogen is recovered through a palladium alloy membrane separation module. The remnants can be sent to a reformer module so that higher purity levels can be achieved, as shown in Fig. 4.7. The operating temperature, by terms of heat recovery, can be reduced before the membrane unit down to 450 °C and then increased before entering the second reactor [21].

The remnants of the membrane separator can be also recycled back into the combustion chamber of an allothermal gasifier, as shown in Fig. 4.8.

Figure 4.8 presents the coupling of an allothermal gasifier called BioHPR (biomass heat pipe reformer) with a separation module. BioHPR is a gasification technology in which the needed heat for the reforming reaction is provided to the gasifier from a combustion chamber by means of liquid metal heat pipes [22]. In the configuration presented in Fig. 4.8, a palladium membrane is used to separate hydrogen from the product gas, whereas the rest of the gases are being combusted and provide the needed heat. Palladium membranes are used in this configuration to achieve high purity hydrogen, which will allow its further use in PEM fuel cells [19].

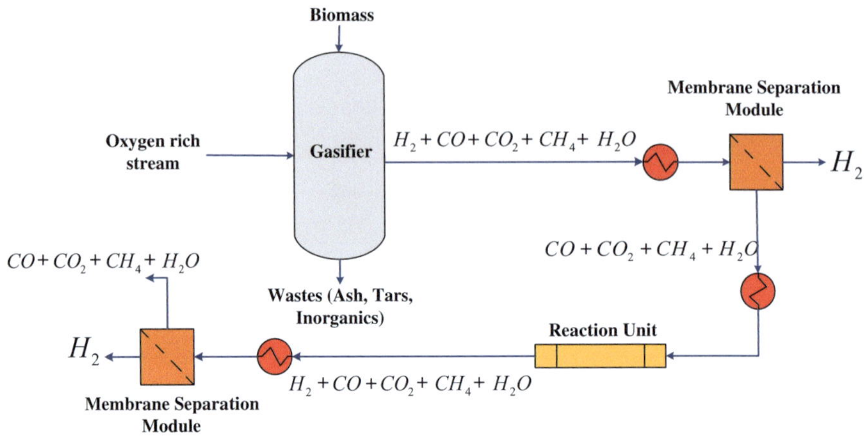

Fig. 4.7 Gasifier and separation module

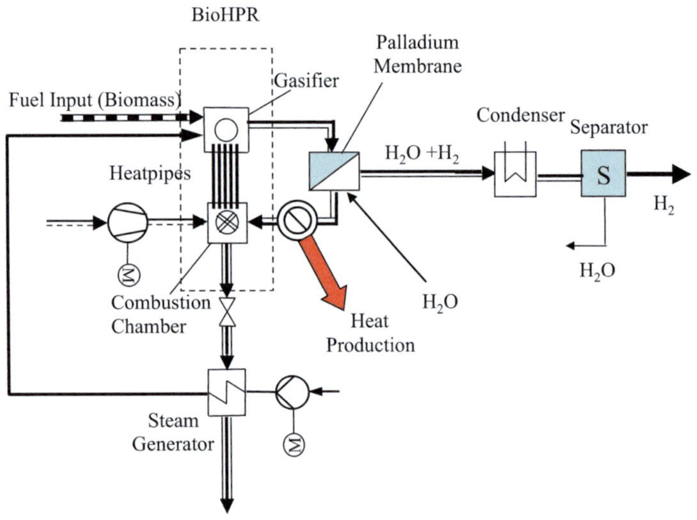

Fig. 4.8 Allothermal gasifier and separation module (Reprinted with permission from [19]. Copyright © 2009, Elsevier)

4.6.4 Water-Gas Shift Reaction

The water-gas shift reaction (WGS) is an exothermic reaction, which is favored at low temperatures and is independent of the operating pressure:

$$CO + H_2O \leftrightarrow CO_2 + H_2, \ \Delta H_{298} = -41.15 \ \text{kJ/mol} \quad (4.8)$$

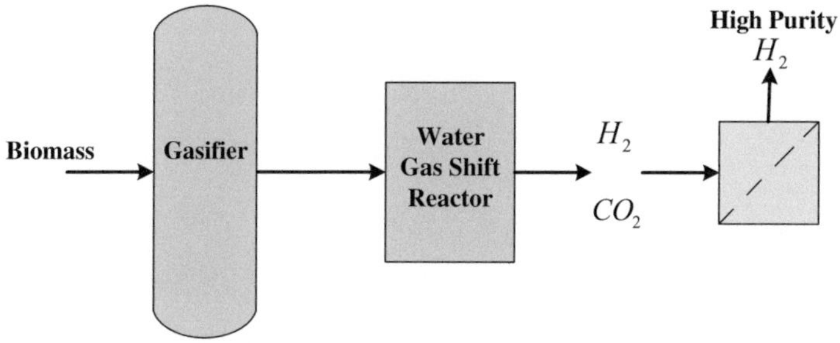

Fig. 4.9 Gasifier with water-gas shift reactor

The main advantage over the steam reforming is that this reaction requires no external heat and as a result no heat exchangers, resulting in a simpler and more compact design and consequently a lower capital cost. The WGS is on most occasions performed in two reactors: a high-temperature reactor at 350–500 °C and a low temperature one at approximately 200 °C [23]. Fe-Cr-based catalysts are the most common catalysts in high-temperature reactors, while in the low-temperature reactors, Cu-based catalysts are used [24, 25] (Fig. 4.9).

For hydrogen production from biomass gasification, a water-gas shift reactor is placed after the biomass gasifier, and the final gas consists mainly of hydrogen and CO_2. Hydrogen can be then very easily separated by means of a membrane. Therefore, ultrapure hydrogen can be produced from this process.

4.6.5 Water-Gas Shift with Pressure Swing Adsorption

Pressure swing adsorption (PSA) is used for the under pressure separation of gas species from a mixture of gases, via molecular sieves, manufactured so that their pore diameter allows the separation of different sized molecules of syngas stream. The most noticeable feature of this technology is the high level of hydrogen purity achieved (99.0–99.9 %) as a result of its ability to remove gas phase impurities to the required level. The operating pressure varies from 18 to 36 bars, while the purity level does not severely affect the hydrogen yield of the PSA system. The main drawbacks of this technology are the relatively modest recovery from the feed-stream hydrogen, approximately 86 % and the low operating temperatures.

Bhattacharya et al. [26] proposed a model for hydrogen production via oxygen-blown biomass gasification followed by a water-gas shift reactor. The required oxygen is produced in low pressure cryogenic air separation unit, in which a nitrogen-rich stream is produced as well. The syngas is produced in a biomass gasifier. It is then further enriched in hydrogen in a shift reactor by the additional injection of water and by controlling the reactor temperature. After the drying process, the two main contents of the gas mixture are H_2 and CO_2. Finally, the

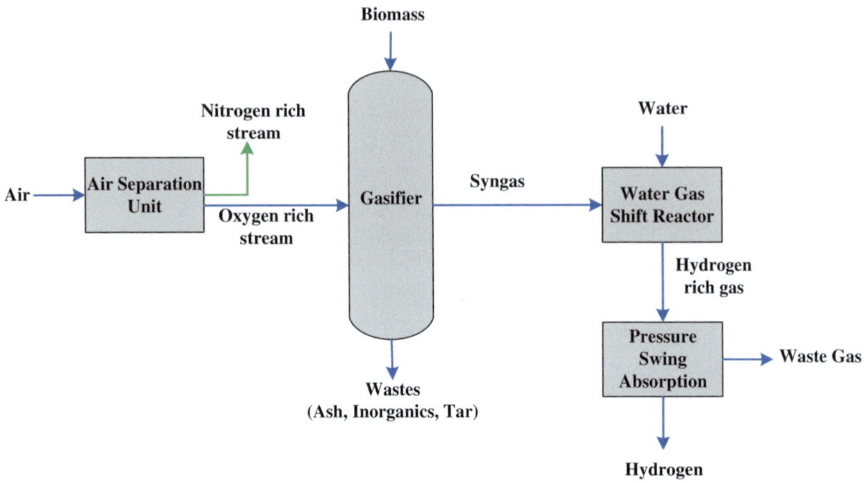

Fig. 4.10 Hydrogen production by means of PSA

pure hydrogen is received via a PSA system, where the CO_2 is separated from the syngas (Fig. 4.10).

4.6.6 Adsorption Enhanced Reforming

A process that improves the hydrogen separation of the product gas from biomass gasification is the use of calcium carbonate as bed material at a dual fluidized bed (DFB) gasifier. The use of CaO causes an increase in H_2 content to a value of up to 70 %. This is done since, parallel to the reforming temperature, the carbonation reaction takes place and therefore, the equilibrium of the shift reaction is moved to the products, according to the Le Chatelier principle.

$$CaO + CO_2 \leftrightarrow CaCO_3 \qquad \text{Carbonation reaction} \tag{4.9}$$

$$CO + H_2O \leftrightarrow CO_2 + H_2 \qquad \text{CO shift reaction} \tag{4.10}$$

The $CaCO_3$ is discharged in the DFB gasifier and in turn creates CaO. This so-called adsorption enhanced reforming (AER) process was demonstrated in a technical scale in the framework of an EU project on a biomass power plant in Güssing [27]. It is vital that all impurities contained in the raw syngas such as dust, alkalis, tars, and sulfur compounds are separated before the CO contained in the syngas is converted into hydrogen in a catalytic two-stage shift reactor.

For the segregation of the remaining CO, this is realized by pressure swing adsorption (PSA), in which the raw synthesis gas flows at high pressure through, e.g., an activated carbon bed, where heavy gases such as CO and CO_2 are heavily adsorbed compared to hydrogen, resulting in receiving high purity hydrogen at the outlet of the adsorber. By switching to a second adsorber, the activated carbon is regenerated by the reduction of pressure.

For the production of high purity hydrogen, membrane processes are usually used. Thereby, hydrogen diffuses at temperatures of around 200 °C through metallic palladium/silver membranes. Hydrogen's partial pressure at the membrane is a contributing factor in the rate of the permeation, a deciding parameter on the required membrane area and consequently on the cost of the process. For this reason, the syngas has to be compressed to high pressures.

4.6.7 *Typical Hydrogen Production Process Integrated in Biomass Gasification Systems*

A process flow sheet for an integrated hydrogen production system with biomass gasification is presented in Fig. 4.11. At first, biomass is dried to 15–30 % moisture content and fed into a fluidized bed gasifier [18]. A high amount of steam is fed into the gasifier, to provide the needed H_2O molecules for the high-temperature shift conversion. Syngas is cleaned from the tars in a reformer, where heavy hydrocarbons are treated catalytically at 800–900 °C. After the first shift reactor, which operates at 350–400 °C, the gas is cooled down to 40 °C via heat exchangers, before being pressurized to 30 bars and fed to an acid gas removal stage. Steam is again added at the low temperature shift step and is removed alongside with CO_2 in a second acid gas removal stage. The final stage consists of a methanation reaction for the conversion of CO to CH_4. The purity of hydrogen with this process can be as high as up to 95 % [18]. Figure 4.11 depicts the process for pure hydrogen production after product gas treatment.

4.7 Hydrogen Production by Reaction Integrated Novel Gasification

A state-of-the-art method for the production of hydrogen via biomass gasification has been proposed by Lin et al. [28–30]. This method, called hydrogen production by reaction integrated novel gasification (HyPr-RING), is a combination of water-hydrocarbon reaction, water-gas shift reaction, and absorption of CO_2 in a single reactor under both sub- and supercritical water conditions. The main reaction is shown by Eq. 4.11:

$$C + 2H_2O + CaO \rightarrow CaCO_3 + 2H_2,\ \Delta H_{298} = -88\ \text{kJ/mol} \tag{4.11}$$

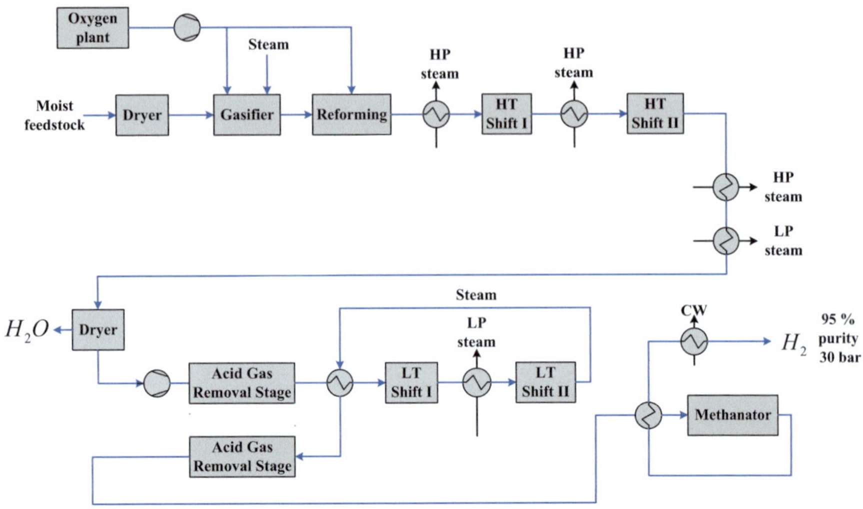

Fig. 4.11 Process flow of a traditional hydrogen production process with high and low temperature shift reactions [18]

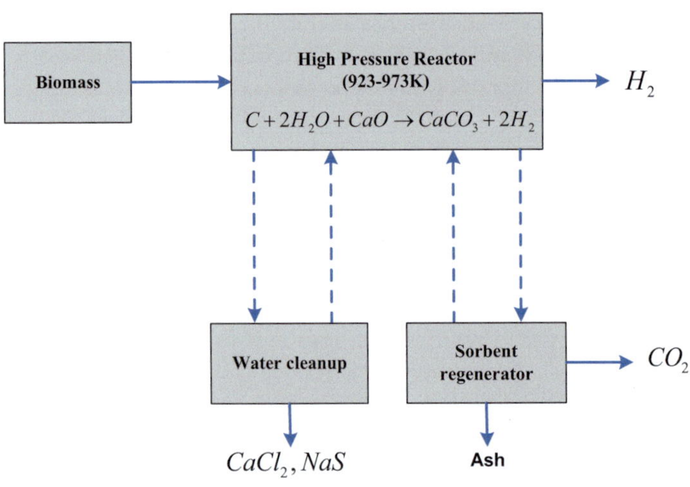

Fig. 4.12 Flowchart of the hydrogen production by reaction integrated novel gasification (*HyPr-RING*) method

Equation 4.11 is exothermic and can produce high-yield hydrogen at relatively low temperatures, of approximately 923–973 K. The advantages of HyPr-RING over conventional gasification are that hydrogen separation and gas separation occur in one reactor at low temperature [31]. Figure 4.12 presents a diagram of the described method.

The basic concept of the HyPr-RING consists of two reactors, the gasifier and the regenerator, as shown in Fig. 4.12. Major products from this process are H_2 and CO_2, while the raw materials supplied are biomass, water, and CaO. In the gasifier, CaO reacts with high-pressure water to form $Ca(OH)_2$, which absorbs CO_2 giving finally $CaCO_3$ and releasing heat. The produced $CaCO_3$ is used, in the regenerator, to regenerate CaO, while CO_2 is also released through this reaction.

4.8 Economics of Hydrogen Production from Biomass Gasification

Due to increasing concerns over environmental impact of fossil fuels and their finite nature, research and development of new energy sources has become inevitable [32]. Interest is growing in hydrogen from both an environmental and a thermodynamic point of view, as it is the cleanest and most efficient combustion fuel. Given that fact, hydrogen contribution to the energy market is expected to increase in order to keep up with this trend [1–3].

The future of hydrogen market is highly dependent on the availability of a low and environmental friendly source [33]. At the moment, the least expensive way to produce hydrogen is via steam methane reforming (SMR), which is used to produce more than half of the global hydrogen production [11]. The costs for the production of hydrogen using this method range between 1.27 €/kg (2007), for large facilities, and 3.18 €/kg (2007), for a smaller 500 kg/d facility [34]. Approximately 52–68 % of the final hydrogen price, in this method, is derived from natural gas feedstock costs, while the remaining costs come from capital costs [35].

On the other hand, the costs of the hydrogen production from biomass range from 1.05 €/kg (2001) to 1.50 €/kg (2001), according to Hamelinck et al. [36]. Based on an assessment, made by Nath et al. [37], the most economic technology of producing hydrogen from biomass is via biomass gasification yet is still rather unfavorable compared to SMR, with Spath et al. reporting that hydrogen by direct gasification of lignocellulosic biomass is about three times more expensive than the one produced by SMR [38]. Ni et al. [28] made a comparison (Fig. 4.13) of the estimated costs of hydrogen production via biomass gasification and natural gas SMR [39].

According to Padro et al. the cost of hydrogen produced via biomass gasification is expected to range between 1.22 €/kg (1999) and 2.42 €/kg (1999), while the facilities that use direct gasification to produce hydrogen are expected to have a 5 % higher cost than the indirect method [39]. A study by Mann determined the cost of hydrogen produced from biomass gasification to be approximately 1.69 €/kg (2004), when the cost of biomass is assumed 39 €/dry ton (2004) and the facility daily output is 139.7 t/day [40]. Another economically competitive, although yet in early development stage, method to produce hydrogen via biomass gasification is supercritical water gasification. Spritzer and Hong [41] calculated the cost of

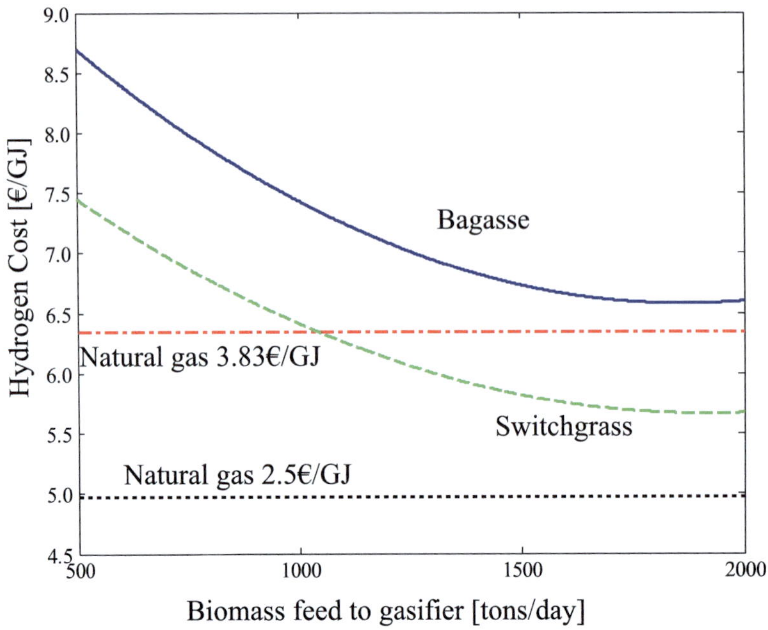

Fig. 4.13 Comparison of estimated costs (2006) of hydrogen production via biomass gasification and natural gas SMR [28]

Table 4.5 Estimation of hydrogen production costs from different technologies in € (2007) [43]

Production category	Method	€/GJ	€/kg
Thermochemical	Steam methane reforming	5.48	0.66
	Coal gasification	9.95	1.20
	Hydrocarbon partial oxidation	6.99	0.84
	Biomass gasification	8.76	1.05
	Biomass pyrolysis	8.93	1.07
Electrolytic	Electrolysis	20.36	2.45
Photolytic	Solar electrolysis	30.37	3.64
	Photobiological production	26.64	3.20

hydrogen produced from supercritical water gasification to be approximately 2.55 €/GJ, about 0.30 €/kg (2002). In general, at the moment, hydrogen production via biomass gasification is not yet considered economically competitive with SMR processes [14], [42].

Norman [43] presented the estimated hydrogen production costs for the main technologies, as listed in Table 4.5.

4.9 Conclusions and Future Outlook

The production of hydrogen from gasification of biomass is a technological concept that has attracted a lot of interest. On one hand, this is because it provides the potential to generate a highly efficient, clean, and environmentally friendly fuel, such as hydrogen, which can be used for power generation but also for the synthesis of biochemicals. On the other hand, the biomass-to-hydrogen conversion pathway is in principle associated to zero net carbon emissions, and therefore, it does not contribute to the greenhouse effect.

In this chapter, a multitude of technological processes allowing the gasification of biomass and the subsequent upgrading of the produced syngas in order to create a stream of pure hydrogen as an end product were presented. Each process exhibits certain technical advantages and disadvantages, while it is associated with different specific costs dependent on its economics and also its efficiency. For the present, the production of hydrogen from biomass is substantially cost competitive compared with the other main technologies, and gasification of biomass has a lot of potential to become gradually more market attractive in the future.

References

1. Rezaiyan J, Cheremisinoff NP. Gasification technologies: a primer for engineers and scientists. Boca Raton: CRC Press; 2005.
2. NNFCC. Review of technologies for gasification of biomass and wastes. E4Tech. 2009.
3. Knoef H. Handbook of biomass gasification. Biomass Technology Group,BTG; Enchede, 2005.
4. Knoef H. State of the art review of biomass gasification in Europe, 5. Internationales Fachsymposium, Entwicklungsstand und Marktreife Holzvergaser-Technik, Karlsruhe. 2002.
5. Karl J. Dezentrale Energiesysteme, Neue Technologien im liberalisierten Energiemarkt. 3rd ed. München: Oldenbourg Verlag; 2012.
6. Karellas S. Online analysis of the composition of biogenous gases and their effect on microturbine and fuel cell systems. Düsseldorf: VDI Verlag; 2006. Reihe 6, Nr. 537.
7. Kaltschmitt M, Hartmann H, Hofbauer H. Energie aus Biomasse, Grundlagen, Techniken und Verfahren. Springer-Verlag, Berlin-Heidelberg, 2001
8. Hofbauer H, Fleck T, Veronik G. Gasification feedstock database, IEA Bioenergy agreement, Task XIII, Activity 3, Technische Universität Wien. 1997.
9. Zhang J. Hydrogen production by biomass gasification in supercritical water. Energeia. 2008;19(6):1–3.
10. Liao B, Guo L, Lu Y, Zhang X. Solar receiver/reactor for hydrogen production with biomass gasification in supercritical water. Int J Hydrog Energy. 2013;38:13038–44.
11. Demirbas A. Hydrogen production from biomass via supercritical water gasification. Energy Source. 2010;32(14):1342–54.
12. Osada M, Sato T, Watanabe M, Shirai M, Arai K. Catalytic gasification of wood biomass in subcritical and supercritical water. Combust Sci Technol. 2006;178:537–52.
13. Lin S, Lu Y, Guo L, Zhang X. Hydrogen production by biomass gasification in supercritical water with bimetallic Ni-M/γAl_2O_3 catalysis (M=Cu, Co and Sn). Int J Hydrog Energy. 2011;36(22):14391–400.
14. Reddy SN, Nanda S, Dalai AK, Kozinski JA. Supercritical water gasification of biomass for hydrogen production. Int J Hydrog Energy. 2014;39(13):6912–26.

15. Matsumura Y, Minowa T, Potic B, Kersten SRA, Prins W, Van Swaij WPM, van de Beld B, Elliott DC, Neuenschwander GG, Kruse A, Jerry Antal Jr M. Biomass gasification in near- and super-critical water: status and prospects. Biomass Bioenergy. 2005;29(4):269–92.
16. Gas Technology Institute. Direct hydrogen production from biomass gasifier using hydrogen-selective membrane. Final Report, contract Number: RD-38, GTI Project 20309. Prepared for Xcel Energy, Minneapolis, 2007.
17. Khan Z, Yusup S, Ahmad MM, Lai Fui Chin B. Hydrogen production from palm kernel shell via integrated catalytic adsorption steam gasification. Energy Convers Manag. 2014;87:1224–30.
18. Hannula I. Hydrogen production via thermal gasification of biomass in near-to-medium term. Espoo: VTT; 2009.
19. Karellas S, Kakaras E, Papadopoulos T, Schäfer C, Karl J. Hydrogen production from allothermal biomass gasification by means of palladium membranes. Fuel Process Technol. 2008;89(6):582–8.
20. De Falco M, Marrelli L, Iaquaniello G. Membrane reactors for hydrogen production processes. Springer; London 2001.
21. De Falco M, Iaquaniello G, Cucchiella G, Marrelli L. Reformer and membrane modules plan to optimize natural gas conversion to hydrogen. Nova Science Publishers; New York 2009.
22. Karl J. Biomass heat pipe reformer—design and performance of an indirectly heated steam gasifier. Biomass Convers Biorefin. 2014;4(1):1–14.
23. Huber GW, Iborra S, Corma A. Synthesis of transportation fuels from biomass: chemistry, catalysts and engineering. Chem Rev. 2006;106:4044–98.
24. Spath P, Aden A, Eggeman T, Ringer M, Wallace B, Jechura J. Biomass to hydrogen production detailed design and economics utilizing the battelle columbus laboratory indirectly-heated gasifier. Golden: US DOE National Renewable Energy Laboratory (NREL); 2005.
25. Iribarren D, Susmozas A, Petrakopoulou F, Dufour J. Environmental and exergetic evaluation of hydrogen production via lignocellulosic biomass gasification. J Clean Prod. 2014;69:165–75.
26. Bhattacharya A, Bhattacharya A, Datta A. Modeling of hydrogen production process from biomass using oxygen blown gasification. Int J Hydrog Energy. 2012;37(24):18782–90.
27. Koppatz S, Pfeifer C, Rauch H, Hofbauer H, Marquard-Moellenstedt T, Specht M. H_2 rich product gas by steam gasification of biomass with in situ CO_2 absorption in a dual fluidized bed system of 8 MW fuel input. Fuel Process Technol. 2009;90(7–8):914–21.
28. Ni M, Leung YC, Leung MKH, Sumathy K. An overview of hydrogen production from biomass. Fuel Process Technol. 2006;87(5):461–72.
29. Lin SY, Suzuki Y, Hatano H, Harada M. Hydrogen production from hydrocarbon by integration of water-carbon reaction and carbon dioxide removal (HyPr-RING) method. Energy Fuels. 2001;15(2):339–43.
30. Lin SY, Harada M, Suzuki Y, Hatano H. Process analysis for hydrogen production by reaction integrated novel gasification (HyPr-RING). Energy Convers Manag. 2005;46(6):869–80.
31. Lin SY, Suzuki Y, Hatano H, Harada M. Developing an innovative method, HyPr-RING, to produce hydrogen from hydrocarbons. Energy Convers Manag. 2002;43(9–12):1283–90.
32. Balat M, Balat M. Political, economic and environmental impacts of biomass-based hydrogen. Int J Hydrog Energy. 2009;34–9:3589–603.
33. Bartels JR, Pate MB, Olson NK. An economic survey of hydrogen production from conventional and alternative energy sources. Int J Hydrog Energy. 2010;35(16):8371–84.
34. Williams RB, Kornbluth K, Erickson PA, Jenkins BM, Gildart MC (2007) Estimates of hydrogen production potential and costs from California landfill gas. Proceeding of 15th European Biomass conference & exhibition, Berlin.
35. Roan V, Betts D, Twining A, Dinh K, Wassink P, Simmons T. Investigation of the feasibility of coal-based methanol for application in transportation fuel cell systems. Georgetown University Advanced Vehicle Development Program; Georgetown, 2004.

36. Hamelinck CN, Faaij APC. Future prospects for production of methanol and hydrogen from biomass. J Power Sources. 2002;111(1):1–22.
37. Nath K, Das D. Hydrogen from biomass. Curr Sci. 2003;85(3):265–71.
38. Spath PL, Mann MK, Amos WA. Update of hydrogen from biomass: determination of the delivered cost of hydrogen. Golden: National renewable Energy Laboratory (NREL) ; 2003.
39. Padro CEG, Putsche V. Survey of the economics of hydrogen technologies. Golden: National renewable Energy Laboratory NREL; 1999.
40. Mann MK. Hydrogen from biomass via gasification and catalytic steam reforming. 2005. http://www.hydrogen.energy.gov/h2a_prod_studies.html.
41. Spitzer MH, Hong GT. Supercritical water partial oxidation, FY 2003 progress report. National Renewable Energy Laboratory, NREL Report, Golden, 2003. p. 1–5.
42. Balat H, Kirtay E. Hydrogen from biomass: present scenario and future prospects. Int J Hydrog Energy. 2010;35(14):7416–26.
43. Norman K. Interim report: feasibility of microscale glucose reforming for renewable hydrogen, SANDIA report no: SAND2007-1713. New Mexico: Sandia National Laboratories; 2007.

Chapter 5
Hydrogen Production from Catalytic Biomass Pyrolysis

Lucía García, Javier Ábrego, Fernando Bimbela, and José Luis Sánchez

Abstract An overview is presented on different catalytic routes for producing hydrogen from biomass via pyrolysis processes. Fundamentals of biomass pyrolysis along with general aspects related to the types of processes and catalysts are discussed. Processes that allow hydrogen production in this field have been divided into single-step and multi-step processes. These processes are reviewed in this chapter, showing the state of the art. In both strategies, a hydrogen-rich product gas is obtained which, conveniently conditioned and purified, may serve for various purposes.

Catalytic pyrolysis of raw biomass feedstocks aiming at producing hydrogen can be carried out by directly contacting the raw material with a catalyst having a high selectivity towards hydrogen production inside the pyrolysis reactor, in a single-step process. Another possibility for producing hydrogen from biomass follows a strategy based on multiple steps. In the majority of multi-step processes, the biomass raw material is subjected to fast pyrolysis for producing a liquid product, denoted as biomass pyrolysis liquids or bio-oil, and afterwards processing the bio-oil or fractions of it in a catalytic steam reforming process with suitable catalysts.

Future trends of catalytic biomass pyrolysis process technologies are described in this chapter.

Keywords Hydrogen • Biomass • Pyrolysis • Catalyst • Nickel • Bio-oil • Steam reforming

---

L. García (✉) • J. Ábrego • F. Bimbela • J.L. Sánchez
Thermochemical Processes Group (GPT), Aragón Institute for Engineering Research (I3A), Universidad de Zaragoza, Zaragoza, Spain
e-mail: luciag@unizar.es; abrego@unizar.es; fbs@unizar.es; jlsance@unizar.es

Z. Fang et al. (eds.), *Production of Hydrogen from Renewable Resources*, Biofuels and Biorefineries 5, DOI 10.1007/978-94-017-7330-0_5

5.1 Introduction

Biomass is a renewable resource that can be used for hydrogen production. Some of the advantages of using biomass are the vast variety of materials and the high dispersion through the planet. This chapter is centered on lignocellulosic biomass since it is the most studied one.

Pyrolysis is a thermochemical process that consists in the thermal decomposition of carbon-containing resources (such as biomass) in a non-oxidizing atmosphere and in the absence of any other reactant agent. Solid, liquid, and gas product fractions are obtained with proportions that are much dependent on the process operating conditions, mainly process temperature, heating rate, and residence time of the vapors.

Pyrolysis is a thermochemical route of biomass conversion that has been explored for hydrogen production. There are a large number of research works focusing on hydrogen production by pyrolysis. These works have identified the catalyst as the key factor in the process. The catalyst increases hydrogen yield working at moderate temperatures, which is favorable from an energy point of view.

Catalytic biomass pyrolysis processes can be divided into one-step and multi-step processes. In one-step process, only one reactor is used. This reactor converts biomass into hydrogen. In a multi-step process, at least two reactors are required. In the first one, biomass pyrolysis takes place, and in the second reactor, pyrolysis products are converted to a high hydrogen yield. In most cases the biomass raw material is subjected to fast pyrolysis for producing a liquid product, bio-oil. Bio-oil is easily transported and different strategies can be employed for hydrogen production, for example, catalytic steam reforming of aqueous fraction of bio-oil. In the 1980s, several technologies of fast pyrolysis were developed to generate bio-oil. Bio-oil is the main intermediate in multi-step processes.

Fundamentals of biomass pyrolysis are presented to understand the main operating variables involved in the process. Some pyrolysis reactors for bio-oil production are included, given the relevance of this intermediate.

In the catalyst section, the main properties and performance of the catalyst in the biomass pyrolysis process are summarized. Also, many experimental works show the vast variety of catalysts used, being nickel-based catalysts preferred due to their price.

The main facts of the one-step processes are presented, analyzing hydrogen yield and the main findings of the literature works. Recent innovative processes such as biomass pyrolysis in molten alkali and microwave plasma are also included.

The multi-step processes section is mainly focused on catalytic steam reforming of bio-oil. This option presents some advantages that can be further implemented at an industrial scale based on economic studies.

5.2 Fundamentals of Biomass Pyrolysis

Pyrolysis is a thermochemical process that consists in the thermal decomposition of biomass in a non-oxidizing atmosphere and in the absence of any other reactant agent. Heating biomass in an inert atmosphere triggers a complex series of chemical reactions that include primary breakdown of biomass constituents, as well as secondary reactions involving the products of direct decomposition of biomass. All these chemical reactions occur coupled to mass and heat transfer processes, and solid, liquid, and gas product fractions are obtained. Usually, the term pyrolysis englobes all these simultaneous processes and not only the thermal decomposition of biomass. The solid fraction that can be obtained from pyrolysis is usually referred to as char, or, more recently, biochar. The liquid fraction, frequently named bio-oil, is a product of great interest for hydrogen production by means of catalytic processes, especially in multi-step processes. The gas fraction is maximized operating at high temperatures.

5.2.1 Composition and Characteristics of Lignocellulosic Biomass

Here, the main characteristics of lignocellulosic biomass will be presented. Note that other types of biomass materials with significant differences in their composition, such as animal residues, municipal solid waste, sewage sludge, and others, are not included in this description; however, they might also have potential for hydrogen production by means of pyrolysis. The three major constituents of lignocellulosic biomass are cellulose (40–50 %), hemicellulose (20–40 %), and lignin (5–30 %) [1]. The structure of cellulose, as well as the main monomers of hemicellulose and lignin, is shown in Fig. 5.1. Some authors have found that interactions between these components are negligible [2]; thus, the pyrolysis products could be considered as the summation of the individual contributions from the three main components.

Apart from the three main components, minor amounts of solvent-extractable compounds (extractives) can be found in lignocellulosic biomass: triglycerides, fatty acids, resin acids, steryl esters, sterols, and lignans [3]. Inorganic constituents can also be found in biomass, with total contents varying in a range from less than 1 % in softwoods to 15 % in herbaceous biomass [4]. Because of the catalytic effects of most of the main inorganic constituents (K, Ca, Na, P), biomass decomposition reactions and char formation can be altered by their presence, especially for cellulose [5].

The original water content of biomass also plays an important role during pyrolysis for various reasons. First, it influences heat requirements because it needs to be evaporated in the pyrolysis reactor and may render the process uneconomical. Second, an excessive water content impedes biomass particle size

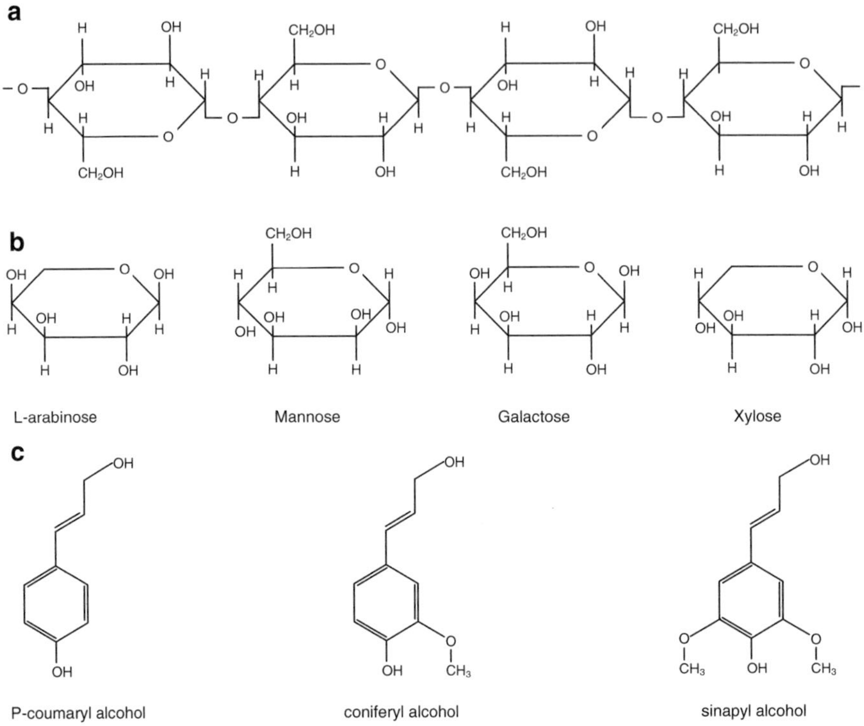

Fig. 5.1 Main components of lignocellulosic biomass: (**a**) cellulose, (**b**) hemicellulose monomers, and (**c**) lignin monomers

reduction, which is needed in several types of pyrolysis reactors. Third, most of the water ends up in the liquid fraction. Finally, the presence of water has been shown to produce higher yields of char [6, 7].

5.2.2 *Reaction Pathways and Types of Pyrolysis*

In most cases, the pyrolysis process takes place at moderate temperatures (300–600 °C) and is driven towards the production of a majority of solid or liquid fraction. If a major gaseous fraction is preferred, then gasification or combustion processes at higher temperatures, involving the presence of oxidizing agents such as air or steam, are more suitable. Bridgwater [8] classified pyrolysis processes and indicated that the main product obtained is gas when fast pyrolysis is carried out at temperatures higher than 700 °C.

Traditionally, a distinction between slow and fast pyrolysis, based on the process heating rate, has been made at the mentioned temperature interval to distinguish processes with major solid, liquid, or gas yields. However, this classification

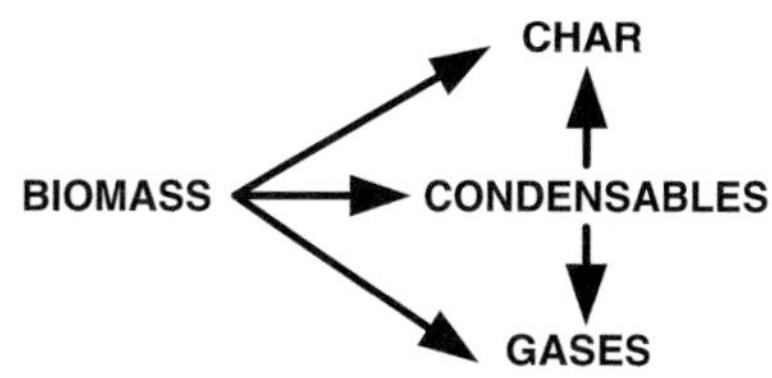

Fig. 5.2 Simplified pyrolysis reaction scheme

oversimplifies the complexity of the pyrolysis process [5]; the residence time of devolatilization products and the conditions in which this residence time goes by are equally relevant. These factors are crucial because primary decomposition condensable products (sometimes called primary tars) can undergo two competitive reaction pathways: cracking to non-condensable gaseous products or formation of secondary solid char depending on the reactor conditions (mainly pressure and gas flow rate) [7, 9] and inorganic content of biomass, as shown in Fig. 5.2. As a result of both competitive pathways, thc final condensable yield is reduced; thus, they should be avoided if liquid yield has to be maximized. Therefore, the so-called fast pyrolysis conditions, i.e., high heating rates, careful temperature control, very low vapor residence times, and rapid cooling of the vapors, are required for maximum bio-oil production [10].

The thermodynamics of the pyrolysis process are also determined by these reaction pathways. For instance, secondary char formation from condensable compounds is an exothermal process [7] that might cause an overall exothermal heat of pyrolysis if operational conditions are directed towards maximum char formation, whereas if bio-oil (condensable products) is preferred, the overall process will be clearly endothermic.

5.2.3 Product Distribution and Characteristics

The proportions of the solid, liquid, and gaseous product fractions in non-catalyzed fast pyrolysis can range between 10–40 %, 20–75 %, and 10–30 %, respectively, and are greatly dependent on the process operating conditions, mainly process temperature and pressure, heating rate, and residence time of the vapors. The biomass feedstock composition and properties (moisture, particle size, or density) also play an important role in product distribution.

The liquid fraction or bio-oil is a complex mixture of water and diverse organic compounds. Water (15–30 %) comes from both the original biomass moisture and devolatilization reactions. A typical composition of the whole bio-oil is (average of different lignocellulosic biomasses, on a dry basis) [11] 56.7 % C, 6.3 % H, 36.8 % O, and 0.2 % N. The maximum amount of obtainable H_2 is limited by both chemical composition and water content of the liquid fraction.

Bio-oil can be easily separated into two distinct phases either by fractionation (water addition) or centrifugation. The aqueous phase is a complex mixture consisting of carboxylic acids, aldehydes and ketones, alcohols, sugars, low molecular weight oligomers, and other more complex carbohydrates. The remaining fraction that is often referred to as pyrolytic lignin or organic phase contains a wide variety of high molecular mass lignin-derived compounds.

As a whole, bio-oil from lignocellulosic biomass is acidic in nature (with pH values between 2 and 4) and contains high amounts of oxygenated compounds. Both characteristics cause chemical instability over time, causing storage problems [12]. The high oxygen content confers bio-oil a low energy density compared to that of petroleum-based fuels. Indeed, it has a heating value of less than half of that of hydrocarbon fuels [13]. Finally, bio-oil might also have high contents of solid particles in suspension.

The origin of some of the individual components of bio-oil can be directly related to the main biomass constituents. For instance, levoglucosan, glycolaldehyde, and cellobiosan are products of cellulose pyrolysis [14], which has been extensively studied. Their amounts are significantly influenced by inorganic content, and they can further react to form gases or char depending on the operational conditions. Lignin pyrolysis has been much less investigated [5]; it produces mainly phenolic compounds such as guaiacols and syringols [15], and "pyrolytic lignin," which is composed of relatively large fractions of the original lignin.

Biomass charcoal, also named char or biochar, is a carbonaceous solid that retains part of the original biomass structure and can have a carbon content as high as 90 % [7]. Lignin from biomass produces comparatively higher yields of char than cellulose and hemicellulose. Lignin also concentrates mostly in the original inorganic content of biomass. Its main uses are as activated carbon precursor and as a fuel. Recent research proposes its use as soil amendment with the additional benefit of long-term carbon capture [16].

The permanent gases from pyrolysis of biomass are mainly composed of CO and CO_2, but also CH_4, H_2, and light hydrocarbons. Additionally, H_2S or NH_3 may be present if the amounts of sulfur or nitrogen in the original biomass are high. The gas mixture can be burned to provide part of the energy needed to drive the pyrolysis process or be subjected to catalytic treatment in a one-step process to improve H_2 yield, as well as to convert the condensable fraction.

5.2.4 Pyrolysis Reactors

Because bio-oil is an important intermediate in hydrogen production by catalytic pyrolysis of biomass using multi-step processes, this subsection examines pyrolysis reactors as they pertain to bio-oil production.

At typical fast pyrolysis conditions (500 °C, hot vapor residence time of ~1 s), a high yield of up to 75 % bio-oil can be obtained. To achieve these operational conditions, a fluidized bed is the preferred choice. Other reactors will be briefly discussed. A more detailed description of the main features of fast pyrolysis reactors can be found in the literature [5, 17].

Fluidized bed reactors allow a very good temperature control and heat transfer to the biomass material, but require careful selection of feedstock particle size distribution. They can be scaled up to maximum throughputs of several tons per hour. Usually, a bed material is required for enhancing heat transfer and fluidization of biomass, if needed. The bed material can be kept bubbling or be transported (circulating fluidized beds) to a secondary reactor where char is burned and hot bed material is recycled to the fluidized bed. In both cases, an inert fluidizing agent (such as nitrogen or recycled combustion gases) is needed.

At temperatures between 400 and 650 °C, Fig. 5.3 shows a typical product distribution of non-catalyzed, fluidized bed pyrolysis of lignocellulosic biomass [18]. The yield of hydrogen is generally very low and increases with temperature [19] without reaching a maximum value within this temperature interval. The addition of catalysts significantly alters these product distributions and may trigger hydrogen formation at lower temperatures, as well as higher concentrations of the desired product.

In fluidized bed reactors, a catalyst of adequate physical properties and fluid dynamic behavior can be continuously added to the bed for hydrogen production. However, catalyst particles are subjected to high attrition rates in fluidized beds, and regeneration of the spent catalyst may require char separation. Additionally, char particles carry-over can cause volatile decomposition, decreasing bio-oil yield.

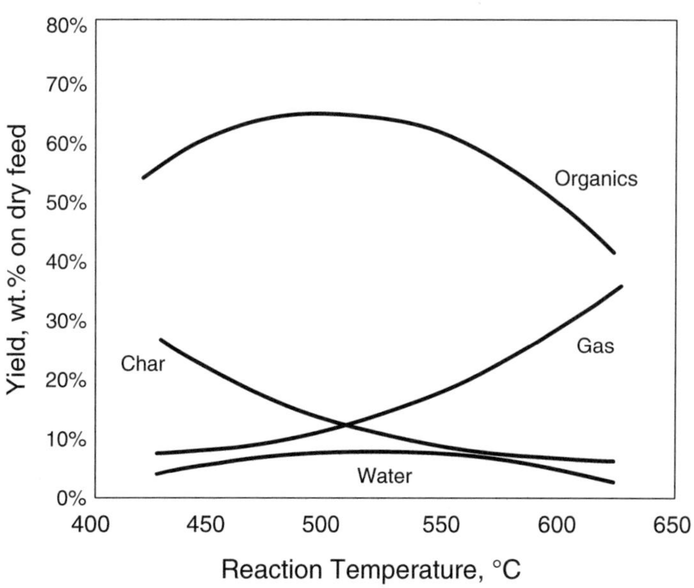

Fig. 5.3 Product distribution of poplar pyrolysis vs. temperature (Reprinted with permission from Bridgwater [17], Copyright © 2011 Elsevier)

The rotating cone reactor, developed at University of Twente [20], uses centrifugal force to mix hot sand and biomass particles. Rapid heating and small vapor residence time are achieved. Combustion of char provides the necessary heat for the process, and no carrier gas is needed. It can be scaled up to several tons per day [5, 21].

In ablative pyrolysis reactors, biomass is pressed against a hot surface, either in motion (rotating cylinder, disk or blade) or static (vortex or cyclone reactors). The main advantages of these systems are the possibilities of processing big biomass particles, using cold gas carriers, and including a catalyst in the hot surface [5]. Its applicability to large-scale systems is limited because scaling is a linear function of heat transfer area [21].

Auger and screw reactors involve mechanically mixing biomass with hot sand and transporting the mixture inside a cylinder. No carrier gas is needed so that auger and screw reactors offer good potential for scale-up [21].

5.3 Catalysts

Like any other hydrogen production process, biomass pyrolysis requires the use of catalysts to produce a hydrogen-rich gas. In fact, the catalyst is considered as a critical factor in the hydrogen production process from biomass to obtain high selectivity and high hydrogen purity. Key properties for a catalyst to be used for hydrogen production are C-C and C-O bond cleavage activity, WGS activity, low coke formation, resistance to deactivation by poisoning, and thermal and mechanical stability. Properties of the catalyst support are also important: usually it also presents an intrinsic catalytic activity that can enhance the reforming or act as a deterrent to coke deposition, which is one of the main drawbacks of the process.

The performance of a catalyst depends on its intrinsic characteristics, such as active metal nature, metal dispersion, and surface area, but also on the chemical nature of the compounds to be reformed. Thus, the process is not only influenced by the biomass being treated, but also by the pyrolysis conditions, as explained in previous sections. Furthermore, the selected process, one-step or multi-step, and the reactors, fixed and fluidized beds, have also a significant influence on catalyst performance.

Regarding the nature of catalysts, one approach is the use of noble metals, which have a high activity towards hydrogen production and lower selectivity to coke formation. In Table 5.1, several noble metal catalysts used in pyrolysis-based hydrogen production are shown. As can be seen, platinum is the most widely used catalyst of this kind.

However, and due to the scarcity in nature and high prices of noble metals, the use of transition metals, especially nickel, has also been studied, in spite of their lower activity and higher tendency to coke deactivation.

Apart from noble metals, the use of several different catalysts such as Na_2CO_3, K_2CO_3, $CaMgCO_3$, La/Al_2O_3, and Cr_2O_3, among others, was reported in a recent

Table 5.1 Noble metal-based catalysts for hydrogen production

Reference	Catalysts	Preparation method	Parameters analyzed
Takanabe et al. [22–24]	0.5 % Pt/ZrO_2	Wet impregnation	Activity tests with model compounds, reaction and deactivation mechanisms
Rioche et al. [25]	Cordierite Pt, Rh, Pd (1 %)/Al_2O_3 Rh, Pd (1 %) /$CeZrO_2$	Incipient wet-ness impregnation	Catalyst screening over different model compounds, reaction temperature
Basagiannis and Verykios [26]	Pt, Pd (1 %)/Al_2O_3 0.5 % Rh/Al_2O_3	Wet impregnation	Catalyst screening, reaction temperature, time on stream
Basagiannis and Verykios [27]	5 % Ru/15 % MgO/Al_2O_3	Wet impregnation	Long-term stability tests with model compounds and the aqueous fraction, reaction temperature, space velocity, structured forms of the catalyst
Basile et al. [28]	Dense Pd/25 % Ag membrane	Cold rolling + diffusion rolling	Permeation tests of the membrane reactor, pressure effect
Iwasa et al. [29]	1–10 % Pt over Al_2O_3, ZrO_2, and other supports	Impregnation over the support	Steam reforming of acetic acid in fixed bed
Dubey et al. [30]	5 % Pt/C	Commercial (Arora-Matthey)	Acetol aqueous reforming, 623–773 K in fixed bed

review [31]. $ZnCl_2$- and Ni-based catalysts are mentioned as having more potential towards maximum gas production. The use of other metals is scarcer in literature, for instance, Shoja et al. [32] used iron fillings for tar cracking in a dual bed reactor, at 850 °C.

Ni-based catalysts constitute the most solid alternative approach to noble metals in the design of suitable catalysts for the steam reforming of bio-oil or its fractions. The advantages of Ni-based catalysts compared to those based on noble metals are principally high activity and selectivity towards H_2 production at a much cheaper cost. However, Ni-based catalysts are more susceptible to carbon formation [33].

Table 5.2 shows both commercially prepared and laboratory-prepared Ni catalysts. Alumina is the preferred support, although many more materials can be found in the available literature. A previous reduction of the active metal is not always used, as the presence of a certain amount of hydrogen in the gas, or its in situ formation, is enough to reduce the Ni oxides at the reactor temperature.

As concluding remarks, and despite the wide range of materials, preparation methods, raw materials, and experimental conditions, it is clear that the presence of an active catalyst and a high temperature (over 1000 K) are needed in order to obtain a hydrogen yield that can reach around 100 g H_2/kg of biomass. Still there is much work to do on the improvement of on-stream catalyst stability over long periods of time, on the grounds that the economy of the process will significantly rely on catalyst performance and durability.

Table 5.2 Ni-based catalysts for H_2 or H_2-rich gas production from biomass pyrolysis

Reference	Catalysts	Parameters analyzed
Wang et al. [34]	UCI G-90C (15 % Ni/70–76 % Al_2O_3/5–8 % CaO)	Reaction temperature, S/C ratio, residence time, activity tests with model compounds, reaction mechanisms
Wang et al. [35, 36]	Commercial catalysts UCI G-90C, ICI 25-4M, ICI 46-1, UCI G-90B, BASF G1-25S, ICI 46-4, UCI G91	Catalyst screening, feedstock screening, reaction temperature, S/C ratio, residence time, space velocity, long-term stability tests, regeneration cycles
Marquevich et al. [37–39]	UCI G-90C, ICI 46-1	Activity and long-term stability tests with model compounds and with bio-oil, reaction temperature, S/C ratio, space velocity, reactor scale
García et al. [33]	UCI G91, ICI 46-1, ICI 46-4, Süd Chemie C11-NK	Catalyst screening in the steam reforming of the aqueous fraction of bio-oil, reaction temperature, S/C ratio, space velocity
Kechagiopoulos et al. [40]	Süd Chemie C11-NK	Activity tests with model compounds and with the aqueous fraction, reaction temperature, S/C ratio, catalyst regeneration
Davidian et al. [41]	Johnson Matthey 4 % Ni/2 % K/carrier La_2O_3-Al_2O_3	Bio-oil cracking activity tests, catalyst regeneration cycles, reaction mechanism
Waheed and Williams [42]	10 % Ni over calcined dolomite, prepared by wet impregnation on the calcined support	Pyrolysis + steam reforming of rice husk, wheat straw, and bagasse. 51 g H_2/kg biomass obtained at 950 °C from rice husk
Zhao et al. [43]	Ni over cordierite (monolith)	Rice husk pyrolysis in a rotary kiln and steam reforming of the resulting vapors. 65 g H_2/kg obtained at 1123 K
Qinglan et al. [44]	NiMo/Al_2O_3 commercial catalyst	36 g H_2/kg was produced at 723 K under nitrogen atmosphere in a one-step fluidized bed process
Zhang et al. [45]	Ni/Al_2O_3, Co/Al_2O_3, NiCo/Al_2O_3	120 g H_2/kg biomass was obtained with the bimetallic catalyst at 825 °C
Ansari et al. [46]	Ni-Fe/Al_2O_3, prepared by impregnation or microemulsion	15 % H_2 in gas, batch experiments with bagasse at 850 °C
Liu et al. [47]	9 % Ni over calcined sepiolite, prepared by wet impregnation	Two-step pyrolysis +reforming, heated by microwave radiation. 100 g H_2/kg biomass (hyacinth)

5.4 One-Step Processes

In one-step processes of catalytic pyrolysis, the catalyst is located in the same vessel where pyrolysis occurs. Considering the schematic representation of catalytic pyrolysis proposed by Garcia et al. [48], both stages, being pyrolysis the first and the action of catalyst the second, occur in the same reactor.

One-step processes have the advantages of process integration and smaller equipment costs compared to multi-step processes. Process integration can achieve some energy savings, as endothermic reactions, such as pyrolysis, occur simultaneously with other exothermic reactions such as water-gas-shift reaction. Heating of pyrolysis products is not required, as it happens, for example, in catalytic steam reforming of bio-oil, one example of multi-step process. The equipment cost diminishes because only one reactor is required in the process.

The drawbacks of this alternative are more severe catalyst deactivation and lower hydrogen content in product gas compared to multi-step processes. Considering a one-step process with a continuously fed fluidized reactor, high liquid yield is produced at relatively low temperature and the catalyst must convert/transform it into gases; as a consequence, more "work" for the catalyst is required [48]. In contrast, considering a multi-step process with two reactors, where in the first pyrolysis is being carried out at high temperature, more cracking occurs and less liquid production is generated (also, liquid conversion by thermal cracking can occur in the piping between the first and the second step, where catalyst is placed); as a consequence, less "work" of the catalyst is needed and then less catalyst deactivation will be observed.

The comparison of one-step process such as catalytic pyrolysis in a continuously fed fluidized reactor [48, 49] with a multi-step process such as catalytic steam reforming of bio-oil [50] indicates that lower hydrogen content in product gas is achieved in one-step catalytic pyrolysis than in catalytic steam reforming of bio-oil. For instance, the content of hydrogen in product gas was 67 vol.% (nitrogen and steam free) at 650 °C and 0.042 g catalyst h/g organics in the steam reforming of the aqueous fraction of bio-oil [50], a multi-step process, while in catalytic pyrolysis of biomass (pine sawdust), a one-step process, the content at the same temperature was 52 vol.% (nitrogen free) at 0.8 g catalyst h/g biomass [48].

In catalytic pyrolysis, one-step process, the moisture of biomass is involved and no more water is added; thus water-gas-shift reaction hardly takes place. In spite of the low content of hydrogen, this process can produce a gas with a H_2/CO ratio useful as synthesis gas for processes such as methanol or Fischer-Tropsch, among others.

In catalytic steam reforming of bio-oil, multi-step process, water is added and water-gas-shift reaction converts CO into CO_2 and H_2, increasing the hydrogen content in product gas. If the purpose is the generation of hydrogen-rich gas in one-step process, the addition of water is required and then the suitable process is steam gasification [51].

Table 5.3 Experimental works of conventional one-step catalytic pyrolysis

Institution	Operating conditions	Experimental installation	Catalyst	Biomass
University of Waterloo (Canada)	500–700 °C atmospheric pressure	Continuous-fed fluidized bed reactor, diameter = 25.4 mm, biomass flow rate <32 g/h	Ni-Al coprecipitated and Ni-Al with Mg or K	Poplar sawdust
University of Zaragoza (Spain) [49, 53]				
University of Zaragoza (Spain) [48, 54, 55]	650 and 700 °C atmospheric pressure	Continuous-fed fluidized bed reactor, diameter = 25.4 mm, biomass flow rate <25 g/h	Ni-Al coprecipitated	Pine sawdust
Karadeniz Technical University (Turkey) [56, 57]	501–752 °C atmospheric pressure	Batch, 1.5 g biomass	$ZnCl_2$, K_2CO_3, and Na_2CO_3 (impregnated biomass)	Cotton cocoon shell, tea factory waste, and olive husks
Delft University of Technology (The Netherlands) [52]	500, 750, and 850 °C	Batch, fixed bed of biomass (70–90 g) + cracking reactor	CaO, FeO, Al_2O_3, MnO, Cr_2O_3, CuO, and Na_2CO_3	Pine sawdust and rice straw
CIRAD-Fôret and Institute Européen des Membranes (France) [58]	700 °C	Batch, 10 g biomass	Ni or Fe nitrates (impregnated biomass)	Oak sawdust
Tianjin University of Science &Technology and Nanjing University of Technology (China) [44]	450, 500, and 590 °C atmospheric pressure	Continuous-fed fluidized bed reactor, diameter = 25 mm, biomass flow rate = 5 g/h	$NiMo/Al_2O_3$	Pine, Alaskan spruce, tropical lauan, and rice husks
University of Tehran and Research Institute of Petroleum Industry (Iran) [46]	850 °C atmospheric pressure	Batch, 1 g biomass (two beds in the same reactor)	$Ni\text{-}Fe/\gamma\text{-}Al_2O_3$	Powdered bagasse

One-step catalytic pyrolysis processes can be divided in two groups: conventional and innovative processes. Table 5.3 presents a summary of experimental works of conventional one-step catalytic pyrolysis processes, carried out at bench scale. In this table it can be observed that catalytic pyrolysis is carried out at atmospheric pressure and temperatures from 450 to 850 °C. There are two different types of experimental installations. In the first type, a batch of biomass is loaded in the reactor, being 90 g the highest amount of biomass reported [52]. The second type uses a continuously fed fluidized bed reactor with a biomass feeding rate lower than 32 g/h [49].

Chen et al. [52] recommended the use of 30 % weight of catalyst load to biomass. Also, they found Cr_2O_3 as the best catalyst with stronger catalytic role than other metal oxides. Cr_2O_3 showed the highest yields to total gas and hydrogen content in gas in their experiments.

In the work of Ansari et al. [46], the reactor contains two beds in a continuous downflow. This is considered as one-step process because only one reactor is used. In the first bed the pyrolysis of bagasse is carried out, and the second bed, below the first one, contains the catalyst and the catalytic cracking of tar is performed. They tested Ni-Fe/γ-Al_2O_3 catalysts prepared by two methods: co-impregnation and microemulsion. Using a microemulsion technique with a water to surfactant ratio of 1, the average metal particle size was decreased to 3.7 nm. Using this catalyst, the gas yield increased from 0.397 to 0.758 m^3/kg and decreased the tar yield from 0.445 to 0.237 g/g biomass compared to the noncatalytic process, while the heating value of the product gas remained almost constant (10–11 MJ/m^3). This last result is a consequence of the decrease in C_nH_m yield due to catalytic cracking.

Qinglan et al. [44] used a fluidized bed, where the primary decomposition and secondary reactions occurred simultaneously. The catalyst selected was a commercial NiMo/Al_2O_3 catalyst. In their study, four biomasses were tested: pine, Alaskan spruce, tropical lauan and rice husks. They found similar results in gas yields of the three woody biomass samples. They attributed this result to their similar compositions. Different results were obtained with rice husks compared to woody biomass. As an example, H_2 yield of woody biomass was higher than that of rice husks under the same operating conditions.

Arauzo et al. employed nickel-based catalysts prepared by coprecipitation in the laboratory [49, 53]. Some catalysts were modified with magnesium or potassium. They concluded that the addition of magnesium in the catalyst did improve resistance to attrition but resulted in a minor loss in gasification activity and increased coke production. The addition of potassium had little effect.

Garcia et al. employed also a Ni-Al coprecipitated catalyst prepared in the laboratory [48, 54, 59]. They studied the influence of calcination and reduction conditions on catalyst performance. They concluded that carrying out the catalytic pyrolysis at 650 and 700 °C, the catalyst calcined at 750 °C can be reduced by H_2 and CO generated during the pyrolysis reaction. The catalyst calcined at 850 °C required more severe reduction conditions to achieve the active phase of the catalyst. In pyrolysis at 650 °C, the most stable catalyst performance is achieved using the catalyst calcined at 850 °C and reduced during 1 h with a hydrogen flow rate of 3080 cm^3 (STP)/min. At a reaction temperature of 700 °C, the highest H_2 and CO yields are obtained using the catalyst calcined at 750 °C without previous reduction.

Garcia et al. [48] also studied the influence of catalyst weight/biomass flow rate (W/mb) ratio on gas production using the Ni-Al coprecipitated catalyst calcined at 750 °C without previous reduction. For W/mb ratios $\geq$0.4 h, no significant modifications were observed on the initial yields of different gases, with a gas composition similar to thermodynamic equilibrium. For W/mb ratios $<$0.4 h, a simple first-order kinetic equation has been suggested for H_2 and CO formation. These authors analyzed gas yield evolution over time. Figure 5.4 shows the results for an experiment carried out at 700 °C with a W/mb ratio of 0.31 h. As can be observed,

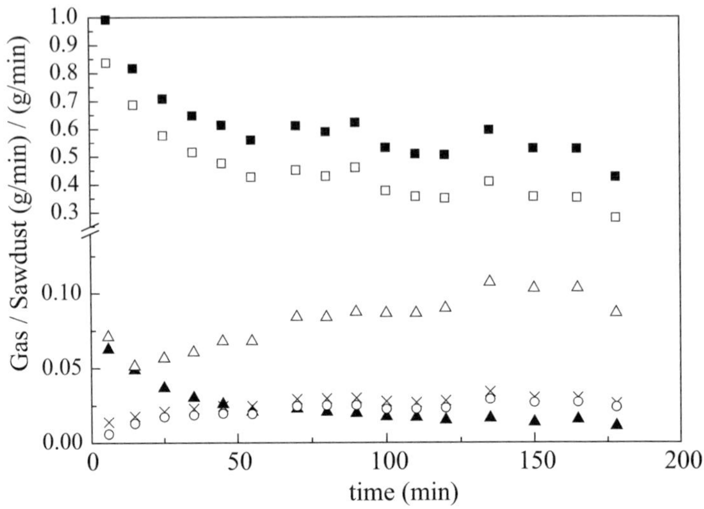

Fig. 5.4 Gas yield evolution versus time for an experiment of catalytic pyrolysis. Temperature = 700 °C, W/mb = 0.31 h. Total gas ■; H_2 ▲; CO □; CO_2 Δ; CH_4 ×; C_2 ○ (Reprinted with permission from García et al. [48], Copyright © 1998 American Chemical Society)

total gas and H_2 and CO yields diminish with experimental time, while CO_2, CH_4, and C_2 yields increase. This evolution is a consequence of the loss of catalyst activity. The deactivation of the catalyst is mainly caused by the formation of carbon deposits on the catalyst surface.

In all the mentioned studies, the effect of the catalyst is the same: increasing gas yield and decreasing liquid yield. Hydrogen yield significantly increases with the presence of the catalyst. In Table 5.4 some results of gas yields and gas composition extracted from some experimental works are shown.

The comparison of these results is difficult. The temperatures of catalytic pyrolysis in the works of Qinglan et al. [44] and Garcia et al. [48] are different, although both studies were performed in an experimental installation with a fluidized bed reactor. Moreover, the catalysts are different. For both studies the maximum hydrogen content showed in Table 5.4 is around 52 vol.%. For a batch installation [58] a smaller hydrogen content was obtained, 28.7 vol.%.

It is worth mentioning that gas yield can be as high as 0.9 g gas/g biomass, for temperatures of 650 and 700 °C obtained in the work of Garcia et al. [48]. The time for these experiments was 49 and 181 min at 650 and 700 °C, respectively, which correspond to 20.4 and 50.6 g biomass fed in, respectively.

The use of a different approach to carry out the pyrolysis of biomass in innovative processes such as the use of a molten alkali reactor or a microwave plasma deserves also to be mentioned.

Jiang et al. [60] studied six biomass feedstocks (fir sawdust, birch sawdust, redwood sawdust, rice stalk, cole stalk, and rice husks) in a stainless steel reactor with about 700 g of molten alkali (NaOH) at temperatures from 350 to 550 °C. The

Table 5.4 Product distribution and gas composition in noncatalytic pyrolysis and conventional one-step catalytic pyrolysis

	Garcia et al. [48]		Qinglan et al. [44]		Bru et al. [58]
Temperature (°C) experimental installation	650 fluidized bed	700 fluidized bed	590 fluidized bed	700 fluidized bed	700 batch
Noncatalytic product yields (g/g biomass)					
Gas	0.338	0.496	0.161	0.343	0.280
Liquid	0.506	0.388	–	–	0.449
Char	0.038	0.036	–	–	0.217
Gas composition (% vol.)					
H_2	16.9	18.8	29.4	30.6	12.5
CO	55.4	55.6	43.5	46.4	46.8
CO_2	12.6	9.7	10.9	6.5	27.3
CH_4	10.6	10.6	11.3	10.0	13.4
C_2	4.5	5.4	4.9	6.5	–
Catalytic product yields (g/g biomass)					
Gas	0.909	0.916	0.541		0.358
Liquid	0.091	0.031	–		0.379
Char	0.058	0.069	–		0.219
Gas composition (% vol.)					
H_2	52.1	47.4	52.9		28.7
CO	40.5	46.1	33.7		35.5
CO_2	5.3	3.0	8.7		28.6
CH_4	2.1	2.4	3.8		7.2
C_2	0	1.1	0.9		–

product gas only contained hydrogen and methane. Redwood sawdust was the biomass with the highest H_2 yield (65.4 g H_2/kg biomass at 450 °C). The increase of temperature caused the increase in H_2 yield with values from 30.7 to 66.5 g H_2/kg biomass at 350 and 550 °C, respectively, using rice stalk as feed. The introduction of additives, especially $NiCl_2$, led to increased H_2 yields. H_2 content in pyrolysis gas was higher than 80 % in most of the studied conditions.

Microwave plasma reactor was used in the study of Spirulina algae pyrolysis by Lin et al. [61]. The pyrolysis was carried out at temperatures of 790, 820, and 848 °C and at atmospheric pressure. 1 g of biomass was loaded in a quartz tube. Although no catalyst was used, significant content of H_2 in product gas and high H_2 yield were obtained, with values of 45 vol.% and 31.5 g H_2/kg biomass, respectively, at 848 °C.

Figure 5.5 shows H_2 yield generated in some one-step pyrolysis processes, both conventional and innovative. It is generated the highest yield of H_2 in molten alkali (NaOH-$NiCl_2$) with a value of 67 g H_2/kg biomass [60].

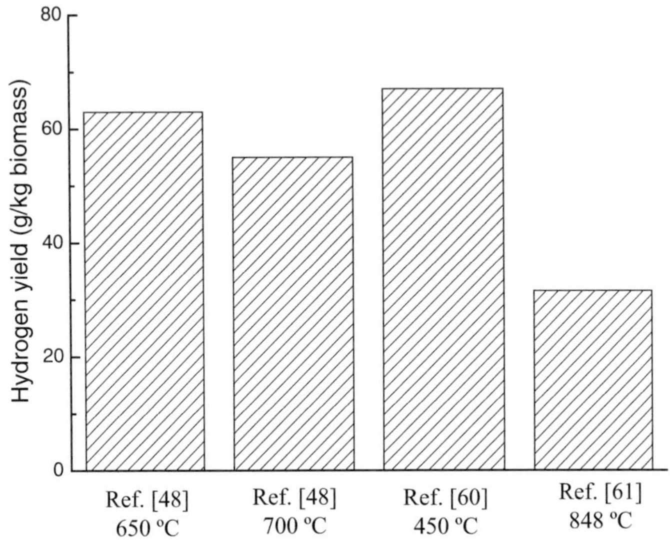

Fig. 5.5 Hydrogen yield generated in one-step pyrolysis processes

5.5 Multi-step Processes

The present section aims at presenting an overview of the different multi-step approaches that have been proposed regarding hydrogen production from catalytic pyrolysis of biomass.

A scheme of the different multi-step routes that lead to hydrogen from catalytic pyrolysis of biomass is presented in Fig. 5.6.

As can be seen, after the pyrolysis stage, three main alternatives have been proposed by various research groups in the literature, which ultimately can be divided into two: catalytic reforming of bio-oil, fractions of it or from its resulting products after a preliminary thermal processing of the bio-oil (pre-reforming), and direct catalytic cracking of the bio-oil. In all these cases, different reactor configurations and designs have been proposed in addition to the significant efforts made by many researchers for developing suitable catalysts for these processes, as previously discussed in this chapter.

All these routes necessarily require a downstream gas conditioning step if high-purity hydrogen is sought. Conventional hydrogen purification processes include further processing downstream of the hydrogen-rich product gas in catalytic water-gas-shift (WGS) reactors [62] and use of hydrogen-selective membranes [63] or pressure-swing adsorption (PSA) equipment [64, 65], among others.

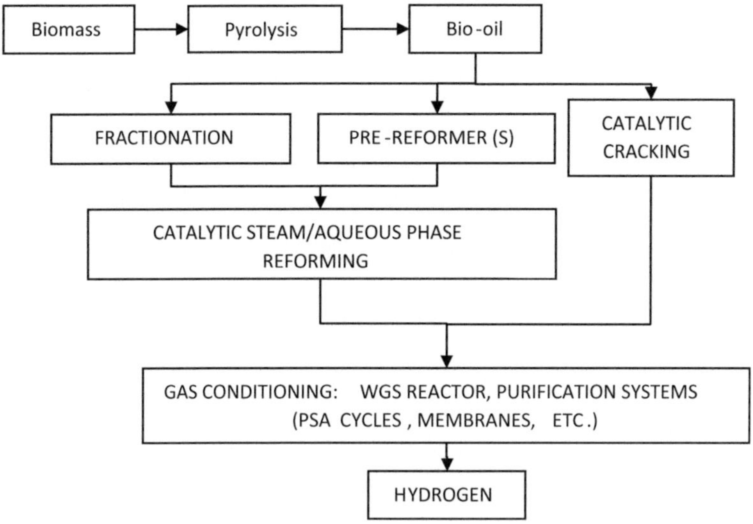

Fig. 5.6 Scheme of multi-step pyrolysis processes for H_2 production

5.5.1 Catalytic Steam Reforming of Bio-Oil

The first studies on hydrogen production from multi-step processes involving biomass pyrolysis were pioneered in 1993 by the National Renewable Energy Laboratory (Colorado, USA) [66, 67]. The strategy proposed by the research group led by Chornet is the catalytic steam reforming of the liquid pyrolysis products or its fractions. A separation of the bio-oil into two phases by water addition was used. The lignin-derived fraction would be devoted to the production of value-added chemicals, benefitting from a high content in phenolic compounds, whereas the carbohydrate-rich aqueous fraction would be subjected to catalytic steam reforming in a process similar to that employed in the production of hydrogen by catalytic steam reforming of natural gas and naphthas [35, 68].

The initial works of this group [34–36] preliminarily explored this process by means of thermodynamic analyses and also conducting experimental tests in a laboratory scale, using a dual fixed-bed quartz microreactor system coupled to a molecular beam mass spectrometer (MBMS). A detailed description of the setup can be found, for example, in the work of Wang and co-workers [34], where the possible reaction mechanisms for producing hydrogen were discussed in depth after conducting the steam reforming of bio-oil model compounds (acetic acid and hydroxyacetaldehyde). Further on, a catalytic fixed-bed reactor setup at bench scale was used in the catalytic steam reforming tests using the aqueous fraction of poplar bio-oil produced at NREL [36]. Screening of catalysts, feedstocks, and operating conditions was initially conducted with different oxygenates used as model compounds [35], representative of the complex composition of bio-oils, and subsequently extended to the aqueous fraction of bio-oil [36]. Both commercial and research Ni-based catalysts were tested, along with a commercial low-temperature shift conversion Cu-based catalyst [35].

In that work, two possible alternatives for hydrogen production from biomass pyrolysis oil were proposed: a regionalized system consisting of small- to medium-scale pyrolysis units coupled to a centralized large-scale catalytic reforming unit, in which the fractionation step would be carried out and subsequently the aqueous fraction of bio-oil would be processed for producing hydrogen. The other envisaged possibility was an integrated system consisting of larger processing plants in which biomass would be directly converted to hydrogen having all the necessary steps: pyrolysis reactor coupled to a catalytic reformer in which pyrolysis vapors would yield hydrogen via steam reforming. Given the intrinsic characteristics of biomass (low energy content, dispersed localization of the feedstock, etc.), from the feasibility analysis carried out, it was concluded that the first alternative was more adequate for producing hydrogen from biomass pyrolysis.

Later on, Czernik et al. [68] proposed a different configuration for the catalytic steam reforming reactor, as a result of the major problem encountered in their previous works, which is catalyst deactivation caused by the formation of carbonaceous deposits. Thus, it was proposed that a fluidized bed reactor could be a more appropriate configuration for the catalytic reformer. As a result, that work and the subsequent works of this group [69–71] focused on the development of suitable catalysts for steam reforming of oxygenates using both a microscale fixed-bed reactor configuration and a 2-in.-diameter fluidized bed reactor setup. A different approach for hydrogen production via distributed bio-oil reforming was further followed by this group, which will be discussed in depth later.

Following the approach of catalytic steam reforming of bio-oil or fractions of it, different research groups have also explored this route ever since. Universidad de Zaragoza (Zaragoza, Spain) worked on the development of suitable catalysts for the process using different process configurations as a result of collaborations with NREL [33].

Afterwards, the research conducted by Universidad de Zaragoza focused on the development of Ni-based catalysts using both a small bench-scale fixed-bed unit [50, 72–76] and a bench-scale fluidized bed setup [76–82]. The fluidized bed setup included a feeding system consisting of a quartz coaxial injection nozzle made of four concentric tubes which helped feeding the liquid feed in the form of spray while avoiding clogging as a result of the refrigeration supplied by the external cooling jacket formed by the outer tubes.

These studies aimed at the development of catalysts with good activity and selectivity to hydrogen, as well as resistant to deactivation by coke deposition and to attrition. The effect of the Ni content on the carbon conversion to gas and the product gas yields, especially H_2, was studied. Furthermore, the incorporation of modifiers to the catalysts was also explored in various works in order to, on the one hand, decrease the formation of carbonaceous deposits on the catalyst surface and, on the other hand, increase the mechanical resistance of the catalysts, thus developing more stable catalysts that ultimately could be used in the fluidized bed reactor during longer operation times. Other works can be found in the literature aiming at the development of suitable catalysts using fixed-bed reactors, both with model compounds [22, 25, 26, 83–86] and with bio-oil or fractions of it [22, 25, 27, 87–91]. Li et al. [92] conducted the catalytic steam reforming of the aqueous fraction of bio-oil using various Ni-based and dolomite catalysts in a stainless steel lab-scale fluidized bed reactor. A similar approach was followed by Xu et al. [93] using sieved particles obtained from a milled commercial NiO/MgO catalyst for the steam reforming of bio-oil from rice husk. Panigrahi et al. at the University of Saskatchewan also conducted steam gasification of bio-oil aiming at hydrogen production in an Inconel tubular reactor containing a noncatalytic fixed bed [94]. The reaction bed was composed of quartz chips, but the possible catalytic wall effect provoked by the presence of active metals such as Ni, Cr, Fe, or Mn in the Inconel alloy was not discussed.

Other alternative configuration is that proposed by the University of Thessaloniki and the Centre for Research and Technology Hellas (CERTH) from Greece. The production of hydrogen was investigated in a spouted bed reactor at a pilot scale [95, 96]. The reactor was made of stainless steel and has two differentiated parts: an inverted conical base followed by a 0.05-m inner diameter cylindrical part. The main advantage of such reactor configuration compared to others is related to the efficient heat recirculation caused by the particular solids dynamics in the spouted bed reactor. On the negative side, spouted bed reactors, similarly to what occurs in fluidized bed configurations, present significant catalyst losses as a result of attrition, which compromises the scale-up and development of such processes at full scale.

These authors explored the steam reforming of model compounds and of the aqueous fraction of bio-oil at atmospheric pressure, varying the usual operating process parameters and also studying the addition of small amounts of oxygen (temperature, steam to carbon molar ratio, and oxygen to carbon ratio). Different catalysts were tested, both commercial and other Ni based prepared over various supports. The authors reported good results in terms of limiting the formation of carbonaceous deposits as a result of the type of reactor used, while particularly a Ni/olivine catalyst presented an adequate mechanical resistance to attrition and yielded the most promising results using model compounds. However, the results of the catalytic steam reforming tests using the aqueous fraction of pine bio-oil reflected a low reforming activity and, subsequently, low hydrogen yields.

The University of Twente (Netherlands) also developed an interesting proof of concept for hydrogen production from catalytic pyrolysis of biomass [97, 98]. Instead of bio-oil fractionation followed by catalytic steam reforming, the group led by Van Swaaij proposed a different approach, aiming at an industrial development of hydrogen production from bio-oil. A two-stage reactor concept consisting of a sand fluidized bed and a catalytic fixed bed was developed. The proposed alternative aimed at overcoming two main problems. On the one hand, large-scale fixed-bed reactors would require a high heat supply at the entrance of the reactor, which could result in a significantly lowered temperature at that point. On the other hand, decoupling of the evaporation, primary pyrolysis oil conversion, and gas conditioning enabled to diminish significantly the catalyst deactivation by deposition of coke on the catalyst surface, since the preliminary cracking of organics from the bio-oil upon heating took place in the inert bed of the fluidized reactor, and simpler molecules could be reformed in the catalytic fixed bed. This also made it unnecessary to develop catalysts with increased mechanical strength.

The two-stage reactor was also proposed by Wu et al. [99], though in this case, both primary and secondary reformers were two catalytic fixed-bed reactors made of stainless steel. The concept proposed by the authors aimed at preserving the steam reforming catalyst operating lifetime by including a low-cost primary reforming catalyst (olivine) that could prevent the Ni-based catalyst in the secondary reformer from rapid deactivation by coke formation. Kan et al. [100] also proposed a dual fixed-bed configuration with two reaction beds in a tubular quartz reactor at lab scale, though in this case the first reaction bed contained quartz sand, thus serving for the vaporization and cracking of the crude bio-oil, and in a second catalytic bed, containing a NiCuZnAl catalyst, steam reforming of the bio-oil vapors took place. More recently, researchers from the University of the Basque Country (Spain) have also explored this approach [101, 102], by using a two-reactor configuration composed of a fixed-bed pre-reformer (thermal step) containing an inert filler (e.g., glass spheres) operating at low temperature (between 200 and 400 °C) and a catalytic fluidized bed reactor for the steam reforming of the bio-oil vapors using Ni-based catalysts.

Another interesting possibility is the cyclic operation of the catalytic fixed-bed reactor by means of developing a chemical looping steam reforming process [103]. This cyclic approach is similar to that initially proposed by the group of Mirodatos for catalytic cracking of the crude bio-oil [41], which will be further discussed. The authors claimed that it is possible to use a Ni-based steam reforming catalyst as an oxygen transfer material operating in a series of cycles of reduction/oxidation. The bio-oil would be steam reformed during the reduction step, in which the catalyst could be reduced during the beginning of the step and afterwards the catalyst could be regenerated in the oxidation step by burning off the carbonaceous deposits formed on the catalyst surface.

Other reactor configurations include a Y-shaped dual catalytic reactor proposed by Hu and Lu that aimed at combining steam reforming of bio-oil and dry reforming of bio-oil using CO_2 [104]. In one branch of the Y-shaped reactor, there is a catalytic fixed-bed reactor (catalytic bed I), in which steam reforming of bio-oil takes place (partial oxidation and oxidative steam reforming are suggested as alternative processes that could take place in this catalytic fixed bed with the pertinent selection of catalysts and reactants). In the second branch acting as inlet, another portion of the bio-oil would directly be fed into the reactor for conducting catalytic dry reforming using the CO_2 from the product gas stream coming out of the catalytic bed I. The catalytic dry reforming takes place in the last section of the reactor, where a second catalytic fixed bed (catalytic bed II) is placed. Rather than focusing on the production of hydrogen, the goal is to produce syngas, though by varying the process conditions the syngas could have different H_2/CO molar ratios.

From an industrial point of view, the most adequate configuration for the bio-oil steam reforming unit is probably that following a similar approach to what the University of Twente proposed. Thus, the industrial unit should have a pre-reformer plus a main catalytic steam reformer, since it would be more flexible in terms of admitting various feedstocks, would mitigate the catalyst deactivation by carbon formation, and could have a better energy efficiency, while benefiting at the same time from a smaller size of the main reformer [105]. This could enhance the economic viability of the process. Trane et al. [105] proposed a simple flow sheet of a two-reactor industrial plant for the steam reforming of bio-oil in which the production of hydrogen takes place in three steps: a low-temperature steam reforming reactor (pre-reformer), a high-temperature reformer, and a low-temperature shift reactor. An optional downstream desulfurization reactor was proposed in two possible locations: after the pre-reformer or right after the water-gas-shift reactor.

5.5.2 Catalytic Cracking of Bio-Oil

A different approach was proposed by the group of Mirodatos from the Catalysis Research Institute in France [41, 106, 107]. The group proposed an alternative strategy, which is to control the carbon deposition during a cracking step and to frequently regenerate the catalyst, in resemblance to the catalytic cracking of methane for hydrogen production. Therefore, the continuous hydrogen production process requires at least two (or more) parallel reactors. In one reactor the catalytic cracking reaction producing hydrogen is taking place, while on the other reactor, the coke is gasified or burned in order to restore the catalytic activity, periodically switching the cracking/regeneration cycle in each reactor. Thus, this approach is claimed to operate in autothermic regime, with the combustion of the coke during the regeneration step in one of the reactors supplying the necessary heat for the cracking hydrogen production step in the other reactor.

This group designed a double-walled stainless steel tubular reactor at laboratory bench scale that had an external cooling jacket using water as refrigerant in order to inject bio-oil at temperatures below 50 °C, thus avoiding undesired polymerization of bio-oil and clogging of the reactor. Bio-oil without any further addition of water was injected by means of a syringe pump through a capillary located on top of the catalytic fixed-bed reactor.

Recently, the research team led by Czernik at the National Renewable Energy Laboratory (USA) has proposed a different strategy, which is based on distributed bio-oil reforming aiming at developing an integrated system that will provide distributed production of hydrogen [108]. The project tackles the challenge of hydrogen production in a dispersed manner by converting bio-oil to H_2 at a targeted total dispensed hydrogen cost between 2 and 4 US$/kg of H_2 (produced, delivered, and dispensed, but untaxed). The rationale of the work lies on the much easier transportation of liquid bio-oil to scattered points in which H_2 may be required (e.g., fueling stations), which could be delivered and stored in order to be processed into hydrogen in the vicinity of the demanding points.

The proof of concept aims at developing a compact, low-capital cost and little-to-none maintenance process. The proposed schematic flow diagram comprises a multi-step approach, in which bio-oil (blended with an alcohol, such as methanol, so as to control the physical and chemical properties of the liquid, namely, viscosity) is firstly volatilized using ultrasonic atomization in a nozzle evaporator operating at temperatures around 400 °C. The vapors would flow through a hot filter at operating temperatures between 400 and 600 °C in order to collect solid char formed during the volatilization process. The next step includes a catalytic partial oxidation/steam reforming packed-bed reactor operating at 800–850 °C in autothermal regime (by adding small amounts of air that will burn part of the feed and thus provide the necessary heat for maintaining the process in autothermal regime). The catalyst would either be a Ni-based or a Pt-based. The authors claim that the best results to date have been obtained in preliminary tests in a bench-scale reactor system using the latter (a Pt-based commercial catalyst supplied by BASF). The other elements from the system include a third step consisting of a catalytic WGS packed-bed reactor operating at 350 °C and further separation and purification of hydrogen from the product gas stream by means of an electrochemical separator.

5.5.3 *Other Approaches*

The production of hydrogen by means of aqueous phase catalytic reforming of bio-oil was explored by the University of Massachusetts [109]. The concept of aqueous phase reforming of biomass oxygenates derives from the preliminary studies on hydrogen production by biomass supercritical water gasification developed by the research group led by Antal [110]. Later on, the research group led by Dumesic at the University of Wisconsin extensively developed the concept of catalytic reforming of oxygenates in liquid phase [111–113].

The idea is to conduct the catalytic reforming in liquid medium by using moderate pressures, typically in the range of 1500–5000 kPa, and thus diminishing the reforming temperature to values around 500 K. The tubular reactor has an upflow packed catalytic bed configuration and is made of stainless steel due to the pressurized conditions. A gas-liquid separator is necessary in order to obtain the H_2-rich product gas. In the work of the research group led by Huber [109], a diluted aqueous fraction of bio-oil (a concentration of around 4–5 % of bio-oil in the solution) was subjected to aqueous phase reforming for producing hydrogen over a 1 wt.% Pt/Al_2O_3 catalyst, with mixed results. The authors reported low conversions, but high hydrogen selectivity. A process flow diagram for the development of a bio-oil aqueous phase reforming process was proposed in the PhD thesis developed by Vispute [114].

Other alternative concepts deal with alternative means for supplying the necessary heat for the reforming reactions, namely, by using an annular Ni-Cr electric wire through which an alternative electric current passes [100, 115] or use of microwave power [47].

5.6 Concluding Remarks and Future Outlook

There are many research works that have studied hydrogen production from catalytic biomass pyrolysis.

Of the two approaches for producing hydrogen from pyrolysis of biomass, the multi-step can benefit from the economy of scale, as lignocellulosic biomass is a naturally dispersed resource, and transporting bio-oil to a central reforming unit, of bigger size, could be economically more favorable. Thus, bio-oil is the most important intermediate in the multi-step processes of catalytic biomass pyrolysis.

Catalysts are a key point in the process, and due to the heterogeneity of biomass available and the different processes, at this moment it has to be tailor-designed for each application. Research in new processes and new catalyst formulations is required to solve the inconveniences of catalyst deactivation, although studies are already being carried out in this direction.

There is lack of economy and energy analyses that can help to spur research and focus efforts on industrial application. Hydrogen is an important raw gas with application both in chemical synthesis and as a fuel in high efficiency systems such as fuel cells. The depletion of oil will increase even more the interest in hydrogen from renewable sources with pilot or demonstration scale plants probably becoming feasible as the price of oil rises.

Acknowledgments The authors wish to express their gratitude to the Spanish MINECO (projects ENE2010-18985 and ENE2013-41523-R). It is also acknowledged the permission for reprinting figures from American Chemical Society and Elsevier.

References

1. McKendry P. Energy production from biomass (part 1): overview of biomass. Bioresour Technol. 2002;83:47–54.
2. Yang H, Yan R, Chen H, Zheng C, Lee DH, Liang DT. In-depth investigation of biomass pyrolysis based on three major components: hemicellulose, cellulose and lignin. Energy Fuel. 2006;20:388–93. doi:10.1021/ef0580117.
3. Kallioinen A, Vaari A, Rättö M, Konn J, Siika-aho M, Viikari L. Effects of bacterial treatments on wood extractives. J Biotechnol. 2003;103:67–76. doi:10.1016/S0168-1656(03)00051-8.
4. Agblevor FA, Besler S. Inorganic compounds in biomass feedstocks. 1. Effect on the quality of fast pyrolysis oils. Energy Fuel. 1996;10:293–8. doi:10.1021/ef950202u.
5. Lédé J. Biomass fast pyrolysis reactors: a review of a few scientific challenges and of related recommended research topics. Oil Gas Sci Technol Rev d'IFP Energ Nouvelles. 2013;68:801–14. doi:10.2516/ogst/2013108.

6. Gray MR, Corcoran WH, Gavalas GR. Pyrolysis of a wood-derived material. Effects of moisture and ash content. Ind Eng Chem Process Des Dev. 1985;24:646–51. doi:10.1021/i200030a020.
7. Antal Jr MJ, Grønli M. The art, science, and technology of charcoal production. Ind Eng Chem Res. 2003;42:1619–40. doi:10.1021/ie0207919.
8. Bridgwater AV. Catalysis in thermal biomass conversion. Appl Catal A Genet. 1994;116:5–47. doi:10.1016/0926-860X(94)80278-5.
9. Di Blasi C. Modeling chemical and physical processes of wood and biomass pyrolysis. Prog Energ Combust Sci. 2008;34:47–90. doi:10.1016/j.pecs.2006.12.001.
10. Bridgwater AV, Meier D, Radlein D. Principles and practice of biomass fast pyrolysis processes for liquids. Org Geochem. 1999;51:3–22. doi:10.1016/S0146-6380(99)00120-5.
11. Sipilä K, Kuoppala E, Fagernäs L, Oasmaa A. Characterization of biomass-based flash pyrolysis oils. Biomass and Bioenergy. 1998;14:103–13. doi:10.1016/S0961-9534(97)10024-1.
12. Oasmaa A, Meier D. Norms and standards for fast pyrolysis liquids. J Anal Appl Pyrolysis. 2005;73:323–34. doi:10.1016/j.jaap.2005.03.003.
13. Czernik S, Bridgwater AV. Overview of applications of biomass fast pyrolysis oil. Energy Fuel. 2004;18:590–8. doi:10.1021/ef034067u.
14. Antal MJJ, Varhegyi G. Cellulose pyrolysis kinetics: the current state of knowledge. Ind Eng Chem Res. 1995;34:703–17. doi:10.1021/ie00042a001.
15. Nowakowski DJ, Bridgwater AV, Elliott DC, Meier D, de Wild P. Lignin fast pyrolysis: results from an international collaboration. J Anal Appl Pyrolysis. 2010;88:53–72. doi:10.1016/j.jaap.2010.02.009.
16. Woolf D, Amonette JE, Street-Perrott FA, Lehmann J, Joseph S. Sustainable biochar to mitigate global climate change. Nat Commun. 2010;1:56. doi:10.1038/ncomms1053.
17. Bridgwater AV. Review of fast pyrolysis of biomass and product upgrading. Biomass Bioenergy. 2012;38:68–94.
18. Meier D, Van De Beld B, Bridgwater AV, Elliott DC, Oasmaa A, Preto F. State-of-the-art of fast pyrolysis in IEA bioenergy member countries. Renew Sustain Energy Rev. 2013;20:619–41. doi:10.1016/j.rser.2012.11.061.
19. Garcia-Perez M, Wang XS, Shen J, Rhodes MJ, Tian F, Lee W-J, Wu H, Li CZ. Fast pyrolysis of oil Mallee woody biomass: effect of temperature on the yield and quality of pyrolysis products. Ind Eng Chem Res. 2008;47:1846–54. doi:10.1021/ie071497p.
20. Wagenaar BM, Kuipers JAM, Prins W, van Swaaij WPM. The rotating cone flash pyrolysis reactor. In: Bridgwater AV, editor. Advances thermochemical biomass conversion. Dordrecht: Springer; 1993. p. 1122–33.
21. Butler E, Devlin G, Meier D, McDonnell K. A review of recent laboratory research and commercial developments in fast pyrolysis and upgrading. Renew Sustain Energy Rev. 2011;15:4171–86. doi:10.1016/j.rser.2011.07.035.
22. Takanabe K, Aika K, Seshan K, Lefferts L. Sustainable hydrogen from bio-oil – steam reforming of acetic acid as a model oxygenate. J Catal. 2004;227:101–8.
23. Takanabe K, Aika K, Seshan K, Lefferts L. Catalyst deactivation during steam reforming of acetic acid over Pt/ZrO2. Chem Eng J. 2006;120:133–7.
24. Takanabe K, Aika K, Inazu K, Baba T. Steam reforming of acetic acid as a biomass derived oxygenate: bifunctional pathway for hydrogen formation over Pt/ZrO 2 catalysts. J Catal. 2006;263:263–9. doi:10.1016/j.jcat.2006.07.020.
25. Rioche C, Kulkarni S, Meunier FC, Breen JP, Burch R. Steam reforming of model compounds and fast pyrolysis bio-oil on supported noble metal catalysts. Appl Catal B Environ. 2005;61:130–9. doi:10.1016/j.apcatb.2005.04.015.
26. Basagiannis AC, Verykios XE. Catalytic steam reforming of acetic acid for hydrogen production. Int J Hydrog Energ. 2007;32:3343–55.

27. Basagiannis AC, Verykios XE. Steam reforming of the aqueous fraction of bio-oil over structured Ru/MgO/Al2O3 catalysts. Catal Today. 2007;127:256–64. doi:10.1016/j.cattod.2007.03.025.
28. Basile A, Gallucci F, Iulianelli A, Borgognoni F, Tosti S. Acetic acid steam reforming in a Pd-Ag membrane reactor: the effect of the catalytic bed pattern. J Memb Sci. 2008;311:46–52.
29. Iwasa N, Yamane T, Takei M, Ozaki J, Arai M. Hydrogen production by steam reforming of acetic acid: comparison of conventional supported metal catalysts and metal-incorporated mesoporous smectite-like catalysts. Int J Hydrog Energ. 2010;35:110–7. doi:10.1016/j.ijhydene.2009.10.053.
30. Dubey VR, Vaidya PD. Kinetics of steam reforming of acetol over a Pt/C catalyst. Chem Eng J. 2012;180:263–9. doi:10.1016/j.cej.2011.11.034.
31. Uddin MN, Daud WMAW, Abbas HF. Effects of pyrolysis parameters on hydrogen formations from biomass: a review. RSC Adv. 2014;4:10467–90. doi:10.1039/C3RA43972K.
32. Shoja M, Babatabar MA, Tavasoli A, Ataei A. Production of hydrogen and syngas via pyrolysis of bagasse in a dual bed reactor. J Energ Chem. 2013;22:639–44. doi:10.1016/S2095-4956(13)60084-4.
33. García L, French R, Czernik S, Chornet E. Catalytic steam reforming of bio-oils for the production of hydrogen: effects of catalyst composition. Appl Catal A Gen. 2000;201:225–39. doi:10.1016/S0926-860X(00)00440-3.
34. Wang D, Montane D, Chornet E. Catalytic steam reforming of biomass-derived oxygenates: acetic acid and hydroxyacetaldehyde. Appl Catal A Gen. 1996;143:245–70.
35. Wang D, Czernik S, Montane D, Mann M, Chornet E. Biomass to hydrogen via fast pyrolysis and catalytic steam reforming of the pyrolysis oil or its fractions. Ind Eng Chem Res. 1997;36:1507–18.
36. Wang D, Czernik S, Chornet E. Production of hydrogen from biomass by catalytic steam reforming of fast pyrolysis oils. Energy Fuel. 1998;12:19–24.
37. Marquevich M, Czernik S, Chornet E, Montané D. Hydrogen from biomass: steam reforming of model compounds of fast-pyrolysis oil. Energy Fuel. 1999;13:1160–6.
38. Marquevich M, Coll R, Montane D. Steam reforming of sunflower oil for hydrogen production. Ind Eng Chem Res. 2000;39:2140–7.
39. Marquevich M, Medina F, Montané D. Hydrogen production via steam reforming of sunflower oil over Ni/Al catalysts from hydrotalcite materials. Catal Commun. 2001;2:119–24.
40. Kechagiopoulos PN, Voutetakis SS, Lemonidou AA, Vasalos IA. Hydrogen production via steam reforming of the aqueous phase of bio-oil in a fixed bed reactor. Energy Fuel. 2006;20:2155–63.
41. Davidian T, Guilhaume N, Iojoiu E, Provendier H, Mirodatos C. Hydrogen production from crude pyrolysis oil by a sequential catalytic process. Appl Catal B Environ. 2007;73:116–27.
42. Waheed QMK, Williams PT. Hydrogen production from high temperature pyrolysis/steam reforming of waste biomass: rice husk, sugar cane bagasse, and wheat straw. Energy Fuel. 2013;27:6695–704. doi:10.1021/ef401145w.
43. Zhao B, Zhang X, Sun L, Meng G, Chen L, Xiaolu Y. Hydrogen production from biomass combining pyrolysis and the secondary decomposition. Int J Hydrog Energ. 2010;35:2606–11. doi:10.1016/j.ijhydene.2009.04.011.
44. Qinglan H, Chang W, Dingqiang L, Yao W, Dan L, Guiju L. Production of hydrogen-rich gas from plant biomass by catalytic pyrolysis at low temperature. Int J Hydrog Energ. 2010;35:8884–90. doi:10.1016/j.ijhydene.2010.06.039.
45. Zhang Y, Li W, Zhang S, Xu Q, Yan Y. Hydrogen production by the catalytic reforming of volatile from biomass pyrolysis over a bimetallic catalyst. Energ Sources A Recover Util Environ Eff. 2013;35:1975–82. doi:10.1080/15567036.2011.580327.
46. Hojjat Ansari M, Jafarian S, Tavasoli A, Karimi A, Rashidi M. Hydrogen rich gas production via nano-catalytic pyrolysis of bagasse in a dual bed reactor. J Nat Gas Sci Eng. 2014;19:279–86. doi:10.1016/j.jngse.2014.05.018.

47. Liu S, Zhu J, Chen M, Xin W, Yang Z, Kong L. Hydrogen production via catalytic pyrolysis of biomass in a two-stage fixed bed reactor system. Int J Hydrog Energ. 2014;39:13128–35. doi:10.1016/j.ijhydene.2014.06.158.
48. García L, Salvador ML, Arauzo J, Bilbao R. Influence of catalyst weight/biomass flow rate ratio on gas production in the catalytic pyrolysis of pine sawdust at low temperatures. Ind Eng Chem Res. 1998;37:3812–9. doi:10.1021/ie9801960.
49. Arauzo J, Radlein D, Piskorz J, Scott DS. A new catalyst for the catalytic gasification of biomass. Energy Fuel. 1994;8:1192–6. doi:10.1021/ef00048a005.
50. Bimbela F, Oliva M, Ruiz J, García L, Arauzo J. Hydrogen production via catalytic steam reforming of the aqueous fraction of bio-oil using nickel-based coprecipitated catalysts. Int J Hydrog Energ. 2013;38:14476–87. doi:10.1016/j.ijhydene.2013.09.038.
51. García L, Salvador ML, Arauzo J, Bilbao R. Catalytic steam gasification of pine sawdust. Effect of catalyst weight/biomass flow rate and steam/biomass ratios on gas production and composition. Energy Fuel. 1999;13:851–9. doi:10.1021/ef980250p.
52. Chen G, Andries J, Spliethoff H. Catalytic pyrolysis of biomass for hydrogen rich fuel gas production. Energy Convers Manag. 2003;44:2289–96. doi:10.1016/S0196-8904(02)00188-7.
53. Arauzo J, Radlein D, Piskorz J, Scott DS. Catalytic pyrogasification of biomass. Evaluation of modified nickel catalysts. Ind Eng Chem Res. 1997;36:67–75. doi:10.1021/ie950271w.
54. Garcia L, Salvador ML, Bilbao R, Arauzo J. Influence of calcination and reduction conditions on the catalyst performance in the pyrolysis process of biomass. Energy Fuel. 1998;12:139–43. doi:10.1021/ef970097j.
55. García L, Salvador M, Arauzo J, Bilbao R. Catalytic pyrolysis of biomass: influence of the catalyst pretreatment on gas yields. J Anal Appl Pyrolysis. 2001;58–59:491–501. doi:10.1016/S0165-2370(00)00114-5.
56. Çağlar A, Demirbaş A. Hydrogen rich gas mixture from olive husk via pyrolysis. Energy Convers Manag. 2002;43:109–17. doi:10.1016/S0196-8904(01)00012-7.
57. Demirbaş A. Gaseous products from biomass by pyrolysis and gasification: effects of catalyst on hydrogen yield. Energy Convers Manag. 2002;43:897–909. doi:10.1016/S0196-8904(01)00080-2.
58. Bru K, Blin J, Julbe A, Volle G. Pyrolysis of metal impregnated biomass: an innovative catalytic way to produce gas fuel. J Anal Appl Pyrolysis. 2007;78:291–300. doi:10.1016/j.jaap.2006.08.006.
59. Garcia L, Salvador ML, Arauzo J, Bilbao R. CO2 as a gasifying agent for gas production from pine sawdust at low temperatures using a Ni/Al coprecipitated catalyst. Fuel Process Technol. 2001;69:157–74. doi:10.1016/S0378-3820(00)00138-7.
60. Jiang H, Wu Y, Fan H, Ji J. Hydrogen production from biomass pyrolysis in molten alkali. AASRI Procedia. 2012;3:217–23. doi:10.1016/j.aasri.2012.11.036.
61. Lin KC, Lin Y-C, Hsiao Y-H. Microwave plasma studies of Spirulina algae pyrolysis with relevance to hydrogen production. Energy. 2014;64:567–74. doi:10.1016/j.energy.2013.09.055.
62. Ratnasamy C, Wagner JP. Water gas shift catalysis. Catal Rev. 2009;51:325–440. doi:10.1080/01614940903048661.
63. Adhikari S, Fernando S. Hydrogen membrane separation techniques. Ind Eng Chem Res. 2006;45:875–81. doi:10.1021/ie050644l.
64. Sircar S, Golden TC. Purification of hydrogen by pressure swing adsorption. Sep Sci Technol. 2000;35:667–87.
65. Silva B, Solomon I, Ribeiro AM, Lee U-H, Hwang YK, Chang J-S, Loureiro JM, Rodrigues AE. H2 purification by pressure swing adsorption using CuBTC. Sep Purif Technol. 2013;118:744–56. doi:10.1016/j.seppur.2013.08.024.
66. Chornet E, Czernik S, Wang D, Gregoire C, Mann MK. Biomass to hydrogen via pyrolysis and reforming. In: Proc. 1994 U.S. DOE Hydrog. Progr. Rev. NREL/CP-470-6431, CONF-9404194. Livermore; 1994. p. 407–32.

67. Milne TA, Elam CC, Evans RJ. Hydrogen from biomass – state of the art research challenges. 2001 IEA Rep. IEA/H2/TR-02/001.
68. Czernik S, French R, Feik C, Chornet E. Hydrogen by catalytic steam reforming of liquid byproducts from biomass thermoconversion processes. Ind Eng Chem Res. 2002;41:4209–15.
69. Czernik S, French R, Magrini-Bair K, Chornet E. The production of hydrogen by steam reforming of trap grease progress in catalyst performance. Energy Fuel. 2004;18:1738–43.
70. Czernik S, Evans R, French R. Hydrogen from biomass-production by steam reforming of biomass pyrolysis oil. Catal Today. 2007;129:265–8. doi:10.1016/j.cattod.2006.08.071.
71. Magrini-Bair KA, Czernik S, French R, Parent YO, Chornet E, Dayton DC, Feik C, Bain R. Fluidizable reforming catalyst development for conditioning biomass-derived syngas. Appl Catal A Gen. 2007;318:199–206.
72. Bimbela F, Oliva M, Ruiz J, García L, Arauzo J. Hydrogen production by catalytic steam reforming of acetic acid, a model compound of biomass pyrolysis liquids. J Anal Appl Pyrolysis. 2007;79:112–20.
73. Bimbela F, Oliva M, Ruiz J, García L, Arauzo J. Catalytic steam reforming of model compounds of biomass pyrolysis liquids in fixed bed: acetol and n-butanol. J Anal Appl Pyrolysis. 2009;85:204–13. doi:10.1016/j.jaap.2008.11.025.
74. Bimbela F, Oliva M, Ruiz J, García L, Arauzo J. Steam reforming of bio-oil aqueous fractions for syngas production and energy. Environ Eng Sci. 2011;28:757–63. doi:10.1089/ees.2010.0367.
75. Bimbela F, Chen D, Ruiz J, García L, Arauzo J. Ni/Al coprecipitated catalysts modified with magnesium and copper for the catalytic steam reforming of model compounds from biomass pyrolysis liquids. Appl Catal B Environ. 2012;119–120:1–12. doi:10.1016/j.apcatb.2012.02.007.
76. Remón J, Medrano J, Bimbela F, García L, Arauzo J. Ni/Al–Mg–O solids modified with Co or Cu for the catalytic steam reforming of bio-oil. Appl Catal B Environ. 2013;132–133:433–44.
77. Galdámez JR, García L, Bilbao R. Hydrogen production by steam reforming of bio-oil using coprecipitated Ni–Al catalysts. Acetic acid as a model compound. Energy Fuel. 2005;19:1133–42. doi:10.1021/ef049718g.
78. Ramos MC, Navascués AI, García L, Bilbao R. Hydrogen production by catalytic steam reforming of acetol, a model compound of bio-oil. Ind Eng Chem Res. 2007;46:2399–406. doi:10.1021/ie060904e.
79. Medrano JA, Oliva M, Ruiz J, García L, Arauzo J. Catalytic steam reforming of acetic acid in a fluidized bed reactor with oxygen addition. Int J Hydrog Energ. 2008;33:4387–96. doi:10.1016/j.ijhydene.2008.05.023.
80. Medrano JA, Oliva M, Ruiz J, García L, Arauzo J. Catalytic steam reforming of model compounds of biomass pyrolysis liquids in fluidized bed reactor with modified Ni/Al catalysts. J Anal Appl Pyrolysis. 2009;85:214–25. doi:10.1016/j.jaap.2008.11.025.
81. Medrano JA, Oliva M, Ruiz J, García L, Arauzo J. Hydrogen from aqueous fraction of biomass pyrolysis liquids by catalytic steam reforming in fluidized bed. Energy. 2011;36:2215–24. doi:10.1016/j.energy.2010.03.059.
82. Medrano JA, Oliva M, Ruiz J, García L, Arauzo J. Catalytic steam reforming of butanol in a fluidized bed and comparison with other oxygenated compounds. Fuel Process Technol. 2014;124:123–33. doi:10.1016/j.fuproc.2014.02.022.
83. Hu X, Lu G. Investigation of steam reforming of acetic acid to hydrogen over Ni–Co metal catalyst. J Mol Catal A Chem. 2007;261:43–8. doi:10.1016/j.molcata.2006.07.066.
84. Vagia EC, Lemonidou AA. Hydrogen production via steam reforming of bio-oil components over calcium aluminate supported nickel and noble metal catalysts. Appl Catal A Gen. 2008;351:111–21. doi:10.1016/j.apcata.2008.09.007.
85. Hu X, Lu G. Investigation of the effects of molecular structure on oxygenated hydrocarbon steam re-forming. Energy Fuel. 2009;23:926–33. doi:10.1021/ef8008647.

86. Matas Güell B, Silva IMT, Seshan K, Lefferts L. Sustainable route to hydrogen – design of stable catalysts for the steam gasification of biomass related oxygenates. Appl Catal B Environ. 2009;88:59–65. doi:10.1016/j.apcatb.2008.09.018.
87. Wang Z, Pan Y, Dong T, Zhu X, Kan T, Yuan L. Production of hydrogen from catalytic steam reforming of bio-oil using C12A7-O- -based catalysts. Appl Catal A Gen. 2007;320:24–34. doi:10.1016/j.apcata.2006.12.003.
88. Yan C-F, Cheng F-F, Hu R-R. Hydrogen production from catalytic steam reforming of bio-oil aqueous fraction over Ni/CeO2–ZrO2 catalysts. Int J Hydrog Energ. 2010;35:11693–9.
89. Ortiz-Toral PJ, Satrio J, Brown RC, Shanks BH. Steam reforming of bio-oil fractions: effect of composition and stability. Energy Fuel. 2011;25:3289–97. doi:10.1021/ef200628q.
90. Seyedeyn-Azad F, Salehi E, Abedi J, Harding T. Biomass to hydrogen via catalytic steam reforming of bio-oil over Ni-supported alumina catalysts. Fuel Process Technol. 2011;92:563–9.
91. Chen T, Wu C, Liu R. Steam reforming of bio-oil from rice husks fast pyrolysis for hydrogen production. Bioresour Technol. 2011;102:9236–40.
92. Li H, Xu Q, Xue H, Yan Y. Catalytic reforming of the aqueous phase derived from fast-pyrolysis of biomass. Renew Energy. 2009;34:2872–7. doi:10.1016/j.renene.2009.04.007.
93. Xu Q, Lan P, Zhang B, Ren Z, Yan Y. Hydrogen production via catalytic steam reforming of fast pyrolysis bio-oil in a fluidized-bed reactor. Energy Fuel. 2010;24:6456–62. doi:10.1021/ef1010995.
94. Panigrahi S, Dalai AK, Chaudhari ST, Bakhshi NN. Synthesis gas production from steam gasification of biomass-derived oil. Energy Fuel. 2003;17:637–42. doi:10.1021/ef020073z.
95. Kechagiopoulos PN, Voutetakis SS, Lemonidou AA, Vasalos IA. Sustainable hydrogen production via reforming of ethylene glycol using a novel spouted bed reactor. Catal Today. 2007;127:246–55. doi:10.1016/j.cattod.2007.05.018.
96. Kechagiopoulos PN, Voutetakis SS, Lemonidou AA, Vasalos IA. Hydrogen production via reforming of the aqueous phase of bio-oil over Ni/olivine catalysts in a spouted bed reactor. Ind Eng Chem Res. 2009;48:1400–8. doi:10.1021/ie8013378.
97. Van Rossum G, Kersten SRA, Van Swaaij WPM. Catalytic and noncatalytic gasification of pyrolysis oil. Ind Eng Chem Res. 2007;46:3959–67.
98. Van Rossum G, Kersten SRA, Van Swaaij WPM. Staged catalytic gasification/steam reforming of pyrolysis oil. Ind Eng Chem Res. 2009;48:5857–66.
99. Wu C, Huang Q, Sui M, Yan Y, Wang F. Hydrogen production via catalytic steam reforming of fast pyrolysis bio-oil in a two-stage fixed bed reactor system. Fuel Process Technol. 2008;89:1306–16. doi:10.1016/j.fuproc.2008.05.018.
100. Kan T, Xiong J, Li X, Ye T, Yuan L, Torimoto Y, Yamamoto M, Li Q. High efficient production of hydrogen from crude bio-oil via an integrative process between gasification and current-enhanced catalytic steam reforming. Int J Hydrog Energ. 2010;35:518–32.
101. Valle B, Remiro A, Aguayo AT, Bilbao J, Gayubo AG. Catalysts of Ni/α-Al2O3 and Ni/La2O3-αAl2O3 for hydrogen production by steam reforming of bio-oil aqueous fraction with pyrolytic lignin retention. Int J Hydrog Energ. 2013;38:1307–18. doi:10.1016/j.ijhydene.2012.11.014.
102. Remiro A, Valle B, Oar-Arteta L, Aguayo AT, Bilbao J, Gayubo AG. Hydrogen production by steam reforming of bio-oil/bio-ethanol mixtures in a continuous thermal-catalytic process. Int J Hydrog Energ. 2014;39:6889–98. doi:10.1016/j.ijhydene.2014.02.137.
103. Lea-Langton A, Zin RM, Dupont V, Twigg MV. Biomass pyrolysis oils for hydrogen production using chemical looping reforming. Int J Hydrog Energ. 2012;37:2037–43. doi:10.1016/j.ijhydene.2011.05.083.
104. Hu X, Lu G. Bio-oil steam reforming, partial oxidation or oxidative steam reforming coupled with bio-oil dry reforming to eliminate CO2 emission. Int J Hydrog Energ. 2010;35:7169–76.
105. Trane R, Dahl S, Skjøth-Rasmussen MS, Jensen AD. Catalytic steam reforming of bio-oil. Int J Hydrog Energ. 2012;37:6447–72. doi:10.1016/j.ijhydene.2012.01.023.

106. Domine ME, Iojoiu EE, Davidian T, Guilhaume N, Mirodatos C. Hydrogen production from biomass-derived oil over monolithic Pt- and Rh-based catalysts using steam reforming and sequential cracking processes. Catal Today. 2008;133–135:565–73.
107. Davidian T, Guilhaume N, Daniel C, Mirodatos C. Continuous hydrogen production by sequential catalytic cracking of acetic acid: part I. Investigation of reaction conditions and application to two parallel reactors operated cyclically. Appl Catal A Genet. 2008;335:64–73.
108. Czernik S, French R, Penev M. Distributed bio-oil reforming. US DOE Hydrog Fuel Cells Progr – FY 2013. Annu Prog Rep. 2013;II:119–23.
109. Vispute TP, Huber GW. Production of hydrogen, alkanes and polyols by aqueous phase processing of wood-derived pyrolysis oils. Green Chem. 2009;11:1433–45. doi:10.1039/B912522C.
110. Yu D, Aihara M, Antal MJ. Hydrogen production by steam reforming glucose in supercritical water. Energy Fuel. 1993;7:574–7. doi:10.1021/ef00041a002.
111. Shabaker J, Davda RR, Huber GW, Cortright RD, Dumesic JA. Aqueous-phase reforming of methanol and ethylene glycol over alumina-supported platinum catalysts. J Catal. 2003;215:344–52. doi:10.1016/S0021-9517(03)00032-0.
112. Cortright RD, Davda RR, Dumesic JA. Hydrogen from catalytic reforming of biomass-derived hydrocarbons in liquid water. Nature. 2002;418:964–7. doi:10.1038/nature01009.
113. Huber GW, Shabaker JW, Dumesic JA. Raney Ni-Sn catalyst for H2 production from biomass-derived hydrocarbons. Science. 2003;300:2075–7. doi:10.1126/science.1085597.
114. Vispute T. Pyrolysis oils: characterization, stability analysis, and catalytic upgrading to fuels and chemicals. Dissertations. Paper 349. 2011. http://scholarworks.umass.edu/open_access_dissertations.
115. Hou T, Yuan L, Ye T, Gong L, Tu J, Yamamoto M, Torimoto Y, Li Q. Hydrogen production by low-temperature reforming of organic compounds in bio-oil over a CNT-promoting Ni catalyst. Int J Hydrog Energ. 2009;34:9095–107. doi:10.1016/j.ijhydene.2009.09.012.

Chapter 6
Low Carbon Production of Hydrogen by Methane Decarbonization

Alberto Abánades

Abstract Hydrogen is one of the energy vectors that is proposed to have an important role in the future. The implementation of the called "hydrogen economy" is a challenge that requires the development of sustainable production technologies for hydrogen. Such technologies are intended to be fed by renewable energy sources such as solar, wind, or biomass. Hydrogen generation from wind and solar photovoltaic cells will be done via water electrolysis. Direct thermal energy sources such as solar thermal are expected to use thermochemical methods, either from water or hydrocarbons as the main raw material. Biomass can be used in this context in two ways: as raw material, through the production of syngas or bio-methane to produce hydrogen by thermochemical processes, and as an electricity producer to generate hydrogen via electrolysis. In this chapter, techniques for hydrocarbon decarbonization will be discussed. The general chemical description of methane pyrolysis will be the starting point to describe concepts for its implementation, mainly applying solar technology. Other alternatives, such as the use bio-methane as raw material and the integration of methane pyrolysis with ammonia production or biofuel synthesis will be discussed. The scientific viability of methane decarbonization has been tested and proven, but viable industrial implementation of the technology still remains. The status of the implementation of this technology on an industrial scale will be discussed analyzing its main technological showstoppers and their potential solutions.

Keywords Low carbon hydrogen production • Methane decarbonization • Methane decomposition • Hydrogen from hydrocarbon

6.1 Introduction

The development of a sustainable energy system is one of the most critical problems that humankind envisages. One of the objectives for such a system is based on the reduction or elimination of the greenhouse gas emissions. The

A. Abánades (✉)
Department of Energy Engineering, School of Industrial Engineering,
Technical University of Madrid, Madrid, Spain
e-mail: abanades@etsii.upm.es

Z. Fang et al. (eds.), *Production of Hydrogen from Renewable Resources*,
Biofuels and Biorefineries 5, DOI 10.1007/978-94-017-7330-0_6

environmental impact of the anthropogenic activity is affecting the natural balance of many ecosystems in the earth. The atmosphere, which has a paramount importance in the global energy balance of our planet, is being exposed to a modification of its greenhouse gases content (CO_2, methane, etc.) due to the combustion of fossil fuels and other industrial processes. Fossil fuels are limited and their use certainly will come to an end. The complete consumption of those available resources will produce huge amounts of greenhouse gases that will dramatically affect the atmospheric environment and produce dramatic changes in the ecosystem.

The deployment of a low-carbon energy system, with the aim of null greenhouse gases emissions (in particular CO_2), is a basic target of the global energy policy, which should be made compatible with the economic and social development. Some ideas are considered to achieve this goal in the future. The development of renewable energies, as the case of biomass, solar, and wind, as main primary energy sources is a step forward toward sustainability, the increase of the energy efficiency is also one of the pillars, but one of the most important facts would be the change in the energy vectors that could make transition to a CO_2-free energy utilization scheme possible, in which current fuels could be substituted by clean liquid fuels, electricity, or hydrogen.

The hydrogen economy, in which hydrogen becomes one of the basic energy vectors, requires the development of technologies that can produce large quantities of hydrogen from any source available in nature, for instance, from water or any other hydrogenated molecule as hydrocarbons. At present, available hydrogen production technologies are mainly fossil fuel reforming and gasification, water electrolysis, biomass fermentation, and thermochemical cycles. In particular, methane steam reforming, naphtha/oil reforming, and coal gasification are nowadays the main source of hydrogen because of their economics and their technological maturity. Nevertheless, those methods imply the transformation of carbon contained in the fossil raw material into carbon dioxide, contributing to greenhouse gas emissions. Their application in the framework of a low-carbon energy system requires their use with carbon capture and sequestration (CCS) techniques, which will become a handicap for their economic viability in the future. Other hydrogen production processes based on electrolysis, which are very mature from the technological point of view, depend on the cost of the technology for the non-fossil electricity input (wind, photovoltaic, solar thermal, geothermal) and its development.

A CO_2-free scheme for the transformation of hydrocarbons into hydrogen is of great interest, not only for its application to fossil resources but also for hydrogenated organic renewable sources. One of the sound alternatives is hydrocarbon decarbonization. The first conceptual proposals and experimental activities for the development of these techniques have been completed for methane, which is the main component of the natural gas and biogas and the hydrocarbon with the highest hydrogen to carbon ratio.

The basic reaction in the methane decarbonization process is

$$CH_4 \rightarrow C + 2H_2 \quad (\Delta h_0 = 74.85\ \text{kJ/mol}) \tag{6.1}$$

The main characteristic of this process is the absence of oxygen, which eliminates CO_2/CO by-products, making the process very suitable to produce hydrogen for fuel cell supply. There have been tested several techniques to achieve reaction (6.1), as direct thermal methane cracking (TMC) or catalytic methane decarbonization (CMD) that will be described in the next sections.

6.2 Socioeconomic Benefits of Methane Decarbonization

The evaluation of the cost/benefits of technological developments is one of the main preliminary tasks that should be done to assess their impact in the long term. This is particularly relevant in the energy sector, which is facing the challenge to reduce environmental impact and provide secure, stabile, and sustainable energy resources. Socioeconomic analysis of an existing or new process is based on its technical performance and its broader environmental, economical, and social effects.

In the particular case of methane decarbonization, the socioeconomic analysis may be done from different aspects: resource availability and environmental, social, economic, and political impacts, even if there is clear interrelation between them.

Natural gas and biogas are called to play a fundamental role in the medium term in the global energy system. Natural gas conventional reserves, and lately unconventional resources, will be one of the pillars to satisfy the energy demand. For instance, the World Energy Outlook [1] estimates that 25 % of the primary energy in 2035 will come from natural gas in sectors like energy production, transportation, industry, and petrochemicals. The use the available natural gas resources in an "as business as usual" scenario will lead to the emission of huge amounts of carbon dioxide into the atmosphere. In fact, such emissions can imply an increase in the atmospheric temperature beyond the limits considered acceptable by the IPCC (Intergovernmental Panel for the Climate Change) of the United Nations [2]. Obviously, the utilization of fossil fuel will drive to their exhaustion as it is a finite resource, but they will be needed in the first steps to the transition toward a future sustainable low-carbon energy system, in which they should be substituted by renewable energy sources.

Low-carbon alternatives to fossils will be the focus in the reduction of the effects of CO_2 emissions with the development of technologies to capture and sequestration (CCS) or utilize (CCU) such combustion products so as to avoid CO_2 accumulation in the atmosphere. With those technologies, the carbon (C/CO_2) is taken from the fossils on the earth and isolated or reused, increasing the amount of carbon in an anthropogenic cycle (production of new fuels or into CO_2 reservoirs). The case of bio-methane, for instance, will imply a cycle in which a constant amount of C/CO_2 is being recycled through biological absorption of CO_2 and its human production for energy production.

The philosophy behind methane decarbonization is that the carbon in the methane molecule is extracted before its oxidation. The result is a C/CO_2 cycle in which the carbon is reduced or maintained, without passing from the earth to the atmospheric cycle. In the case of using bio-methane methane decarbonization, the net carbon dioxide in the atmosphere may be reduced or applied to cycles containing CO_2 as, for instance, biodiesel production without increasing the long-term carbon inventory.

In the long-term sustainable future, a pillar of an energy system without CO_2 emission might be the development of the hydrogen economy, in which no carbon is involved. Several processes are available or under study to produce hydrogen. Many of them intend to treat hydrogenated materials, as biomass hydrocarbons and water, to obtain hydrogen with electrolysis, fermentation, or other thermochemical processes. Those processes are certainly of great interest and many of them will be part of the technological portfolio of the hydrogen economy in the future. Nevertheless, they have to face enormous scientific, technological, and economic challenges to achieve viability. The fact is that 95 % of the current hydrogen production worldwide comes from natural gas and oil/naphtha steam reforming and coal gasification [3].

The production of hydrogen by available techniques is a must to develop its utilization technology. Without the development and future availability of hydrogen applications and machines, there will be no hydrogen demand and the existence of hydrogen production methods will not be practical. Therefore, the development of the hydrogen economy needs the current contribution of the fossil fuels to provide the hydrogen for the development of its end-use technologies. The production of hydrogen by low-carbon methods, such as fossil reforming and gasification with carbon capture, or fossil decarbonization will help to mitigate the greenhouse gases emissions that could be produced during this transition phase.

The main reason for the current massive utilization of fossil steam reforming and gasification is its competitive cost with respect to other mature technologies, such as electrolysis. Nowadays, natural gas is the preferred raw material for hydrogen production, but this may change in the future, as the availability and competitive market price of coal will imply the substitution of natural gas by coal in industrial processes, or in ammonia production due to mobility. The result could be an increase in the global greenhouse gas emissions in the medium term from hydrogen production.

The development of cost-effective and low-emission technologies for hydrogen production such as methane decarbonization will have a positive economic and environmental impact as it can make available a technology that could change the described tendency. The cost estimates of methane decarbonization will be discussed in another section of this chapter. In Table 6.1, a comparison between methane steam reforming, coal gasification, and methane decarbonization is shown.

From the theoretical analysis of those technologies, it can be seen that the production of 100 Mtoe (millions tonnes of oil equivalent) of hydrogen may produce 103.8 Mton (millions metric tonnes) of carbon, avoiding the production of 255.18 Mton (million metric tonnes) CO_2 with steam methane reforming (SMR), or 626.85 Mton CO_2 in comparison with coal gasification. Moreover, the density of supercritical CO_2 is 0.47 kg/m^3. Using such a value as a reference for its storage or

Table 6.1 Comparison between mature hydrogen production techniques from fossils and methane pyrolysis/decarbonization

	Methane steam reforming	Coal gasification	Methane pyrolysis
Reaction	$CH_4 + 2H_2O = CO_2 + 4H_2$	$C + 2H_2O = CO_2 + 2H_2$	$CH_4 = C + 2H_2$
Heat of reaction (kJ/mol C)	235.00	178.15	74.85
Heat of reaction (kJ/mol H_2)	63.25	89.08	37.43
Energy efficiency in transformation (%)	74	60	55
Energy efficiency with CCS (%)	54	43	55
CO_2 emission (mol CO_2/mol H_2)	0.34	0.83	0.05
CO_2 emission with CO_2-free energy source (mol CO_2/mol H_2)	0.25	0.5	0
CO_2 production with CO_2-free energy source (kg/GJ)	61.1	150.9	0
Carbon production (kg/GJ)	0	0	24.8

Extracted with permission from [4]. Copyright © 2012, Elsevier

sequestration in liquid form, it can be estimated that if the production of 100 Mtoe of hydrogen via steam reforming is integrated with CCS (carbon capture and sequestration) processes, the volume of CO_2, which would require an appropriate storage system, should be on the order of 0.54 billion (10^9) of cubic meters (bcm) in contrast to the 0.054 bcm that would be needed to store carbon (approx. density 1.9 kg/m^3) produced via the methane-cracking process. The volume difference would amount to a factor of 10, which arises from the difference in density (approx. 4) and mass (2.47) between carbon and CO_2.

6.3 Methane Pyrolysis Reaction

Methane is a chemically stable molecule in which a considerable research effort has been devoted to due to its abundance in nature and its vast range of practical applications as a raw material in industrial processes or as an energy carrier. The characterization of the methane pyrolysis reaction has been extensively studied. The thermal dissociation of methane is an endothermic reaction that has previously described as $CH_4 \rightarrow C$ (solid) + H_2 (gas). Such a reaction is initiated forming a free radical of the form:

$$\mathrm{CH4} \rightarrow \mathrm{CH_3}^* + \mathrm{H}^* \tag{6.2}$$

This reaction has the highest activation energy and dominates the kinetics of the overall reaction. It is followed by the formation of gaseous products via a chain mechanism of the form:

$$\begin{aligned} 2(\mathrm{CH_3}^* + \mathrm{H}^*) &\rightarrow \mathrm{C_2H_6} + \mathrm{H_2} \rightarrow \mathrm{C_2H_4} + 2\mathrm{H_2} \rightarrow \mathrm{C_2H_2} + 3\mathrm{H_2} \\ &\rightarrow 2\mathrm{C} + 4\mathrm{H_2} \end{aligned} \tag{6.3}$$

The activation of the methane molecule to initiate methane cracking requires the splitting of the strong C-H bonds (dissociation energy: 104 kcal/mol). This fact implies the need of high temperatures for this endothermic reaction in spite of any catalyst that could weaken this bond. From theoretical thermodynamic analysis, the Gibbs energy of the methane-cracking reaction has been reported as

$$\begin{aligned} \Delta G^0\left(\frac{\mathrm{J}}{\mathrm{mol}}\right) &= 8,9658.88 - 102.27 \cdot T(K) - 0.00428 \cdot T(K)^2 \\ &\quad - \frac{249,9358.99}{T(K)} \end{aligned} \tag{6.4}$$

When the Gibbs energy of the reaction equals zero, the temperature conditions to develop the reaction take place. According to this expression, around 819 K (545 °C), the reaction will theoretically start to develop. The presence of additional components in the reaction, such as a catalyst, can reduce this temperature, as they will reduce the required activation energy to break the C-H bonds.

An additional evaluation of the thermodynamic equilibrium of the methane-cracking reaction obtained with the Cantera library [6] is represented versus temperature in Fig. 6.1, showing how around 1200 °C, the full conversion of methane into hydrogen is theoretically feasible. Nevertheless, the full conversion in a practical device or experimental mock-up strongly depends on the kinetics of the reaction. The understanding of the kinetics is already an open issue.

Most kinetic analyses of the methane-cracking reaction have a first-order kinetics with an Arrhenius model of the type:

$$K(T) = k_0 \cdot \exp\left(-\frac{E_a}{R \cdot T}\right) \tag{6.5}$$

The k_0 corresponds to a pre-exponential factor, and E_a is the activation energy. From the work of several authors with different types of reactors and temperatures, those parameters have been ranged experimentally from 10^{12} to 10^{16} for the case of the pre-exponential factor and between 330 and 450 kJ/mol for the activation energy without catalyst [7–15].

Evaluation of several options to catalyze the methane decarbonization has been made in some studies. Their main challenge is the improvement of the reaction kinetics so that it is viable to implement a methane-cracking process at temperatures below 1000 °C.

Metallic catalysts have been studied very intensively for methane catalytic decarbonization (MCD) with Ni, Pt, Co, Pd, or Cu as base materials in several supports [16]. In general, Ni-based catalysts give better results, and the carbon formation and diffusion into nickel become an important issue to understand the growth of the filamentous carbon formation, whose balance with the carbon production by methane gasification determines the deactivation of the catalyst by carbon saturation and coke formation at its surface. Some examples of the activity of some metallic catalysts are shown in the Table 6.2, obtained from several research works in the field, showing the maximum conversion achieved and the reaction temperature.

The possibility of using carbonaceous materials as catalyst for the reaction has been studied by several researchers. This fact has clear interest, as it might be possible the autocatalysis of the reaction, using self-produced materials for the enhancement of the reaction. Table 6.3 reports some data provided by Mudarov et al. [26] that constitutes a good summary of the initial catalytic activity (K_0) of carbonaceous materials. Among the main carbonaceous catalyst, activated carbon (AC) and carbon black (CB) are the most interesting. A higher catalytic initial activity corresponds to higher initial surface area.

The deactivation of carbonaceous materials is connected to the growing of carbon aggregates in the surface of the catalyst, reducing its surface area. Some analysis has been done to evaluate the behavior of this catalyst with time. The group

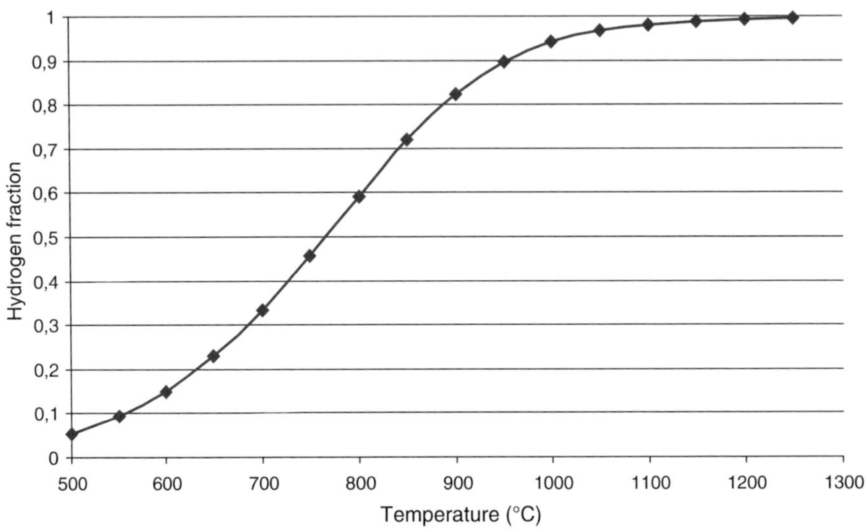

Fig. 6.1 Conversion rate at the thermodynamic equilibrium of the methane-cracking reaction

Table 6.2 Examples of metallic catalysts for methane decarbonization with carbon yields and conversions (Conv.) for the given temperatures (*T*)

Catalyst	*T* (°C)	Conv. (%)	Carbon yield (g C/g catalyst)	Ref
2Ni-1Cu-1Al (atomic)	750	70	191 g C/g Ni	[19]
15Ni-3Cu-2Al (atomic)	690	60	465 g C/g Ni	[19]
23Ni-77La_2O_3 (w%)	700	75	78.2 g C/g Ni	[21]
50Fe-50Al_2O_3 (w%)	625	4	53 g C/g Fe	[22]
19.5Ni-80.5SiO_2 (w%)	500	35.2	61.3 g C/g Ni	[23]
70 Ni-10Cu-10Fe-10Al_2O_3 (mole)	700	60	17.14 g C/g Ni	[25]
NiO	700	10	398 g C/g Ni	[18]
90Ni-10SiO_2 (w%)	700	70	200 g C/g Ni	[24]
60 Ni-25Cu-15SiO_2 (mole)	647	60	801 g C/g Ni	[20]

Extracted with permission from [17]. Copyright © 2011, Elsevier

of Serrano [27] (Fig. 6.2) shows the evolution of the catalytic activity of black carbon (CB black pearls and vulcan), activated carbon (AC), and carbon nanotubes (MWNT). Apart of the surface area, the pore volume is another important factor. A high pore volume implies a lower deactivation of the pore as the pore mouths are not so easily closed by the growing of carbon aggregates.

A graphical summary of applications of the most popular catalysts for the methane decarbonization reaction is depicted in Fig. 6.3. The thermal methane cracking without the presence of a catalyst may be practically implemented above 1000 °C. Nevertheless, at this temperature, the conversion and kinetics of the reaction are rather low. Most authors explore temperatures above 1500 °C for its practical implementation. The carbon produced is mainly amorphous with an

Table 6.3 Initial catalytic activity of a selected set of carbonaceous materials

Material	Surface area (m^2/g)	k_0 (mmol/min-g)
Acetylene black	80	0.22
AC, hardwood	1500	2.04
AC, lignite	650	1.77
AC, petroleum coke	2570	1.43
AC, phenol resin	2260	1.53
CB, black pearls 120	25	0.22
CB, black pearls 2000	1500	1.15
CB, vulcan XC72	254	0.48
Graphite, natural	4–6	0.02

Extracted with permission from [26]. Copyright © 2005, Elsevier

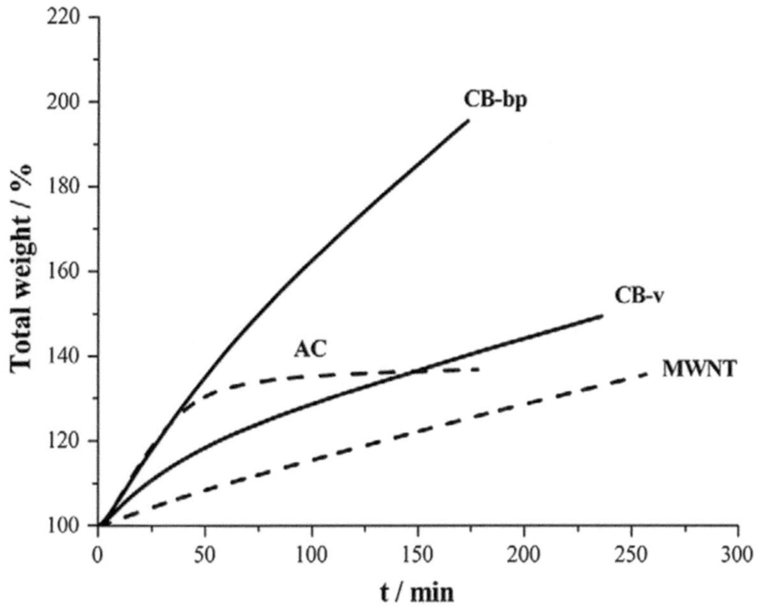

Fig. 6.2 Evolution of the activity and catalyst weight for some carbonaceous catalysts. *AC* activated carbon, *CB* carbon black, *v* (vulcan)-*bp* (black pearls), *MWNT* (carbon nanotubes) (Reprinted with permission from [27]. Copyright © 2009, Elsevier)

increasing agglomeration strength obtained at the higher temperatures. Figure 6.4 shows the microstructure of the carbon formed at 1000 °C from the methane pyrolysis reaction during the experiments done at CIEMAT [28]. It can be observed the evolution of the carbon formed with the temperature in comparison with the graphite produced at 1500 °C shown in Fig. 6.5. The carbon particles are tinier at 1500 °C, producing a stronger agglomeration. The surface characteristics at higher temperatures seem to be more favorable for certain catalytic effects.

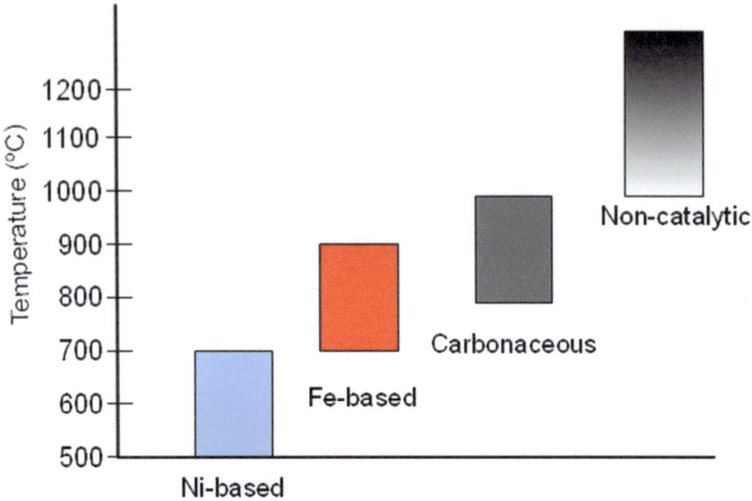

Fig. 6.3 Temperature range of applicability of catalysts for methane decarbonization

Fig. 6.4 Electron microscopy of the carbon formed from methane cracking at 1000 °C (Reprinted with permission from [28]. Copyright © 2011, Elsevier)

The most interesting catalyst from the techno-economical point of view may be based on Ni compounds, as the methane conversion is very significant between temperatures of 500 and 700 °C for reasonable reaction times (seconds) as shown in Table 6.2. Cost of Ni compounds is lower with respect to other catalysts such as Pd,

Fig. 6.5 Electron microscopy of the graphite formed from methane cracking at 1500 °C (Courtesy of Abánades in cooperation with CIEMAT)

Pt, Ru, Mo, or W, and they have an acceptable selectivity. Fe-based catalysts are a cost-effective alternative, but they require higher temperatures than Ni compounds for an acceptable conversion. The formation mechanisms of carbon in the catalyst particles produce carbon filaments in this case.

Carbonaceous catalysts such as activated carbon and black carbon are an option with a lot of potential for a good techno-economical viability. Their catalytic effect is less important than metallic catalysts but they require higher reaction temperatures, up to 1000 °C. The carbon produced is concentrated on the surface of the catalyst, and, therefore, it is highly influenced by its surface characteristics, reaching the production of graphitic carbon in some cases.

6.4 Technical Options for Methane Decarbonization

The practical implementation of methane pyrolysis might be done through three main paths: direct thermal cracking at very high temperature, catalyzed thermal decarbonization, and plasma-torch driven methane pyrolysis.

The direct thermal cracking is exclusively based on the heating of methane up to temperatures in which the kinetics of the reaction produces very high conversions in a reasonable time. To achieve these requirements, high temperatures are needed (above 1300 °C). Experimental work has reported that almost 100 % conversion of methane is possible.

A set of designs proposed for direct thermal cracking are based on the utilization of concentrated solar irradiation to provide the energy required to sustain the endothermic methane pyrolysis reaction. Those designs use tubular structures in which the natural gas is flowing and heated to develop its decomposition. The next section of this chapter will be devoted to these promising concepts.

Catalyzed thermal decarbonization has been widely studied and a lot of experimental work has been reported in relation with metallic and carbonaceous catalysts. Several catalysts of this type have been tested. Apart of experimental tests regarding the analysis of the heterogeneous catalysis and supports, the challenge is to design a continuous device that could be used on the industrial scale. As pointed out, nickel could be applied as catalyst for the reaction between 500 and 700 °C, which is a convenient temperature for an engineering industrial device. Nevertheless, as a consequence of the process, part of the carbon is deposited on the tip of the catalyst causing its deactivation.

The design of a continuous operating reactor requires a regeneration section for the catalyst, generally based on the oxidation of the encapsulating carbon that produced the catalyst deactivation. Different reactor designs have been successfully used alternating cracking and regeneration cycles in continuous mode, as in parallel fixed-bed reactors and fluidized-bed reactors. For the case of a fixed-bed reactor, carbon deposition constitutes an operational problem that leads to the blockage of the reactor or an elevated pressure drop through the bed. In addition, the low heat transfer in the fixed bed may lead to high temperatures during the regeneration phase, which may cause catalyst sintering. The fluidized-bed reactors overcome some of those problems, but they have high operation costs, as demonstrated for the case of the HYPRO process [29].

Plasma-torch driven methane pyrolysis has been implemented industrially and is based on the combined utilization of electricity and heat for the reaction. Similar methods based on the production of carbonaceous catalysts by plasma arc and heat have been proposed as well. Among the concepts for this technique, that of Mudarov [30] shown in the next section and that of the Kvaerner CB&H process [32] are discussed. In this last case, an industrial plant was built, producing 500 kg/h of pure carbon and 2000 Nm^3/h of hydrogen, equivalent to 10 MW, with a consumption of 2.1 MW of electricity and 1 MW of high-temperature steam.

6.5 Concept Proposals

Apart of the industrial implementation of the plasma-torch method of Hvaerner [32], the most active developments in the field, at least from the point of view of industrial conceptual design, are connected with the utilization of solar energy to produce hydrogen. Sound initiatives have been proposed either in the United States or in Europe.

Table 6.4 Experimental solar devices that implement methane decarbonization according to reaction temperature (*T*)

T (°C)	Catalyst or particle feeding	CH_4 flow (l/min)	Max. conversion (%)	Ref
1327	Black carbon BP2000	8.6–15.6	95	[33]
1450	No, $Fe(CO)_5/Fe(C_5H_5)_2$	5–9.7	Almost 100 %	[34]
1550	No	0.1	75 %	[35]
1800	No			[36]
1400	Reactor with Rh	3.8		[37]
1700	No	2–10	42	[37]

A summary of the solar devices that have been studied for solar cracking is depicted in Table 6.4. Most of these experiments aim to reach temperatures above 1300 °C and are mainly linked to research centers in Europe, such as the CNRS (France) and ETH (Switzerland), but also in Israel (the Weizmann Institute) and in the United States (NREL, National Renewable Energy Laboratory).

As a clear example of the solar methane cracking, Steinfeld et al. [33] tested a 5 kW prototype of a solar reactor seeded with black carbon particles as solar irradiation thermal absorber and carbonaceous catalyst for thermal cracking of methane. The cylindrical reactor was 200 mm in length and 100 mm in diameter, with 60 mm hole, illuminated trough a 240 mm-diameter quartz window (Fig. 6.6). They tested the reactor in the temperature range from 1300 to 1600 K. The methane flow rate was between 8.6 and 15.6 L/min, with carbon volume fraction until 7.10^{-5}. They reported maximum methane-to-hydrogen conversion of 95 % at a residence time less than 2 s and an experimental solar-to-chemical energy conversion efficiency of 16 %.

In Europe, the most important studies have been concentrated in the participants of the SOLHYCARB project, which was founded by the European Commission, in which a 10 kW reactor has been tested [36]. A scheme of the reactor is depicted in Fig. 6.7. The methane decarburation process was tested in the temperature range 1740–2070 K (1466–1796 °C), showing how increasing temperature permits enhancement of methane conversion, and long residence times reduce the amount of other hydrocarbon production. It was also reported that a typical carbon mass balance showed that about half of the initial carbon content in the methane feed was found on the form C_2H_2 at the reactor exit, 2 % as particles in the filter, 34 % in the tubes as thermophoretic carbon deposits, or 3 % in the non-converted CH_4. The large amount of acetylene shows how the decarbonization reaction is not complete.

A practical approach to use black carbon produced in the process as catalyst is based on a decarburation reactor with a catalyst production on-line by plasma-arc device [30]. The catalyst in this case is a carbon aerosol obtained from methane plasma-arc pyrolysis that is flowed in the reactor to serve as a seed for thermal methane-cracking reaction. A schematic view of the process is shown in Fig. 6.8. The electrical input to that process is estimated as 14 % with respect to the total energy required, with an overall energy efficiency of 83.9 %, although practical

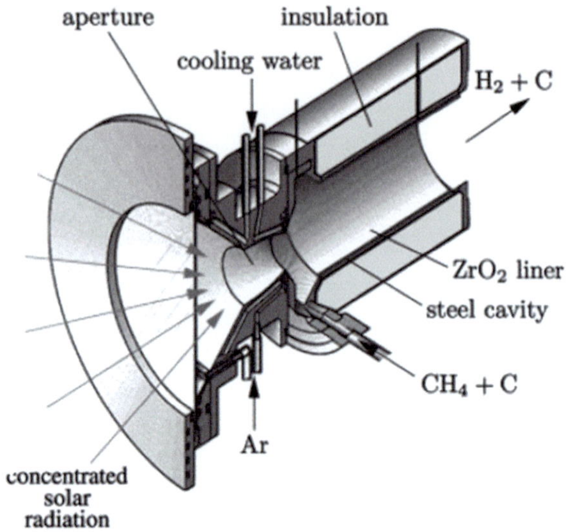

Fig. 6.6 Design of Steinfeld et al. (Reprinted with permission from [33]. Copyright © 2009, Elsevier)

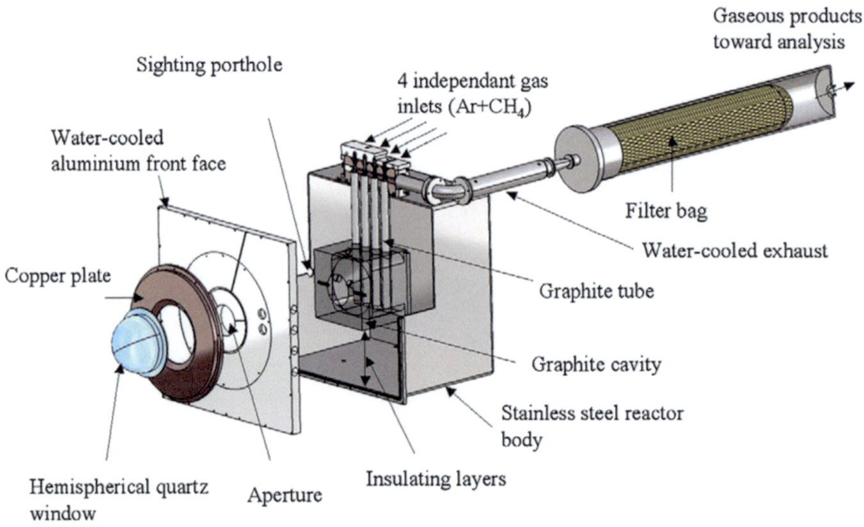

Fig. 6.7 Indirect heated tubular reactor proposed in the SOLHYCARB project (Reprinted with permission from [36]. Copyright © 2011, Elsevier)

implementation should include other losses in filters and pumping and include the heating value of carbon as an output of the process. The experimental data reported show stable continuous operation of 8 h at 870 °C in a once-through utilization of the electric arc-produced carbon catalyst.

Another industrial scale proposal is depicted in Fig. 6.9. Concentrated solar sunlight is used in this case to provide the necessary heating to drive the

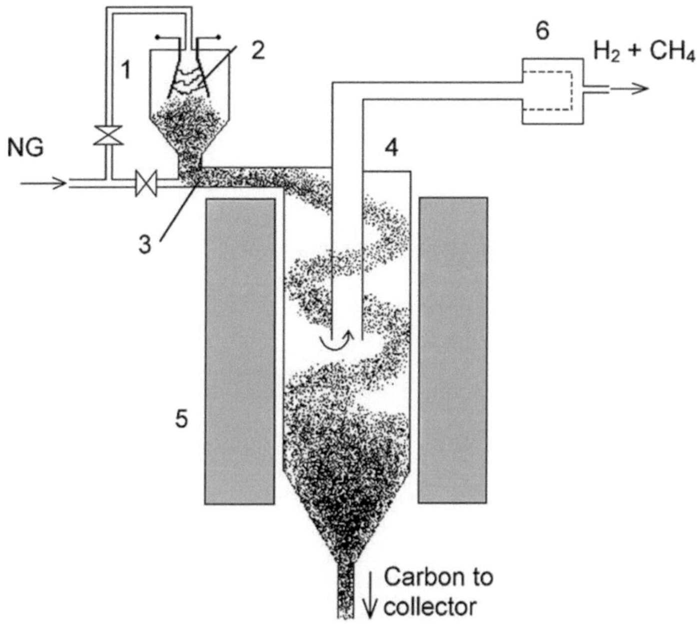

Fig. 6.8 Schematic diagram of a vortex-flow reactor couple with the nonthermal plasma device for generating carbon aerosol. *1* Nonthermal plasma reactor, *2* arc discharge, *3* carbon aerosol, *4* vortex-flow reactor, *5* heater, *6* bag filter (Reprinted with permission from [31]. Copyright © 2009, Elsevier)

endothermic methane-cracking reaction. The core of the concept is a fluid-wall reactor, in which hydrogen itself as sweeping gas is used to avoid carbon deposition on the walls of the tubular reactor. The practical implementation of this type of reactor has encountered some difficulties due to the effectiveness of the sweeping gas. In fact, attempts to reproduce this operational method have found many difficulties to work in a continuous mode [37].

Industrial initiatives toward the implementation of methane decarbonization have been quite active from the beginning of research on the technology. Table 6.5 lists some of the main patents that have a direct practical relationship with direct methane decarbonization. The methods are directly derived from the experimental work in the field and include different tube dimensions and gas feeding. The use of a mixed gas feed includes H_2 or argon to dilute carbon particles or avoid its clogging on the reactor walls.

Other more general patents have been released related to methane decarbonization. For instance, Dahl and Weimer in 2003 (US 6872378) patented a solar thermal aerosol flow reactor as an environmentally beneficial process using concentrated sunlight to heat radiation absorbing particles to drive endothermic gas phase reactions for hydrogen production or other hydrogen synthesis gases. The scheme of the patent is provided in Fig. 6.10. An intermediate heat exchanger to preheat the natural gas and a set of bags to catch the carbon produced are a basic

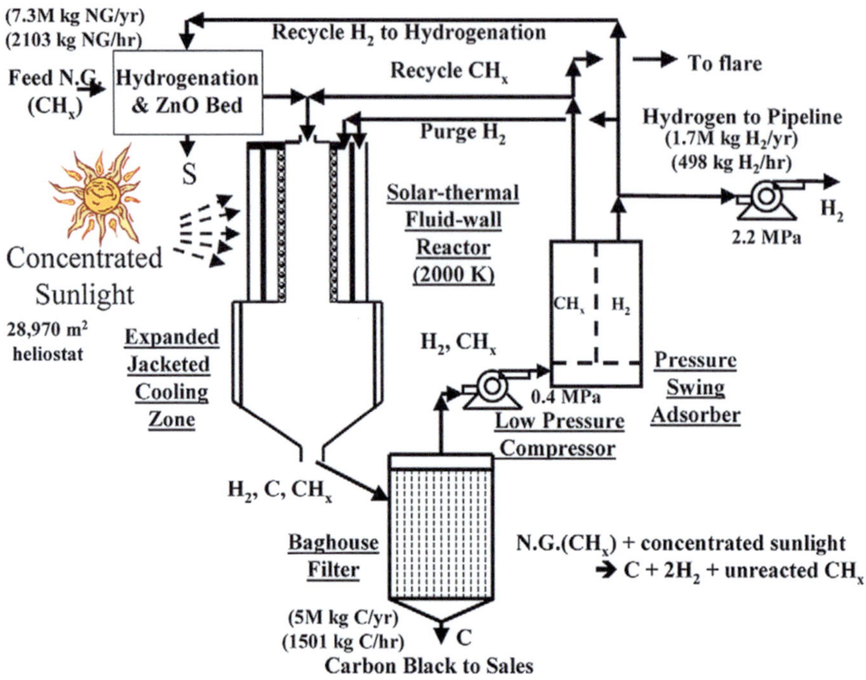

Fig. 6.9 Scheme for industrial methane cracking suggested by Dahl et al. (Reprinted with permission from [38]. Copyright © 2004, Elsevier)

part of this process. The cogenerated carbon and hydrogen are the main products of the process.

Along the same line, in 2012, Muradov (US 8147755) patented the conceptual design of a proposal as "a novel process and apparatus disclosed for sustainable, continuous production of hydrogen and carbon by catalytic dissociation or decomposition of hydrocarbons at elevated temperatures using in-situ generated carbon particles." In that patent, the energy input may be produced by a high-temperature energy source, either plasma (thermal, nonthermal, microwave, corona discharge, glow discharge, dielectric barrier discharge) or radiation sources (such as electron beam, gamma, ultraviolet). Muradov referred to oxidative as oxygen, air, ozone, nitrous oxide (NO_2), and other oxidizing agents.

6.6 Economic Analysis

The implementation of a technology is mainly dependent on the economic benefits that are able to be provided for society. Those benefits are of different natures. For instance, social benefits are of paramount importance and the creation of quality jobs is one of its main indicators. This fact is connected to the operation and

Table 6.5 Main patents related to direct methane decarbonization from solar irradiation

Patent	Dimensions	Graphite characteristics	Flow rates	Problems
Matovich 1977 (US 4056602)	I.D./O.D. 76.2/ 101.6 mm	Fluid-wall system	H_2:141 l/min	Difficult to filter the carbon particles
	L = 914.4 mm	Porous graphite tube, 20 μm average pore radius	CH_4:28–141 l/min	
Schramm 1987 (US 4643890)	I.D./O.D. 324/330 mm	Fluid-wall system	–	–
	L = 1200 mm	Nonporous graphite tube with perforations of 0.76 mm as diameter		
Weimer 2003 (WO 03/ 076334)	I.D./O.D. 15/18 mm	Fluid-wall system	H_2:1 l/min	No carbon deposition on the wall
	L = 440 mm	300 mm porous (49 % porosity), 70 mm of both ends nonporous	CH_4:4 l/min	Without fluid-wall system, clogging in 8 min
Promes (Odeillo)	I.D. 10 mm	Fluid-wall system	Ar + CH_4:1–5 l/min	Clogging in 10–60 min
	L = 200 mm	Two concentric graphite tubes		
SOLHYCARB	I.D. 8 mm	Non-fluid-wall system	–	Clogging at the end of the tube
	L = 320 mm	Alumina tube		

maintenance costs (OPEX) and to a lesser extent to the temporal capital costs (CAPEX). On the other hand, the environmental benefits are connected as well, as are the resources being exploited in a sustainable way, and that also implies sustainable activity of the society.

The economic analysis is focused in the evaluation of the cost of the product provided by the process and its comparison with the mature and future alternatives. In the Table 6.6, the production cost of hydrogen is summarized without including distribution costs. Coal gasification and natural gas steam reforming are the most competitive technologies for producing hydrogen with biomass gasification in some specific cases. The centralized production systems benefit from the economy of scale. Wind and photovoltaic (PV) methods depend strongly on the costs of the renewable installation. The cost of methane-cracking processes has been evaluated by some of the proposals previously described.

Methane cracking has been developed as a hydrogen production technique driven by solar thermal technology so far. In those facilities, the largest cost share arises from the solar field, which depends strongly on the evolution of the heliostat cost. In addition, the receiver is also a very important part of the costs. For instance, in the economic evaluation of the SOLHYCARB [45] project, the capital cost for the heliostat field was set up to 230 $/m^2. Previous analysis done by NREL established a heliostat cost of 250 $/m^2, in the same range of the SOLHYCARB project cost estimation. In such analysis, the heliostat and solar-related equipment

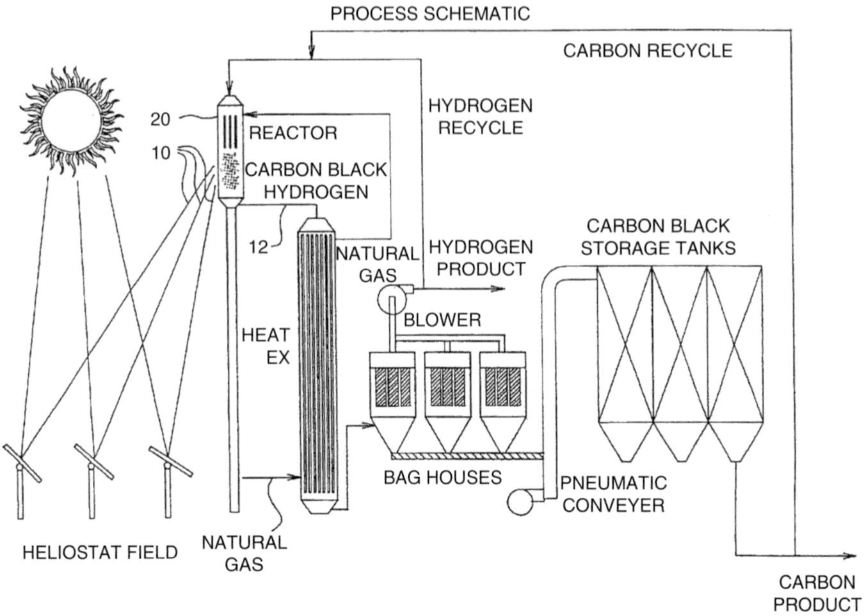

Fig. 6.10 Scheme for industrial methane cracking suggested in the patent by Weimer, Dahl et al. (Patent US 6872378)

Table 6.6 Cost of hydrogen production with various technologies

Technology	Fuel	Prod. cost ($\$_{2008}$/kg)	Ref
Central steam reforming	Natural gas	1.5	[40]
Distrib. steam reforming	Natural gas	2.6	[43]
Gasification	Coal	1.2	[43]
Gasification with CCS	Coal	1.8	[40]
Gasification	Biomass	1.4	[43]
Distributed electrolysis	Grid electricity	6.8	[43]
Central electrolysis	Wind	3.8	[43]
Distributed electrolysis	Wind	7.3	[43]
Thermochemical cycle	Nuclear	1.4	[43]
Pyrolysis/cracking	Natural gas + solar	3.0	[41]
Pyrolysis/cracking	Natural gas + solar	3.6	[44]
Pyrolysis/cracking	Natural gas + solar	4.5	[38, 39]
Steam reforming	Natural gas + solar	2.2	[42]
PV electrolysis	Solar	9.1	[40]

Extracted with permission from [4]. Copyright © 2012, Elsevier

amount to 60 % of the total capital costs. These values are in agreement with the current state of the art in solar tower technology. Nevertheless, those costs have a large potential to be improved in the future, with a foreseeable target of 100 $/m^2

for the heliostat field, which would reduce the overall capital costs of 30 % leading to an hydrogen cost of 2 \$/kg H_2, close to the competitiveness with fossil fuel gasification.

The cost estimates previously shown are based on the exploitation of solar power as the main primary energy source to enable the methane-cracking reaction. The solar technology will most likely undergo further development, leading to a sensitive cost reduction, through the reduction of the heliostat cost, as stated, and improvements in the scale economy, plant size, and receiver design. Another option to provide energy for the endothermic methane pyrolysis may be the burning of a relatively small portion of the H_2, which is produced by the process. Such a solution would eliminate the capital cost of the solar equipment purchase and installation, as a consequence the H_2 production cost can be expected to be reduced approximately 50 %, hence achieving an indicative value of 2 \$/kg, which is competitive with steam methane-reforming (SMR) systems without including carbon capture and sequestration (CCS). It is also likely competitive in the short term with other alternatives for hydrogen production based on wind or solar PV. That energy may be also obtained with heat produced by biogas or by any other type of biofuel that could be used at the temperatures required for the development of the reaction.

The economic viability of the process can be affected by the selling of the carbon that is eventually produced. This cost effect is taken into account in many of the economic analyses. Different hypotheses have been evaluated with respect to internal rate of return (IRR, %) and the initial investment, as well as the price of natural gas, showing how cost estimates for solar methane cracking are between 2.5 and 4.5 \$/kg without taking into account sales of black carbon, being competitive with standard methane steam reforming from a carbon price on the order of 0.7 \$/kg [35]. Nevertheless, the massive industrial implementation of methane cracking will reduce the value of carbon to a low level, and its contribution to the economics of the process will be marginal unless new technologies and applications for the carbon product are developed.

6.7 Application to Industrial Processes

The utilization of hydrogen has one of its more important markets in its application to industrial processes. The production of many chemicals requires the generation of hydrogen to be combined with other molecules to manufacture commodities such as ammonia or methanol. Nowadays, a significant amount of hydrogen is used for upgrading heavy oil as well. At the present state of the technology, those hydrogen production processes are based on fossil fuel reforming or gasification, leading to a significant amount of CO_2 production. Other techniques to produce carbon-free hydrogen are based on electrolysis, which is economically viable in rather limited cases.

6.7.1 Ammonia Production

One of the most relevant industrial commodities is the production of ammonia. This chemical is the starting feedstock for nitrogenous fertilizer synthesis, for which is dedicated to 80 % of its production. The remaining 20 % is used in the manufacturing of fibers, explosives, amines, amides, plastics, and organic nitrogen components, as nitric acid and urea. Since the required H_2 is currently extracted from fossil fuels, for instance, natural gas reforming, fertilizer production constitutes one of the main industrial CO_2 sources from worldwide chemical industry.

A fraction (30 %) of the CO_2 produced in the ammonia manufacturing process is currently removed from the reformer using amines scrubbing and reused in the production of urea. According to the fertilizer industry (www.fertilizer.org), natural gas is the dominant feedstock for ammonia production, as it has the highest hydrogen content. Nevertheless, some other fossil fuel feedstocks, such as naphtha, fuel oil, and coal, are being used in developing countries (e.g., India and China) and have a higher CO_2 release.

According to the International Energy Agency [1], natural gas-based ammonia production contributes up to 72 % of the total amount, with an energy intensity of 41.6 GJ/t NH_3 and a CO_2 emission level in the order of 2.1 Mton CO_2/Mt NH3, leading to the order of 0.3 GtonCO_2/year. The world production of ammonia in 2011 was 164 Mt, with China being the most important producer, with 75 % of its hydrogen being produced from coal. Huge efforts are being put in place to increase the energy efficiency of the processes, which can be improved by as much as 20 %. Nevertheless, the development of a CO_2-free process to provide hydrogen may reduce the greenhouse gases emissions by a relevant factor and can contribute greatly to the abatement of global greenhouse gas emission.

As a general statement, the introduction of methane cracking in the ammonia production process (Fig. 6.11) would allow a massive reduction in the associated CO_2 emissions. From the previous estimation of the sector emissions, assuming a methane-cracking ammonia market penetration of 50 %, the CO_2 abatement as a result of the application of this technology would be on the order of 0.15 Gton/y.

6.7.2 Biofuel Production

Liquid fuels are hydrogenated chemicals produced by industry from raw materials containing carbon and hydrogen. The raw materials for their production are fossil sources such as carbon, natural gas or oil, and biomass. The hydrogen to carbon ratio of a fuel is one of its most important characteristics. As a general rule, the highest the H/C ratio has the highest heating value and the lowest the carbon dioxide emissions. Therefore hydrogen is being added to low H/C fuels to improve their performance and reduce their specific emissions.

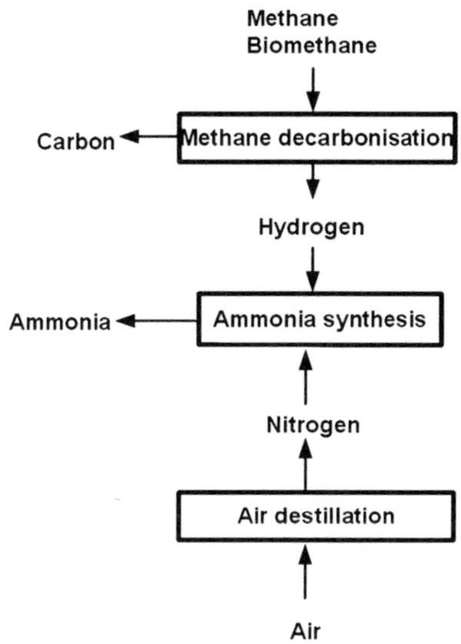

Fig. 6.11 Simplified scheme for the integration of methane decarbonization in ammonia synthesis

As an example, the production of biodiesel is based on the transesterification of vegetable oils and fats with the addition of methanol. A simplified scheme for the integration of methane pyrolysis in this process is depicted in Fig. 6.12. Methane or bio-methane may be used as main raw material for the hydrogen production, which is combined with carbon dioxide to produce methanol for the transesterification process. The methanol is made from carbon-free hydrogen. The carbon dioxide can be obtained from carbon capture, closing the carbon cycle with a null net CO_2 emission. Alternative advanced processes may be based on hydrogenation of oil and fats, in which hydrogen should be provided to triglycerides to saturate their carbon bonds.

In both cases that use either methanol or hydrogen, these chemicals are commonly obtained from fossils, mainly natural gas. For the case of methanol, some industrial initiatives produce methanol from the combination of hydrogen from geothermal electrolysis and a CO_2 source in special locations, such as Iceland [46]. Additional methods may be developed with other low-carbon hydrogen sources as methane cracking. For the case of hydrogenation, hydrogen is commonly produced by natural gas steam reforming, and a direct environmental improvement of the technology might be done by technologies such as methane pyrolysis.

6.8 Main Technological Problems

From the analysis of previous work on the topic, it emerges that the technological development of the methane decarbonization reaction faces some limitations in its industrial deployment. The most important source of technical problems is the

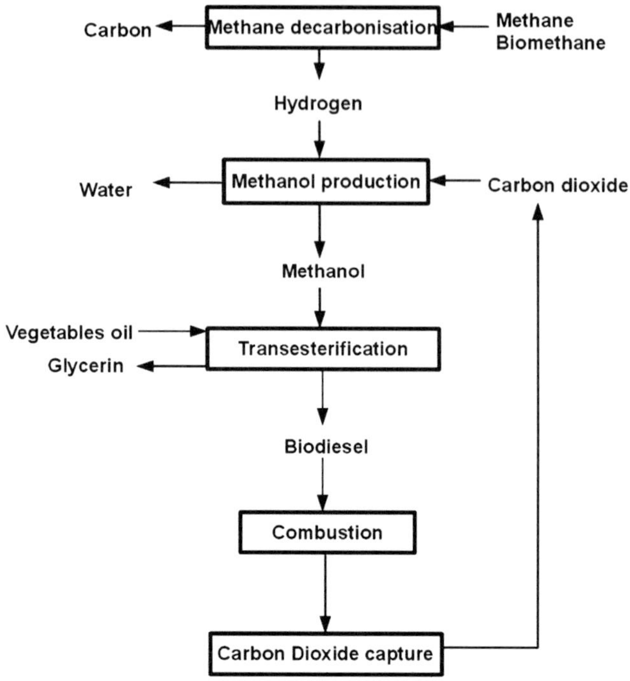

Fig. 6.12 Simplified scheme for the integration of methane decarbonization in a biodiesel production process

production of carbon particles, which are deposited in the equipment or on the catalysts producing reactor clogging and catalysis deactivation. An example of this problem is depicted in the Fig. 6.13, in which the carbon agglomeration is shown, in which the carbon seeds attached to the walls of the gas-flowing tube and grew. The most applied treatment to remove such deposits is based on the use of oxidizing agents such as steam, oxygen, and CO_2 that produce undesirable carbon oxides. The mechanical strength of the carbon clogging depends on the temperature of the process. Temperatures above 1300 °C produce very hard agglomerations that make it unlikely to extract the graphitic carbon by mechanical methods without damaging the reactor.

Practical industrial implementation of the process in the case of catalyzed reactions is limited by the deactivation of the catalyst and its cost. The deactivation of the catalyst is a process derived from the formation of coke in the surface of the catalyst that prevents the contact of the methane molecules with the catalyst surface. In the case of the carbonaceous catalyst, the coke fills the pores, thus reducing the active surface. In the case of metallic catalysts, the coke covers the active metallic atoms, at a rate higher than the carbon diffusion trough the metallic structure and support, thus producing complete deactivation of the catalysts

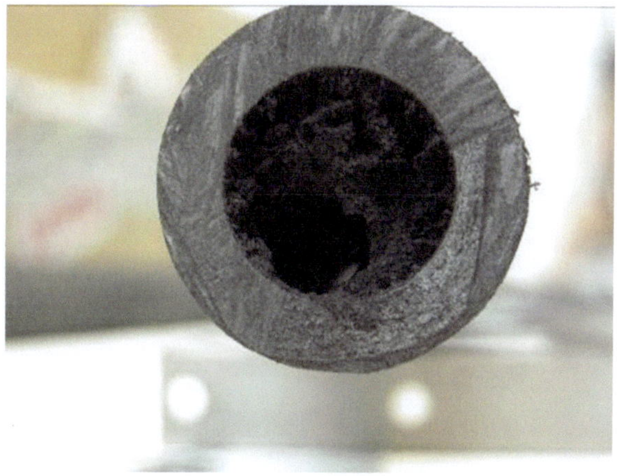

Fig. 6.13 Carbon agglomeration in a methane-cracking test tube (Reprinted with permission from [28]. Copyright © 2011, Elsevier)

(Fig. 6.14). The result for catalysts such as Ni is the formation of carbon nanotubes [30].

There have been some projects that have attempted to implement industrial deployment of the decarburation reaction. The first one was the HYPRO project by Universal Oil Products in the 1960s, which was based on a fluidized-bed reactor with Ni-Fe-Co catalyst in Al_2O_3 support [29]. The system with two fluidized beds was too complex and expensive and its techno-economical viability was not successful.

Metallic catalysts have demonstrated their suitability for use at reasonable reaction temperatures (below 800 °C), but their performance must be increased an order of magnitude to ensure their viability in a once-through process [4]. Multiple step processes (deactivation/regeneration) are based on oxidation of the coke layer around the catalyst producing CO_2. Catalyst deactivation is produced when there is a mismatch between the coke layer formation on the catalyst surface and the coke diffusion through the metallic particle, leading to the complete encapsulation of the catalyst and the elimination of its contact surface with the methane gas.

Carbon diffusion through the metallic particle generally produces carbon filaments between the catalyst and its support. The control of carbon diffusion and filament growth is the key issue for catalyst deactivation. Operating conditions that are required to effectively produce a steady-state and stable catalytic effect should be adopted under very strict set points, which make it very difficult to implement a practical long-term continuous process. Up to know, long-term operation of a powerful catalyst has not been demonstrated.

As pointed out, the regeneration of spent metallic catalyst is required for an environmentally friendly and economic process and constitutes a real challenge for the present state of the art. The regeneration method should be chosen according to the energy balance of the process, the regeneration time, and the performance of the regenerated catalyst. None of the state-of-the-art methods can fulfill these conditions [4]. For instance, air regeneration is able to produce energy that will cover part

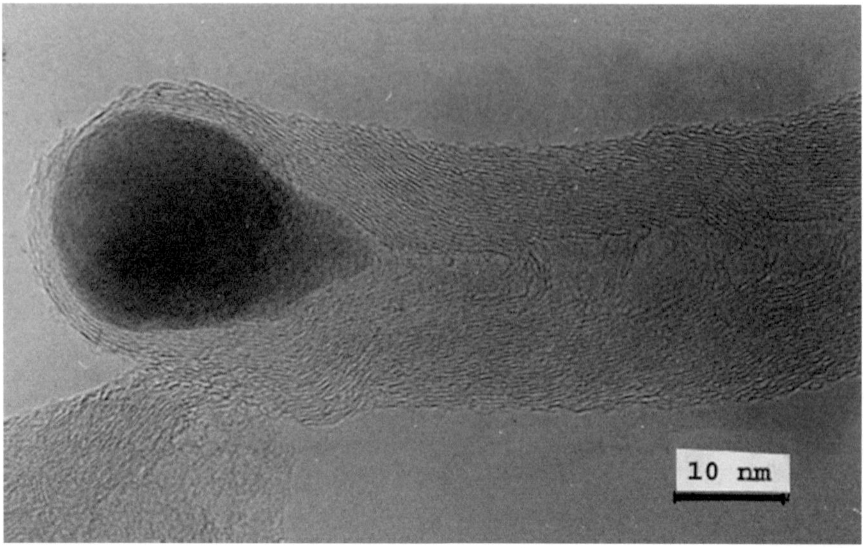

Fig. 6.14 Metallic catalyst on the tip of a coke filament produced by catalytic methane cracking (Reprinted with permission from [47]. Copyright © 1999, Elsevier)

of the energy demand for the methane-cracking process, but it has the drawback of producing hot spots, catalyst oxidation, sintering, and even disintegration of the catalyst [46]. Another suitable technique, such as steam regeneration, is more promising for the catalyst integrity and quality. Many steam regeneration cycles have been reported for Ni-based catalysts [49], although they tend to reduce catalyst performance. The main drawback of catalyst regeneration is high complexity of the chemical process.

Another technological problem associated to the current technological proposals could come from the scale-up of the concepts. The step forward from the experimental device, based generally on tubular reactor, to an industrial scale system is not trivial. The industrial system might not be a result of a simple multiplication of tubes to reach a reasonable amount of hydrogen production, for instance, 10 MW as in the case of the Kvaerner process. The economic viability of the system may require the design of reactors with higher volume. A device based on a gaseous volume is limited if it is intended to keep a stable high temperature as a result of the thermophysical properties of the gas itself. Thermal conductivity, for instance, and thermal diffusivity will limit the volume of a temperature controlled process in which an endothermic reaction takes place. The utilization of carbon particles for a better thermal homogeneity is potentially a good option to overcome this problem.

6.9 Conclusions and Future Outlook

The development of methane cracking for hydrogen production requires a better understanding of some basic information and the design of reactors that can operate without problems such as equipment clogging, catalysis deactivation, continuous operation, and economy.

About the basic information, a lot of uncertainties are still pending in relation to the kinetic parameters of the methane-cracking reaction. There are several orders of magnitude in parameters as reported for the pre-exponential factor for Arrhenius-type modeling. Even the activation energy has a lot of uncertainty. Without this information, it will be very unreliable to scale experimental devices to the industrial facility level.

A lot of work has been done on heterogeneous catalysis. Efforts should be made to synthesize new catalysts that can provide high conversion with high thermal and chemical stability and high carbon capacity [50]. Fluidized bed reactors with cyclic methane-cracking reaction and catalyst regeneration sections are a promising concept that should be developed further so that continuous operation can be achieved.

The methods based on direct methane decarbonization are mainly searching for high temperatures provided by solar furnaces. Its main technological problem is the graphitic carbon produced and how to remove it to avoid clogging of the reactor. Continuous operation of solar furnaces, generally based on 3-D concentration systems for the illumination of a central tower or parabolic dishes, will require research and development to remove such carbon by mechanical or chemical cleaning. Some attempts have been proposed that use gas stream turbulence or gas sweeping through porous media, although there are no successes on the reproducibility and control of these features.

The use of liquid media has been proposed in some patents, mainly the one by Steinberg and Dong (US 5767165), in which molten iron, tin, or molten salts (NaCl, NaF) are used. A scheme of this proposal is depicted in Fig. 6.15. The preferred option in most of the cases for the liquid media is molten metals, as their density is very different from that of the carbon produced and separation seems to be achievable by physical means. The use of molten tin or copper has been proposed [5], in a scheme in which it could be possible to separate the carbon skimming off the surface of molten metal, as slag is skimmed off the surface of molten iron in a blast surface. Some experimental work has been done as well to implement the methane-cracking reaction in the context of nuclear generation IV reactor development, to extend liquid metal reactors to hydrogen production [51] using tin and lead as reaction media. Nevertheless, the reproducibility of those results must be confirmed. The hydrogen flow rate was quite small in this last case, below 10 ml/min, and the scale-up of the process will require a device with a capacity of several orders of magnitude higher.

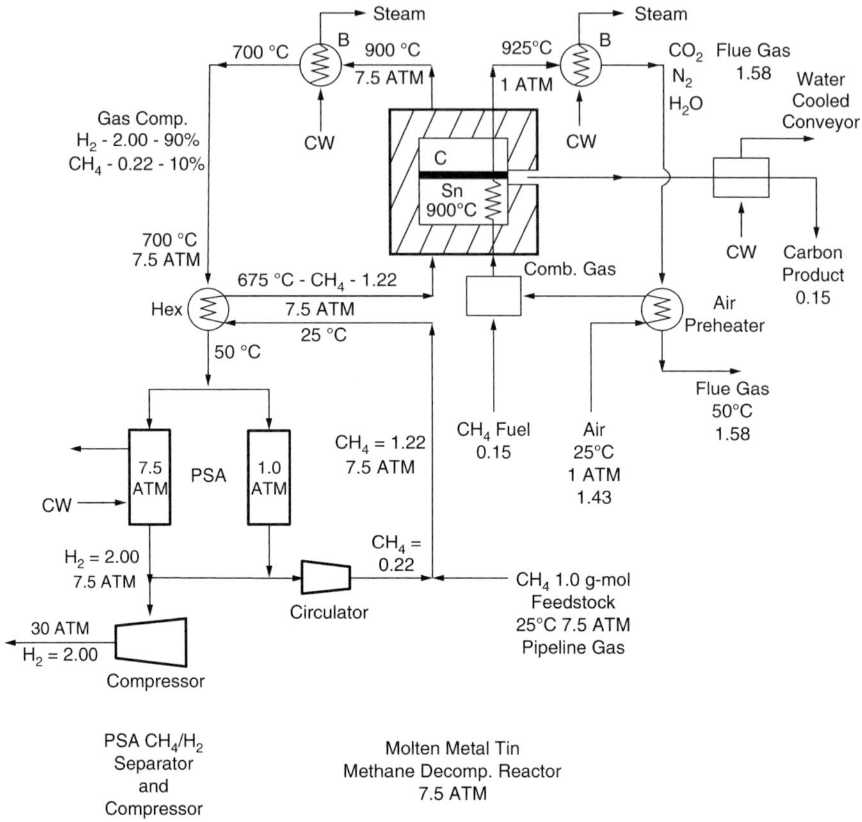

Fig. 6.15 Conceptual design of a methane decarbonization process developed on liquid media (Reprinted with permission from [5]. Copyright © 1999, Elsevier)

The use of molten media (liquid metals and molten salts) to host the methane-cracking reactions is promising, although a lot of experimental work should be done to confirm if a continuous operation scheme is feasible on an industrial scale. A reasonable range for the reaction temperature for a methane-cracking process that applies liquid media should be determined. The main source of problems might be the compatibility between hydrogen, carbon, and the molten media and their thermal stability should be evaluated at the reaction conditions, as well as the kinetics and rates versus temperature of the reaction. The application of molten media avoids problems associated with gas flow stagnant and blocking due to carbon agglomeration, but opens new questions related to corrosion, as most of the liquid metals are quite aggressive with steel materials. Nevertheless, carbon separation processes from liquid media should be developed from additional experimental evaluations.

References

1. IEA-World Energy Outlook. Special report 'Are we entering a golden age of gas'. 2011.
2. IPCC. Climate Change 2014 Synthesis Report. 2014. http://www.ipcc.ch.
3. Kalamaras CM, Efstathiou AM. Hydrogen production technologies: current state and future developments. Conference papers in energy 2013. New York: Hindawi Publishing Corporation; 2013.
4. Abánades A, Rubbia C, Salmieri D. Thermal cracking of methane into hydrogen for a CO2-free utilization of natural gas. Int J Hydrog Energy. 2012;38(20):8491–6.
5. Steinberg M. Fossil fuel decarbonization technology for mitigating global warming. Int J Hydrog Energy. 1999;24(8):771–7.
6. Goodwin DG, Moffat HK, Speth RL. Cantera. An object-oriented software toolkit for chemical kinetics, thermodynamics, and transport processes. 2014. http://www.cantera.org. Version 2.1.2.
7. Kassel LS. The thermal decomposition of methane. J Am Chem Soc. 1932;54(10):3949–61.
8. Skinner GB, Ruehrwein RA. Shock tube studies on the pyrolysis and oxidation of methane. J Phys Chem. 1959;63(10):1736–42.
9. Kevorkian V, Heath CE, Boudart M. The decomposition of methane in shock waves. J Phys Chem. 1960;64(8):964–8.
10. Kozlov GI, Knorre VG. Single-pulse shock tube studies on the kinetics of the thermal decomposition of methane. Combust Flame. 1962;6:253–63.
11. Palmer HB, Hirt TJ. The activation energy for the pyrolysis of methane. J Phys Chem. 1963;67 (3):709–11.
12. Steinberg M. Production of hydrogen and methanol from natural gas with reduced CO2 emission. Int J Hydrog Energy. 1998;23(6):419–25.
13. Rodat S, Abanades S, Coulio J, Flamant G. Kinetic modelling of methane decomposition in a tubular solar reactor. Chem Eng J. 2009;146(1):120–7.
14. Shantarovic PS, Pavlov BV. Thermal cracking of methane. Int Chem Eng. 1962;2(3):415–8.
15. Chen CJ, Back MH, Back RA. The thermal decomposition of methane. I. kinetics of the primary decomposition to C2H6 + H2; rate constant for the homogeneous unimolecular dissociation of methane and its pressure dependence. Can J Chem. 1975;53(23):3580–90.
16. Abbas HF, Wan Daud WMA. Hydrogen production by methane decomposition: a review. Int J Hydrog Energy. 2010;35:1160–90.
17. Li Y, Li D, Wang G. Methane decomposition to COx-free hydrogen and nano-carbon material on group 8–10 base metal catalyst: a review. Catal Today. 2011;162:1–48.
18. Li Y, Zhang BC, Xie XW, Liu JL, Xu YD, Shen WJ. Novel Ni catalysts for methane decomposition to hydrogen and carbon nanofibers. J Catal. 2006;238:412–24.
19. Chen JL, Li XM, Li YD, Qin YN. Production of hydrogen and nanocarbon from direct decomposition of undiluted methane on high-nickeled Ni–Cu–Alumina catalysts. Chem Lett. 2003;32:424–5.
20. Ashok J, Reddy PS, Raju G, Subrahmanyam M, Venugopal A. Catalytic decomposition of methane to hydrogen and carbon nanofibers over Ni–Cu–SiO2 catalysts. Energy Fuel. 2009;23:5–13.
21. Gallego GS, Barrault J, Batiot-Dupeyrat C, Mondragon F. Production of hydrogen and MWCNTs by methane decomposition over catalysts originated from LaNiO3 perovskite. Catal Today. 2010;149:365–71.
22. Avdeeva LB, Reshetenko TV, Ismagilov ZR, Likholobov VA. Iron-containing catalysts of methane decomposition: accumulation of filamentous carbon. Appl Catal A Gen. 2002;228:53–63.
23. Ishihara T, Miyashita Y, Iseda H, Takita Y. Decomposition of methane over Ni/SiO2 catalysts with membrane reactor for the production of hydrogen. Chem Lett. 1995;2:93–4.

24. Ermakova MA, Ermakov DY, Kuvshinov GG, Plyasova LM. New nickel catalysts for the formation of filamentous carbon in the reaction of methane decomposition. J Catal. 1999;187:77–84.
25. Chesnokov VV, Chichkan AS. Production of hydrogen by methane catalytic decomposition over Ni–Cu–Fe/Al2O3 catalyst. Int J Hydrog Energy. 2009;34:2979–85.
26. Muradov N, Smith F, T-Raissi A. Catalytic activity of carbons for methane decomposition reaction. Catal Today. 2005;102–103:225–33.
27. Serrano DP, Botas JA, Gui-Lópex R. H_2 production from methane pyrolysis over commercial carbon catalysts: kinetic and deactivation study. Int J Hydrog Energy. 2009;34:4488–94.
28. Abánades A, et al. Experimental analysis of direct thermal methane cracking. Int J Hydrog Energy. 2011;36:12877–86.
29. Poblenz J, Scott N. Methods for hydrogen production by catalytic decomposition of a gaseous hydrocarbon stream. U.S. Patent No 3,284,161. 1966.
30. Villacampa JI, et al. Catalytic decomposition of methane over Ni-Al2O3 coprecipitated catalysts. Reaction and regeneration studies. Appl Catal Gen. 2003;252:363–83.
31. Muradov N, Smith F, Bockerman G, Scammon K. Thermocatalytic decomposition of natural gas over plasma-generated carbon aerosols for sustainable production of hydrogen and carbon. Appl Catal Gen. 2009;365:292–300.
32. Gaudernack, Lynum S. Hydrogen from natural gas without release of CO2 to the atmosphere. Int J Hydrog Energy. 1998;23(12):1087–93.
33. Maag G, Zanganeh G, Steinfeld A. Solar thermal cracking of methane in a particle-flow reactor for the co-production of hydrogen and carbon. Int J Hydrog Energy. 2009;34(18):7676–85.
34. Yeheskel J, Epstein M. Thermolysis of methane in a solar reactor for mass-production of hydrogen and carbon nano-materials. Carbon. 2011;49(14):4695–703.
35. Abanades S, Flamant G. Solar hydrogen production from the thermal splitting of methane in a high temperature solar chemical reactor. Sol Energy. 2006;80:1321–32.
36. Rodat S, Abanades S, Flamant G. Co-production of hydrogen and carbon black from solar thermal methane splitting in tubular reactor prototype. Sol Energy. 2011;85:645–52.
37. Maag G, Rodat S, Flamant G, Steinfeld A. Heat transfer model and scale-up of an entrained-flow solar reactor for the thermal decomposition of methane. Int J Hydrog Energy. 2010;35 (24):13232–41.
38. Dahl JK, et al. Rapid solar-thermal dissociation of natural gas in an aerosol flow reactor. Energy. 2004;29:715–25.
39. Dahl JK, Buechler KJ, Weimer AW, Lewandowski A, Bingham C. Solar-thermal dissociation of methane in a fluid-wall aerosol flow reactor. Int J Hydrog Energy. 2004;29:725–36.
40. Pregger T, Graf D, Krewitt W, Sattler C, Roeb M, Möller S. Prospects of solar thermal hydrogen production processes. Int J Hydrog Energy. 2009;34:4256–67.
41. Weimer AW, et al. Thermal dissociation of methane using a solar coupled aerosol flow reactor. NREL USA, Proceedings of the 2000 DOE Hydrogen Programme Review. 2000.
42. EU project SOLREF. Hydrogen production via solar reforming of hydrocarbons. 2008. http://www.pre.ethz.ch/research/projects/?id=solref.
43. U.S. Energy Information Administration. The impact of increased use of hydrogen on petroleum consumption and carbon dioxide emissions. 2008. Report#:SR-OIAF-CNEAF/2008-04. http://www.eia.gov/oiaf/servicerpt/hydro/hydrogen.html. Available May 2015.
44. Muradov N, Veziroglu TN. "Green" path from fossil-based to hydrogen economy: an overview of carbon-neutral technologies. Int J Hydrog Energy. 2008;33:6804–39.
45. Sattler C. From a pilot solar reactor to an industrial plant. Process analysis and cost issues. SOLHYCARB Event, Odeillo, 28 Sep 2009.
46. C.R.I. Carbon Recycling International. http://www.carbonrecycling.is/. Available Dec 2014.
47. Avdeeva LB, Kochubey DI, Shaikhutdinov SK. Cobalt catalysts of methane decomposition: accumulation of the filamentous carbon. Appl Catal A. 1999;177:43–51.
48. Rahman M, Croiset E, Hudgins R. Catalytic decomposition of methane for hydrogen production. Top Catal. 2006;37:137–45.

49. Choudhary VR, Banerjee S, Rajput AM. Continuous production of H2 at low temperature from methane decomposition over Ni-containing catalyst followed by gasification by steam of the carbon on the catalyst in two parallel reactors operated in cyclic manner. J Catal. 2001;198 (1):136–41.
50. Amin M, Croiset E, Epling W. Review of methane catalytic cracking for hydrogen production. Int J Hydrog Energy. 2011;36:2904–35.
51. Serban M, Lewis MA, Marshall CL, Doctor RD. Hydrogen production by direct contact pyrolysis of natural gas. Energy Fuel. 2003;17:705–13.

Chapter 7
Hydrogen Production by Supercritical Water Gasification of Biomass

Ekin Kıpçak and Mesut Akgün

Abstract It is widely accepted that hydrogen energy can sustainably provide the world's growing energy needs. Currently, hydrogen is produced from mainly nonrenewable feedstocks through biochemical and thermochemical technologies. However, to achieve a genuinely sustainable, economic, viable, and environmentally benign technology, hydrogen will need to be produced from renewable energy sources through innovative production processes. This chapter focuses on one of these technologies that provide a novel approach for hydrogen production: supercritical water gasification of biomass. The process has significant potential for the conversion of biomass to produce hydrogen and other combustible gases. It also has major advantages when compared with other processes, such as eliminating the necessity for drying of the feedstock, providing high gasification efficiency and hydrogen selectivity, enabling the formation of clean gaseous products, and producing much lower amounts of tars and chars. To increase the hydrogen selectivity, the use of catalysts is common, with the preferred catalysts being alkaline salts, some metals, and metal oxides. The supercritical water gasification process can exhibit different gas compositions or activities with respect to the feedstock, reaction conditions, or catalyst used. Therefore, in this chapter, hydrogen production from various biomass sources by supercritical water gasification is comparatively discussed with examples from the literature. The term biomass covered in this chapter includes model compounds (such as glucose, cellulose, and lignin), alcohols, and real biomass (such as industrial wastewaters and sewage sludge). The effects of reaction time, system temperature and pressure, biomass concentration, oxidant concentration, catalyst use, and the kind of catalyst on the hydrogen yield are investigated.

Keywords Biofuel • Biomass • Catalyst • Hydrogen • Supercritical water • Supercritical water gasification

---

E. Kıpçak (✉) • M. Akgün
Chemical Engineering Department, SCFT-Supercritical Fluid Technologies Research Group, Yildiz Technical University, Davutpaşa Campus, Istanbul 34210, Turkey
e-mail: eyildir@yildiz.edu.tr; akgunm@yildiz.edu.tr

Z. Fang et al. (eds.), *Production of Hydrogen from Renewable Resources*, Biofuels and Biorefineries 5, DOI 10.1007/978-94-017-7330-0_7

7.1 Introduction

Hydrogen is the simplest and most abundant element on earth. It is a prominent feedstock in chemical, petrochemical, metallurgical, electronics, and food processing industries. For instance, the syntheses of methanol and ammonia, the processes of hydrotreatment and hydrocracking, the production of reformulated gasoline, and the hydrogenation of oils and fats, along with fuel cell applications, benefit the utilization of hydrogen [1–3].

A great deal of interest has been attributed to hydrogen as an alternative fuel. Currently, the majority of the world's energy demand is supplied from fossil fuels (coal, petroleum, and natural gas). However, regarding the gradual increase of the energy use and the present consumption amounts, it is foreseen that these nonrenewable energy reserves will be depleted in the near future. What is more, the use of fossil fuels causes serious environmental problems due to the emission of air pollutants and greenhouse gases. These conditions encourage scientists to investigate renewable, sustainable, and environment-friendly alternative energy sources [4, 5].

Hydrogen is one of these potential candidates that have the ability to minimize humankind's dependence on fossil fuels and reduce the severe impact of environmental pollution. First and foremost, hydrogen is abundant, as it is the most plentiful element in the universe. Since the reaction of hydrogen with oxygen is very fast, it is considered to be an efficient fuel. Another point of emphasis should be made on the prevention of carbon emissions, as during the utilization of hydrogen as a fuel, the only combustion product is water [6, 7]. Moreover, hydrogen can be used in fuel cells for the purposes of transportation or electricity generation. Such applications with fuel cells result in considerably higher thermodynamic efficiencies than those of conventional internal combustion engines [2, 8]. At the same time, hydrogen is not an energy source, but an energy carrier. Though it cannot be found in its free molecular form, hydrogen is combined with other elements to form water, biomass, or hydrocarbons that make up fuels such as natural gas, petroleum, and coal [5, 6]. Therefore, hydrogen should be obtained from these aforementioned compounds via some biological, electrochemical, and thermochemical processes.

Biological hydrogen production processes include photosynthesis, biological water-gas shift reactions, and fermentative hydrogen production. The latter involves the employment of anaerobic (dark fermentation) or photoheterotrophic (light fermentation) microorganisms, during which carbohydrate-rich biomass is used. Hydrogen can also be produced via photosynthesis, with the aid of solar energy. The biological water-gas shift reaction, on the other hand, requires the employment of some specific heterotrophic bacteria that are able to perform water-gas shift reaction at ambient conditions [5]. The main advantages of these processes are their environmentally friendly nature and the lower amount of energy consumption when compared with other routes. However, biological hydrogen production processes are not adequate yet to meet the large amount of hydrogen demands in a hydrogen-oriented economy [2].

Electrolytic processes have the advantage of being simple and producing hydrogen with a high purity. However the energy consumption is very high and the production rate is limited in such processes, which in turn lead to high operation costs [6, 9].

Hydrogen production through thermochemical (reforming) processes includes pyrolysis, conventional gasification, and supercritical water gasification. Pyrolysis is the thermal degradation of carbonaceous matter into char, liquid, and gaseous products. The process is carried out in the absence of air or oxygen, at a temperature above 500 °C. The endothermic reaction proceeding during this process generates a hydrogen-rich gaseous product. Catalysts may also be used during the process [5]. Gasification, on the other hand, results in the formation of combustible gaseous products, with the use of air, oxygen, or steam as the gasifying agents. The process occurs at higher temperatures than pyrolysis, generally above 800 °C. However, the traditional gasification technology that involves the use of coal results in lower hydrogen amounts in the product gas and unwanted tar formations. What is more, the process necessitates the use of feedstock with moisture content generally less than 35 %. This, in consequence, leads to the increase of drying costs [5, 6]. On the other hand, the main advantage of the process is its flexibility with various feedstocks. Nevertheless, the steam reforming of natural gas is the most widely used and well-established process for hydrogen production in today's world [2]. But the process has significant drawbacks, as the catalysts used are usually overheated, sintered, or deactivated during the reforming reaction. Generally low thermal conductivities within the catalyst bed, the sulfurous or carbonyl groups present in the fuel, tar, and char formations, are the main reasons for these drawbacks [9].

Supercritical water gasification (SCWG) is a thermochemical hydrogen production process that has been attracting significant attention for the last three decades. The SCWG technology involves the employment of supercritical water (water above its critical point of 374.1 °C and 22.1 MPa) as the gasifying agent. The process benefits from the unique properties of water at these conditions, such as its low viscosity and dielectric constant, tunable dissolving power, increased thermal conductivity and diffusivity, and lower mass transfer resistance [9–11]. The SCWG process owns a significant potential for the conversion of hydrocarbons or biomass to produce hydrogen and other combustible gases. It also has major advantages when compared with other thermochemical hydrogen production processes. For instance, SCWG eliminates the necessity for the drying of the feedstock used in the process. The gasification efficiency and hydrogen selectivity are high, the formation of clean gaseous products is enabled, and lower amounts of tars and chars are produced [6, 10, 12].

All of the aforementioned hydrogen production processes have specific advantages of their own. Furthermore, both renewable and nonrenewable energy sources are important for the further development of these processes. However, to achieve a genuinely sustainable, economic, viable, and environmentally benign technology, hydrogen should be produced from renewable energy sources and through innovative production processes. One of these important energy sources is biomass. Biomass comprises all the living matter present on earth; hence it is considered as a renewable and abundant energy source. It offsets the environmental impact of greenhouse gas emissions, since biomass consumes atmospheric carbon dioxide during growth. Since biomass contains low amounts of sulfur, nitrogen, and metals, detrimental gas emissions during its combustion are in negligible amounts [8, 10, 12]. On account of these desirable characteristics of biomass, this chapter focuses on its conversion by an immense innovative thermochemical technology: supercritical water gasification. The oncoming sections include detailed information on the properties of supercritical water, and SCWG processes and studies published in the literature regarding hydrogen production through supercritical water gasification of biomass.

7.2 Supercritical Fluids and Supercritical Water

The idea of supercritical fluids first came forward at the beginning of nineteenth century, with the discovery of the critical point by Baron Charles Cagniard de la Tour. His studies led him to the determination of critical temperatures and pressures of various compounds. These investigations were carried forward, and the nature of the supercritical state was further debated by scientists such as Michael Faraday, Dmitri Mendeleev, and Thomas Andrews [13]. With the pioneering work of these scientists, the concepts of the critical point and the supercritical phase are elucidated in today's world.

A supercritical fluid is defined as any substance at a temperature and pressure above its critical point. The liquid and gaseous phases merge together and become indistinguishable at these conditions. At the same time, the properties of supercritical fluids are intermediate between those of liquid and gas phases. For instance, supercritical fluids have gas-like diffusivities and viscosities and liquid-like densities. Moreover, these properties are tunable, which means that they can be adjusted by changing the operating temperature and pressure to the desired values.

Owing to such benign properties, a wide range of applications benefit from supercritical fluids. These applications include pharmaceuticals, cosmetics, polymers, microelectronics, food sciences, powders, nano-systems, biotechnology, environment, and energy. The employment of supercritical fluids in these areas brings significant advantages, since most of these substances are nontoxic, noncarcinogenic, nonflammable, thermodynamically stable, and inexpensive [13]. Although most of the supercritical fluids have the aforementioned desirable properties, the most widely used one in industrial processes is supercritical water.

7.2.1 The Physical Properties of Supercritical Water

Being ecologically safe, abundantly available, and as the most important solvent in nature, water has very interesting properties at supercritical conditions. Below the

Table 7.1 Some physical properties of water at standard and supercritical conditions

	At standard conditions [17]	At subcritical conditions [17, 18]	At supercritical conditions [17, 18]	
Temperature (°C)	25	300	374	500
Pressure (MPa)	0.10	25	25	25
Density (kg/m^3)	997.05	743	322	89.86
Static dielectric constant	78.50	21.48	5.90	1.46
pK_w	14.00	11.12	13.84	16.97
Dynamic viscosity (Pa·s)	$89 \cdot 10^{-5}$	$9.2 \cdot 10^{-5}$	$5.1 \cdot 10^{-5}$	$3 \cdot 10^{-5}$

critical point, the liquid and gaseous phases are separated by the vapor pressure curve. However, the properties of these phases become increasingly similar and completely identical toward and at the critical point, respectively. The critical temperature (T_c) of water is 374.14 °C and its critical pressure (P_c) is 22.064 MPa [14, 15]. These relatively high values are due to the strong hydrogen bonds and consequently the strong interaction between the water molecules. Nevertheless, the transition from ambient to supercritical conditions increases both the kinetic energy of the water molecules and their intermolecular distances. Hence, it becomes very difficult to maintain the hydrogen bonds between the molecules, which in consequence make water to show a solvent behavior comparable to those of nonpolar fluids [16].

Due to these structural changes, supercritical water (SCW) has properties that are very different from those of ambient water. Some physical properties of water at standard and supercritical conditions are comparatively presented in Table 7.1. For instance, SCW has a much lower density than that of liquid water. The breaking of the hydrogen bond network reduces the barrier for the translational and rotational motions of the molecules. This triggers the self-diffusivity of water to increase with increasing temperature and decreasing density. A decrease in the density of water from 1000 to 100 kg/m^3 increases the diffusivity of water by an order of magnitude [16]. Moreover, SCW has low viscosity. Hence, these favorable transport properties make SCW an excellent medium for homogenous, fast, and efficient reactions [15].

An interesting change observed above the critical point is the great decrease of SCW's dielectric constant. The dielectric constant, or the static relative permittivity (ε) of water, determines its solution properties [2]. At standard conditions, water is poorly miscible with hydrocarbons and gases. On the contrary, due to its relatively high dielectric constant of 78.5, it is a good solvent for salts [15]. With increasing temperature and decreasing density, the dielectric constant of water begins to decrease, being in the range of 10 at the vicinity of the critical point. Consequently, SCW is completely miscible with organic compounds and gases, whereas it becomes a poor solvent for ionic species such as inorganic salts. This important property allows SCW to be a significant solvent for the homogenous reactions of organic compounds and gases. Moreover, due to the tunable solvent power of SCW with temperature and density, different operating conditions can be chosen to precipitate particles of a controlled size or structure [15, 16, 19].

Another unique property of SCW is its ionic product (K_w). At 25 °C, the ionic product of water is 10^{-14} mol^2/L^2 [20]. At the vicinity of the critical point, this value increases by almost three orders of magnitude, and the reason for this tremendous increase is explained by the self-dissociation of water to be endothermic [15]. Hence, higher concentrations of H_3O^+ and OH^- ions are formed in subcritical water and in supercritical water at high pressures, making water an acidic or basic catalyst precursor. This phenomenon, along with the not too low dielectric constant values, reinforces ionic reactions. In consequence, this region is generally preferred for synthesis or hydrolysis reactions [16, 20, 21]. Contrarily, above the critical point, since the water density and dielectric constant are low, the solvation power for ionic components decreases [15]. This causes a decrease in the ionic product. The prominent change in water properties at supercritical conditions causes the reaction pathways to change in character from ionic to free radical. In other words, free radical reactions dominate at high-temperature and low-density conditions [16, 21]. In epitome, due to the evident change of the ionic product, SCW supports either ionic or free radical reactions. This makes SCW an adjustable solvent, with respect to different conditions of temperature and pressure.

7.2.2 *The Role of Supercritical Water in Chemical Reactions*

The previously mentioned physical properties make SCW to have a significant role in chemical reactions. For instance, due to its low density at supercritical conditions, the intermolecular distances increase. Consequently, it becomes much easier for molecules to relocate and react with other molecules. The low viscosity and high diffusivity of SCW make it an excellent medium for fast and highly efficient chemical reactions. The low dielectric constant causes organic materials to have a very high solubility and complete miscibility in SCW. Thus, the reactions involving these materials proceed in a single homogenous phase and without being subjected to any interfacial transport limitations. Moreover, due to the tunable ionic product with respect to temperature and pressure, the selectivity of ionic or free radical reactions taking place in SCW can be adjusted. The work of Kruse and Dinjus [21] sets a good example for this phenomenon. The authors' study involving glycerol at a temperature of 395 °C and a pressure of 25–45 MPa revealed that at high pressures, the yields of acetaldehyde and formaldehyde as reaction products increased. On the contrary, during the reactions performed at low pressures, the yields of methanol and allyl alcohol increased. It was suggested that the latter products were formed through free radical reactions, whereas acetaldehyde and formaldehyde were formed by ionic reactions. In other words, the reason for the change in the yields of different products at different reaction conditions is the tunable ionic product of SCW.

During the reactions taking place in SCW, water is not only a reaction medium but also a reactant or a catalyst [15, 16, 21]. For example, water as a reactant leads to hydrolysis reactions. In such reactions, the breakdown of the polymeric structures of the reactants takes place, like the hydrolysis of cellulose to sugars. The main advantage of these hydrolysis reactions is the high solubility of intermediate products in SCW. Consequently, the reactions proceed with a high efficiency and unwanted formations such as chars and tars are inhibited [21–23]. Another example of water being a reactant is the water-gas shift reaction, in which carbon monoxide and water react to form carbon dioxide and hydrogen. In SCW media, the excess amounts of water may drive the equilibrium of the reactions to the forward direction. Thus, the reaction can proceed without the use of any catalysts and hydrogen generation is promoted by the water-gas shift reaction [22, 24].

Due to the high ionic product of water at the near-critical region and at supercritical region for high pressures, greater concentrations of H_3O^+ and OH^- ions are formed. This condition makes SCW an acidic or basic catalyst precursor. Hence, many chemical reactions that necessitate the use of a catalyst at normal conditions proceed without any catalysts in SCW. The dehydration of cyclohexene, the dehydration of lactic acid to acrylic acid, the Friedel-Crafts reactions that involve the reaction of aromatic compounds with some alcohols and alkenes, and the Cannizzaro-type reactions such as the production of ethanol and formic acid from formaldehyde can be given as examples for this phenomenon [2, 21, 25].

7.2.3 Gasification Reactions in Supercritical Water Media

About four decades ago, Sanjay Amin, who was a graduate student working at the Massachusetts Institute of Technology, was investigating the decomposition of organic compounds in hot water. Amin observed that if the experiments were performed in subcritical water, the reaction produced great amounts of chars and tars, in addition to hydrogen and carbon dioxide. However, when the experiments were conducted at conditions of supercritical water, the chars and tars entirely disappeared [26]. This uttermost discovery was further developed by Michael Modell from the same university, who provided the basis for the gasification reactions at supercritical water media.

Supercritical water gasification (SCWG) is a very efficient process during which biomass is converted to flammable gaseous products at conditions exceeding the critical temperature and critical pressure of water. Since hydrothermal gasification processes are performed at both the subcritical and supercritical conditions of water, SCWG is also called hydrothermal gasification. The SCWG process provides a novel and efficient approach, due to the very high solubility of organic materials in supercritical water. Consequently the reactions proceed in a single homogenous phase, without being subjected to any mass transfer limitations.

The gaseous product obtained in SCWG process involves hydrogen and carbon dioxide, along with methane, C_2–C_4 hydrocarbons (such as ethane, propane, and propylene), and carbon monoxide. The product gas composition is generally determined by the operating conditions. In this respect, depending on the temperature at which the process is employed, the SCWG process is generally categorized in two groups: high-temperature gasification and low-temperature gasification. High-temperature gasification is performed above a temperature of 500 °C and hydrogen production is favored during the process. Low-temperature gasification, on the other hand, is performed below 500 °C. Here, a methane-rich gaseous product is obtained. Hydrogen may also be formed by the aid of catalyst use, though the yields are lower due to the endothermic nature of the reforming reaction [2, 27, 28]. Even though the SCWG process is carried out at milder reaction conditions when compared with conventional steam reforming and pyrolysis reactions, the gasification efficiency is much higher [29].

Especially when low-temperature gasification is employed, the use of catalysts is very common in order to lower the activation energy and therefore reach the desired gasification efficiency, to increase the amount of the gaseous product, or to enhance the selectivity of the desired product. Catalysts can also be used to inhibit the polymerization of oily products or to increase the conversion of reaction intermediates to gaseous products, consequently reducing char and tar formations [27]. For these reasons, the use of homogenous catalysts such as KOH, NaOH, K_2CO_3, and Na_2CO_3 or heterogeneous catalysts such as activated carbon and metal-based catalysts like nickel, ruthenium, platinum, rhodium, and palladium used with various supports are very common in SCWG processes.

The advantages of SCWG can be summarized as below:

- Unlike conventional gasification or pyrolysis, the water content of the biomass feedstock used in SCWG does not constitute a problem. Therefore, the need for expensive and energy-consuming drying pretreatment steps is eliminated. During gasification reactions carried out in SCW media, biomass with water content up to 80 % can be employed [29–31].
- The formation of tars and chars is reduced, due to the increased solubility of reaction intermediates in supercritical water. Instead of polymerizing into such unwanted formations, these intermediates are transformed to gaseous products [21, 22, 30].
- The gaseous product has a very high solubility and complete miscibility in supercritical water. This enables the occurrence of single-phase reactions [2].
- Due to the much reduced mass transfer limitations and the occurrence of single-phase reactions, the reactions taking place in SCWG process are fast [2].
- Heteroatoms like sulfur and nitrogen leave the process in the aqueous effluent. This way, expensive gas cleaning steps are prevented [21, 31].
- A hydrogen-rich gaseous product is obtained during the process [21, 22].
- The gaseous product is not diluted with inert gases [31].
- The selectivity of the process toward a specific gaseous product, such as methane or hydrogen, can be controlled by adjusting the operating conditions [24, 31].
- The produced carbon dioxide can easily be separated, due to its high solubility in water [21, 31].
- The acquirement of a highly pressurized product gas prevents costly gas compression steps [31].
- Because of the high operating pressures, compact gasification systems can be developed. Moreover, the short reaction times lead to the utilization of smaller-sized reactors [2].

However, apart from these significant advantages, SCWG process has some disadvantages as well [2, 24, 31]:

- It may be hard to pump some biomass solutions and sludges in continuous systems.
- The employment of high temperatures and pressures may cause corrosion. Hence, the regarding constraint of using corrosive-resistant reactor materials increases the cost of the SCWG equipment.
- Apart from the high risk of corrosion, the severe operating conditions also require attentive operational safety throughout the gasification system.
- The endothermic nature of the reforming reactions leads to high energy consumptions. Therefore, a very effective heat exchanger design should be made to increase the energy efficiency of the system.
- The heating process should be made with extreme care, so that biomass solutions do not decompose and cause clogging before they reach supercritical conditions.
- The low solubility of inorganic substances may also cause clogging of the SCWG system.
- If supported heterogeneous catalysts are used in the system, the most significant drawback is the catalyst's lifetime. Such catalysts can be deactivated after several uses due to the char/tar formations on their surface or the presence of nitrogen/sulfur-containing compounds in the biomass feedstock. Moreover, the use of these catalysts may cause plugging of the SCWG reactor, since the reactor's cross-sectional area becomes much narrower than that of noncatalytic reactors.
- The problem of reactor plugging may also arise because of homogenous catalyst use, as the alkali salts have a low solubility in supercritical water. The high pH of the environment due to the presence of alkali also yields the possibility of reactor corrosion.

However, scientists in today's world continue to carry out their studies to solve these aforementioned problems and benefit from the significant advantages of the SCWG process.

7.3 Hydrogen Production by Supercritical Water Gasification

For the last three decades, scientists have been conducting and publishing studies on supercritical water gasification processes. The studies generally focus on energy production from biomass and biodegradable organic wastes, the simultaneous energy recovery and treatment of wastewaters, utilization of efficient catalysts, and challenges to overcome due to the previously mentioned disadvantages of SCWG. In this sense, with the current prospects of the hydrothermal processes, the pioneering work of scientists at Xi'an Jiaotong University (China), Institut für

Technische Chemie Forschungszentrum Karlsruhe (Germany), Hamburg University of Technology (Germany), Tohoku University (Japan), Hiroshima University (Japan), Korea Institute for Science and Technology (Republic of Korea), Paul Scherrer Institute (Switzerland), University of Twente (The Netherlands), Yıldız Technical University (Turkey), University of Leeds (UK), Pacific Northwest National Laboratory (USA), University of Michigan (USA), Auburn University (USA), and University of Hawaii (USA) is gratefully acknowledged. The studies conducted at the aforementioned institutions are elaborately discussed in the valuable reviews by Matsumura et al. [32] and Peterson et al. [33].

The biomass used during the investigations conducted at these institutions is generally in the form of model solutions. Such model biomass solutions include the building blocks of plant cells like cellulose, hemicellulose, lignin, and glucose; alcohols like methanol, ethanol, and glycerol; and ketones like acetone. Apart from the aqueous solutions of model biomass compounds, though smaller in number, there are also some studies regarding the SCWG of real biomass materials. These studies involve tobacco, corn, cotton, and sunflower stalks; plant residues such as corn cob, rice husk, and nut shells; industrial wastes such as sawdust, tannery wastes, and paper pulp sludge; domestic wastes such as sewage sludge; and industrial wastewaters such as winery wastes, cheese whey wastewater, textile wastewater, and olive mill wastewater.

These generally laboratory-scaled studies intend to investigate the gasification of biomass solutions at supercritical conditions and to elucidate the effect of different parameters on the reaction mechanism, biomass conversion, gasification efficiency, and gaseous product composition. The investigations revealed that the main parameters effecting SCWG of biomass are temperature, pressure, residence time, feedstock concentration, oxidant concentration, and catalyst use. Detailed information regarding these parameters and feedstocks used will be given in the oncoming sections. However, the complex and not fully illuminated nature of supercritical water gasification is the major challenge associated with this process. It is still not entirely possible to predict the effects of various parameters on the reactions and reaction products. Nevertheless, possible reactions taking place during SCWG of biomass are summarized in Table 7.2. Accordingly, the primary reactions taking place during SCWG processes are endothermic reforming reactions, exothermic methanation reactions, and the water-gas shift reaction, along with coke gasification and Boudouard coking.

7.3.1 *Influence of Process Parameters on Hydrogen Production*

As shown in Table 7.2, some of the reactions taking place during the SCWG of biomass are reversible reactions. This condition necessitates the optimization of the process parameters to obtain the maximum hydrogen yield. The main parameters

Table 7.2 Possible reactions taking place during SCWG of biomass

Reaction	$\Delta H_{298\ K}$ (kJ/mol)
Decomposition of organics: $C_nH_mO_p + \frac{n}{2}(H_3O^+ + OH^-) \rightarrow (n+p)CO + \left(\frac{m}{2}+n\right)H_2$ + intermediate products(C_nH_m)	
Water-gas shift reaction: $CO + H_2O \leftrightarrow CO_2 + H_2$	−41.1
Methanation of CO: $CO + 3H_2 \leftrightarrow CH_4 + H_2O$	−206.2
Methanation of CO_2: $CO_2 + 4H_2 \leftrightarrow CH_4 + 2H_2O$	−165.0
Methanation of C: $C + 2H_2 \leftrightarrow CH_4$	−75.0
Reforming of methane: $CH_4 + H_2O \leftrightarrow 3H_2 + CO$	206.0
Boudouard coking (CO disproportionation): $2CO \leftrightarrow C + CO_2$	−172.4
Coke gasification: $CO + H_2 \leftrightarrow C + H_2O$	−131.3

affecting SCWG of biomass are temperature, pressure, residence time, feedstock concentration, oxidant concentration, and catalyst use. Hence, this section evaluates the effect of each parameter on hydrogen production, comparatively referring to the studies published in literature.

7.3.1.1 Temperature

Temperature is one of the most important factors affecting the gasification reactions taking place in SCW media. The relationship of the reaction rate to temperature in Arrhenius equation causes higher temperatures to be more suitable for thermal degradation reactions. For instance, in their study regarding methanol reforming in supercritical water, Boukis et al. [34] reported that a methanol conversion of 22 % at 400 °C increased up to 99.5 % at 600 °C. Another study with methanol yielded similar results, stating that a temperature increase from 500 to 700 °C caused high methanol conversions [35]. Goodwin and Rorrer [36] investigated the SCWG of xylose as a form of hemicellulose between 450 and 650 °C, at a pressure of 250 bar. The authors reported that a xylose conversion of about 30 % at 450 °C increased above 99 % at 650 °C.

The combined steam reforming and water-gas shift reactions have a highly endothermic nature. Thus, a very high external energy is needed to drive the equilibrium reaction to form hydrogen. According to Le Chatelier's principle, high reaction temperatures shift the equilibrium of the reaction to the forward direction, thus promoting the production of hydrogen. On the other hand, low reaction temperatures favor the production of methane [2, 26]. A lot of published papers give countenance to this fact. Susanti et al. [9] examined the SCWG of isooctane at 601–676 °C. The authors stated that the hydrogen gas yield increased from 0.93 to 2.31 mol/mol isooctane with an increase in temperature from 601 to 676 °C and that this was related with the combined steam reforming and water-gas shift reaction to be endothermic. Ding et al. [10] obtained similar results regarding their work on the hydrothermal gasification of cellulose at 400–550 °C. For a water to biomass ratio of 3:1, the hydrogen yields obtained at 400 °C and 550 °C were 0.50 mmol/g biomass and 0.95 mmol/g biomass, respectively. In their study on the SCWG of methanol at 550–700 °C and 27.6 MPa, Taylor et al. [35] stated that the amount of hydrogen in the dry gas composition increased with increasing temperature. In another study investigating the SCWG of xylose at 450–650 °C and 250 bar, per each mol of xylose fed, the amount of hydrogen generated was reported to increase [36]. Susanti et al. [27] investigated the hydrothermal gasification of glucose at a pressure of 25 MPa. The experimental results unveiled that as the reaction temperature was increased from 650 to 767 °C, the amount of hydrogen in the gaseous effluent increased from 68 % to 71.4 % and the hydrogen yield increased from 7.9 to 11.5 mol/mol glucose. In another study with glucose at 25 MPa conducted by Jin et al. [37], a temperature increase from 550 to 650 °C resulted in an increase of hydrogen fraction from 14.68 % to 29.30 % and hydrogen yield from 4.425 to 8.939 mol/kg. Therdthianwong et al. [38] examined hydrogen production by ethanol reforming in SCW. The authors stated that both ethanol decomposition and ethanol reforming were favorable at high temperatures. Accordingly, the reactions carried out at 600 °C showed greater hydrogen selectivities than those carried out at 500 °C. Guo et al. [39] investigated the reactions of indole, which is a product from hydrothermal processing of algal biomass, under SCW conditions. The experiments conducted between 550 and 700 °C showed that the highest hydrogen molar yields were achieved at 700 °C. Liu et al. [40] obtained similar results on their paper regarding the hydrothermal gasification of glycerol between 650 and 800 °C.

The literature studies performed with real biomass present congenerous hydrogen tendencies with temperature. For instance, Zhang et al. [41] investigated energy recovery from sewage sludge with SCW treatment at 400–550 °C. Accordingly the experiment conducted at 550 °C showed an almost tenfold increase in the yield of hydrogen, reaching 14.5 mol/kg dried sludge, when compared to that at 400 °C. Kıpçak et al. [42] found similar results in their work regarding the hydrothermal gasification of olive mill wastewater in SCW conditions. The authors' experiments were performed at five different temperatures (400, 450, 500, 550, and 600 °C), for five different reaction times (30, 60, 90, 120, and 150 s), at a constant pressure of 25 MPa. The change in the hydrogen content of the gaseous product with respect to reaction temperature and time is presented in Fig. 7.1. As it can be seen from Fig. 7.1, the hydrogen content of the gaseous product increased dramatically with temperature, especially for low reaction times, and the maximum amount of hydrogen (about 11 %) was encountered at 600 °C.

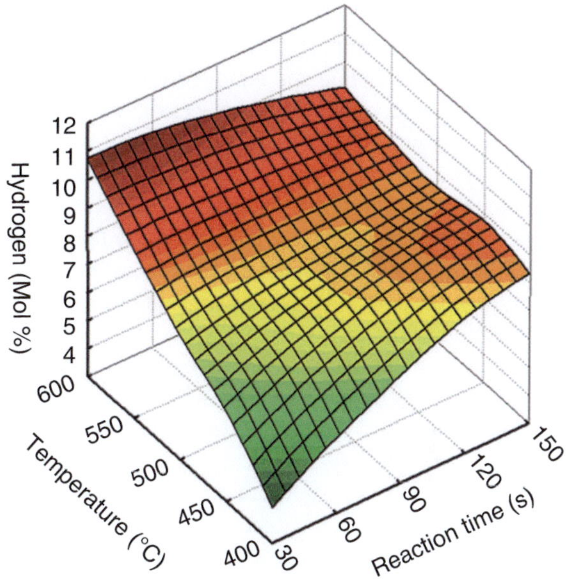

Fig. 7.1 Change in the hydrogen content of gaseous product with respect to reaction temperature and time for the hydrothermal gasification olive mill wastewater (Reprinted with permission from Kipçak et al. [42], copyright 2011, Elsevier)

At the same temperature range and pressure as the aforementioned paper, Cao et al. [43] investigated hydrogen production from SCWG of black liquor in a continuous flow system and drew likewise conclusions. Their experimental results showed that temperature had an enormous influence on the SCWG of black liquor. Both the total gas and hydrogen yields were almost doubled, hydrogen yield increasing from 6.82 to 11.26 mol/kg. The authors suggested that the alkali present in black liquor promoted the completion of the water-gas shift reaction, thus increasing the generation of hydrogen. Lu et al. [44] reported that the most effective operating parameter on hydrogen yield during the SCWG of corn cob was temperature. Another study on hydrogen production by sewage sludge gasification in SCW with a fluidized bed reactor, performed by Chen et al. [45], had similar conclusions. The regarding experiments were made at 480–540 °C and at a pressure of 25 MPa. The authors reported that as the temperature was increased from 480 to 540 °C, the molar fraction of hydrogen was increased from 40.13 % to 43.6 %, and the hydrogen yield was increased from 6.74 to 9.26 mol/kg.

7.3.1.2 Pressure

The effect of pressure on SCWG is mainly influenced by the role of water during chemical reactions. The previously mentioned physical properties of SCW, such as the density, dielectric constant, and ionic product, increase with increasing pressure. Hence, at elevated pressures, the rates of ionic reactions like hydrolysis are enhanced. The increased density at high pressures causes SCW to be an efficient energy transfer agent and a collision partner in intermolecular energy transfer steps, thus promoting such reactions. Moreover, the major steps that generate a typical chemical reaction are the diffusion of the reactant molecules to each other and their chemical transformation and the diffusion of the reaction products away from each other. The primary factor that affects the diffusion of a solute in a solvent is its diffusivity, which is highly dependent on the solvent's viscosity. And the solvent's viscosity is a parameter that is highly dependent on pressure. Hence, the chemical reaction rates are strongly dependent on the operating pressures, since the diffusion taking place at high pressures is hindered by the formation of a solvent cage around the solute molecules. This phenomenon is called the "cage effect." As a result of the cage effect, the reaction products are kept in a cage formed by the solvent molecules and this condition inhibits fission-type reactions like decomposition. The product molecules that cannot leave the cage show a tendency to either merge together or transform into reactants. The cage created by the solvent molecules also isolates the reactant molecules. This condition promotes solute-solvent interactions (such as hydrolysis or water-gas shift reaction) rather than solute-solute interactions. Consequently, high pressures are favorable for reactions like the water-gas shift reaction; but at the same time, decomposition reaction rates are reduced [46].

The influence of the cage effect is used by some papers in order to explain the effect of operating pressure on hydrogen yield during the SCWG of biomass. For instance, Sricharoenchaikul [46] studied black liquor gasification in supercritical water. The experiments performed at 220, 300, and 400 atm resulted with hydrogen fractions of 15.53 %, 13.37 %, and 10.69 % in the gaseous effluent, respectively. On the other hand, the amount of the total gaseous products and the carbon conversion remained relatively stable irrespective of pressure, indicating the hindrance of decomposition reactions due to the cage effect. Karakuş et al. [47] investigated the effect of operating pressure in their study on the catalytic SCWG of 2-propanol. The experiments performed at 400 °C and 5–25 MPa showed an increasing hydrogen trend with pressure, as presented in Fig. 7.2. Accordingly, the hydrocarbon content of gaseous product such as methane, ethane, and propane decreased with pressure, whereas the hydrogen content increased. This increase was especially distinctive for the experiments performed with Pt/Al_2O_3 catalyst, yielding an increase of hydrogen from 45 % at 5 MPa to 95 % at 25 MPa. The authors attributed this increment to two significant factors: firstly, high pressures favoring the water-gas shift reaction and, secondly, the increase in the ionic product of water with pressure. They stated that the high ionic product of SCW formed greater concentrations of H_3O^+ and OH^- ions, which enhanced the hydrolysis of 2-propanol to yield carbon monoxide and hydrogen.

Gadhe and Gupta [48] also obtained similar hydrogen tendencies with pressure in their study on methanol reforming in supercritical water. The experiments performed at 700 °C showed that as the pressure was increased from 69 to 276 bar, the molar hydrogen yield (mol of hydrogen formed per mol of methanol fed) increased from 0.35 to 1.23.

On the other hand, the rather complex and not fully illuminated nature of SCW prevents the creation of a general conclusion on pressure effects. With reference to this fact, some published papers mention that pressure has a negligible effect on the

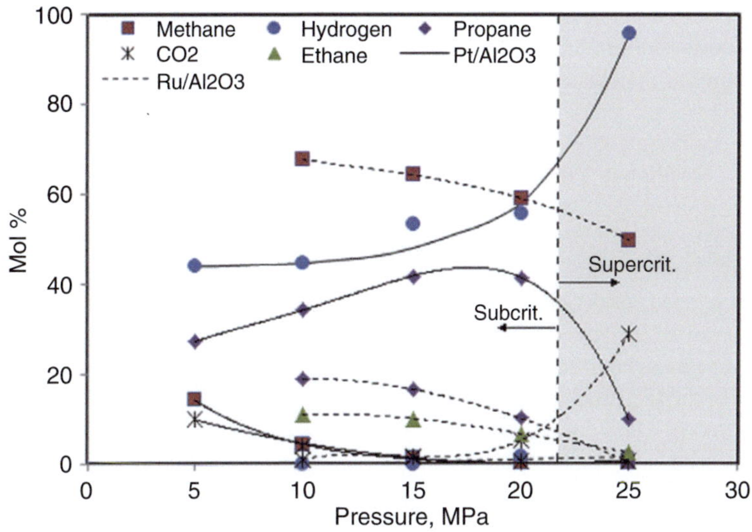

Fig. 7.2 Influence of reaction pressure on gaseous effluent contents for the catalytic supercritical water gasification of 2-propanol (temperature, 400 °C; reaction time, 10 s; feed concentration, 0.5 M) (Reprinted with permission from Karakuş et al. [47], copyright 2013, Hydrogen Energy Publications)

formation of hydrogen. For example, Kıpçak et al. [42] found similar results in their work on the hydrothermal gasification of olive mill wastewater. To understand the effect of pressure, a series of experiments were performed at 550 °C and for a residence time of 60 s. As the pressure was increased from 100 to 300 bar, the hydrogen content of the gaseous effluent did not change significantly. But the authors preferred not to the use low pressures, since smut formations due to carbonization were observed at such conditions. They reported that the carbon content of the olive mill wastewater settled to the reactor inner wall as smut, instead of transforming into gaseous products, which caused system clogging with time. Boukis et al. [34] noted that the variation of pressure from 25 to 45 MPa showed a negligible effect on the gas composition during the SCWG of methanol at 600 °C. In another paper on the hydrothermal gasification of glucose at 23, 25, and 27 MPa, it was concluded that the employed pressures had no significant effect on the hydrogen yields [37]. However, it was mentioned that subcritical pressures had an adverse effect on hydrogen production, due to the properties of water at these conditions. At the same time, too high pressures caused trouble for the design and maintenance of the system. Hence, the paper remarked that the ideal operation pressure was about 25 MPa due to the experimental evidence. In their study on ethanol reforming in SCW, Byrd et al. [49] reported likewise hydrogen tendencies with pressure. The experiments conducted at 700 °C showed that an increase of pressure from 221 to 276 bar resulted in a nearly constant hydrogen yield. D'Jesus et al. [50] investigated the influence of process variables during the gasification of corn silage in SCW. The authors pointed out that the amount of gaseous product did not vary with an increase of process pressure from 250 to 400 bar. In another study of the same authors on the gasification of corn and clover grass in supercritical water, similar conclusions were obtained [51].

7.3.1.3 Residence Time

The effect of residence time on hydrogen production during SCWG depends explicitly on other process parameters like temperature, pressure, and the kind of feedstock used. Short-chained hydrocarbons require shorter residence times for their complete conversion than those of longer chained hydrocarbons. Hence, the residence times required to obtain maximum hydrogen production vary greatly with respect to the feedstock properties, such as the chain length, the chemical structure, and the oxygen content [2]. If excessively long residence times are used, hydrogen may undergo subsequent reactions such as methanation, which in consequence can reduce its eventual yield. Moreover, the cited literature studies unveil that model biomass solutions are gasified in shorter residence times, whereas much longer residence times are required for the complete conversion of real biomass with complex compositions. Thus, in conclusion, the residence times employed during SCWG processes may increase or decrease the hydrogen yields.

The findings of the literature studies that investigate the effect of residence time support the fact that the residence time may increase or decrease the hydrogen yields. Kıpçak et al. [42] investigated the effects of five different reaction times during the hydrothermal gasification of olive mill wastewater at 400–600 °C and 25 MPa. As shown in Fig. 7.1, at low reaction temperatures, the increase of reaction time from 30 to 150 s promoted the hydrogen fraction in the gaseous product. On the contrary, the fraction of hydrogen was inversely proportional with reaction time at higher temperatures. The authors interpreted this condition as the degradation of the formed hydrogen to the final reaction products at elevated temperatures.

Another study on the catalytic gasification of 2-propanol at 400–600 °C and 25 MPa showed similar results [3]. Figure 7.3 presents the change in the hydrogen yield with respect to reaction time for the employed catalysts at different temperatures. Generally, the hydrogen yield exhibits an increasing trend as the reaction time increases from 10 to 30 s. It should be noted that for high temperatures and for the Fe-Cr catalyst that showed higher hydrogen selectivity, there was a prominent increment in the hydrogen yield as the reaction time was increased from 10 to 20 s. However, as the reaction time was further increased to 30 s, the hydrogen yields attained an almost constant value, as an evidence for the decomposition of the formed hydrogen to other gases with the enhancement of temperature.

There are some other studies in literature that show increasing hydrogen formation with residence time. Susanti et al. [9] investigated the effect of residence time for the SCWG of isooctane at a fixed temperature of 632 °C and a fixed pressure of 25 MPa. As the residence time was increased from 6 to 33.3 s, almost a fivefold increase in the hydrogen gas yield from 1.14 to 5.52 mol/mol isooctane was encountered. According to the authors, the increase in hydrogen gas yield at longer

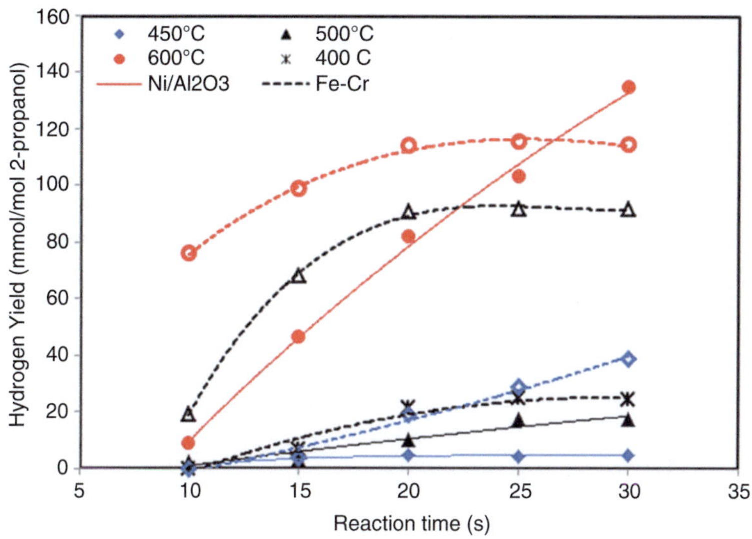

Fig. 7.3 Change of hydrogen yield with reaction time for the catalytic supercritical water gasification of 2-propanol (Reprinted with permission from Akgün and Kıpçak [3], copyright 2014, Elsevier)

residence times indicated the need of an extended period of time for the completion of the reaction. Goodwin and Rorrer [36] also reported an increased hydrogen yield with residence time in their work on the SCWG of xylose at 450–650 °C and 250 bar. Lu et al. [44] investigated the influence of different process parameters on hydrogen production by SCWG of corn cob. The experimental results exhibited an increased hydrogen yield as the residence time was increased from 20 to 40 s. The authors state that a more complete gasification of biomass would be realized with longer residence times and that when compared with the effects of temperature and pressure, residence time had no significant effect on hydrogen yield, since the SCWG reaction rate was very rapid in the range of experimental operating parameters. In another article on black liquor gasification at SCW conditions, the effect of residence time was analyzed at 650 °C and 300 bar [46]. It was reported that the increase of residence time from 5 to 120 s enhanced the hydrogen yields. In their study on ethanol reforming in SCW conditions, Arita et al. [52] examined the time dependence of the gaseous product yields at 500 °C. The results showed that the molar yield of hydrogen increased as the residence time was increased from 10 to 30 min.

On the other hand, there are some studies that report a first increasing and then decreasing hydrogen yield with residence time. For instance, Byrd et al. [49] investigated the effect of residence time on hydrogen production by ethanol reforming in SCW. During the experiments performed at 800 °C and 221 bar, an increase of residence time from 1 to 4 s promoted the hydrogen yields. However, further increase of the residence time had an adverse effect on the hydrogen yields.

The authors noted that at high temperatures, low residence times could prevent the methanation reactions from reaching equilibrium. This made it possible to significantly limit the extent of methanation and the consequent hydrogen loss by operating at low residence times. Boukis et al. [53] also obtained similar hydrogen trends with residence time. In their work on methanol reforming in SCW for hydrogen production, the effect of residence times varying between 3 and 100 s was investigated at 600 °C and 25 MPa. The initial increase in the hydrogen yield was followed by a steady decrease, as the residence times were elongated. This effect was attributed to the methanation reaction of carbon monoxide and hydrogen, as methane formations exhibited a linear increase as a function of time. Susanti et al. [54] analyzed the noncatalytic gasification of isooctane in SCW. The experiments were conducted at 763 °C and 25 MPa, for a residence time of 60–120 s. The authors reported that when the residence time was increased from 60 to 91 s, a significant increase in the hydrogen gas yield from 7.7 mol/mol isooctane to 12.4 mol/mol isooctane was observed. However, the hydrogen yield decreased to 11.8 mol/mol isooctane when the residence time was 120 s. On the contrary, just like the case in the previously mentioned article, long residence times were seen to favor methane formations. Thus, the decrease in the hydrogen yield and the increase in the methane yield at long residence times were interpreted as the indication of hydrogen consumption by the methanation reactions.

7.3.1.4 Feedstock Concentration

Feedstock concentration is one of the important operating parameters that have a major impact on the yield and composition of the SCWG products. This impact is generally attributed to the water-gas shift reaction. If low feedstock concentrations are used, the excess amount of water may shift the water-gas shift reaction to the forward direction. This, in consequence, promotes the hydrogen and carbon dioxide yields in the gaseous product. Moreover, excess amount of water may also shift the methanation reactions of carbon monoxide and carbon dioxide (Table 7.2) to the backward direction, thus decreasing the methane yield in the gaseous product.

In the literature, especially studies performed with model biomass solutions support the aforesaid phenomena. Many articles point out the greater gas yields and organic matter conversions when low feedstock concentrations are used. On the other hand, at high feedstock concentrations, a tendency to form char and tar is also emphasized due to the polymerization reactions. For example on their work on the SCWG of isooctane, Susanti et al. [9] reported that as the feedstock concentration was increased from 5.7 to 32.7 wt.%, the total gas yield decreased from 2.13 to 1.27 L/g isooctane and the total organic carbon content of the liquid product increased significantly from 311.3 to 1600.5 ppm. In another study of Susanti et al. [27], it was stated that as the glucose feed concentration was increased from 2 to 15 wt.%, the total gas yield decreased from 2.04 to 1.41 g/L and the total

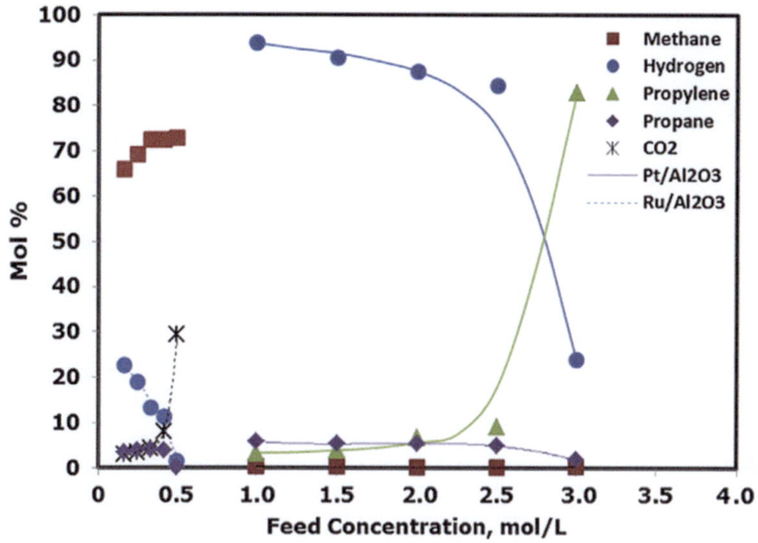

Fig. 7.4 Influence of feedstock concentration on the gaseous effluent contents during the catalytic supercritical water gasification of 2-propanol (temperature, 400 °C; reaction time, 10 s; pressure, 25 MPa) (Reprinted with permission from Karakuş et al. [47], copyright 2013, Hydrogen Energy Publications)

organic carbon present in the liquid effluent increased from 20 to 41 ppm. The authors also stated that gasification experiments could not be performed for glucose concentrations higher than 20 wt.%, due to the plugging of the feed preheater line. In their study on methanol reforming in SCW, Boukis et al. [34] reported that the methanol conversion decreased dramatically for feedstock concentrations greater than 50 wt.%. In another study performed with methanol, while methanol was completely reacted for 15 and 25 wt.% feeds, incomplete conversion was observed for a feedstock concentration of 45 wt.% [35]. Therdthianwong et al. [38] examined the effect of water to ethanol ratio on ethanol conversions during SCWG. Accordingly, an increase in ethanol conversion was encountered for increasing water to ethanol ratios. The authors also observed that at high ethanol concentrations, the formation of coke, char, and high-molecular-weight carbonaceous species plugged at the sensor of the reactor's back pressure regulator.

Literature studies also mention lower hydrogen formation with increasing feedstock concentration. For instance, Karakuş et al. [47] examined the catalytic SCWG of 2-propanol and a series of experiments were performed at 400 °C, 25 MPa and 10 s, in order to comprehend the effect of feedstock concentration. The tested 2-propanol concentrations ranged from 0.17 to 0.5 mol/L and 1 to 3 mol/L. The experimental results are presented in Fig. 7.4. Accordingly, the fraction of hydrogen in the gaseous product decreased with increasing 2-propanol concentrations for each of the catalysts used. This decrease was attributed to the inhibition of the water-gas shift reaction due to the high concentrations of the reactants in the reaction medium and the authors suggested the employment of low feedstock concentrations in order to obtain a gaseous effluent rich in hydrogen.

Susanti et al. [9] investigated the SCWG of isooctane at 630 °C and 25 MPa, for a residence time of 18 s. As the feed concentration increased from 5.7 to 32.7 wt.%, the amount of hydrogen in the produced gas decreased from 56.5 % to 40.1 %. Moreover, an almost threefold decrease in the hydrogen gas yield was encountered from 4.90 to 1.72 mol/mol isooctane. Based on equilibrium point of view, the authors stated that the lower amount of water in the feed shifted the isooctane steam reforming and the water-gas shift reactions to the backward direction, thus decreasing the hydrogen and increasing the carbon monoxide yields. The equilibrium of the carbon monoxide methanation reaction then shifted to the forward direction, increasing the methane yields. Hence, the authors pointed out decreased hydrogen formations for higher feedstock concentrations. Susanti et al. [27] also investigated the SCWG of glucose solutions at 740 °C and 25 MPa, with a reaction time of 60 s. The increase of glucose feed concentrations from 2 to15 wt.% caused a decrease in the hydrogen gas yield from 11.2 to 5.7 mol/mol glucose. Ding et al. [10] examined the SCWG of cellulose for different water to biomass ratios ranging from 3 to 7 and for different reaction temperatures ranging from 400 to 550 °C. When the water to biomass ratio was increased from 3 to 7, the hydrogen yield increased by 44 %, 11 %, and 22 % at 400, 470, and 550 °C, respectively. The authors mentioned that at higher gasification temperatures, the structure of water changed and led to a reduction in both intramolecular and intermolecular hydrogen bonding. Therefore, they suggested that water could provide hydrogen and consequently increase the hydrogen yield. During their study on methanol reforming in SCW, Boukis et al. [34] mentioned that hydrogen production declined significantly for methanol concentrations greater than 50 wt.% for experiments carried out at 600 °C, 25 MPa, and 15 s. Another article on methanol reforming at 700 °C had similar results, stating that as the methanol concentration increased from 15 to 45 wt.%, the hydrogen production notably decreased [35]. Therdthianwong et al. [38] investigated the SCW reforming of another alcohol, ethanol, at 500 °C and 25 MPa. The experimental results indicated an enhanced water-gas shift reaction with increasing water to ethanol ratios, thus greater amount of hydrogen yields. Another study on the SCWG of glycerol pointed out a dramatic decrease in the hydrogen yield as the glycerol feedstock concentration increased from 5 to 45 wt.% [40]. Cao et al. [43] studied the influence of black liquor concentration ranging from 3 to 9.5 wt.% at 550 °C, 25 MPa with a 7.42 s residence time. Accordingly, the dilution of black liquor favored hydrogen formations. During the SCWG of 3 wt.% black liquor, maxima in both the hydrogen fraction (61.02 %) and the hydrogen yield (27.55 mol/kg) were achieved. Chen et al. [45] had similar conclusions in their work regarding the SCWG of sewage sludge. The experiments performed at 540 °C and 25 MPa showed that as the concentration of sewage sludge was increased from 4 to 12 wt.%, the molar fraction of hydrogen decreased gradually. The article specified that high reaction temperatures and low feedstock concentrations were favorable for hydrogen production, which could enhance the steam reforming and water-gas shift reactions better. Yu et al. [55] investigated the effect of glucose concentration at 600 °C and 34.5 MPa, for a reaction time of 34 s. The authors reported that the hydrogen yield decreased by a factor of two as the reactant concentration increased from 0.1 to 0.8 M.

7.3.1.5 Oxidant Concentration

Some published articles investigate the effect of oxidant use, which is in concentrations less than the theoretical oxygen amount required for the complete decomposition of the organic matter content of a biomass. Generally hydrogen peroxide is the employed oxidant in such partial oxidation reactions. For instance, Susanti et al. [9] investigated the effect of partial oxidation during the SCWG of isooctane, with the aid of hydrogen peroxide as the oxidant. The experiments were performed at a temperature of 637 °C, pressure of 25 MPa, residence time of 18 s, and feedstock concentration of 9.9 wt.%. As the hydrogen peroxide concentration was increased from 0 to about 2.7 mol/L, the total gas yield increased from 2.08 to 2.94 L/g isooctane. Hydrogen and carbon monoxide gas yields were seen to reach their maximum values at medium hydrogen peroxide concentrations. As the oxidant concentration was increased from 0 to 2.7 mol/L, the hydrogen gas yield also increased from 4.00 to 6.13 mol/mol isooctane, which then decreased to 4.56 mol/mol isooctane with a further increment of oxidant concentration to about 4.5 mol/L. It was concluded that high oxidant concentrations in SCW conditions resulted in complete oxidation of isooctane to form carbon dioxide and water. Williams and Onwudili [29] examined the impact of hydrogen peroxide in various concentrations on the SCWG of glucose. The authors reported that in the absence of any hydrogen peroxide, there was a large amount of carbonization. However, an increase in the concentration of hydrogen peroxide decreased the amounts of char, oil, and water-soluble products while increasing the gas yields. Due to the presence of oxygen in the system, an increment in carbon dioxide concentration was observed. On the other hand, the amount of hydrogen increased to a point and then started decreasing for oxidant concentrations above 6 wt.%. It was pointed out that partial oxidation was selective for hydrogen production, whereas excess amounts of oxidant caused direct oxidation to carbon dioxide. Jin et al. [37] had similar observations on the SCWG of glucose and lignin at 600 °C and 25 MPa. The employed oxidant was hydrogen peroxide, used in concentrations of 0–40 % of the chemical oxygen demand of the biomass solution. Accordingly, the gasification efficiencies increased with respect to the increase in the added oxidant amount. Moreover, the hydrogen yields showed a first increasing and then decreasing trend with oxidant concentration. It was reported that hydrogen peroxide in SCW formed $OH^{\bullet}$ free radicals and the presence of these free radicals greatly decreased char and tar formations, thereby increasing the gasification efficiencies.

Another study with isooctane at 764 °C and 25 MPa, with a residence time of 105 s, investigated the effect of various oxygen equivalent ratios (ER) ranging from 0 to 0.3 [54]. With the aid of hydrogen peroxide as the oxidant, it was seen that the total gas yield decreased from 4.22 to 3.73 L/g when the ER increased from 0 to 0.3, indicating that partial oxidation did not promote gas production. The hydrogen gas yield on the other hand showed a very slight increase from 12.4 to 12.6 mol/mol isooctane as the ER increased from 0 to 0.1. Further increase of the ER to 0.3 decreased the hydrogen gas yield to 10.1 mol/mol isooctane. The authors attributed this to the high reaction temperatures and stated that it may therefore be unnecessary to use partial oxidation reactions to enhance hydrogen gas yields when high temperatures, long residence times, and low feedstock concentrations are used. Xu et al. [56] studied the influence of oxidation coefficient on product properties in sewage sludge treatment by SCW. The experiments performed at 400 °C and 25 MPa, with a residence time of 2.5 min examined hydrogen peroxide oxidation coefficients (practically added oxidant amount/theoretically required oxidant amount, denoted with "n" in the article) ranging from 0 to 2.5. Hydrogen gas yields were seen to initially increase and then decrease for $0 \leq n \leq 1$. The maximum hydrogen yield was encountered for the oxidation coefficient of 0.6. Hence, it was concluded that small amounts of oxidant helped to generate hydrogen, but excessive amounts caused the gaseous products to be converted to carbon dioxide and water. It was also concluded that the partial oxidative gasification processes at supercritical conditions could be used to produce hydrogen while simultaneously and harmlessly dealing with sewage sludge. A similar conclusion was also drawn by Kıpçak and Akgün [57] in their work on the oxidative gasification of olive mill wastewater in supercritical water. The partial oxidative gasification experiments were performed with hydrogen peroxide as the oxygen source, used in concentrations ranging between about 9 and 182 mmol/L. The experiments were carried out between 400 and 700 °C, at a constant pressure of 25 MPa and for a reaction time of 30 s. It was seen that the use of an oxidant caused an almost twofold increase in the gasification yields when compared with the results obtained for hydrothermal gasification experiments [42]. While the maximum gasification yield was 7.1 mL/mL wastewater for the case of hydrothermal gasification, it was about 15 mL/mL wastewater for partial oxidative gasification. On the other hand, increasing oxidant concentrations were seen to have an adverse effect on hydrogen production, which was more prominent at high reaction temperatures (Fig. 7.5).

As it can be seen from Fig. 7.5, the increase of oxidant concentration from about 9 to 182 mmol/L decreased the hydrogen fraction from 8.43 % to 4.34 % at 650 °C and from 9.42 % to 0.14 % at 700 °C. The authors explained this phenomenon by the degradation of the formed hydrogen into the final reaction products of carbon dioxide and water, both through the impacts of temperature and oxidant concentration. It was concluded that the use of oxidant concentrations corresponding to 1–2 % of the chemical oxygen demand of the wastewater was sufficient for the energy recovery and simultaneous treatment of olive mill wastewater.

7.3.1.6 Use of Catalyst

In SCWG processes the use of catalysts is very common for several significant purposes: to lower the activation energy to reach complete gasification, to convert biomass feedstock into intermediates that can be more readily gasified, to reduce char and tar formations, to reach the desired gasification efficiency, and to enhance the selectivity of the desired gaseous product. For these reasons, a lot of studies have

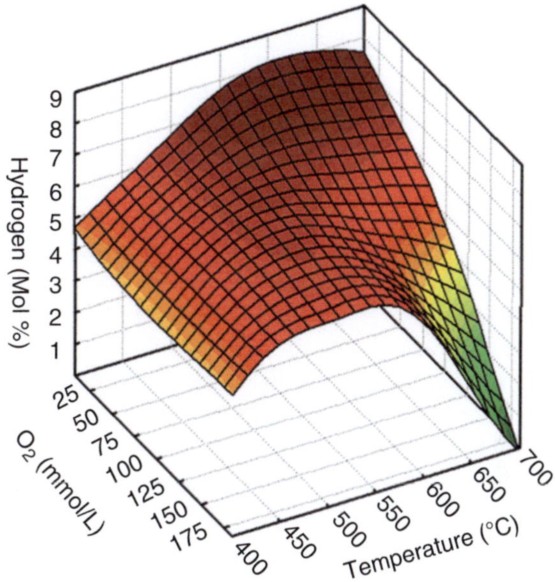

Fig. 7.5 Change in the hydrogen content of gaseous product with respect to oxidant concentration and temperature during the partial oxidative gasification of olive mill wastewater (reaction time, 30 s; pressure, 25 MPa) (Reprinted with permission from Kıpçak and Akgün [57], copyright 2012, Elsevier)

been made with the aim of determining efficient, stable, and inexpensive catalysts that can be used in SCWG processes. The commonly investigated catalysts are either homogenous alkali catalysts or heterogeneous metal-based catalysts.

7.3.1.6.1 Alkali Catalysts

The main homogenous alkali catalysts employed during the SCWG of biomass are NaOH, KOH, Na_2CO_3, and K_2CO_3 [24, 58]. The primary importance of these alkali catalysts arises from the enhancement of hydrogen yield through their use by promoting the water-gas shift reaction [24, 31, 59]. Furthermore, the use of alkali catalysts may lead to the formation of carbonates from their reaction with carbon dioxide. The decrease of carbon dioxide in the gaseous effluent may also cause the equilibrium of the water-gas shift reaction to the forward direction, enhancing hydrogen formation. Moreover, these catalysts commonly promote the splitting of the C-C bonds, thereby increasing the gasification efficiency and reducing char and tar formations [60]. The alkali addition may also increase the reactor wall activity to catalyze the SCWG reactions. However, the use of these catalysts has three main drawbacks. Firstly, it is difficult to recover these catalysts from the reactor's effluent, which is a considerable problem in terms of economic point of view [59, 60]. Secondly, the increased pH of the solution due to alkali addition may corrode the reactor walls [2, 60]. And thirdly, due to their low solubility in SCW, alkali catalysts have the possibility to precipitate in the reactor system, which in turn may cause plugging issues [24, 31]. Hence, the aforementioned drawbacks require greater utilization and investigation of alkali catalyst use.

NaOH

NaOH is one of the most commonly employed alkali catalysts. Watanabe et al. [61] investigated the effect of NaOH addition on the SCWG of glucose and cellulose at 673 K and for a reaction time of 15 min. Accordingly, NaOH improved the hydrogen yield significantly for both biomass solutions. While the hydrogen yields were about 2 % for experiments performed without any catalysts, they increased above 20 % for glucose and 10 % for cellulose with the use of NaOH. Moreover, the yields of carbon monoxide were negligibly small. Hence, the authors suggested that carbon monoxide rapidly reacted with water to produce hydrogen and carbon dioxide with the enhancement of NaOH. Another article by Watanabe et al. [62] showed similar results for lignin gasification at 673 K. It was seen that NaOH addition significantly enhanced the formations of hydrogen and carbon dioxide. While the hydrogen yield was about 2.3 % for the noncatalytic runs, it increased to 15 % for NaOH-added runs. Moreover, the carbon monoxide yields were very low. With these results, it was stated that the rate of water-gas shift reaction was remarkably promoted by alkali. Osada et al. [63] investigated the noncatalytic and NaOH-catalyzed SCWG of lignin and cellulose at 400 °C. For the noncatalytic experiments made with lignin, the hydrogen fraction in the gaseous effluent was 7 mol%. This amount increased to 22 mol% with the use of NaOH. For the case of cellulose gasification, the experiments performed without any catalysts resulted in a hydrogen fraction of 14 mol%. This value was further increased to 43 mol% with the use of NaOH. Onwudili and Williams [64] also investigated the role of NaOH in hydrogen production during hydrothermal gasification of glucose. The experiments performed at 450 °C and 34 MPa showed that NaOH addition improved the hydrogen formations.

KOH

Ding et al. [10] performed the noncatalytic and KOH added SCWG of cellulose and pinewood at 550 °C, with a water to biomass ratio of 7:1 and for 30 min. For cellulose, with the presence of KOH, the gas yield increased from 16.4 wt.% without catalyst to 38.4 wt.% with catalyst. Moreover, the solid product formation was reduced from 16.9 wt.% without catalyst to 1.0 wt.% with catalyst. The results also showed that KOH had a high selectivity for hydrogen, which was explained by the shift of the water-gas shift reaction's equilibrium toward hydrogen production. Experiments made with pinewood as the feedstock showed similar results. Accordingly, the hydrogen yield increased from 0.83 to 5.55 mmol/g pinewood with KOH catalyst. Chen et al. [45] investigated the effects of various alkali catalysts (KOH, NaOH, K_2CO_3, and Na_2CO_3) during the gasification of sewage sludge at 540 °C and 25 MPa. KOH showed the highest catalytic activity for improving the hydrogen yield. With its employment, a maximum hydrogen yield of 15.49 mol/kg and a maximum hydrogen molar fraction of 55.96 % were achieved. Kruse et al. [65] studied the gasification of pyrocatechol in SCW with the presence of KOH. The experiments were performed at 500 °C, 250 bar, and 1 h of reaction time. The authors stated that an increase in the amount of KOH from 0 to 5 wt.% caused a decrease in the carbon monoxide content of the gaseous effluent and a corresponding increase in the hydrogen and carbon dioxide yields. This was explained by the improvement of the water-gas shift reaction by the addition of the catalyst. Guo et al. [66] examined the effects of various alkali catalysts, including KOH, on glycerol gasification. Studies performed at 526 °C and 6.5 s residence time showed that the use of KOH increased the hydrogen and carbon dioxide yields while decreasing the carbon monoxide yield. For instance, the molar fraction of hydrogen for the noncatalytic experiment was 57.5 %, which further increased to 67.4 % with the use of KOH.

Na_2CO_3

Xu et al. [60] investigated the effect of Na_2CO_3 catalyst on the SCWG of glycine. The initial hydrogen yield was considerably low without catalyst use, but increased sharply with the employment of Na_2CO_3. Guo et al. [66] analyzed the effect of Na_2CO_3 use on the SCWG of glycerol at 526 °C and for a residence time of 6.5 s. It was seen that Na_2CO_3 improved the gasification efficiency and increased the molar fraction of hydrogen in the gaseous effluent from 57.5 % obtained in the noncatalytic run to 67.0 %. In their work on the hydrothermal gasification of glucose at 923.15 K and 25 MPa, Hao et al. [67] investigated the influence of Na_2CO_3 addition. The experiments performed without any catalysts resulted in a hydrogen yield of 16.8 %, methane yield of 14.2 %, and a carbon monoxide yield of 39.1 %. The addition of 0.01 M Na_2CO_3 increased the hydrogen and methane yields to 23.1 % and 19.3 %, respectively, while lowering the carbon monoxide yield to 2.7 %.

K_2CO_3

Guo et al. [66] investigated the effects of various alkali catalysts on the gasification of glycerol at 526 °C, 25 MPa, and a residence time of 6.5 s, the results of which are presented on Fig. 7.6. It can be seen that the use of alkali catalysts significantly increased the hydrogen and carbon dioxide yields while decreasing the carbon monoxide yield. The activity of the catalysts with respect to hydrogen yields decreased in the order of $NaOH > Na_2CO_3 > KOH > K_2CO_3$. Although K_2CO_3 exhibited the lowest activity for hydrogen formation, it still caused a notable increase in the hydrogen fraction. The molar fractions of hydrogen, methane, and carbon monoxide in the gaseous product for the noncatalytic conditions were 57.5 %, 3.4 %, and 19.2 %, respectively. When K_2CO_3 catalyst was employed, the molar fraction of hydrogen increased to 67.4 %, methane decreased to 3.2 %, and carbon monoxide decreased to 0.31 %. Sınağ et al. [68] reported a similar hydrogen enhancement in their work on the SCWG of glucose with K_2CO_3 at 500 °C and 30 MPa. With the use of K_2CO_3, the recovered amount of hydrogen per mol of glucose was approximately twice as much as without catalyst. This, along with low amounts of carbon monoxide, was interpreted as the enhancement of the

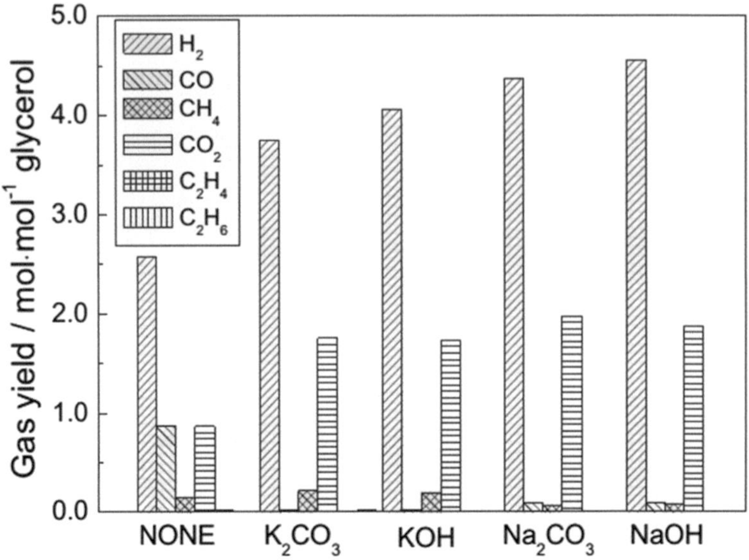

Fig. 7.6 The gas yields in SCWG of glycerol with different alkali catalysts (temperature, 526 °C; pressure, 25 MPa; residence time, 6.5 s; feedstock concentration, 10 wt.%; catalyst concentration, 0.5 wt.%) (Reprinted with permission from Guo et al. [66], copyright 2012, Hydrogen Energy Publications)

water-gas shift reaction with the catalyst effect. Madenoğlu et al. [69] investigated the gasification of glucose at temperatures ranging from 400 to 600 °C and pressures ranging from 20 to 42.5 MPa, with and without the use of K_2CO_3 catalyst. At 600 °C and 20 MPa, the hydrogen yield was nearly doubled from 2.25 to 3.87 mol/mol glucose, with the addition of the catalyst. K_2CO_3 was seen to improve hydrogen and carbon dioxide yields, by lowering the carbon monoxide yield through the water-gas shift reaction. The influence of K_2CO_3 was also discussed in another article on the SCWG of real biomass samples (cauliflower residue, tomatoes residue, and hazelnut shell) at 600 °C and 35 MPa [70]. In the absence of the catalyst, the hydrogen yield (mol hydrogen/kg C in the feed) was found as 20.2, 17.9, and 11.7 for cauliflower residue, tomatoes residue, and hazelnut shell, respectively. With the employment of K_2CO_3 catalyst, these values increased to 32.1 for cauliflower residue, 30.9 for tomatoes residue, and 18.5 for hazelnut shell.

7.3.1.6.2 Metal-Based Catalysts

The use of metal catalysts such as nickel, ruthenium, platinum, palladium, rhodium, and cobalt employed with various supports is frequently encountered in SCWG processes. These catalysts generally have a high activity, enabling efficient carbon conversions even at low reaction temperatures. Their easy separation and recovery, along with their noncorrosive nature, are among the other primary advantages of metal catalysts [24]. However, the catalyst lifetime is a significant challenge associated with their use. There is a possibility that they can be oxidized in the hydrothermal operation environment. Moreover, metal-based catalysts may be deactivated due to tarry compound deposits on their surface or the presence of sulfur/nitrogen containing compounds in the feedstock [2].

Nickel

Nickel-based catalysts are generally preferred for their lower cost, convenience for use at high temperatures, prevention of char and tar formations, and efficiency in accelerating the carbon conversions [24, 58]. What is more, these catalysts generally increase the rates of water-gas shift and methanation reactions [71, 72]. Literature studies show that the activities of nickel catalysts usually depend on the operating temperature and the nature of the support material [28]. However, the most significant drawback of these catalysts is their easy deactivation, mostly due to the adsorption of the reaction intermediates onto the catalyst surface [24, 28]. Hence, it is very common to use compounds like CeO_2, La_2O_3, SiO_2, and ZrO_2, called promoters, in order to enhance the activity and stability of the nickel catalysts [59, 73].

Kruse et al. [65] investigated the influence of Raney nickel on the SCWG of pyrocatechol solutions. The addition of the catalyst led to an increase in the hydrogen and carbon dioxide yields, but not a significant increase in the methane yield. The authors reported that, although the effect of Raney nickel was lower than that of alkali catalysts, it still acted on the water-gas shift reaction by increasing the hydrogen fraction by 170 % and decreasing carbon monoxide fraction by 70 % when compared with the experiments without catalyst addition. In another article on the effect of Raney nickel on glucose gasification at 500 °C and 30 MPa, similar observations were made [68]. It was reported that Raney nickel addition had a positive effect on hydrogen, methane, and carbon dioxide formations.

Akgün and Kıpçak [3] studied the behavior of Ni/Al_2O_3 catalyst on the gasification of 2-propanol at 400–600 °C and 25 MPa, in the reaction time range of 10–30 s. To understand the effect of catalyst addition, noncatalytic experiments were also performed at 450 °C. The highest gasification yield and the hydrogen content of the gaseous product in the regarding blank experiments were recorded as 0.15 L/L feed and 2.5 %, respectively. On the other hand, the use of Ni/Al_2O_3 catalyst improved the gasification yield to 9.5 L/L feed and the hydrogen content to 62 %. Lu et al. [28] investigated hydrogen production by glucose gasification in SCW over Ni/Al_2O_3 catalyst at 400 °C and 24.5 MPa. The hydrogen and methane yields were 1.3 mol/kg and 0.8 mol/kg, respectively, without the addition of the catalyst. With the addition of Ni/Al_2O_3, the hydrogen yield improved significantly to 10.5 mol/kg, while the methane yield was increased by 257.5 %. Osada et al. [63] studied the effect of Ni/Al_2O_3 on the catalytic gasification of cellulose and lignin. The experimental results unveiled that the use of Ni/Al_2O_3 enhanced hydrogen production for both biomass compounds. For the case of cellulose, at 400 °C, and for a reaction time of 15 min, the hydrogen fraction obtained at the noncatalytic run was 14 %. This value increased up to 23 % when Ni/Al_2O_3 was employed. At the same experimental conditions, the use of Ni/Al_2O_3 improved the hydrogen fraction from 7 % for without catalyst to 17 %. In another study of Osada et al. [74], lignin gasification at 673 K was performed. Accordingly the molar fractions of hydrogen, methane, carbon monoxide, and carbon dioxide obtained were 3.4 %, 69.5 %, 7.9 %, and 19.0 %, respectively, for the experiment without any catalyst use. Ni/Al_2O_3 improved the hydrogen molar fraction to 18.4 % and increased carbon dioxide to 58.4 % while decreasing the methane fraction to 18.1 % and reducing carbon monoxide to 4.0 %. Therdthianwong et al. [75] also reported similar results in their work regarding ethanol reforming in SCW at 500 °C and 25 MPa.

As it was mentioned before, nickel catalysts are sometimes used with promoters in order to maintain their stability and prevent their deactivation. For instance, Li et al. [59] studied SCWG of glucose over $Ni\text{-}Cu/Al_2O_3$, $Ni\text{-}Co/Al_2O_3$, and $Ni\text{-}Sn/Al_2O_3$ catalysts at 673 K. The hydrogen selectivity with no catalyst addition was 19.2 %. This value was further improved to 50.2 % for $Ni\text{-}Cu/Al_2O_3$ and 61.1 % for $Ni\text{-}Co/Al_2O_3$. On the other hand, it was reported that Sn additive retrograded the catalytic activity of the nickel-based catalyst for hydrogen production. In another gasification study with glucose at 673 K and 24.5 MPa, Lu et al. [73] investigated the effect of $Ni/CeO_2/Al_2O_3$ catalyst. It was reported that catalyst use enhanced hydrogen, methane, and carbon dioxide formations. Therdthianwong et al. [75] examined the reforming of ethanol over $Ni/CeZrO_2/Al_2O_3$ catalyst in SCW at 500 °C and 25 MPa. It was seen that the catalyst significantly improved ethanol conversion, along with hydrogen and methane yields. Lee [76] studied the influence of Ni-Y, Ni-Fe, and Ni-Co catalysts on activated carbon supports for the SCWG of glucose. It was reported that all three of the catalysts improved carbon conversions along with hydrogen, methane, and carbon dioxide yields. On the contrary, the use of these promoted nickel catalysts caused a decrease in carbon monoxide formations.

Ruthenium

Ruthenium catalysts are especially used in low-temperature gasification processes. The most significant advantage of these catalysts is the long-term maintainability of their activities and stabilities [24, 59]. They are usually preferred for their enhancement of methanation reactions. On the other hand, ruthenium catalysts may be subject to catalyst poisoning in the presence of sulfur-containing feedstock. Moreover, their high cost is another disadvantage [11, 59].

Osada et al. [63] investigated the effect of Ru/TiO_2 catalyst on the gasification of lignin and cellulose at 400 °C. For the case of lignin, when compared with the noncatalytic results, the employment of Ru/TiO_2 increased the hydrogen fraction of the gaseous effluent from 7 % to 14 % while enhancing the methane fraction from 33 % to 41 %. On the other hand, the results for cellulose gasification were quite different. Ru/TiO_2 decreased the hydrogen fraction from 14 % to 9 %, but an enormous increase in the methane fraction was observed. While it was 3 % for

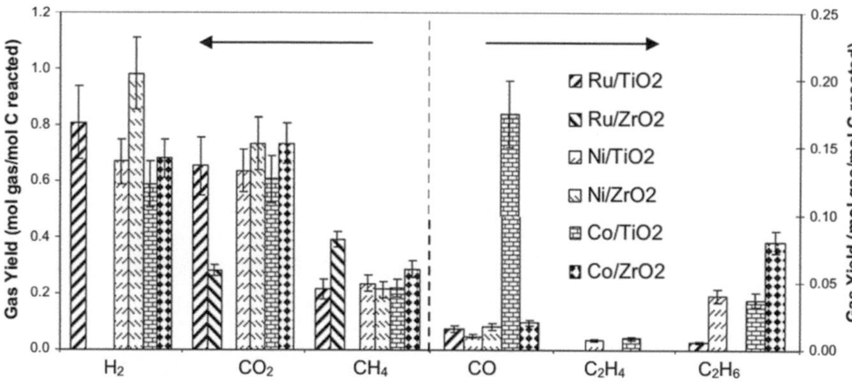

Fig. 7.7 The product gas yields for various TiO_2- and ZrO_2-supported metal catalysts used in the SCWG of switchgrass biocrude (temperature, 650 °C; pressure, 250 bar) (Reprinted with permission from Byrd et al. [78], copyright 2011, Hydrogen Energy Publications)

noncatalytic gasification, it increased to 44 % with Ru/TiO_2 use. In another study with Ru/TiO_2 on the SCWG of lignin at 673 K, similar results with those of cellulose in the previously mentioned article were obtained [77]. For the noncatalytic experiments, the molar fractions of hydrogen and methane in the gas product were 21.5 % and 30.6 %, respectively. With the use of Ru/TiO_2, the fraction of hydrogen decreased to 4.5 % and the fraction of methane increased to 49.0 %. Byrd et al. [78] studied the influence of various titania and zirconia supported metal catalysts on the gasification of switchgrass biocrude at 600 °C and 250 bar. Their experimental results are presented in Fig. 7.7. The figure clearly presents the impact of support material on gas yields. For instance, the second highest hydrogen yield was seen for Ru/TiO_2, but the gas product contained very little hydrogen for Ru/ZrO_2. This catalyst showed the highest methane activity instead. The authors also mention the poor closure of carbon balance for this catalyst, which was due to the extensive char formation during its use. On the other hand, the highest hydrogen yield was encountered for Ni/ZrO_2 catalyst (0.98 mol hydrogen/mol carbon). For the case of Co, while the hydrogen yields were not as high as those of Ni and Ru catalysts, a significant yield of carbon monoxide was observed, especially for Co/TiO_2. The authors explain this with cobalt's low activity for the water-gas shift reaction.

Al_2O_3 is one of the most commonly used supports for ruthenium. For instance, Karakuş et al. [47] studied the catalytic gasification of 2-propanol over Ru/Al_2O_3 at 400–550 °C, 25 MPa and a reaction time of 10–30 s. The employment of Ru/Al_2O_3 catalyst significantly increased the gasification yield (amount of gaseous product obtained per the amount of 2-propanol solution fed) from about 1 to 27.2 L/L feed. Moreover, it increased the hydrogen and methane yields. The authors reported that Ru/Al_2O_3 showed a very high gasification activity toward methane, favoring the water-gas shift reaction and methanation of carbon monoxide. In a study made by Byrd et al. [49], ethanol reforming in SCW with Ru/Al_2O_3 catalyst was investigated. It was reported that the use of Ru/Al_2O_3 significantly increased the hydrogen yield. Besides, a major reduction in carbon monoxide and methane yields was encountered. Byrd et al. [79] also studied hydrogen production from glucose using Ru/Al_2O_3 catalyst in SCW. The experiments were performed at 700 °C and 248 bar, with a glucose concentration of 1 wt.%, for both the noncatalytic and catalytic runs. The use of Ru/Al_2O_3 increased the hydrogen fraction from 54.5 % to 68.9 %, decreased the methane fraction from 8.5 % to 1.3 %, and lowered the carbon monoxide fraction from 5.6 % to 0.1 %. The authors suggest that glucose initially underwent dehydrogenation on the catalyst surface to give adsorbed intermediates. The subsequent cleavage of C-C bonds led to the formation of hydrogen and carbon monoxide, the latter then reacting with water to form carbon dioxide and further hydrogen through the water-gas shift reaction. Onwudili and Williams [80] investigated the gasification of glucose at 550 °C and 36 MPa, with and without the employment of Ru/Al_2O_3 catalyst. Accordingly, the use of Ru/Al_2O_3 increased the gas production significantly. During the noncatalytic gasification, the obtained gaseous product contained 12 % hydrogen, 3.38 % methane, 3.16 % C_2–C_4 hydrocarbons, 15.4 % carbon monoxide, and 66 % carbon dioxide by mol. For the catalytic experiments, on the other hand, the gaseous product was comprised of 29 % hydrogen, 21.6 % methane, 2.02 % C_2–C_4 hydrocarbons, 2.02 % carbon monoxide, and 48.3 % carbon dioxide. The authors pointed out the evident selectivity of Ru/Al_2O_3 catalyst toward hydrogen and methane formations.

Other Metal Catalysts

Akgün and Kıpçak [3] compared the hydrogen yields of Ru/Al_2O_3, Pt/Al_2O_3, Ni/Al_2O_3, and Fe-Cr in their article regarding the catalytic hydrogen production from 2-propanol in SCW. Figure 7.8 presents the highest hydrogen yields obtained for each catalyst. While the highest hydrogen yields for Ru/Al_2O_3, Ni/Al_2O_3, and Fe-Cr were obtained between 100 and 150 mmol/mol 2-propanol, it was about 1050 mmol/mol 2-propanol for Pt/Al_2O_3, for the experiment conducted at 400 °C. Consequently, the authors concluded that Pt/Al_2O_3 was much more selective for hydrogen production, especially for low reaction temperatures.

Osada et al. [74] investigated the effects of various metal catalysts (Ru, Rh, Pt, Pd, and Ni) used with various supports (TiO_2, Al_2O_3, and C) on the gasification of lignin at 673 K. The authors reported that the catalyst selectivities to hydrogen were in the order of nickel > palladium > platinum > rhodium > ruthenium, while the selectivities for methane were in the order of ruthenium > rhodium > platinum > palladium > nickel. Yamaguchi et al. [81] also studied the gasification of lignin with the same catalysts and at the same temperature as the previously mentioned article. It was reported that the catalyst selectivities for hydrogen were in the following order: palladium > ruthenium > platinum > rhodium > nickel.

Youssef et al. [82] investigated AC (activated carbon), Pd/AC, Ru/AC, Ru/Al_2O_3, and NaOH activities during the gasification of hog manure at 500 °C and 28 MPa. The gas yield distributions of these catalysts are presented in Fig. 7.9. The major components of the product gas yields for all experiments were hydrogen,

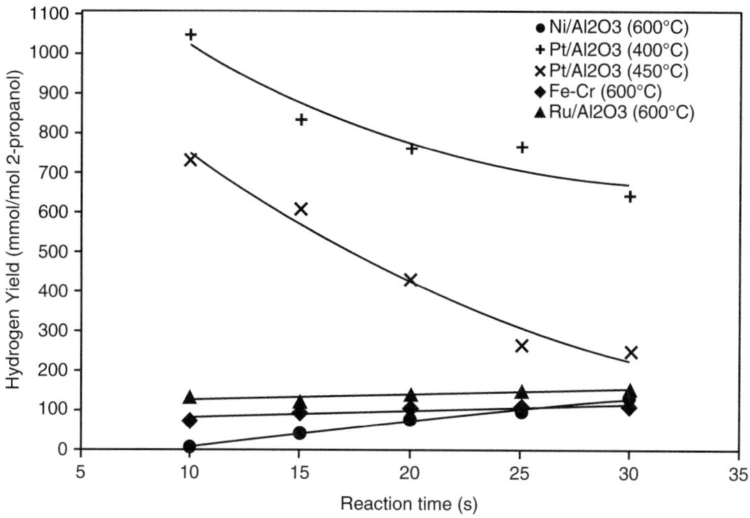

Fig. 7.8 A comparison of the highest hydrogen yields obtained for various catalysts during the SCWG of 2-propanol (pressure: 25 MPa) (Reprinted with permission from Akgün and Kıpçak [3], copyright 2014, Elsevier)

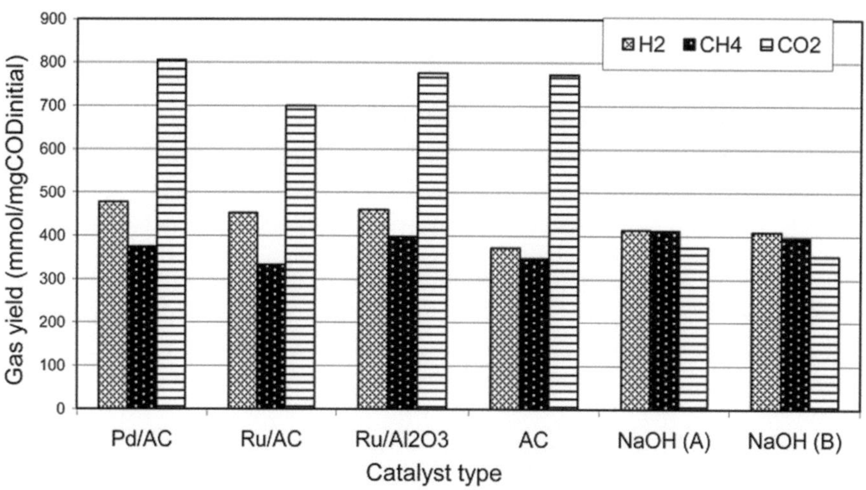

Fig. 7.9 The gas yields in SCWG of hog manure with different catalysts (temperature, 500 °C; pressure, 28 MPa; catalyst amount, 2.5 g) (Reprinted with permission from Youssef et al. [82], copyright 2010, Hydrogen Energy Publications)

methane, and carbon dioxide. Trace amounts of carbon monoxide were encountered only for the experiments made with NaOH. Similar to the results of [81], palladium showed greater hydrogen selectivity than ruthenium. The order of hydrogen production in catalytic gasification was reported as Pd/AC > Ru/Al_2O_3 > Ru/AC >

AC > NaOH. On the other hand, the order of feedstock's chemical oxygen demand reduction was reported as NaOH > Ru/AC > AC > Ru/Al_2O_3 > Pd/AC.

As a result, it can be said that catalysts exhibit different activities or gas compositions depending on the feedstock and the operating conditions used. Hence it is not possible to affirm that a specific catalyst is superior over another for a specific selectivity by rigid boundaries. This necessitates additional studies regarding hydrogen and biofuel production from biomass by supercritical water gasification.

7.3.2 Literature Studies

In the literature, a lot of valuable studies have been made regarding supercritical water gasification of biomass for the last three decades. A summary of these studies, which were elaborately discussed in the previous sections, is presented in Table 7.3 for noncatalytic hydrothermal gasification and in Table 7.4 for catalytic gasification. The tables include the feedstock used, the operating conditions, the reactor type, the kind of catalyst, if used, and the major findings. In the findings section, the lowest and highest amounts of hydrogen, methane, C_2–C_4 hydrocarbons, and carbon monoxide are presented.

7.4 Conclusions and Future Outlook

Hydrogen is a promising energy source, which can sustainably provide the world's growing energy needs, minimize the dependence on fossil fuels, and reduce the impact of environmental pollution. Among the biological, thermochemical, and electrochemical processes that are employed for hydrogen production, supercritical water gasification owns a significant potential. Due to the unique properties of supercritical water, SCWG offers many advantages for hydrogen production from biomass. For instance, since biomass with water content up to 80 % can be used, the process does not necessitate expensive drying pretreatment steps. A hydrogen-rich gaseous product is obtained in SCWG. Moreover, the selectivity of the process toward a specific gaseous product can be controlled by adjusting the operating conditions. Reduced mass transfer limitations, homogenous and fast reactions, high levels of conversions, reduced amounts of chars and tars, short residence times, and smaller-sized reactors are the other advantages that the SCWG process offers.

However, despite these numerous advantages, there are several challenges that should be overcome for the commercialization of SCWG. The first challenge is associated with the energy efficiency of the process. The high operating temperatures and the regarding heating requirements necessitate the use of an efficient heat recovery unit. Secondly, the reactor design should be made with extreme care, involving adequate safety measures and noncorrosive materials, due to the severe

Table 7.3 Hydrothermal gasification studies 1993–2013

Feedstock	Temperature (°C)	Pressure (MPa)	Res. time (min)	Reactor	H_2	CH_4	CO	C_2–C_4	Unit	Ref.
Ethanol	550–700	27.6	0.05 and 0.10	Flow	48 %	26 %	3 %	–	mol%	[35]
Ethanol	500–600	25	50–290	Flow	18.6–60.1 %	8.2–19.4 %	12.6–23.2 %	–	mol%	[38]
Ethanol	450–500	–	10–60	Batch	1.2–14 %	0.06–10 %	0–2 %	–	mol%	[52]
Ethylene glycol	550–700	27.6	0.05 and 0.10	Flow	61 %	6 %	11 %	–	mol%	[35]
Glucose	600–767	25	0.25–1	Flow	5.7–11.5	0.20–1.15	0–0.25	C_2H_6: 0.02–0.25	mol/mol glucose	[27]
Glucose	330–380	9.3–22.5	0–120	Flow	0.20–1.27 %	0.11–1.40 %	3.60–8.89 %	–	wt.%	[29]
Glucose	500–650	23–27	~25 g/min (flow rate)	Flow	10–36 %	1–4 %	17–38 %	–	mol%	[37]
Glucose	650	34.5	0.5	Flow	1–8.5	0.90–0.96	0–3	–	mol/mol glucose	[55]
Glucose	750–800	22–33	0.067–0.108	Flow	23–40 %	12–26 %	1–48 %	C_2H_6: 2–8 %	mol%	[84]
Glycerol	300–430	30	5–120	Batch	0–0.077	0.010–0.041	0.060–0.110	0.006–0.036	mol/mol glycerol	[30]
Isooctane	601–676	25	0.10–0.55	Flow	35–60 %	14–26 %	0.5–2.1 %	3–26 %	mol%	[9]
Isooctane	763	25	1–2	Flow	55–68 %	8.6–17.7 %	1.3–2.5 %	C_2H_6: 0.5–1.5 %	mol%	[54]
Lignin	500–650	23–27	~25 g/min (flow rate)	Flow	28–57 %	1.4–3 %	0–2 %	–	mol%	[37]
Methanol	400–600	25–45	0.05–1.67	Flow	56–75 %	1–30 %	2–24 %	–	vol.%	[34]
Methanol	550–700	27.6	0.05 and 0.10	Flow	72 %	4 %	6 %	–	mol%	[35]
Methanol	400–600	25–45	0.05–1.67	Flow	0.64–2.40	0–0.52	0.08–0.60	–	mol/mol methanol	[53]

(continued)

Table 7.3 (continued)

Feedstock	Temperature (°C)	Pressure (MPa)	Res. time (min)	Reactor	H_2	CH_4	CO	C_2–C_4	Unit	Ref.
Xylose	450–650	25	0–0.67	Flow	36–62 %	2.6–7.1 %	1.6–26.0 %	–	mol%	[36]
Black liquor	400–600	25	0.082–0.228	Flow	6.82–27.55	1.87–6.25	0–0.22	C_2: 0–0.83	mol/kg	[43]
Black liquor	375–650	22–40	0.083–2	Flow	5.0–23.8 %	1.6–19.3 %	1.2–12.6 %	C_2: 2.5–30.1 %; C_4: 5.2–26.8 %	wt.%	[46]
Corn cob	500–650	23–27	~25 g/min (flow rate)	Flow	24–39 %	3.8–8.8 %	3.8–7.5 %	–	mol%	[37]
Corn cob	550–650	22.5–27.5	0.33–0.67	Flow	5.12–19.60	1.03–5.68	0.05–0.99	C_2H_4: 0.08–0.41; C_2H_6: 0.06–0.78	mol/kg	[44]
Corn silage	300–700	25–40	0.6–10	Flow	0–33 %	0–27 %	0.5–16 %	C_2H_6: 0–5 %	vol.%	[50]
Microalgae	450–550	24	0–77	Batch	0.4–10.2	0.2–7.7	0.1–0.9	C_2H_4: 0–0.5; C_2H_6: 0–2.3	mmol/g	[85]
Paper mill sludge	400–550	23.8–45.5	20–120	Batch	1.5–14.5	0.3–3.0	0.2–0.3	–	mol/kg sludge	[41]
Sewage sludge	800	38	7–14	Batch	19.6–33.6 %	0.7–1.6 %	36.4–46.1 %	–	mol%	[86]
Textile wastewater	450–650	25	0.5–2.5	Flow	5.88–15.46 %	17.11–40.14 %	0–53.84 %	C_2H_6: 1.61–6.23 %; C_3H_8: 0–0.50 %	mol%	[83]
Winery waste	500	25	5	Batch	33.94 %	8.64 %	0.14 %	–	mol%	[87]

Table 7.4 Catalytic gasification studies 2000–2013

Feedstock	Catalyst	Temperature (°C)	Pressure (MPa)	Res. time (min)	Reactor	H_2	CH_4	CO	C_2–C_4	Unit	Ref.
Cellulose	NaOH	400–440	30–35	10–15	Batch	2.2–23.1 %	–	0–10.0 %	–	mol%	[61]
Cellulose	Ru/TiO_2	400	–	15	Flow	9 %	44 %	–	1 %	mol%	[63]
Cellulose	Ni/Al_2O_3	400	–	15	Flow	23 %	3 %	7 %	1 %	mol%	[63]
Cellulose	NaOH	400	–	15	Flow	43 %	3 %	–	–	mol%	[63]
Cellulose	NaOH	450	34	–	Batch	0.043	0.033	–	0.049	g	[64]
Ethanol	Ru/Al_2O_3	600–800	22.1–27.6	0.083–0.333	Flow	53.2–73.0 %	2.8–18.0 %	0.2–5.6 %	–	mol%	[49]
Ethanol	Ni/Al_2O_3	500	25	0.833	Flow	1.5	0.6	0.4	–	mol/mol ethanol	[75]
Ethanol	$Ni/CeZrO_2/Al_2O_3$	500	25	0.833	Flow	1.1	1.2	0.05	–	mol/mol ethanol	[75]
Glucose	Ni/Al_2O_3	400	24.5	20	Batch	10.5	3.0	–	–	mol/kg	[28]
Glucose	Ni/CeO_2–Al_2O_3	400	24.5	20	Batch	12.7	2.8	–	–	mol/kg	[28]
Glucose	Ni/Al_2O_3	400	–	20	Batch	10.5	2.9	–	–	mol/kg	[59]
Glucose	Ni-Cu/Al_2O_3	400	–	20	Batch	10.9	1.7	–	–	mol/kg	[59]
Glucose	Ni-Co/Al_2O_3	400	–	20	Batch	8.7	0.9	–	–	mol/kg	[59]
Glucose	Ni-Sn/Al_2O_3	400	–	20	Batch	3.7	0.9	–	–	mol/kg	[59]
Glucose	ZrO_2	400–440	30–35	10–15	Batch	2.0–24.3 %	–	0–5.6 %	–	mol%	[61]
Glucose	NaOH	450	34	–	Batch	0.043	0.050	–	0.047	g	[64]
Glucose	KOH	650	25	1.2–3.7	Flow	25.7–27.4 %	16.4–19.2 %	0.5 %	C_2H_4: 0.7–1.1 %; C_2H_6: 2.8–3.7 %	mol%	[67]
Glucose	Na_2CO_3	650	25	1.8–3.0	Flow	23.1–24.1 %	19.3–20.0 %	0.9–2.7 %	C_2H_4: 1.0 %; C_2H_6: 3.0–3.2 %	mol%	[67]
Glucose	K_2CO_3	500	30	60	Batch	1.43–1.61	0.51–0.74	–	C_2H_6: 0.06–0.09	mol/mol glucose	[68]
Glucose	Raney nickel	500	30	60	Batch	1.63–1.94	0.49–0.54	–	C_2H_6: 0.09–0.11	mol/mol glucose	[68]
Glucose	K_2CO_3	400–600	20–42.5	60	Batch	0.4–3.9	0.1–1.5	0–0.1	0–2.6	mol/mol glucose	[69]

(continued)

Table 7.4 (continued)

Feedstock	Catalyst	Temperature (°C)	Pressure (MPa)	Res. time (min)	Reactor	H_2	CH_4	CO	C_2–C_4	Unit	Ref.
Glucose	$Ni/CeO_2/Al_2O_3$	400	24.5	20	Batch	6.5–12.8	0.9–2.6	–	–	mol/kg	[73]
Glucose	Ni/AC	550–700	28	0–390	Flow	3.5–4.1	0.7–0.9	0.1–0.2	–	mol/mol	[76]
Glucose	Ni-Y/AC	550–700	28	0–390	Flow	1.1–6.9	0.2–1.3	0.1–0.4	–	mol/mol	[76]
Glucose	Ni-Co/AC	550–700	28	0–390	Flow	2.5–3.6	0.7–0.9	0.1–0.2	–	mol/mol	[76]
Glucose	Ni-Fe/AC	550–700	28	0–390	Flow	2.4–3.0	0.6–0.7	1.0–1.5	–	mol/mol	[76]
Glucose	Ru/Al_2O_3	700–800	24.8	0.017–0.100	Flow	68.9 %	1.3 %	0.1 %	–	mol%	[79]
Glucose	Ru/Al_2O_3	550	36	10	Batch	12.0–94.0 %	3.38–43.1 %	0–15.4 %	0.84–4.57 %	mol%	[80]
Glucose	Ni/Al_2O_3	400–500	28	30	Batch	0.6–2.1	0.1–1.2	0.8–4.1	–	mol/mol glucose	[92]
Glucose	$RuNi/Al_2O_3$	600–750	24	180–1440	Flow	30.9–63.3 %	1.8–17.8 %	0.5–28.1 %	–	vol.%	[93]
Glycerol	Na_2CO_3	380–550	25	–	Flow	225–1300	63–175	0–38	C_{2+}: 0–38	mL/g glycerol	[60]
Glycerol	Ru/Al_2O_3	700–800	24.1	0.017–0.067	Flow	42.2–70.0 %	3.6–19.6 %	0–4.3 %	–	mol%	[89]
Glycerol	Ru/ZrO_2	510–550	35	0.033–0.167	Flow	0.2–0.6	0.1–0.3	0.3–0.5	–	mol/mol glycerol	[91]
Glycine	Na_2CO_3	380–550	25	–	Flow	250–1050	0–25	–	C_{2+}: 75–100	mL/g glycine	[60]
Lignin	Ru/TiO_2	400	–	15	Flow	14 %	41 %	–	1 %	mol%	[63]
Lignin	Ni/Al_2O_3	400	–	15	Flow	17 %	19 %	4 %	1 %	mol%	[63]
Lignin	NaOH	400	–	15	Flow	22 %	15 %	–	1 %	mol%	[63]
Lignin	Ru/C	400–450	30	60	Batch	3.1–7.2 %	47.4–48.1 %	0–0.7 %	0–0.6 %	mol%	[81]
Lignin	Rh/C	400–450	30	60	Batch	6.6–7.4 %	38.7–47.3 %	–	0.3–2.1 %	mol%	[81]
Lignin	Pt/C	400–450	30	60	Batch	7.6–9.7 %	25.5–28.7 %	0–0.2 %	2.5–3.6 %	mol%	[81]
Lignin	Pd/C	400–450	30	60	Batch	11.1–15.4 %	16.6–27.8 %	0–7.0 %	1.7–3.3 %	mol%	[81]

Lignin	Ni/C	400–450	30	60	Batch	6.0–9.5 %	23.1–25.4 %	1.2–2.2 %	2.0–3.2 %	mol%	[81]
Lignin	Ru/TiO_2	400	37.1	0–180	Batch	3–13 %	41–47 %	0–1.5 %	–	mol%	[88]
Lignin	Ru/C	400	37.1	0–180	Batch	3–10 %	41–51 %	0–2.5 %	–	mol%	[88]
Lignin	Ru/Al_2O_3	400	37.1	0–180	Batch	3–12 %	42–51 %	0–1 %	–	mol%	[88]
Lignin	Ru/MgO	400	–	120	Batch	0.0001–0.0005	2.9×10^{-5}–0.0005	–	C_2H_6: 0–1.7×10^{-4}	mol	[90]
Methanol	Nano-Cu	500–700	6.9–27.6	0.25	Flow	0.09–1.23	0.01–0.02	0.02–0.13	–	mol/mol methanol	[48]
Pyrocatechol	KOH	600–700	20–40	0.25–4	Flow	2.7–7.8	0.32–1.44	0.07–0.17	–	mol/mol pyrocatechol	[65]
Corn silage	$KHCO_3$	550–700	25	3–25	Flow	29.6–34.4 %	15.0–20.5 %	0.6–2.8 %	C_2H_6: 2.6–4.8 %	vol.%	[51]
Potato	NaOH	450	34	–	Batch	0.041	0.051	–	0.037	g	[64]
Sawdust	Raney nickel	300–410	12–34	10–100	Batch	3–24 %	14–49 %	0.1–9 %	–	vol.%	[94]
Starch	NaOH	450	34	–	Batch	0.043	0.037	–	0.044	g	[64]
Whey	NaOH	300–390	9.5–24.5	0–120	Batch	5.2–147	0.2–14.2	–	0.2–16.5	mg	[95]

reaction conditions employed in the process. Another challenge arises due to the wall effects of the reactors. Generally, the reactors used for the SCWG processes are constructed of alloys to avoid corrosion. Apart from being expensive, these alloys also contain high amounts of nickel. The nickel-containing reactor walls may act as a catalyst, which may in turn lead to uncertainties in the experimental results. There are also some challenges associated with the feedstock used. Commercial SCWG systems will also have to deal with real biomass, which contain inorganic materials and impurities. These inorganics and impurities may precipitate in the reactor and cause clogging issues. The formation of chars and tars may also block the reactors. Hence, the heating of the reactors should be very rapid in order to minimize such formations. Finally, catalyst use in commercial SCWG systems also creates some challenges. The use of homogenous alkali catalysts may cause corrosion of the reactor. What is more, since alkali salts have a very low solubility in supercritical water, some problems may arise from their precipitation in the reactor. Heterogeneous catalysts, on the other hand, are expensive and can be deactivated after several uses due to char and tar formations on their surface. Catalyst pores are also prone to being blocked by the impurities and inorganic materials found in the feedstock. The supports of these heterogeneous catalysts also have the possibility to oxidize and degrade at supercritical conditions.

In conclusion, once the aforementioned challenges are handled during its commercialization, owing to its unique and significant advantages, supercritical water gasification will be a sustainable, viable, and environmentally benign technology for hydrogen production from biomass.

References

1. Mattos LV, Noronha FB. Partial oxidation of ethanol on supported Pt catalysts. J Power Sources. 2005;145:10–5.
2. Susanti RF, Kim J, Yoo K. Supercritical water gasification for hydrogen production: current status and prospective of high-temperature operation. Supercritical fluid technology for energy and environmental applications, chapter 6. Elsevier:Poland, 2014.
3. Akgün M, Kıpçak E. Catalytic hydrogen production from 2-propanol in supercritical water: comparison of some metal catalysts. J Supercrit Fluids. 2014;90:101–9.
4. Alauddin ZABZ, Lahijani P, Mohammadi M, Mohamed AR. Gasification of lignocellulosic biomass in fluidized beds for renewable energy development: a review. Renew Sust Energ Rev. 2010;14:2852–62.
5. Kalinci Y, Hepbasli A, Dincer I. Biomass-based hydrogen production: a review and analysis. Int J Hydrog Energy. 2009;34:8799–817.
6. Zhang J. Hydrogen production by biomass gasification in supercritical water. Energeia. 2008;19(6):1–6.
7. Reddy SN, Nanda S, Dalai AK, Kozinski JA. Supercritical water gasification of biomass for hydrogen production. Int J Hydrog Energy. 2014;39:6912–26.
8. Kırtay E. Recent advances in production of hydrogen from biomass. Energy Convers Manag. 2011;52:1778–89.
9. Susanti RF, Veriansyah B, Kim J, Kim J, Lee Y. Continuous supercritical water gasification of isooctane: a promising reactor design. Int J Hydrog Energy. 2010;35:1957–70.

10. Ding N, Azargohar R, Dalai AK, Kozinski JA. Catalytic gasification of cellulose and pinewood to H_2 in supercritical water. Fuel. 2014;118:416–25.
11. Azadi P, Farnood R. Review of heterogeneous catalysts for sub- and supercritical water gasification of biomass and wastes. Int J Hydrog Energy. 2011;36:9529–41.
12. Yanik J, Ebale S, Kruse A, Saglam M, Yüksel M. Biomass gasification in supercritical water: part 1. Effect of the nature of biomass. Fuel. 2007;86:2410–5.
13. Jessop PG, Leitner W. Chemical synthesis using supercritical fluids. Germany: WILEY-VCH; 1999.
14. Poling BE, Prausnitz JM, O'Connell JP. The properties of gases and liquids. 5th ed. New York: McGraw-Hill Higher Education; 2001.
15. Kruse A, Dinjus E. Hot compressed water as reaction medium and reactant 1. Properties and synthesis reactions. J Supercrit Fluids. 2007;39:362–80.
16. Akiya N, Savage PE. Roles of water for chemical reactions in high-temperature water. Chem Rev. 2002;102:2725–50.
17. Krammer P, Vogel H. Hydrolysis of esters in subcritical and supercritical water. J Supercrit Fluids. 2000;16:189–206.
18. Zhou N, Krishnan A, Vogel F, Peters WA. A computational model for supercritical water oxidation of toxic organic wastes. Adv Environ Res. 2000;4:79–95.
19. Brunner G. Near critical and supercritical ware. Part I. Hydrolytic and hydrothermal processes. J Supercrit Fluids. 2009;47:373–81.
20. He C, Chen CL, Giannis A, Yang Y, Wang JY. Hydrothermal gasification of sewage sludge and model compounds for renewable hydrogen production: a review. Renew Sust Energ Rev. 2014;39:1127–42.
21. Kruse A, Dinjus E. Hot compressed water as reaction medium and reactant 2. Degradation reactions. J Supercrit Fluids. 2007;41:361–79.
22. Gasafi E, Reinecke MY, Kruse A, Schebek L. Economic analysis of sewage sludge gasification in supercritical water for hydrogen production. Biomass Bioenergy. 2008;32:1085–96.
23. Kruse A, Gawlik A. Biomass conversion in water at 330–410°C and 30–50 MPa. Identification of key compounds for indicating different chemical reaction pathways. Ind Eng Chem Res. 2003;42:267–79.
24. Guo Y, Wang SZ, Xu DH, Gong YM, Ma HH, Tang XY. Review of catalytic supercritical water gasification for hydrogen production from biomass. Renew Sust Energ Rev. 2010;14:334–43.
25. Anikeev VI, Yermakova A, Manion J, Huie R. Kinetics and thermodynamics of 2-propanol dehydration in supercritical water. J Supercrit Fluids. 2004;32:123–35.
26. Basu P. Biomass gasification and pyrolysis: practical design and theory. United States: Elsevier Academic Press; 2010.
27. Susanti RF, Dianningrum LW, Yum T, Kim Y, Lee BG, Kim J. High-yield hydrogen production from glucose by supercritical water gasification without added catalyst. Int J Hydrog Energy. 2012;37:11677–90.
28. Lu Y, Li S, Guo L, Zhang X. Hydrogen production by biomass gasification in supercritical water over $Ni/\gamma Al_2O_3$ and Ni/CeO_2-γAl_2O_3 catalysts. Int J Hydrog Energy. 2010;35:7161–8.
29. Williams PT, Onwudili J. Composition of products from the supercritical water gasification of glucose: a model biomass compound. Ind Eng Chem Res. 2005;44:8739–49.
30. Müller JB, Vogel F. Tar and coke formation during hydrothermal processing of glycerol and glucose. Influence of temperature, residence time and feed concentration. J Supercrit Fluids. 2012;70:126–36.
31. Van Rossum G, Potic B, Kersten SRA, Van Swaaij WPM. Catalytic gasification of dry and wet biomass. Catal Today. 2009;145:10–8.
32. Matsumura Y, Minowa T, Potic B, Kersten SRA, Prins W, Van Swaaij WPM, Van de Beld B, Elliott DC, Neuenschwander GG, Kruse A, Antal Jr MJ. Biomass gasification in near- and super-critical water: status and prospects. Biomass Bioenergy. 2005;29:269–92.

33. Peterson AA, Vogel F, Lachance RP, Fröling M, Antal Jr MJ, Tester JW. Thermochemical biofuel production in hydrothermal media: a review of sub- and supercritical water technologies. Energy Environ Sci. 2008;1:32–65.
34. Boukis N, Diem V, Habicht W, Dinjus E. Methanol reforming in supercritical water. Ind Eng Chem Res. 2003;42:728–35.
35. Taylor JD, Herdman CM, Wu BC, Wally K, Rice SF. Hydrogen production in a compact supercritical water reformer. Int J Hydrog Energy. 2003;28:1171–8.
36. Goodwin AK, Rorrer GL. Reaction rates for supercritical water gasification of xylose in a micro-tubular reactor. Chem Eng J. 2010;163:10–21.
37. Jin H, Lu Y, Guo L, Cao C, Zhang X. Hydrogen production by partial oxidative gasification of biomass and its model compounds in supercritical water. Int J Hydrog Energy. 2010;35:3001–10.
38. Therdthianwong S, Srisiriwat N, Therdthianwong A, Croiset E. Hydrogen production from bioethanol reforming in supercritical water. J Supercrit Fluids. 2011;57:58–65.
39. Guo Y, Wang S, Huelsman CM, Savage PE. Products, pathways, and kinetics for reactions of indole under supercritical water gasification conditions. J Supercrit Fluids. 2013;73:161–70.
40. Liu Q, Liao L, Liu Z, Dong X. Hydrogen production by glycerol reforming in supercritical water over Ni/MgO-ZrO_2 catalyst. J Energy Chem. 2013;22:665–70.
41. Zhang L, Xu C, Champagne P. Energy recovery from secondary pulp/paper-mill sludge and sewage sludge with supercritical water treatment. Bioresour Technol. 2010;101:2713–21.
42. Kıpçak E, Söğüt OÖ, Akgün M. Hydrothermal gasification of olive mill wastewater as a biomass source in supercritical water. J Supercrit Fluids. 2011;57:50–7.
43. Cao C, Guo L, Chen Y, Guo S, Lu Y. Hydrogen production from supercritical water gasification of alkaline wheat straw pulping black liquor in continuous flow system. Int J Hydrog Energy. 2011;36:13528–35.
44. Lu Y, Guo L, Zhang X, Ji C. Hydrogen production by supercritical water gasification of biomass: explore the way to maximum hydrogen yield and high carbon gasification efficiency. Int J Hydrog Energy. 2012;37:3177–85.
45. Chen Y, Guo L, Cao W, Jin H, Guo S, Zhang X. Hydrogen production by sewage sludge gasification in supercritical water with fluidized bed reactor. Int J Hydrog Energy. 2013;38:12991–9.
46. Sricharoenchaikul V. Assessment of black liquor gasification in supercritical water. Bioresour Technol. 2009;100:638–43.
47. Karakuş Y, Aynacı F, Kıpçak E, Akgün M. Hydrogen production from 2-propanol over Pt/Al_2O_3 and Ru/Al_2O_3 catalysts in supercritical water. Int J Hydrog Energy. 2013;38:7298–306.
48. Gadhe JB, Gupta RB. Hydrogen production by methanol reforming in supercritical water: catalysis by in-situ-generated copper nanoparticles. Int J Hydrog Energy. 2007;32:2374–81.
49. Byrd AJ, Pant KK, Gupta RB. Hydrogen production from ethanol by reforming in supercritical water using Ru/Al_2O_3 catalyst. Energy Fuel. 2007;21:3541–7.
50. D'Jesus P, Boukis N, Kraushaar-Czarnetzki B, Dinjus E. Influence of process variables on gasification of corn silage in supercritical water. Ind Eng Chem Res. 2006;45:1622–30.
51. D'Jesus P, Boukis N, Kraushaar-Czarnetzki B, Dinjus E. Gasification of corn and clover grass in supercritical water. Fuel. 2006;85:1032–8.
52. Arita T, Nakahara K, Nagamib K, Kajimoto O. hydrogen generation from ethanol in supercritical water without catalyst. Tetrahedron Lett. 2003;44:1083–6.
53. Boukis N, Diem V, Galla U, Dinjus E. Methanol reforming in supercritical water for hydrogen production. Combust Sci Technol. 2006;178:467–85.
54. Susanti RF, Nugroho A, Lee J, Kim Y, Kim J. Noncatalytic gasification of isooctane in supercritical water: a strategy for high-yield hydrogen production. Int J Hydrog Energy. 2011;36:3895–906.
55. Yu D, Aihara M, Antal Jr MJ. Hydrogen production by steam reforming glucose in supercritical water. Energy Fuel. 1993;7:574–7.

56. Xu D, Wang S, Tang X, Gong Y, Guo Y, Zhang J, Wang Y, Ma H, Zhou L. Influence of oxidation coefficient on product properties in sewage sludge treatment by supercritical water. Int J Hydrog Energy. 2013;38:1850–8.
57. Kıpçak E, Akgün M. Oxidative gasification of olive mill wastewater as a biomass source in supercritical water: effects on gasification yield and biofuel composition. J Supercrit Fluids. 2012;69:57–63.
58. Guo LJ, Lu YJ, Zhang XM, Ji CM, Guan Y, Pei AX. Hydrogen production by biomass gasification in supercritical water: a systematic experimental and analytical study. Catal Today. 2007;129:275–86.
59. Li S, Lu Y, Guo L, Zhang X. Hydrogen production by biomass gasification in supercritical water with bimetallic Ni-M/γAl_2O_3 catalysts (M = Cu, Co and Sn). Int J Hydrog Energy. 2011;36:14391–400.
60. Xu D, Wang S, Hu X, Chen C, Zhang Q, Gong Y. Catalytic gasification of glycine and glycerol in supercritical water. Int J Hydrog Energy. 2009;34:5357–64.
61. Watanabe M, Inomata H, Arai K. Catalytic hydrogen generation from biomass (glucose and cellulose) with ZrO_2 in supercritical water. Biomass Bioenergy. 2002;22:405–10.
62. Watanabe M, Inomata H, Osada M, Sato T, Adschiri T, Arai K. Catalytic effects of NaOH and ZrO_2 for partial oxidative gasification of n-hexadecane and lignin in supercritical water. Fuel. 2003;82:545–52.
63. Osada M, Sato T, Watanabe M, Adschiri T, Arai K. Low-temperature catalytic gasification of lignin and cellulose with a ruthenium catalyst in supercritical water. Energy Fuel. 2004;18:327–33.
64. Onwudili JA, Williams PT. Role of sodium hydroxide in the production of hydrogen gas from the hydrothermal gasification of biomass. Int J Hydrog Energy. 2009;34:5645–56.
65. Kruse A, Meier D, Rimbrecht P, Schacht M. Gasification of pyrocatechol in supercritical water in the presence of potassium hydroxide. Ind Eng Chem Res. 2000;39:4842–8.
66. Guo S, Guo L, Cao C, Yin J, Lu Y, Zhang X. Hydrogen production from glycerol by supercritical water gasification in a continuous flow tubular reactor. Int J Hydrog Energy. 2012;37:5559–68.
67. Hao XH, Guo LJ, Mao X, Zhang XM, Chen XJ. Hydrogen production from glucose used as a model compound of biomass gasified in supercritical water. Int J Hydrog Energy. 2003;28:55–61.
68. Sinag A, Kruse A, Rathert J. influence of the heating rate and the type of catalyst on the formation of key intermediates and on the generation of gases during hydropyrolysis of glucose in supercritical water in a batch reactor. Ind Eng Chem Res. 2004;42:502–8.
69. Madenoglu TG, Saglam M, Yuksel M, Ballice L. Simultaneous effect of temperature and pressure on catalytic hydrothermal gasification of glucose. J Supercrit Fluids. 2013;73:151–60.
70. Madenoglu TG, Boukis N, Saglam M, Yuksel M. Supercritical water gasification of real biomass feedstocks in continuous flow system. Int J Hydrog Energy. 2011;36:14408–15.
71. Taylor AD, DiLeo GJ, Sun K. Hydrogen production and performance of nickel based catalysts synthesized using supercritical fluids for the gasification of biomass. Appl Catal B Environ. 2009;93:126–33.
72. Yoshida T, Oshima Y, Matsumura Y. Gasification of biomass model compounds and real biomass in supercritical water. Biomass Bioenergy. 2004;26:71–8.
73. Lu Y, Li S, Guo L. Hydrogen production by supercritical water gasification of glucose with $Ni/CeO_2/Al_2O_3$: effect of Ce loading. Fuel. 2013;103:193–9.
74. Osada M, Sato T, Watanabe M, Arai K, Shirai M. Water density effect on lignin gasification over supported noble metal catalysts in supercritical water. Energy Fuel. 2006;20:930–5.
75. Therdthianwong S, Srisiriwat N, Therdthianwong A, Croiset E. Reforming of bioethanol over Ni/Al_2O_3 and $Ni/CeZrO_2/Al_2O_3$ catalysts in supercritical water for hydrogen production. Int J Hydrog Energy. 2011;36:2877–86.
76. Lee I. Effect of metal addition to Ni/activated charcoal catalyst on gasification of glucose in supercritical water. Int J Hydrog Energy. 2011;36:8869–77.

77. Osada M, Hiyoshi N, Sato O, Arai K, Shirai M. Reaction pathway for catalytic gasification of lignin in presence of sulfur in supercritical water. Energy Fuel. 2007;21:1854–8.
78. Byrd AJ, Kumar S, Kong L, Ramsurn H, Gupta RB. Hydrogen production from catalytic gasification of switchgrass biocrude in supercritical water. Int J Hydrog Energy. 2011;36:3426–33.
79. Byrd AJ, Pant KK, Gupta RB. Hydrogen production from glucose using Ru/Al_2O_3 catalyst in supercritical water. Ind Eng Chem Res. 2007;46:3574–9.
80. Onwudili JA, Williams PT. Hydrogen and methane selectivity during alkaline supercritical water gasification of biomass with ruthenium-alumina catalyst. Appl Catal B Environ. 2013;132–133:70–9.
81. Yamaguchi A, Hiyoshi N, Sato O, Bando KK, Osada M, Shirai M. Hydrogen production from woody biomass over supported metal catalysts in supercritical water. Catal Today. 2009;146:192–5.
82. Youssef EA, Elbeshbishy E, Hafez H, Nakhla G, Charpentier P. Sequential supercritical water gasification and partial oxidation of hog manure. Int J Hydrog Energy. 2010;35:11756–67.
83. Kıpçak E, Akgün M. In-situ gas fuel production during the treatment of textile wastewater at supercritical conditions. Water Sci Technol. 2013;67(5):1058–67.
84. Hendry D, Venkitasamy C, Wilkinson N, Jacoby W. Exploration of the effect of process variables on the production of high-value fuel gas from glucose via supercritical water gasification. Bioresour Technol. 2011;102:3480–7.
85. Guan Q, Savage PE, Weib C. Gasification of alga Nannochloropsis Sp. in supercritical water. J Supercrit Fluids. 2012;61:139–45.
86. Wilkinson N, Wickramathilaka M, Hendry D, Miller A, Espanani R, Jacoby W. Rate determination of supercritical water gasification of primary sewage sludge as a replacement for anaerobic digestion. Bioresour Technol. 2012;124:269–75.
87. Escot Bocanegra P, Reverte C, Aymonier C, Loppinet-Serani A, Barsan MM, Butler IS, Kozinski JA, Gökalp I. Gasification study of winery waste using a hydrothermal diamond anvil cell. J Supercrit Fluids. 2010;53:72–81.
88. Osada M, Sato O, Arai K, Shirai M. Stability of supported ruthenium catalysts for lignin gasification in supercritical water. Energy Fuel. 2006;20:2337–43.
89. Byrd AJ, Pant KK, Gupta RB. Hydrogen production from glycerol by reforming in supercritical water over Ru/Al_2O_3 catalyst. Fuel. 2008;87:2956–60.
90. Furusawa T, Sato T, Sugito H, Miura Y, Ishiyama Y, Sato M, Itoh N, Suzuki N. Hydrogen production from the gasification of lignin with nickel catalysts in supercritical water. Int J Hydrog Energy. 2007;32:699–704.
91. May A, Salvado J, Torras C, Montane D. Catalytic gasification of glycerol in supercritical water. Chem Eng J. 2010;160:751–9.
92. Youssef EA, Chowdhury MBI, Nakhla G, Charpentier P. Effect of nickel loading on hydrogen production and chemical oxygen demand (COD) destruction from glucose oxidation and gasification in supercritical water. Int J Hydrog Energy. 2010;35:5034–42.
93. Zhang L, Champagne P, Xu CC. Supercritical water gasification of an aqueous by-product from biomass hydrothermal liquefaction with novel Ru modified Ni catalysts. Bioresour Technol. 2011;102:8279–87.
94. Waldner MH, Vogel F. Renewable production of methane from woody biomass by catalytic hydrothermal gasification. Ind Eng Chem Res. 2005;44:4543–51.
95. Muangrat R, Onwudili JA, Williams PT. Alkaline subcritical water gasification of dairy industry waste (whey). Bioresour Technol. 2011;102:6331–5.

Part III
Electrochemical and Solar Conversions

Chapter 8
Hydrogen Production from Water and Air Through Solid Oxide Electrolysis

Kongfa Chen, Dehua Dong, and San Ping Jiang

Abstract High-temperature solid oxide electrolyzers (SOEs) or solid oxide electrolysis cells (SOECs) are electrochemical devices for the efficient production of hydrogen or syngas as feedstock for liquid fuels such as methanol, gasoline, and diesel using electricity and unused heat from nuclear plants, steelmakers, or renewable energy sources. This chapter aims to review the principles, status, and progress in the electrochemical hydrogen production process by water electrolysis such as low-temperature alkaline electrolysis cells and polymer exchange membrane electrolysis cells with particular emphasis on the new electrolysis technologies of high-temperature SOECs. The material and material degradation issues associated with high-temperature electrolysis processes are reviewed and discussed.

Keywords Hydrogen generation • Solid oxide electrolyzer • Materials • Principles

8.1 Introduction

Hydrogen plays an important role in the energy sector and is one of the most important feedstocks for the production of hydrocarbon fuels and chemicals. Hydrogen, which has high energy density and environmental friendliness, is considered to be an ideal energy carrier for renewable energy storage [1, 2]. However, hydrogen does not exist in its pure state in nature, like oxygen, and has to be produced from hydrogen-containing resources such as natural gas, coal, biomass, and water by reforming, gasification, electrolysis, or thermal decomposition [3]. Currently, about 96 % of hydrogen is produced from fossil fuels [4, 5]. The industrial processes produce significant amounts of CO_2, which is a major greenhouse gas (GHG) and has a strong impact on global warming. Due to the depletion of fossil fuels, hydrogen production from non-renewable hydrocarbon sources is not sustainable.

K. Chen • D. Dong • S.P. Jiang (✉)
Department of Chemical Engineering, Fuels and Energy Technology Institute,
Curtin University, Perth, WA 6102, Australia
e-mail: S.Jiang@curtin.edu.au

Z. Fang et al. (eds.), *Production of Hydrogen from Renewable Resources*,
Biofuels and Biorefineries 5, DOI 10.1007/978-94-017-7330-0_8

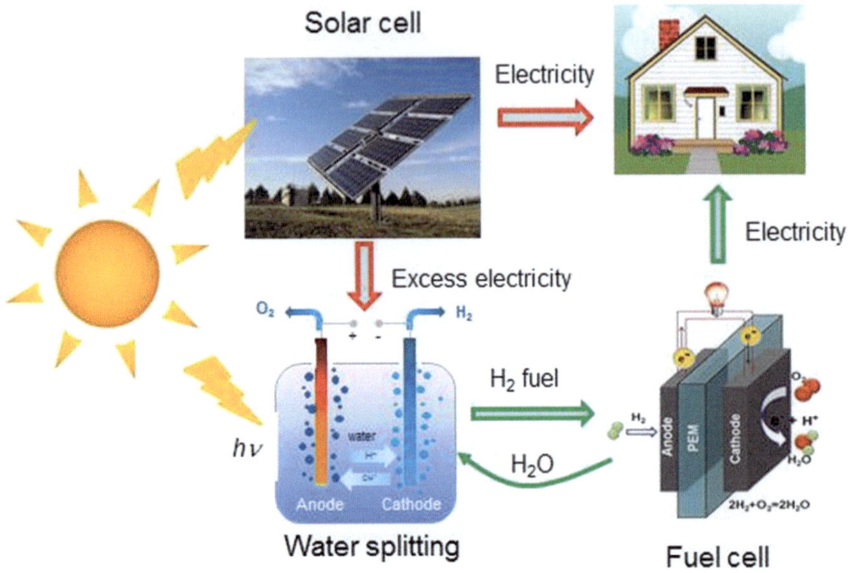

Fig. 8.1 Schematic representation of the energy cycle using water splitting to store excess solar electrical energy and fuel cells to provide the electricity during the low peak period

Hydrogen production from water splitting or electrolysis derived from renewable energy, such as hydro, solar (photovoltaic), or wind energy, is sustainable and provides an environmentally friendly pathway to contribute toward meeting the constantly growing demand for energy supply and storage [6, 7]. For example, conversion of intermittent or excess solar (photovoltaic, PV) electrical energy into chemical energy by water electrolysis into hydrogen fuels can be used to store surplus solar energy during peak generation periods. During low-generation periods (e.g., the night), these H_2 fuels can then be used to efficiently regenerate electricity via fuel cells. Fuel cells are energy conversion devices that electrochemically convert fuels such as hydrogen to electricity with high power density, high efficiency, and low GHG emissions [8]. Using H_2 as a fuel, the only by-product of the fuel cell reaction is water, which can be fed back into the water electrolysis process. Figure 8.1 shows a schematic of the role of water electrolysis (electrochemical or photoelectrochemical types) and fuel cells in such environmentally friendly energy solutions.

Water electrolysis for hydrogen production has many advantages such as high purity, simple process, and no GHG emission. Industrial water electrolyzers are mutual technology and have been established for more than 100 years. However, the energy requirement in conventional industrial electrolyzers is high, 4.5–5.0 kWh/ m^3 H_2. Thus, there are urgent needs to enhance the efficiency of water electrolyzers and reduce energy consumption. The main focus of this chapter is to briefly introduce the principle of water electrolysis and various types of water electrolyzers with particular emphasis on the new electrolysis systems of high-temperature solid oxide electrolyzers (SOEs) or solid oxide electrolysis cells (SOECs). Fundamental and materials development of SOECs are reviewed and discussed.

8.2 Water Electrolysis

Water electrolysis uses electricity to drive the reactions and produces hydrogen and oxygen at cathode and anode, respectively. Charges carried by ionic species such as OH^-, H^+, or O^{2-}, move through an ionic conducting electrolyte to maintain the electrical neutrality of the system. According to the ionic conducting electrolyte, there are three types of electrolyzers as shown in Fig. 8.2, namely, alkaline electrolyzer, proton exchange membrane (PEM) electrolyzer, and high-temperature SOECs.

Cell voltage, U, is an important parameter to represent energy consumption of water electrolyzers. The thermodynamic decomposition voltage of water is 1.23 V at 25 °C. According to Faraday's law, the electric quantity, Q, to produce 1 mol H_2 (i.e., 22.4 L at standard condition) is 2 F. Thus, theoretical energy consumption, W_t, to produce 1 m^3 H_2 is

$$\begin{aligned} W_{\mathrm{t}} = UQ &= 1.23 \times \left(2 \times \frac{1000}{22.4} \times 96,485 \times \frac{1}{1000} \times \frac{1}{3600}\right) \\ &= 2.94 \ \ \mathrm{kWh/m^3H_2} \end{aligned} \tag{8.1}$$

However, in practical cells, gas evolution reaction occurs at 1.65–1.7 V, and in industrial cells the cell voltages of water electrolysis are typically in the range of

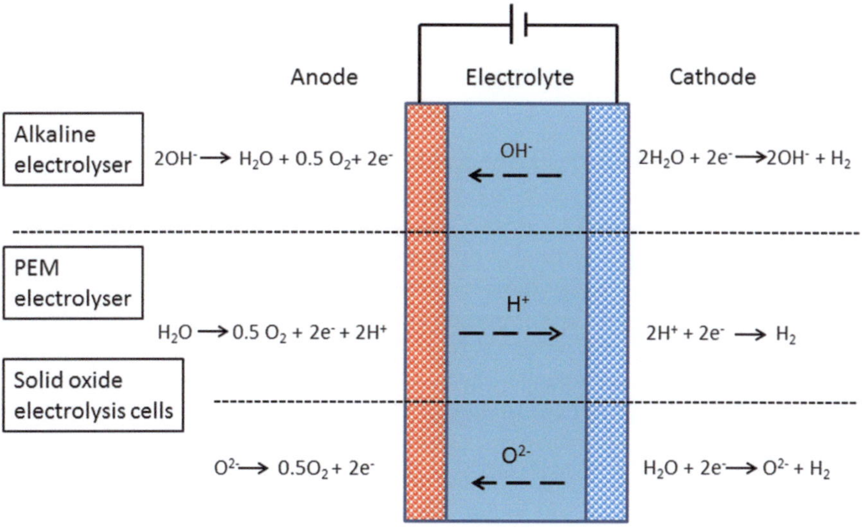

Fig. 8.2 Classification of water electrolyzers

1.8–2.6 V, due to high overpotential and large ohmic voltage drop. If the cell voltage is 2.0 V, practical energy consumption, W_p, is

$$\begin{aligned} W_\mathrm{t} = UQ = 2.0 \times \left(2 \times \frac{1000}{22.4} \times 96,485 \times \frac{1}{1000} \times \frac{1}{3600}\right) \\ = 4.79\ \mathrm{kWh/m^3H_2} \end{aligned} \tag{8.2}$$

The energy efficiency, η_e, of water electrolysis for hydrogen production is

$$\eta_\mathrm{e} = \frac{2.94}{4.79} \times 100 = 61.4\ \% \tag{8.3}$$

Thus, increasing energy efficiency of water electrolysis by reducing the overpotentials for the H_2 evolution reaction at the cathode, O_2 evolution reaction at the anode, and overall cell resistance will significantly enhance the utilization efficiency of renewable primary energy.

8.2.1 Alkaline Electrolyzer

Alkaline electrolyzers are the dominant type of units in commercial operation today. The electrolyte is typically 20–30 wt.% KOH and operating temperatures are in the range of 70–100 °C. Porous Raney nickel is the most common electrode used in alkaline electrolysis cells. The electrode reactions are

$$\text{Cathode}: 2\mathrm{H_2O}(l) + 2\mathrm{e}^- \rightarrow \mathrm{H_2}(g) + 2\mathrm{OH}^-(aq) \tag{8.4}$$

$$\text{Anode}: 2\mathrm{OH}^-(aq) \rightarrow \mathrm{H_2O}(l) + 1/2\mathrm{O_2}(g) + 2\mathrm{e}^- \tag{8.5}$$

Under a polarization potential between cathode and anode, water is dissociated into OH^- and H_2 on the cathode side, and OH^- moves through electrolyte and reaches the anode side, where OH^- releases oxygen molecules and electrons with a by-product of water. The electrolysis technology has been developed for commercial applications after being discovered by Troostwijk and Diemann in 1789 [9]. The system is simple and low-cost, and its efficiency is in the range of 47–82 % [3]. The main disadvantage of the technology is the low purity of hydrogen product due to the cross diffusion of hydrogen and oxygen between electrodes, which also causes safety issues related to hydrogen explosion [10].

During water electrolysis, bubbles cannot be removed rapidly from the electrolytic system. Bubble coverage on the electrode surface and bubble dispersion in the electrolyte, i.e., the bubble effect, can lead to high ohmic voltage drop and large reaction overpotential. The bubble effect is one of the key factors for high energy consumption and has been extensively reviewed recently [11]. The hydrogen production rate is also limited by low current load as the electrolysis cells have high ohmic resistances.

8.2.2 PEM Electrolyzer

PEM electrolyzers use proton-conducting polymeric membranes as electrolytes and operate at a similar temperature range. In PEM cells, protons produced at the anode are selectively conducted across a polymer membrane and the following electrode reactions occur:

$$\text{Cathode}: 2\text{H}^+(aq) + 2\text{e}^- \rightarrow \text{H}_2(g) \tag{8.6}$$

$$\text{Anode}: \text{H}_2\text{O}(l) \rightarrow 2\text{H}^+(aq) \ + \ 1/2\text{O}_2(g) + 2\text{e}^- \tag{8.7}$$

Compared with alkaline electrolyzers, PEM cells can be operated at high current densities to increase hydrogen production rate. In addition, solid polymer membrane can effectively prevent gas diffusion, thus producing high purity of hydrogen. The technology is also well developed and efficiency ranges from 48 % to 65 % [3]. However, the commercialization is limited by the high cost and low stability of noble metal-based electrocatalysts. The use of expensive noble metal electrocatalysts (typically Pt particles) and expensive membranes greatly increases capital costs of hydrogen production. Noble metals can be poisoned by impurities such as CO and lose catalytic activity. Another issue is the oxidation of carbon-based supports and the gas diffusion layer. Less expensive materials are needed.

8.2.3 Solid Oxide Electrolysis Cells

As the splitting of water is an endothermic reaction, electricity demand and decomposition voltage decrease with an increase in temperature. Therefore it is beneficial to operate at elevated temperatures for water electrolysis. Overpotentials and ohmic voltage drops decrease considerably at high operation temperatures. High-temperature SOECs are reversibly operated solid oxide fuel cells (SOFCs) as in Fig. 8.3. SOFC is an energy conversion device to electrochemically convert chemical energy of fuels such as hydrogen, hydrocarbons, and biofuels to electricity and is the most efficient among various fuel cells [8]. Driven by the difference in oxygen chemical potential, oxygen ions, O^{2-}, generated at the oxygen electrode migrate through an oxygen ion-conducting solid electrolyte such as yttria-stabilized zirconia or YSZ to the fuel electrode where they are consumed by oxidation of fuels. The same device can be operated under SOEC mode. In SOEC mode, steam (i.e., H_2O) is oxidized at the hydrogen electrode side to produce H_2 and O^{2-}, which migrate through YSZ to the oxygen electrode side, producing pure O_2 [12, 13]. The SOECs can reach an efficiency of ~90 % [14]. SOECs can also use proton-conducting electrolytes, which have the same reaction on electrodes as PEM electrolyzers [15]. Compared with commercial alkaline electrolyzers and PEM electrolyzers, the SOECs are still in the early stage of development. However, it is a promising technology for large-scale hydrogen production and attracts wide

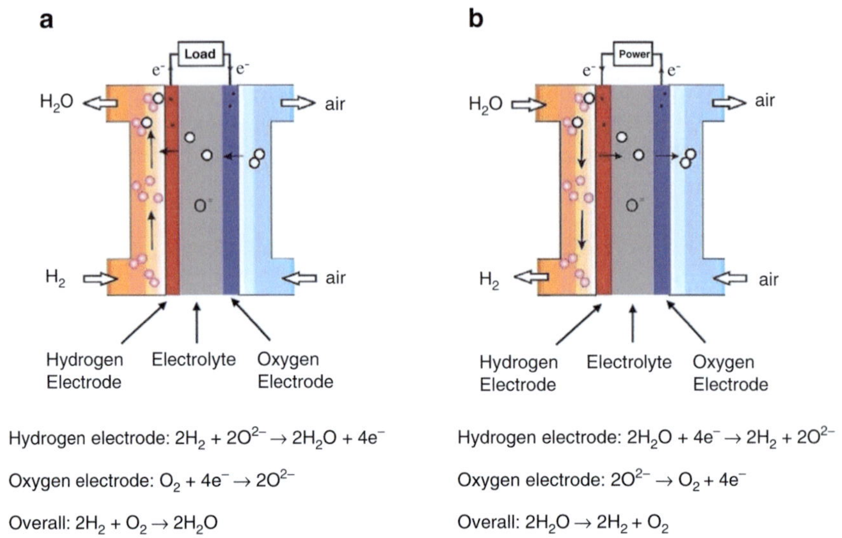

Fig. 8.3 Schematic diagrams of the operation principles of (**a**) solid oxide fuel cell (SOFC) and (**b**) solid oxide electrolysis cell (SOEC)

research interests. High-temperature SOECs have significant advantages over low-temperature water electrolysis, including high efficiency and no requirement of expensive noble metal electrocatalysts due to the high operation temperature of 800–900 °C. At such high temperatures, transition metals such as Ni and metal oxides can be used as the electrodes.

The electrode reactions of an SOEC are shown as below:

$$\text{Cathode}: \ H_2O(g) + 2e^- \rightarrow H_2(g) + O^{2-} \tag{8.8}$$

$$\text{Anode}: O^{2-} \rightarrow 1/2O_2(g) + 2e^- \tag{8.9}$$

Thus, the overall water electrolysis reaction is

$$H_2O(g) + \text{heat} + \text{electricity} \rightarrow H_2(g) + 1/2O_2(g) \tag{8.10}$$

According to reaction thermodynamics, the theoretical energy required for water electrolysis (ΔH) is the sum of thermal energy demand (Q_t) and electrical energy demand, i.e., the Gibbs free energy change (ΔG), and can be expressed as

$$\Delta H = \Delta G + Q_t = \Delta G + T\Delta S \tag{8.11}$$

The thermal energy demand is equal to the product of the absolute temperature (T) and the entropy change (ΔS).

The overall energy requirement is equal to the summary of electrical energy and thermal energy. Figure 8.4 shows the energy requirements of the electrolysis as it varies with temperature. At 25 °C, electrical energy demand (ΔG) and theoretical decomposition or reversible voltage (U^θ) are 474 kJ/mol and 1.23 V, respectively.

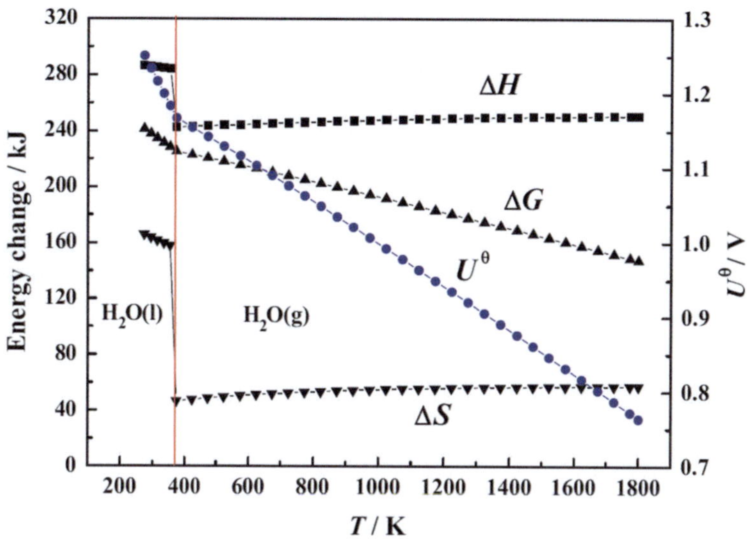

Fig. 8.4 Effect of temperature on thermodynamic properties of the water electrolysis process. For the meanings of symbols, please see text (*Source*: After Ref. [11]. Reproduced with permission from Elsevier)

As temperature increases to 900 °C, ΔG and U^θ decrease to 366 kJ/mol and 0.95 V, respectively. Since the most electricity is produced by converting thermal energy into electrical energy, operating at high temperatures decreases the overall energy requirement. As compared to that at 25 °C, the theoretical energy savings is as much as 23 %. Moreover, waste heat, such as that from nuclear reactions, can be used to maintain high-temperature electrolysis [5]. Therefore, high-temperature SOEC electrolysis can achieve a much higher energy efficiency than low-temperature electrolysis.

In practical SOECs, the cell voltage required is always larger than the theoretical reversible voltage because extra voltage is needed to overcome the voltage losses due to the overpotentials of electrochemical reactions at the anode and cathode, the internal resistance of the electrodes and electrolyte, and the external resistance of current collectors.

8.3 Assessment and Application Status

8.3.1 Technical Assessment

As SOECs are essentially reversibly operated SOFCs, electrode and electrolyte materials developed for SOFCs have been adopted for the development of high-performance SOECs. For example, both tubular and planar configurations have been employed for SOEC stacks [16]. Nuclear power plants produce high-quality steam and electricity simultaneously, which can be coupled with SOECs to produce valuable hydrogen fuel [17]. Therefore, from a technical point of view, high-temperature electrolysis via SOECs is a promising technology for storage of renewable electricity [18]. Very different from the hydrogen production from fossil fuels or biomass, water electrolysis does not produce GHGs such as CO_2, SO_x, and NO_x. As shown above, high-temperature electrolysis requires the least amount of electricity and has the highest energy efficiency as compared to low-temperature electrolysis. Moreover, waste heat from chemical, metallurgical, and thermal power generation industries can be used to preheat steam to operating temperature, thus achieving better economy. SOECs can also be coupled with other high-temperature reaction processes, such as methane reforming. The oxygen produced on the anode side can be used for syngas production via the partial oxidation of methane, leading to high energy efficiency and eliminating the cost of pure oxygen supply for the methane reforming by coupling the two processes.

However, there are some challenges needed to be targeted to realize the commercialization of SOEC technologies.

(a) *High-performance materials* – Power consumption for the electrolysis depends on electrolysis cell resistances. To reduce the power consumption, high-performance electrode and electrolyte materials are required. For example, cell ohmic resistance is mainly determined by the ionic conductivity of the electrolyte. Electrode polarization resistances are related to the microstructure and catalytic activities of electrodes. The conductivities (ionic and/or electronic) of electrodes and interconnect also affect cell performance.
(b) *Performance durability of electrode and electrolyte materials* – Some materials show high performance, but they are not stable under SOEC operating conditions. For example, $Ba_{0.5}Sr_{0.5}Co_{0.2}Fe_{0.8}O_3$-based electrodes show very low electrode polarization resistances owing to high catalytic activity but degrade rapidly during operation in air because Ba can readily react with CO_2 in the air stream to form carbonate. Apart from material stability, electrode microstructure change during high-temperature operation also can cause performance degradation. High temperatures induce the agglomeration and sintering of fine electrode microstructures and reduce the effective reaction areas. For example, Ni is commonly used in the cathode and acts as an electronic conductor and catalyst. During electrolysis, Ni aggregation results in reduced conductivity and catalytic activity. In addition, impurities from input can cause rapid degradation of performance [19].
(c) *Capital cost* – The issues of the high capital cost of the SOECs are the same as those facing the development of SOFCs. The SOEC cost includes material costs and in particular the manufacturing cost and assembly cost. The manufacture of the cell components involves many steps, and some steps require special control or equipment, such as thin electrolyte film deposition techniques. Apart from the materials for the cell components, interconnections and current collectors are required to construct cell stacks. The limited cyclability of high-temperature SOEC stacks adds significantly to the overall capital and maintenance cost of the technology.

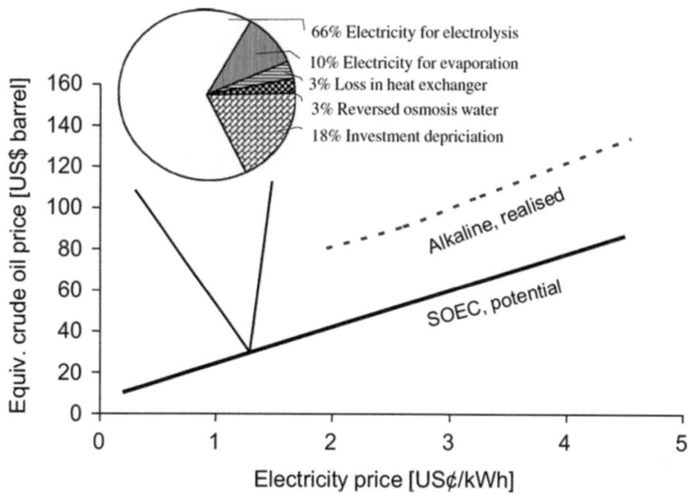

Fig. 8.5 H_2 production vs. electricity price (*Source*: After Ref. [2]. Reproduced with permission from Elsevier)

8.3.2 Economic Assessment

The major issue associated with the commercial viability of water electrolysis is the high electricity and capital cost. As shown in Fig. 8.5, the production cost of hydrogen is mainly determined by electricity cost, which depends on electricity price. As there is no commercial case of hydrogen production by SOECs, the estimated hydrogen cost based on the data in the laboratory has large uncertainties. For example, Jensen et al. from Risø National Laboratory (Denmark) calculated the hydrogen cost of 5US$/GJ (0.71US$/kg) [2]. The hydrogen production cost was calculated based on performance data of current density of -3.6 A/cm^2 at the cell voltage of 1.48 V.

Based on the hydrogen price of 1€/kg, it was concluded that high-temperature electrolysis becomes competitive with H_2 production from fossil fuels when the electricity price is below 0.02–0.03 €/KWh [14]. Brisse et al. reported similar results [20]. The hydrogen production cost is particularly interesting when electricity price is close to or below 50 €/MWe. With this electricity price the hydrogen production cost is close to and even lower than the cost obtained with alkaline electrolyzers (~3.5 €/kg for large H_2 capacity of electrolyzers).

The development of the SOECs also brings significant environmental benefit. In the consideration of the shortage of fossil fuels, renewable energy will be the main energy resource in the future. It is anticipated that hydrogen produced by SOEC is renewable and will play an important role in renewable energy supply and storage industries.

8.3.3 Co-electrolysis of Steam and CO_2

CO_2 is one of the most important GHGs. To reduce CO_2 emissions, carbon capture and sequestration (CCS) is developed as a short- to midterm solution to mitigate environmental impact, and this allows continued use of fossil fuels until renewable energy technology reaches maturity. However, current CCS technology is considered to be too expensive [21], and the alternative is to develop CO_2 utilization technologies. High-temperature co-electrolysis of CO_2 and steam via SOECs makes very efficient use of electricity and heat, provides high reaction rates, and directly produces syngas (CO/H_2 mixture) [22, 23]. Syngas is feedstock of liquid fuel such as gasoline or diesel production by Fischer-Tropsch synthesis route. The reaction of CO_2 electrolysis at the cathode is

$$CO_2(g) + 2e^- \rightarrow CO(g) + O^{2-} \tag{8.12}$$

Compared with water electrolysis, CO_2 electrolysis consumes more power since CO_2 is highly stable. Moreover, CO_2 electrolysis shows rapid degradation due to carbon formation on cathode. Ni-based cathodes are commonly used in SOECs because it is cheaper than noble metals and very active for the electrode reaction. However, CO can be dissociated into carbon over Ni catalyst via Boudouard reaction:

$$2CO \rightarrow CO_2 + C \tag{8.13}$$

The carbon species, amorphous or fibrous, cover the Ni catalyst, deactivating the Ni-based cathode. In the presence of H_2O and CO_2, co-electrolysis of steam and CO_2 occurs via reactions (8.8) and (8.12), producing hydrogen and CO. Compared with CO_2 electrolysis, the co-electrolysis requires less power due to lower cell resistance. One possible reaction is that CO is formed via reverse water-gas shift (RWGS) reaction, and the required hydrogen for the RWGS reaction is produced by electrolysis, as shown below:

$$CO_2 + H_2 \rightarrow CO + H_2O \tag{8.14}$$

Another advantage of the co-electrolysis is that carbon formation can be diminished due to the presence of steam.

Graves et al. [24] did a detailed process and energy balance analysis of a pathway based on CO_2 capture from the atmosphere, high-temperature co-electrolysis of H_2O and CO_2, and Fischer-Tropsch synthesis (Fig. 8.6). From the energy balance, electrolysis is the major energy-consuming step. Assuming the process is entirely driven by electricity and the waste heat is utilized to preheat the CO_2 and H_2O, the net electricity-to-fuel efficiency was calculated as ~70 % [24]. The same group also studied the co-electrolysis of H_2O and CO_2 and found that the performance of SOECs for co-electrolysis of H_2O and CO_2 is higher than for CO_2 electrolysis but slightly lower than H_2O electrolysis, indicating that RWGS plays a role during co-electrolysis [25].

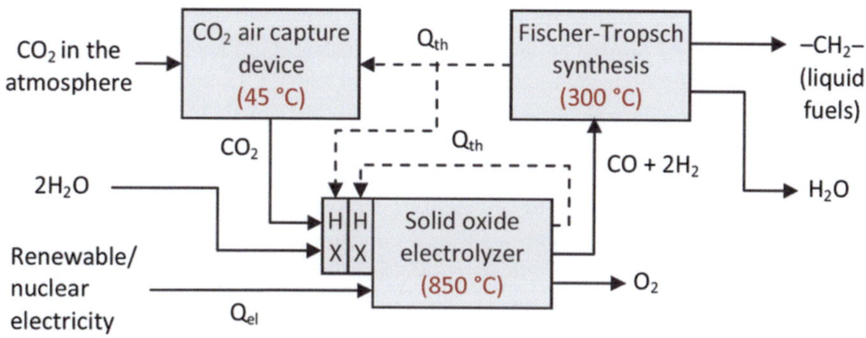

Fig. 8.6 Schematic of the proposed CO_2-recycled synthetic fuel production process. $-CH_2-$ represents a hydrocarbon. *HX* heat exchanger (*Source*: After Ref. [24]. Reproduced with permission from Elsevier)

8.4 Key Materials for SOECs

8.4.1 Electrolyte

The 8 mol% Y_2O_3-stabilized ZrO_2 (YSZ) is a commonly used high-temperature electrolyte material due to its good oxygen ion conductivity and high stability under oxidizing and reducing atmosphere [16, 26]. Alternative electrolyte materials such as scandia and ceria co-doped zirconia (10Sc1CeSZ) [27], rare earth metals doped ceria [28], and $La_{0.9}Sr_{0.1}Ga_{0.8}Mg_{0.2}O_3$ (LSGM) [29, 30] with higher ionic conductivity have been investigated, so as to reduce the internal ohmic loss and maintain high efficiency at a reduced temperature. However, the drawback of Ce-containing electrolytes is that the reduction of Ce^{4+} to Ce^{3+} at high temperatures and under low oxygen partial pressures on the hydrogen electrode side leads to current leak, efficiency loss, and even breakdown under the SOEC operating conditions [31–33]. Therefore, doped ceria is considered not applicable for SOEC. LSGM is chemically incompatible with Ni, and a doped ceria buffer layer is required. Proton conducting oxides such as $BaCe_{0.5}Zr_{0.3}Y_{0.2}O_{3-\delta}$ and $SrZr_{0.9}Yb_{0.1}O_{3-\delta}$ have also been studied [15, 34, 35], though substantial efforts are needed to demonstrate cell performance and stability during long-term use.

8.4.2 Oxygen Electrode

Sr-doped $LaMnO_3$ (LSM) perovskite oxide is a common oxygen electrode for high-temperature SOECs [36–38], because of its good catalytic activity at elevated temperatures, reasonably high electrical conductivity, and excellent thermal and chemical compatibility with YSZ electrolyte. However, LSM is a predominantly electronic conductor with negligible ionic conductivity, and the oxygen oxidation reaction is primarily limited at the electrode/electrolyte interface [39, 40]. Addition of ion-conducting oxides such as YSZ to form an LSM-YSZ composite electrode enlarges the three-phase boundaries (TPBs) to the bulk of the electrode and thus substantially improves the electrode activity [41, 42].

Mixed ionic-electronic conductor (MIEC) $La_{1-x}Sr_xCo_{1-y}Fe_yO_{3-\delta}$, typically $La_{0.6}Sr_{0.4}Co_{0.2}Fe_{0.8}O_{3-\delta}$ (LSCF), is one of the most investigated oxygen electrodes of SOECs. LSCF possesses oxygen vacancy concentration and ion mobility several orders of magnitude higher than LSM and therefore shows much better electrocatalytic activity for the oxygen evolution reaction than the LSM [43, 44]. LSCF-doped ceria composite electrodes are also investigated to enhance the electrode activity and/or reduce the TEC of electrode [45, 46]. Other MIEC oxygen electrodes have been reported including (La,Sr)XO_3 (X=Co, Fe, Cu) [44], $Ba_{0.5}Sr_{0.5}Co_{0.8}Fe_{0.2}O_{3-\delta}$ [47], $Sm_{0.5}Sr_{0.5}CoO_{3-\delta}$ [48], $SmBaCo_2O_{5+\delta}$ [49], $Ba_{0.9}Co_{0.5}Fe_{0.4}Nb_{0.1}O_{3-\delta}$ [50], $SrCo_{1-x}Mo_xO_{3-\delta}$ [51], $La_{2-x}Sr_xCo_{0.5}Ni_{0.5}O_{4\pm\delta}$ [52], $Sr_2Fe_{1.5}Mo_{0.5}O_{6-\delta}$ [53], and $Ln_2NiO_{4+\delta}$ (Ln=Nd, La, Pr) [54, 55].

Dispersion of electrocatalytically active nanoparticles such as GDC and Pd into LSM oxygen electrodes by infiltration substantially enhances the electrode activity for oxygen evolution reaction [56–58]. Another type of effective nanostructured oxygen electrode is prepared by incorporating catalytic nanoparticles such as LSM, LSCF, (La,Sr)CoO_3 (LSC) and (La,Sr)FeO_3 (LSF), and Pd into porous YSZ scaffold [59–62]. The thermal expansion coefficient (TEC) of the infiltrated electrode is largely dependent on the YSZ scaffold [63], which is beneficial to mitigate the large thermal mismatch between cobaltite containing perovskite oxygen electrode and electrolyte.

8.4.3 Hydrogen Electrode

Ni cermet is a state-of-the-art hydrogen electrode of SOECs, due to its high electrical conductivity, high electrocatalytic activity, and low price. Addition of Fe to form a $Ni_{0.9}Fe_{0.1}$ alloy is reported to restrict the sintering and agglomeration of Ni phase, enhances the electrocatalytic activity, and reduces the ohmic resistance [30]. Ion conducting phase such as YSZ is usually added to form a Ni-YSZ composite hydrogen electrode, leading to enlarged TPBs and better thermal and structural stability. The incorporation of electrocatalytically active nanoparticles such as gadolinium-doped ceria (GDC), SDC, and $Mo_{0.1}Ce_{0.9}O_2$ oxides in the Ni-YSZ hydrogen electrode enhances the ionic conductivity and TPBs and thus substantially boosts the electrocatalytic activity and/or operating stability for the water-splitting reaction [64–66]. Nanostructured Ni-infiltrated samaria-doped ceria (SDC) is also prepared to maximize the reaction area [67, 68].

Oxide materials such as $(La_{0.75}Sr_{0.25})_{0.95}Cr_{0.5}Mn_{0.5}O_3$ (LSCM) [69, 70], (La,Sr)TiO_3 [44, 71], Sr_2FeNbO_6 [72], and $Sr_{0.94}Ti_{0.9}Nb_{0.10}O_{3-\delta}$ [73] are investigated as alternative anode materials due to the high phase and structural stability under reducing and inert environment. A small amount of active metal catalyst such as Ni and Fe nanoparticles is added to enhance the electrocatalytic activity and increase the current efficiency of the electrode [74, 75]. In situ exsolution under reducing conditions is an alternative means to uniformly disperse active metal nanoparticles in the hydrogen electrode. Yang et al. developed nanosized CoFe alloy dispersed K_2NiF_4-type $Pr_{0.8}Sr_{1.2}(Co,Fe)_{0.8}Nb_{0.2}O_{4+\delta}$ by annealing $Pr_{0.4}Sr_{0.6}Co_{0.2}Fe_{0.7}Nb_{0.1}O_{3-\delta}$ perovskite in H_2 at 900 °C [76]. The cell showed a high current density of 1.57 A cm^{-2} at electrolysis voltage of 1.3 V, 60 vol.% absolute humidity at 900 °C, and good stability during the electrolysis test for 200 h. Tsekouras et al. reported that adoption of A-site deficiency of $La_{0.4}Sr_{0.4}M_{0.06}Ti_{0.94}O_{3-\delta}$ (M=Fe^{3+} or Ni^{2+}) perovskite oxide facilitates the exsolution of Fe or Ni nanoparticles [77]. Exsolution of Ni and Cu metals is also found in $(La_{0.75}Sr_{0.25})_{0.95}(Cr_{0.5}Ni_{0.5})_{0.95}Ni_{0.05}O_{3-\delta}$, $NbTi_{0.5}M_{0.5}O_4$ (M=Ni,Cu), and $La_{0.5}Sr_{0.5}Ti_{0.75}Ni_{0.25}O_3$ [78–80].

8.5 Performance Degradation of SOEC Electrodes

A major challenge of developing SOECs is the significant performance degradation of electrodes during long-term operation. The SOECs have a considerably high degradation rate varying from 5.6 %/1,000 h for an LSCF oxygen electrode operating for ~4,000 h [81] to as high as ~20 %/1,000 h for an LSM oxygen electrode operating for 2,000 h [82]. This is much higher than that of the state-of-the-art SOFCs, such as 0.5 %/1,000 h for an LSM oxygen electrode running for 19,000 h [83] and 1.4 %/1,000 h for an LSCF oxygen electrode operating for 10,000 h [84]. The structural, chemical, and thermal stability of electrodes play a key role in determining the durability of SOECs.

8.5.1 Oxygen Electrodes

8.5.1.1 LSM

With the influence of anodic current passage under the SOEC operating conditions, electrode activity and structural deterioration tend to occur on the LSM oxygen electrodes. Figure 8.7 shows the typical polarization curves of an LSM oxygen electrode under an anodic current of 500 mA/cm^2 [85]. It is clear that at the final stage of test, stage III, the anodic potential, electrode ohmic resistance, and polarization resistance increase dramatically. Complete electrode delamination from the electrolyte occurred after the test.

Several degradation/failure modes of LSM oxygen electrodes have been proposed. The most common mode of failure is the occurrence of delamination at the oxygen electrode/electrolyte interface [86]. It is generally considered that the

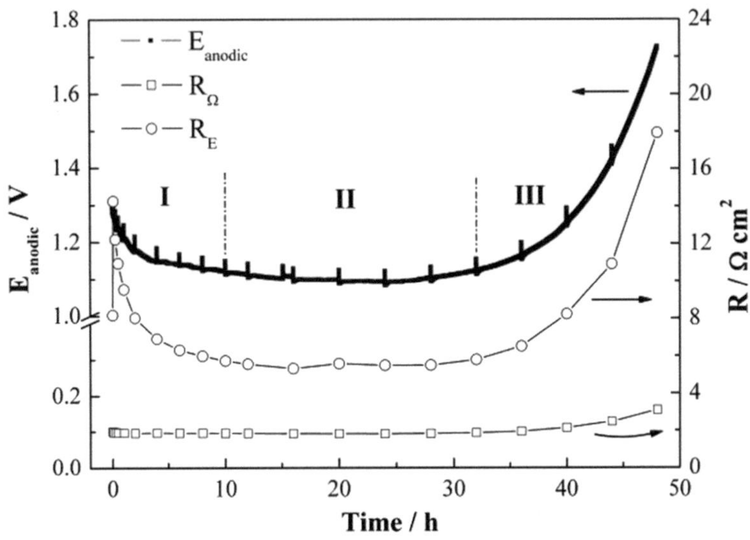

Fig. 8.7 Polarization curves of an LSM oxygen electrode as a function of anodic current passage time at 500 mA/cm^2 and 800 °C in air. E_{anodic}, R_{Ω}, and R_E are anodic potential, electrode ohmic resistance, and polarization resistance, respectively (*Source*: After Ref. [85]. Reproduced with permission from Elsevier)

delamination is due to the high oxygen chemical potential produced under an anodic overpotential, the penetration of evolved oxygen gas into the closed pores/defects at the electrode/electrolyte interface, and the high oxygen partial pressure buildup at the interface [86, 87]. On the other hand, Kim et al. [88] suggested that the densification of LSM-YSZ oxygen electrodes caused by the cation migration is responsible for the delamination of oxygen electrodes.

Virkar presented an electrochemical model for the delamination of oxygen electrode [89]. The model predicts the occurrence of delamination of oxygen electrode is a result of the formation of high internal oxygen pressure within the electrolyte close to the oxygen electrode. Knibbe et al. [90] observed hole/pore formation along the grain boundaries of YSZ electrolyte close to the LSM-YSZ oxygen electrodes under high current densities of 1–2 A/cm^2. The degradation in cell performance is suggested to be related to the nucleation and growth of oxygen clusters in the YSZ grain boundaries near the oxygen electrode.

Keane et al. [91] observed $La_2Zr_2O_7$ formation at the electrolyte surface in contact with the LSM oxygen electrode, and the amount of $La_2Zr_2O_7$ increases with the increase of anodic bias. The formation of resistive $La_2Zr_2O_7$ phase is suggested to weaken the contact at the electrode/electrolyte interface and delamination of LSM electrodes.

Chen et al. [42, 85] studied the degradation mechanism of LSM oxygen electrodes. The delamination of LSM oxygen electrodes is related to the formation of nanoparticles within LSM grains at the electrode/electrolyte interface (Fig. 8.8a). These nanoparticles can be removed by HCl acid treatment (Fig. 8.8b). The

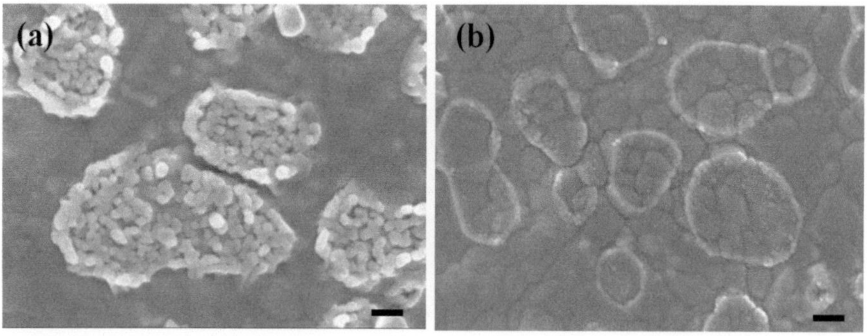

Fig. 8.8 SEM micrograph of YSZ electrolyte surface in contact with the LSM oxygen electrode after the anodic current passage at 500 mA/cm^2 and 800 °C for 48 h: (**a**) before and (**b**) after acid treatment. Scale bar = 200 nm (*Source*: After Ref. [85]. Reproduced with permission from Elsevier)

nanoparticle formation is due to the excess local tensile strains caused by the lattice shrinkage of LSM grains, during the migration or incorporation of oxygen ions from the electrolyte to the LSM grains under the anodic polarization potential driving force [85]. A related kinetics model was thus developed to predict the delamination of LSM oxygen electrodes [92].

8.5.1.2 MIEC Oxygen Electrodes

LSCF is the most commonly investigated MIEC oxygen electrode. Compared with the LSM oxygen electrode, the LSCF electrode exhibits much more structural and performance stability under the SOEC operating conditions [93, 94]. However, the degradation of the oxygen electrode cannot be eliminated. Nguyen et al. reported that the SOEC with LSCF oxygen electrode has no voltage degradation under a low current density of 0.3 A/cm^2 for 2,000 h, though the electrode degradation increases at high current density [95]. In the case of a longer duration of test for 9,000 h, Schefold et al. observed a cell voltage loss of 3.8 %/1,000 h at a higher current density of 1 A/cm^2 [96–98]. The LSCF electrode is partly demixed and Co_3O_4 is formed. There is distinct microstructure change on LSCF grains close to the electrolyte by forming crystal facets and edges (Fig. 8.9a). The high mobility of surface segregated SrO on the LSCF grain surface and its long-range transport is attributed to the formation of a dense $SrZrO_3$ layer at the GDC interlayer/YSZ electrolyte interface region (Fig. 8.9b).

For LSCF oxygen electrode, another cause of degradation is the gradual separation of GDC buffer layer (between LSCF oxygen electrodes and YSZ electrolyte) after SOEC operations [99–101]. Kim et al. [100] studied the effect of co-sintering temperature of GDC/YSZ bilayer and found that co-sintering the YSZ/GDC bilayer at an optimum temperature such as 1,400 °C ensures the intact contact at the YSZ/GDC interface after the electrolysis operation.

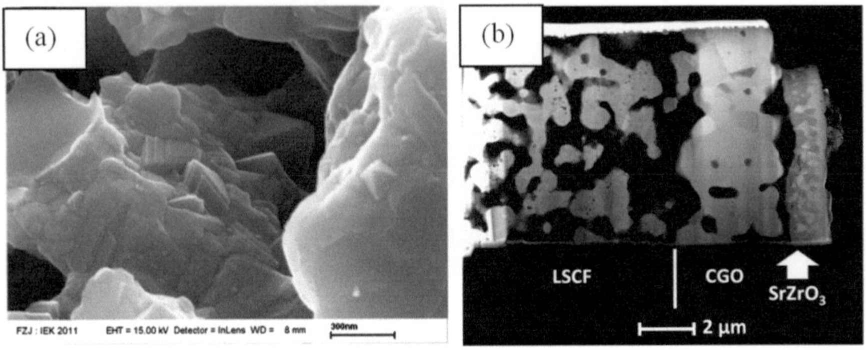

Fig. 8.9 (**a**) SEM micrograph of an LSCF oxygen electrode close to the electrolyte and (**b**) significant $SrZrO_3$ formation in CGO pores after the SOEC was running at 1 A/cm^2, 780 °C for 9,000 h. Scale bar in (**a**) is 300 nm (*Source*: After Ref. [97]. Reproduced with permission from Elsevier)

The concern of using high cobalt-containing MIEC electrodes is the thermal incompatibility due to their high TECs. The TECs of $LaCoO_3$ (LC), LSCF, and $Ba_{0.5}Sr_{0.5}Co_{0.8}Fe_{0.2}O_{3-\delta}$ (BSCF) are 21.5×10^{-6}/K [102], 15.3×10^{-6}/K [103], and 19.95×10^{-6}/K [104], respectively, much higher than 9.9×10^{-6}/K of YSZ [105] and 11.5–11.9×10^{-6}/K of GDC [106]. Thus, it is not surprising that LC and BSCF were separated from the electrolyte even after a short-term test [107, 108]. The use of composite oxygen electrode such as LSCF-GDC, rather than pristine LSCF, is reported to improve the thermal compatibility and electrode stability [109].

8.5.1.3 Degradation by Contaminants

Fe-Cr stainless steel alloy is a leading interconnect material for planar SOECs due to its ease of fabrication, high thermal conductivity, and low cost [110, 111]. The purpose of adding Cr is to form Cr_2O_3 protective scales on the stainless steel surface at high temperatures. However, gaseous Cr species such as CrO_3 and $CrO_2(OH)_2$ from the Cr_2O_3 scale are readily generated at the oxygen electrode side, which can poison the oxygen electrodes. For a manganite oxygen electrode after an electrolysis test for 2,000 h, most gaseous Cr species are captured by the $La_{0.8}Sr_{0.2}CoO_3$ (LSC) current collection layer [82]. The Cr deposition is attributed to a long-range transport of Sr and Co cations to the electrode outmost surface and formation of La-Cr-O phases [112].

Detailed studies have been carried out on Cr deposition and poisoning of LSCF oxygen electrodes under the SOEC operating conditions [113]. After polarization in the presence of Fe-Cr interconnect at 900 °C for 20 h, electrode polarization resistance and overpotential for the O_2 evolution reaction on LSCF electrode are 0.413 $\Omega \cdot cm^{-2}$ and 127 mV, respectively, which is nearly 7 and 18 times of the initial values of the electrode before the polarization. XRD and XPS analysis clearly identified the deposition of $SrCrO_4$, $CrO_{2.5}$, and Cr_2O_3 phases on the surface of LSCF oxygen electrodes, and their formations are closely related to the increased segregation of SrO species under anodic polarization conditions. Sr segregation leads to the Sr deficient at the A-site, thus deteriorating the electrocatalytic activity of the LSCF oxygen electrodes for O_2 oxidation reaction. The presence of gaseous Cr contaminants poisons and degrades the performance of LSCF oxygen electrodes.

Cr deposition and poisoning of LSM oxygen electrodes have been investigated under the SOEC operating conditions [114]. The presence of chromia-forming alloy metallic interconnect significantly degrades and poisons the electrocatalytic activity of LSM oxygen electrodes for the O_2 oxidation reaction and accelerates the delamination of the electrodes. Cr deposition occurs in the LSM electrode bulk and in particular on the YSZ electrolyte surface and on LSM electrode inner surface close to the electrode/electrolyte interface. The results show the formation of Cr, $CrO_{0.87}$, Cr_2O_3, Cr_2O_5, and $SrCrO_4$ on the YSZ electrolyte surface, within the contact rings and on the LSM inner surface. Very different from the predominant formation of $(Cr,Mn)_3O_4$ spinels for the LSM electrodes under the SOFC operation conditions, $(Cr,Mn)_3O_4$ spinel phase could not be found on LSM oxygen electrodes under the SOEC operation conditions. The fundamental reason for the observation of $SrCrO_4$ formation instead of $(Cr,Mn)_3O_4$ on LSM electrode under SOEC operation conditions is that anodic polarization promotes Sr segregation and depresses Mn segregation.

It is well known that segregated SrO plays an important role in the formation of $SrCrO_4$ on the LSCF surface in the presence of an Fe-Cr alloy interconnect under SOFC operating conditions [115]. Backhaus-Ricoult et al. [116] studied the LSM electrode/YSZ electrolyte interface using in situ photoelectron microscopy and observed the increased manganese segregation/migration under cathodic polarization at the interface. This implies that under anodic polarization potential, manganese segregation would be suppressed. Thus, Cr deposition would be dominated by the interaction between the gaseous Cr species and segregated SrO in the case of LSM oxygen electrodes under SOEC operating conditions, leading to Cr deposition and $SrCrO_4$ formation in the bulk of the electrode. The substantially higher Cr deposition at the electrode/electrolyte interface region is most likely related to high Sr segregation. The high activity between the Sr and Cr can also lead to the spread of Sr species from the LSM electrode to the YSZ surface. The accelerated migration and segregation of SrO from the bulk to the free surface and subsequent reaction between SrO and gaseous Cr species based on the nucleation theory [115] are schematically shown in Fig. 8.10 [114].

8.5.1.4 Improved Durability via Reversible Operations

An interesting operational approach to mitigate the degradation of LSM oxygen electrodes is to cycle reversibly between electrolysis and fuel cell modes. Rashkeev et al. studied the oxygen electrode delamination of SOEC and found that

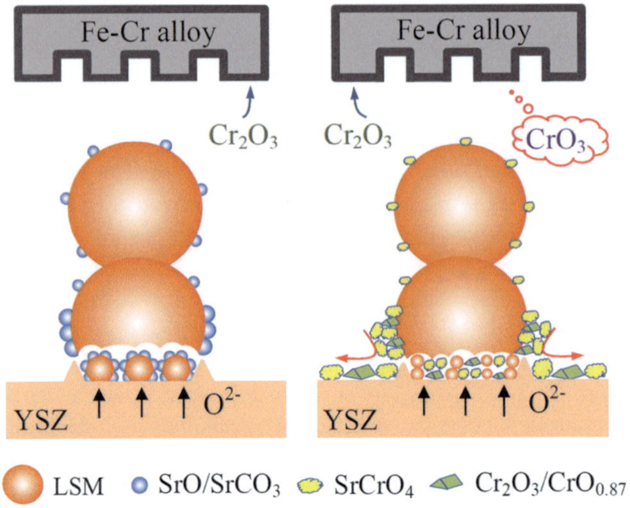

Fig. 8.10 Schemes of Cr deposition on LSM oxygen electrodes under the SOEC operation conditions. (**a**) LSM nanoparticle formation at the electrode/electrolyte interface and accelerated SrO segregation from the bulk to the LSM surface under anodic polarization; (**b**) $Co_2O_3/CrO_{0.87}$ and $SrCrO_4$ formation on the LSM surface, at the electrode/electrolyte interface, and on the YSZ electrolyte surface. The *red arrows* indicate the migration of $SrO/SrCO_3$ from the LSM surface to the YSZ electrolyte surface (*Source*: After Ref. [114]. Reproduced with permission from The Royal Society of Chemistry)

modification of operational regime by applying alternating voltage pulses can suppress the oxygen electrode delamination and increase the lifetime of SOECs [117]. Barnett et al. [118] applied alternating current on symmetric LSM-YSZ composite oxygen electrodes. The reversing current operation at 0.5 A/cm^2 with a current cycle period of 1 and 12 h of cathodic current passage is beneficial to reduce the electrode degradation caused by the anodic polarization, and a shorter cycle period is effective to avoid electrode delamination. The mitigation effect of reversible operation is believed to avoid the occurrence of structural damage at the electrode/electrolyte interface during the incubation time before the delamination occurs [118], though the exact reaction mechanism has not been clarified.

8.5.1.5 Development of Robust Oxygen Electrodes

Design and development of nanostructured electrode are an effective means to manipulate the stability and durability of oxygen electrodes of SOECs. Chen et al. [56, 57] developed a high activity GDC nanoparticles infiltrated LSM oxygen electrode. The infiltrated GDC-LSM electrode retained its high activity and structural stability for the test at 800 °C for 100 h. The promotion of stability is believed to be related to the expansion of reaction sites to the bulk of electrode, thus mitigating the delamination of LSM electrode at the electrode/electrolyte interface. The dispersion of Pd nanoparticles also enhances the structural and performance stability of LSM-GDC composite oxygen electrodes [58, 60].

Another strategy to mitigate the degradation and delamination of LSM electrode is to incorporate LSM nanoparticles into rigid, well-sintered porous YSZ scaffold. The strong and intimate contacted YSZ network provides fast ion-conducting path for rapid transport of oxygen ions from the electrolyte to the electrode and a highly structural stability and integrity. The infiltrated LSM-YSZ electrode is very stable after the electrolysis test for 50 h [59]. More importantly, the grain growth of LSM nanoparticles due to the thermal sintering effect can be offset by LSM lattice shrinkage by the anodic electrolysis polarization, and therefore infiltrated LSM-YSZ composite oxygen electrodes have a high potential as highly stable and active electrodes for long-term electrolysis operation [61].

8.5.2 Ni Cermet Hydrogen Electrodes

Under SOEC operating conditions, a high steam concentration in the hydrogen electrode can pose a detrimental effect on the structural stability of Ni hydrogen electrodes. Growth of Ni particles and oxidation of Ni have been reported after the electrolysis [119–121]. Thus, H_2 is needed in the cathode side to maintain the reducing environment for Ni-based hydrogen electrodes.

There are reports regarding relocation of Ni particles within the hydrogen electrodes under extreme conditions. Hauch et al. reported the formation of a 2–4 μm dense Ni-YSZ layer close to the electrolyte with high current and high steam concentration for 68 h [122]. The possible cause is reduction of gaseous $Ni(OH)_2$ to form Ni at the electrochemically active electrolyte/electrode interface. In contrast, a long-term electrolysis test showed that Ni was depleted in the first 10 μm region of hydrogen electrode adjacent to the electrolyte, while Ni was substantially agglomerated in the outer part of electrode [98]. Chen et al. observed the formation of ZrO_2 nanoparticles on the surface of Ni particles near the YSZ electrolyte by electrolysis at high current [120].

Impurities such as silicon from glass sealant and raw cell materials have an adverse effect on the performance stability of hydrogen electrodes [90, 123]. Under the SOEC operation conditions of $p(H_2O)$ of 0.7 atm and temperature of 850 °C, volatile Si species $Si(OH)_4$ with a vapor pressure of ~10^{-7} atm would come out of the silicate glass sealant and be reduced to glassy phase silica under the reducing environment in the electrochemically active electrode/electrolyte interface region [123]. The deposition of Si can block the active TPBs and pose current constrictions for the current passage from the electrolyte to the Ni-YSZ hydrogen electrode.

H_2S of ppm level can attack the Ni electrode of SOFC by adsorption of sulfur onto Ni grain surface followed by forming nickel sulfide (Ni_2S_3), which significantly reduces the reaction site as well as electrocatalytic activity of the electrode [124]. In the case of SOEC, ppb level of trace sulfur in the inlet gases also significantly poisons the Ni hydrogen electrodes. Ebbesen et al. [125] electrolyzed industrial-grade CO_2 with a degradation rate of 0.22–0.44 mV/h and found that the degradation is independent of current density and is irreversible. After cleaning the CO_2 and/or H_2 inlet gases, a highly stable operation without notable degradation was achieved [19, 126, 127].

8.6 Conclusions and Future Outlook

Hydrogen production by high-temperature SOECs coupled with renewable energy sources is a promising route for the sustainability of energy in the future. The combination of co-electrolysis of H_2O and CO_2 to produce syngas in an SOEC as feedstock with Fischer-Tropsch synthesis to convert syngas to liquid fuels such as gasoline or diesel is a particularly attractive and most promising route for sustainable energy, as liquid transportation fuels are a high-value form of energy. The high-temperature SOECs have significant advantages over the low-temperature electrolysis. The dominant costs of the SOEC process are the electricity cost and capital cost of the electrolyzer. The key challenges in the development of highly efficient, durable, and commercially competitive SOECs technologies are the performance, reliability, and stability of the electrode and electrolyte materials under the electrolysis operation conditions. SOECs are still at the R&D stage, and substantial efforts are required to enhance the performance and durability of the SOEC technologies.

Acknowledgments This project is supported by Curtin University Research Fellowships and the Australian Research Council Discovery Project funding scheme (DP150102044).

References

1. Gahleitner G. Hydrogen from renewable electricity: an international review of power-to-gas pilot plants for stationary applications. Int J Hydrogen Energy. 2013;38:2039–61.
2. Jensen SH, Larsen PH, Mogensen M. Hydrogen and synthetic fuel production from renewable energy sources. Int J Hydrogen Energy. 2007;32:3253–7.
3. Ursua A, Gandia LM, Sanchis P. Hydrogen production from water electrolysis: current status and future trends. Proc IEEE. 2012;100:811.
4. Balat M. Potential importance of hydrogen as a future solution to environmental and transportation problems. Int J Hydrogen Energy. 2008;33:4013–29.
5. Manage MN, Hodgson D, Milligan N, et al. A techno-economic appraisal of hydrogen generation and the case for solid oxide electrolyser cells. Int J Hydrogen Energy. 2011;36:5782–96.
6. Christopher K, Dimitrios R. A review on exergy comparison of hydrogen production methods from renewable energy sources. Energy Environ Sci. 2012;5:6640–51.
7. Abe R. Recent progress on photocatalytic and photoelectrochemical water splitting under visible light irradiation. J Photochem Photobiol C-Photochem Rev. 2010;11:179–209.

8. Jiang SP, Wang X. Chapter 5. Fuel cells: advances and challenges. In: Kharton VV, editor. Handbook of solid state electrochemistry. Berlin:Wiley-VCH. 2011. p. 179–264.
9. Trasatti S. Water electrolysis: who first? J Electroanal Chem. 1999;476:90–1.
10. Carmo M, Fritz DL, Merge J, et al. A comprehensive review on PEM water electrolysis. Int J Hydrogen Energy. 2013;38:4901–34.
11. Wang MY, Wang Z, Gong XZ, et al. The intensification technologies to water electrolysis for hydrogen production – a review. Renew Sustain Energy Rev. 2014;29:573–88.
12. Bierschenk DM, Wilson JR, Barnett SA. High efficiency electrical energy storage using a methane-oxygen solid oxide cell. Energy Environ Sci. 2011;4:944–51.
13. Xie K, Zhang Y, Meng G, et al. Direct synthesis of methane from CO2/H2O in an oxygen-ion conducting solid oxide electrolyser. Energy Environ Sci. 2011;4:2218–22.
14. Hauch A, Ebbesen SD, Jensen SH, et al. Highly efficient high temperature electrolysis. J Mater Chem. 2008;18:2331–40.
15. Bi L, Boulfrad S, Traversa E. Steam electrolysis by solid oxide electrolysis cells (SOECs) with proton-conducting oxides. Chem Soc Rev. 2014;43:8255–70.
16. Ni M, Leung MKH, Leung DYC. Technological development of hydrogen production by solid oxide electrolyzer cell (SOEC). Int J Hydrogen Energy. 2008;33:2337–54.
17. Yildiz B, Kazimi MS. Efficiency of hydrogen production systems using alternative nuclear energy technologies. Int J Hydrogen Energy. 2006;31:77–92.
18. Kasai S. Hydrogen electrical energy storage by high-temperature steam electrolysis for next-millennium energy security. Int J Hydrogen Energy. 2014;39:21358–70.
19. Ebbesen SD, Mogensen M. Exceptional durability of solid oxide cells. Electrochem Solid State Lett. 2010;13:D106–8.
20. Brisse A, Schefold J, International JP. High temperature electrolysis at EIFER, main achievements at cell and stack level. WHEC 2012 conference proceedings – 19th World Hydrogen Energy conference. Energy Procedia. 2012;29:53–63.
21. Kaithwas A, Prasad M, Kulshreshtha A, et al. Industrial wastes derived solid adsorbents for CO2 capture: a mini review. Chem Eng Res Des. 2012;90:1632–41.
22. Zhan ZL, Zhao L. Electrochemical reduction of CO_2 in solid oxide electrolysis cells. J Power Sources. 2010;195:7250–4.
23. Li YX, Zhou JE, Dong DH, et al. Composite fuel electrode $La_{0.2}Sr_{0.8}TiO_{3-\delta}$-$Ce_{0.8}Sm_{0.2}O_{2-\delta}$ for electrolysis of CO_2 in an oxygen-ion conducting solid oxide electrolyser. Phys Chem Chem Phys. 2012;14:15547–53.
24. Graves C, Ebbesen SD, Mogensen M, et al. Sustainable hydrocarbon fuels by recycling CO2 and H2O with renewable or nuclear energy. Renew Sustain Energy Rev. 2011;15:1–23.
25. Graves C, Ebbesen SD, Mogensen M. Co-electrolysis of CO2 and H2O in solid oxide cells: performance and durability. Solid State Ion. 2011;192:398–403.
26. Yu B, Zhang W, Xu J, et al. Preparation and electrochemical behavior of dense YSZ film for SOEC. Int J Hydrogen Energy. 2012;37:12074–80.
27. Laguna-Bercero MA, Skinner SJ, Kilner JA. Performance of solid oxide electrolysis cells based on scandia stabilised zirconia. J Power Sources. 2009;192:126–31.
28. Chaubey N, Wani BN, Bharadwaj SR, et al. Physicochemical properties of rare earth doped ceria Ce(0.9)Ln(0.1)O(1.95) (Ln = Nd, Sm, Gd) as an electrolyte material for IT-SOFC/SOEC. Solid State Sci. 2013;20:135–41.
29. Elangovan S, Hartvigsen JJ, Frost LJ. Intermediate temperature reversible fuel cells. Int J Appl Ceram Technol. 2007;4:109–18.
30. Ishihara T, Jirathiwathanakul N, Zhong H. Intermediate temperature solid oxide electrolysis cell using $LaGaO_3$ based perovskite electrolyte. Energy Environ Sci. 2010;3:665–72.
31. Laguna-Bercero MA, Orera VM. Micro-spectroscopic study of the degradation of scandia and ceria stabilized zirconia electrolytes in solid oxide electrolysis cells. Int J Hydrogen Energy. 2011;36:13051–8.

32. Eguchi K, Hatagishi T, Arai H. Power generation and steam electrolysis characteristics of an electrochemical cell with a zirconia- or ceria-based electrolyte. Solid State Ion. 1996;86–8:1245–9.
33. Zhu S, Wang Y, Rao YY, et al. Chemically-induced mechanical unstability of samaria-doped ceria electrolyte for solid oxide electrolysis cells. Int J Hydrog Energy. 2014;39:12440–7.
34. He F, Song D, Peng RR, et al. Electrode performance and analysis of reversible solid oxide fuel cells with proton conducting electrolyte of BaCe0.5Zr0.3Y0.2O3-delta. J Power Sources. 2010;195:3359–64.
35. Kobayashi T, Abe K, Ukyo Y, et al. Performance of electrolysis cells with proton and oxide-ion conducting electrolyte for reducing nitrogen oxide. Solid State Ion. 2002;154:699–705.
36. Brisse A, Schefold J, Zahid M. High temperature water electrolysis in solid oxide cells. Int J Hydrogen Energy. 2008;33:5375–82.
37. Hauch A, Jensen SH, Ramousse S, et al. Performance and durability of solid oxide electrolysis cells. J Electrochem Soc. 2006;153:A1741–7.
38. Yang CH, Coffin A, Chen FL. High temperature solid oxide electrolysis cell employing porous structured (La0.75Sr0.25)(0.95)MnO3 with enhanced oxygen electrode performance. Int J Hydrogen Energy. 2010;35:3221–6.
39. Ye YM, He TM, Li Y, et al. Pd-promoted $La_{0.75}Sr_{0.25}Cr_{0.5}Mn_{0.5}O_3$/YSZ composite anodes for direct utilization of methane in SOFCs. J Electrochem Soc. 2008;155:B811–8.
40. Badwal SPS, Jiang SP, Love J, et al. Chemical diffusion in perovskite cathodes of solid oxide fuel cells: the Sr doped $LaMn_{1-x}M_xO_3$ (M=Co, Fe) systems. Ceram Int. 2001;27:419–29.
41. Liang MD, Yu B, Wen MF, et al. Preparation of LSM-YSZ composite powder for anode of solid oxide electrolysis cell and its activation mechanism. J Power Sources. 2009;190:341–5.
42. Chen K, Ai N, Jiang SP. Performance and stability of (La, Sr)MnO3–Y2O3–ZrO2 composite oxygen electrodes under solid oxide electrolysis cell operation conditions. Int J Hydrogen Energy. 2012;37:10517–25.
43. Laguna-Bercero MA, Kilner JA, Skinner SJ. Performance and characterization of (La, Sr) MnO3/YSZ and La0.6Sr0.4Co0.2Fe0.8O3 electrodes for solid oxide electrolysis cells. Chem Mater. 2010;22:1134–41.
44. Marina OA, Pederson LR, Williams MC, et al. Electrode performance in reversible solid oxide fuel cells. J Electrochem Soc. 2007;154:B452–9.
45. Tao Y, Nishino H, Ashidate S, et al. Polarization properties of La0.6Sr0.4Co0.2Fe0.8O3-based double layer-type oxygen electrodes for reversible SOFCs. Electrochim Acta. 2009;54:3309–15.
46. Choi MB, Singh B, Wachsman ED, et al. Performance of La0.1Sr0.9Co0.8Fe0.2O3−δ and La0.1Sr0.9Co0.8Fe0.2O3−δ–Ce0.9Gd0.1O2 oxygen electrodes with Ce0.9Gd0.1O2 barrier layer in reversible solid oxide fuel cells. J Power Sources. 2013;239:361–73.
47. Yu B, Zhang WQ, Chen J, et al. Advance in highly efficient hydrogen production by high temperature steam electrolysis. Sci China Ser B-Chem. 2008;51:289–304.
48. Jiang W, Lü Z, Wei B, et al. Sm0.5Sr0.5CoO3–Sm0.2Ce0.8O1.9 composite oxygen electrodes for solid oxide electrolysis cells. Fuel Cells. 2014;14:76–82.
49. Wei B, Chen K, Zhao L, et al. SmBaCo2O5+δ as high efficient oxygen electrode of solid oxide electrolysis cells. ECS Trans. 2013;57:3189–96.
50. Yang Z, Jin C, Yang C, et al. Ba0.9Co0.5Fe0.4Nb0.1O3−δ as novel oxygen electrode for solid oxide electrolysis cells. Int J Hydrogen Energy. 2011;36:11572–7.
51. Aguadero A, Pérez-Coll D, Alonso JA, et al. A new family of mo-doped SrCoO3−δ perovskites for application in reversible solid state electrochemical cells. Chem Mater. 2012;24:2655–63.
52. Laguna-Bercero MA, Kinadjan N, Sayers R, et al. Performance of La2-xSrxCo0.5Ni0.5O4 +/-delta as an oxygen electrode for solid oxide reversible cells. Fuel Cells. 2011;11:102–7.
53. Liu Q, Yang C, Dong X, et al. Perovskite Sr2Fe1.5Mo0.5O6-[delta] as electrode materials for symmetrical solid oxide electrolysis cells. Int J Hydrogen Energy. 2010;35:10039–44.

54. Chauveau F, Mougin J, Bassat JM, et al. A new anode material for solid oxide electrolyser: the neodymium nickelate Nd2NiO4+delta. J Power Sources. 2010;195:744–9.
55. Ogier T, Bassat JM, Mauvy F, et al. Enhanced performances of structured oxygen electrodes for high temperature steam electrolysis. Fuel Cells. 2013;13:536–41.
56. Chen KF, Ai N, Jiang SP. Development of (Gd, Ce)O2-impregnated (La, Sr)MnO3 anodes of high temperature solid oxide electrolysis cells. J Electrochem Soc. 2010;157:P89–94.
57. Chen K, Ai N, Jiang SP. Performance and structural stability of Gd0.2Ce0.8O1.9 infiltrated La0.8Sr0.2MnO3 nano-structured oxygen electrodes of solid oxide electrolysis cells. Int J Hydrogen Energy. 2014;39:10349–58.
58. Chen K, Ai N, Jiang SP. Enhanced electrochemical performance and stability of (La, Sr) MnO3–(Gd, Ce)O2 oxygen electrodes of solid oxide electrolysis cells by palladium infiltration. Int J Hydrogen Energy. 2012;37:1301–10.
59. Yang CH, Jin C, Coffin A, et al. Characterization of infiltrated (La0.75Sr0.25)(0.95)MnO3 as oxygen electrode for solid oxide electrolysis cells. Int J Hydrogen Energy. 2010;35:5187–93.
60. Ai N, Chen K, Liu S, et al. Performance and stability of nano-structured Pd and Pd0.95M0.05 (M = Mn, Co, Ce, and Gd) infiltrated Y2O3–ZrO2 oxygen electrodes of solid oxide electrolysis cells. Int J Hydrogen Energy. 2013;38:16569–78.
61. Chen K, Ai N, Jiang SP. Reasons for the high stability of nano-structured (La, Sr)MnO3 infiltrated Y2O3–ZrO2 composite oxygen electrodes of solid oxide electrolysis cells. Electrochem Commun. 2012;19:119–22.
62. Wang WS, Huang YY, Jung SW, et al. A comparison of LSM, LSF, and LSCo for solid oxide electrolyzer anodes. J Electrochem Soc. 2006;153:A2066–70.
63. Vohs JM, Gorte RJ. High-performance SOFC cathodes prepared by infiltration. Adv Mater. 2009;21:943–56.
64. Hanifi AR, Laguna-Bercero MA, Etsell TH, et al. The effect of electrode infiltration on the performance of tubular solid oxide fuel cells under electrolysis and fuel cell modes. Int J Hydrogen Energy. 2014;39:8002–8.
65. Chen Y, Bunch J, Jin C, et al. Performance enhancement of Ni-YSZ electrode by impregnation of Mo0.1Ce0.9O2+delta. J Power Sources. 2012;204:40–5.
66. Kim-Lohsoontorn P, Kim Y-M, Laosiripojana N, et al. Gadolinium doped ceria-impregnated nickel-yttria stabilised zirconia cathode for solid oxide electrolysis cell. Int J Hydrogen Energy. 2011;36:9420–7.
67. Osada N, Uchida H, Watanabe M. Polarization behavior of SDC cathode with highly dispersed Ni catalysts for solid oxide electrolysis cells. J Electrochem Soc. 2006;153: A816–20.
68. Uchida H, Osada N, Watanabe M. High-performance electrode for steam electrolysis mixed conducting ceria-based cathode with highly-dispersed Ni electrocatalysts. Electrochem Solid State Lett. 2004;7:A500–2.
69. Yang X, Irvine JTS. (La0.75Sr0.25)(0.95)Mn0.5Cr0.5O3 as the cathode of solid oxide electrolysis cells for high temperature hydrogen production from steam. J Mater Chem. 2008;18:2349–54.
70. Xing R, Wang Y, Liu S, et al. Preparation and characterization of La0.75Sr0.25Cr0.5Mn0.5O3−δ-yttria stabilized zirconia cathode supported solid oxide electrolysis cells for hydrogen generation. J Power Sources. 2012;208:276–81.
71. Tsekouras G, Irvine JTS. The role of defect chemistry in strontium titanates utilised for high temperature steam electrolysis. J Mater Chem. 2011;21:9367–76.
72. Ge B, Ma JT, Ai D, et al. Sr2FeNbO6 applied in solid oxide electrolysis cell as the hydrogen electrode: kinetic studies by comparison with Ni-YSZ. Electrochim Acta. 2015;151:437–46.
73. Bernuy-Lopez C, Knibbe R, He Z, et al. Electrochemical characterisation of solid oxide cell electrodes for hydrogen production. J Power Sources. 2011;196:4396–403.
74. Xu S, Chen S, Li M, et al. Composite cathode based on Fe-loaded LSCM for steam electrolysis in an oxide-ion-conducting solid oxide electrolyser. J Power Sources. 2013;239:332–40.

75. Gan Y, Qin Q, Chen S, et al. Composite cathode La0.4Sr0.4TiO3−δ–Ce0.8Sm0.2O2−δ impregnated with Ni for high-temperature steam electrolysis. J Power Sources. 2014;245:245–55.
76. Yang CH, Yang ZB, Jin C, et al. High performance solid oxide electrolysis cells using Pr0.8Sr1.2(Co, Fe)0.8Nb0.2O4+δ–Co–Fe alloy hydrogen electrodes. Int J Hydrogen Energy. 2013;38:11202–8.
77. Tsekouras G, Neagu D, Irvine JTS. Step-change in high temperature steam electrolysis performance of perovskite oxide cathodes with exsolution of B-site dopants. Energy Environ Sci. 2013;6:256–66.
78. Li SS, Qin QQ, Xie K, et al. High-performance fuel electrodes based on NbTi0.5M0.5O4 (M = Ni, Cu) with reversible exsolution of the nano-catalyst for steam electrolysis. J Mater Chem A. 2013;1:8984–93.
79. Arrive C, Delahaye T, Joubert O, et al. Exsolution of nickel nanoparticles at the surface of a conducting titanate as potential hydrogen electrode material for solid oxide electrochemical cells. J Power Sources. 2013;223:341–8.
80. Xu SS, Dong DH, Wang Y, et al. Perovskite chromates cathode with resolved and anchored nickel nano-particles for direct high-temperature steam electrolysis. J Power Sources. 2013;246:346–55.
81. Schefold J, Brisse A, Zahid M, et al. Long term testing of short stacks with solid oxide cells for water electrolysis. ECS Trans. 2011;35:2915–27.
82. Mawdsley JR, Carter JD, Kropf AJ, et al. Post-test evaluation of oxygen electrodes from solid oxide electrolysis stacks. Int J Hydrogen Energy. 2009;34:4198–207.
83. De Haart LGJ, Vinke IC. Long-term operation of planar type SOFC stacks. ECS Trans. 2011;35:187–94.
84. Schuler JA, Wuillemin Z, Hessler-Wyser A, et al. Cr-poisoning in (La, Sr)(Co, Fe)O3 cathodes after 10,000 h SOFC stack testing. J Power Sources. 2012;211:177–83.
85. Chen KF, Jiang SP. Failure mechanism of (La, Sr)MnO3 oxygen electrodes of solid oxide electrolysis cells. Int J Hydrogen Energy. 2011;36:10541–9.
86. Momma A, Kato T, Kaga Y, et al. Polarization behavior of high temperature solid oxide electrolysis cells (SOEC). J Ceram Soc Jpn. 1997;105:369–73.
87. Brichzin V, Fleig J, Habermeier HU, et al. The geometry dependence of the polarization resistance of Sr-doped LaMnO3 microelectrodes on yttria-stabilized zirconia. Solid State Ion. 2002;152:499–507.
88. Kim J, Ji H-I, Dasari HP, et al. Degradation mechanism of electrolyte and air electrode in solid oxide electrolysis cells operating at high polarization. Int J Hydrogen Energy. 2013;38:1225–35.
89. Virkar AV. Mechanism of oxygen electrode delamination in solid oxide electrolyzer cells. Int J Hydrogen Energy. 2010;35:9527–43.
90. Knibbe R, Traulsen ML, Hauch A, et al. Solid oxide electrolysis cells: degradation at high current densities. J Electrochem Soc. 2010;157:B1209–17.
91. Keane M, Mahapatra MK, Verma A, et al. LSM–YSZ interactions and anode delamination in solid oxide electrolysis cells. Int J Hydrogen Energy. 2012;37:16776–85.
92. Zhang Y, Chen K, Xia C, et al. A model for the delamination kinetics of La0.8Sr0.2MnO3 oxygen electrodes of solid oxide electrolysis cells. Int J Hydrogen Energy. 2012;37:13914–20.
93. Elangovan S, Hartvigsen J, Larsen D, et al. Materials for solid oxide electrolysis cells. ECS Trans. 2011;35:2875–82.
94. Minh NQ. Development of Reversible Solid Oxide Fuel Cells (RSOFCs) and stacks. ECS Trans. 2011;35:2897–904.
95. Nguyen VN, Fang Q, Packbier U, et al. Long-term tests of a Jülich planar short stack with reversible solid oxide cells in both fuel cell and electrolysis modes. Int J Hydrogen Energy. 2013;38:4281–90.

96. Schefold J, Brisse A, Tietz F. Nine thousand hours of operation of a solid oxide cell in steam electrolysis mode. J Electrochem Soc. 2012;159:A137–44.
97. Tietz F, Sebold D, Brisse A, et al. Degradation phenomena in a solid oxide electrolysis cell after 9000 h of operation. J Power Sources. 2013;223:129–35.
98. The D, Grieshammer S, Schroeder M, et al. Microstructural comparison of solid oxide electrolyser cells operated for 6100 h and 9000 h. J Power Sources. 2015;275:901–11.
99. Fan H, Keane M, Singh P, et al. Electrochemical performance and stability of lanthanum strontium cobalt ferrite oxygen electrode with gadolinia doped ceria barrier layer for reversible solid oxide fuel cell. J Power Sources. 2014;268:634–9.
100. Kim SJ, Choi GM. Stability of LSCF electrode with GDC interlayer in YSZ-based solid oxide electrolysis cell. Solid State Ion. 2014;262:303–6.
101. Hjalmarsson P, Sun X, Liu Y-L, et al. Influence of the oxygen electrode and inter-diffusion barrier on the degradation of solid oxide electrolysis cells. J Power Sources. 2013;223:349–57.
102. Minh NQ. Ceramic fuel-cells. J Am Ceram Soc. 1993;76:563–88.
103. Tai LW, Nasrallah MM, Anderson HU, et al. Structure and electrical properties of La1 − xSrxCo1 − yFeyO3. Part 2. The system La1 − xSrxCo0.2Fe0.8O3. Solid State Ion. 1995;76:273–83.
104. Wei B, Lu Z, Huang XQ, et al. Crystal structure, thermal expansion and electrical conductivity of perovskite oxides BaxSr1-xCo0.8Fe0.2O3-delta (0.3 <= x <= 0.7). J Eur Ceram Soc. 2006;26:2827–32.
105. Phillipps MB, Sammes NM, Yamamoto O. Gd(1-x)A(x)Co(1-y)Mn(y)O(3) (A = Sr, Ca) as a cathode for the SOFC. Solid State Ion. 1999;123:131–8.
106. Kharton VV, Figueiredo FM, Navarro L, et al. Ceria-based materials for solid oxide fuel cells. J Mater Sci. 2001;36:1105–17.
107. Hino R, Haga K, Aita H, et al. R & D on hydrogen production by high-temperature electrolysis of steam. Nucl Eng Des. 2004;233:363–75.
108. Kim-Lohsoontorn P, Brett DJL, Laosiripojana N, et al. Performance of solid oxide electrolysis cells based on composite La0.8Sr0.2MnO3-[delta] − yttria stabilized zirconia and Ba0.5Sr0.5Co0.8Fe0.2O3-[delta] oxygen electrodes. Int J Hydrogen Energy. 2010;35:3958–66.
109. Choi M-B, Singh B, Wachsman ED, et al. Performance of La0.1Sr0.9Co0.8Fe0.2O3−δ and La0.1Sr0.9Co0.8Fe0.2O3−δ–Ce0.9Gd0.1O2 oxygen electrodes with Ce0.9Gd0.1O2 barrier layer in reversible solid oxide fuel cells. J Power Sources. 2013;239:361–73.
110. Zhu WZ, Deevi SC. Opportunity of metallic interconnects for solid oxide fuel cells: a status on contact resistance. Mater Res Bull. 2003;38:957–72.
111. Zhang X, O'brien JE, O'brien RC, et al. Improved durability of SOEC stacks for high temperature electrolysis. Int J Hydrogen Energy. 2013;38:20–8.
112. Sharma VI, Yildiz B. Degradation mechanism in La0.8Sr0.2CoO3 as contact layer on the solid oxide electrolysis cell anode. J Electrochem Soc. 2010;157:B441–8.
113. Wei B, Chen KF, Zhao L, et al. Chromium deposition and poisoning at La0.6Sr0.4Co0.2Fe0.8O3-delta oxygen electrodes of solid oxide electrolysis cells. Phys Chem Chem Phys. 2015;17:1601–9.
114. Chen KF, Hyodo J, Dodd A, et al. Chromium deposition and poisoning of La0.8Sr0.2MnO3 oxygen electrodes of solid oxide electrolysis cells. Faraday Discuss. 2015. doi:10.1039/C1035FD00010F.
115. Jiang SP, Chen XB. Chromium deposition and poisoning of cathodes of solid oxide fuel cells – a review. Int J Hydrogen Energy. 2014;39:505–31.
116. Backhaus-Ricoult M, Adib K, Clair TS, et al. In-situ study of operating SOFC LSM/YSZ cathodes under polarization by photoelectron microscopy. Solid State Ion. 2008;179:891–5.
117. Rashkeev SN, Glazoff MV. Control of oxygen delamination in solid oxide electrolyzer cells via modifying operational regime. Appl Phys Lett. 2011;99:173506.

118. Hughes GA, Yakal-Kremski K, Barnett SA. Life testing of LSM-YSZ composite electrodes under reversing-current operation. Phys Chem Chem Phys. 2013;15:17257–62.
119. Schiller G, Ansar A, Lang M, et al. High temperature water electrolysis using metal supported solid oxide electrolyser cells (SOEC). J Appl Electrochem. 2009;39:293–301.
120. Chen M, Liu Y-L, Bentzen JJ, et al. Microstructural degradation of Ni/YSZ electrodes in solid oxide electrolysis cells under high current. J Electrochem Soc. 2013;160:F883–91.
121. Kim S-D, Seo D-W, Dorai AK, et al. The effect of gas compositions on the performance and durability of solid oxide electrolysis cells. Int J Hydrogen Energy. 2013;38:6569–76.
122. Hauch A, Ebbesen SD, Jensen SH, et al. Solid oxide electrolysis cells: microstructure and degradation of the Ni/yttria-stabilized zirconia electrode. J Electrochem Soc. 2008;155: B1184–93.
123. Hauch A, Jensen SH, Bilde-Sorensen JB, et al. Silica segregation in the Ni/YSZ electrode. J Electrochem Soc. 2007;154:A619–26.
124. Gong MY, Liu XB, Trembly J, et al. Sulfur-tolerant anode materials for solid oxide fuel cell application. J Power Sources. 2007;168:289–98.
125. Ebbesen SD, Mogensen M. Electrolysis of carbon dioxide in solid oxide electrolysis cells. J Power Sources. 2009;193:349–58.
126. Ebbesen SD, Graves C, Hauch A, et al. Poisoning of solid oxide electrolysis cells by impurities. J Electrochem Soc. 2010;157:B1419–29.
127. Ebbesen SD, Høgh J, Nielsen KA, et al. Durable SOC stacks for production of hydrogen and synthesis gas by high temperature electrolysis. Int J Hydrogen Energy. 2011;36:7363–73.

Chapter 9
Bioelectrochemical Production of Hydrogen from Organic Waste

In S. Kim, Euntae Yang, Mi-Jin Choi, and Kyu-Jung Chae

Abstract Bioelectrochemical hydrogen production is a new technology that uses electrochemically active bacteria under an applied voltage to convert organic matter into hydrogen. This technology is generally referred to as a microbial electrolysis cell (MEC). MECs have gained attention as a novel alternative hydrogen production method because of their high hydrogen conversion efficiency, low energy requirement, and their applicability to many organic substrates. Consequently, the technology has been rapidly advanced. However, various technical challenges remain prior to scale-up and their practical application. This chapter deals with development of MEC technology and includes the following sections: definition and history of hydrogen production, principles and advantages, critical factors affecting MEC performance, anodic biocatalysts and technical challenges, and perspectives and outlooks of hydrogen production from organic waste.

Keywords Bioelectrochemical system • Biohydrogen • Catalyst • Electrochemically active bacteria • Microbial electrolysis cell

9.1 What Is Bioelectrochemical Production of Hydrogen?

Bioelectrochemical hydrogen is produced in a bioelectrochemical cell (BEC) that uses electrochemically active bacteria. The electrochemically active bacteria are microorganisms that have the capability to extracellularly release electrons and act as biocatalysts to convert organic matter into hydrogen [1]. A microbial fuel cell is a biocatalytic conversion system that produces net electrical energy. According to the

I.S. Kim (✉) • E. Yang
School of Environmental Science and Engineering, Gwangju Institute of Science and Technology, Gwangju, South Korea
e-mail: iskim@gist.ac.kr

M.-J. Choi
Ulsan Development Institute, Ulsan, South Korea

K.-J. Chae
Department of Environmental Engineering, Korea Maritime and Ocean University, Busan, South Korea

Z. Fang et al. (eds.), *Production of Hydrogen from Renewable Resources*, Biofuels and Biorefineries 5, DOI 10.1007/978-94-017-7330-0_9

final electron acceptors (e.g., oxygen, ferricyanide, and nitrate), bioelectrochemical cells can be applied to biohydrogen production [2–4], wastewater treatment [5], environmental sensors [6], and bioremediation [7]. Specifically, systems for bioelectrochemical hydrogen production have been referred to using terms such as biocatalyzed electrolysis [8], bioelectrochemically assisted microbial reactor process [9], bioelectrochemical cells [2], bioelectrochemical systems [10], and microbial electrolysis cells [11]. However, microbial electrolysis cells (MECs) are the most commonly used term in these days [12]; hence, MEC is used as the basic terminology in this chapter for bioelectrochemical hydrogen production.

9.2 MEC Principles and Advantages

MECs are a modification of microbial fuel cells that generate hydrogen from organic matter using biocatalyzed electrolysis [3, 4]. Biocatalyzed electrolysis is also carried out by electrochemically active bacteria [2, 13–15]. Figure 9.1 presents a typical two-chambered MEC and the principle of hydrogen production in the

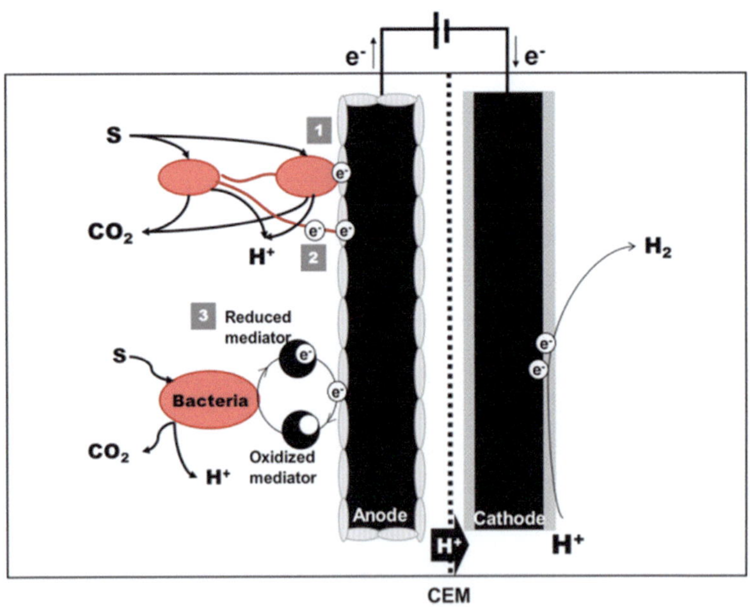

Fig. 9.1 Conceptual diagram of a microbial electrolysis cell (MEC). This configuration is a typical two-chambered type of MEC that generates hydrogen in the cathode compartment. The electrons retrieved from the microbial oxidation of the substrate are transferred to the anode in a number of ways, including (1) direct electron transfer from the attached bacteria, (2) highly conductive nanowires produced by specific bacteria as a long distance transfer method, and (3) using an exogenous mediator or bacterial-origin mediator. The use of a CEM is optional. *S* substrate, *CEM* cation exchange membrane (Reprinted with permission from [22]. Copyright © 2008, Environmental Engineering Research)

MEC. A two-chambered MEC consists of an anode chamber and a cathode chamber, separated by an ion exchange membrane. Electrons retrieved from the microbial oxidation of the substrate are subsequently transferred to the anode in a number of ways, including direct electron transfer from the attached bacteria, highly conductive nanowires produced by specific bacteria as a long distance transfer, and using an exogenous mediator or bacterial-origin mediator [16–20]. The electrons produced from anodic organic oxidation then flow through an external circuit to produce a current, whereas protons are transported through an ion exchange membrane. The electrons and protons combine with protons at the cathode to form hydrogen. Here, electrons and protons are retrieved at the anode in the same manner as for the microbial fuel cell process. However, external power boost is required in order to overcome thermodynamic barriers since the production of hydrogen from acetate oxidation is not thermodynamically spontaneous. For example, if acetate is used as the substrate, the reaction in each chamber is as follows:

$$\text{Anode}: CH_3COO^- + 4H_2O \rightarrow 2HCO_3^- + 9H^+ + 8e^- \qquad E_{anode} = -0.28\text{V (NHE)}$$
$$\text{Cathode}: 8H^+ + 8e^- \rightarrow 4H_2 \qquad E_{cathode} = -0.42\text{ V (NHE)}$$

The potential for the oxidation of acetate (1 M) at the anode and for the reduction of protons to hydrogen at the cathode is −0.28 V and −0.42 V (NHE), respectively. Therefore, hydrogen can be produced theoretically at the cathode by applying a voltage >0.14 V (i.e., −0.42 – (−0. 28) V) [4]. Note that the required voltage is lower than that needed to produce hydrogen from direct water electrolysis (1.23 V at pH 7). In practical terms, however, the applied voltage must be supplied that is higher than 0.14 V due to the high internal resistance in MEC systems; nevertheless, the potential required remains lower than for water electrolysis.

Biohydrogen production from MECs has some important advantages over other technologies (photo fermentation, dark fermentation, and water biophotolysis) in generating hydrogen from organic matter. First, high conversion efficiency to hydrogen is achievable. For example, Cheng and Logan reported an efficiency of 8.55 mol H_2/mol glucose at 0.6 V compared with the typical 4 mol H_2/mol glucose obtained by dark fermentation [21].

Second, high purity of hydrogen is produced in the cathode chamber, such that no further expensive hydrogen purification processes are required. In contrast, for direct biophotolysis, hydrogen purification processes are necessary, since hydrogen and oxygen are concurrently generated by the light-driven dissociation of water ($2H_2O$ + light $\rightarrow 2H_2 + O_2$).

9.3 MEC Architecture

To improve the performance of MECs, many researchers have studied ways to alter their designs to overcome barriers to electrons to promote proton transport and electrode reactivity (Table 9.1).

Table 9.1 Reactor designs of microbial electrolysis cells introduced in previous studies

Reactor design	Reactor volume (L)	Hydrogen yield ($m^3/(m^3$ d))	Ref.
H-shaped two chamber (cation exchange membrane)	Anode 0.31, cathode 0.31	2.9 mol/mol	[3]
Two chamber (cation exchange membrane)	Anode 0.20, cathode 0.20	0.052	[2]
Two chamber (cation exchange membrane)	Anode 3.3, cathode 3.3	0.02	[4]
Cuboid two chamber (anion exchange membrane)	Anode 0.014, cathode 0.028	1.10 $m^3/(m^3$ d)	[23]
Disk-shaped membrane electrode assembly single chamber with gas diffusion electrode	3.3	0.33	[8]
Single chamber with gas diffusion electrode (anion exchange membrane)	3.3	0.31	[8]
Cuboid single chamber lacking membrane	0.028	3.12	[24]
Rectangular MEC with serpentine-shaped flow channels	Anode 0.28, cathode 0.28	0.63	[27]
Upflow single chamber	0.161	59 %	[28]
Cathode-on-top single chamber	0.3	1.58	[29]
Continuous flow single chamber	0.050	6.32	[30]
Concentric tubular two chamber	Anode 0.33, cathode 1.29	1.0 mol/mol	[31]
Multi-electrode continuous flow	2.5	0.53	[32]
Tubular single chamber	2	55.6 mL/d	[33]
Sleeve-shaped single chamber	0.8	2.36	[34]
Submersible	Anode 1, cathode 0.009	0.032	[37]
Anaerobic-baffled reactor	3.46	20.7 %	[35]
Upflow anaerobic sludge blanket type	2	–	[36]
Pilot scale	1000	0.19	
Cassette style design (pilot scale)	Anode 120, cathode 2.6	0.015	[39]

The first MEC configuration was adapted from that of microbial fuel cells. A typical microbial fuel cell is a two-chambered microbial fuel cell that consists of an anode and a cathode chamber separated by a cation exchange membrane. At the early stage of MEC technology, in an H-type MEC reactor, hydrogen generation was 2.9 mol H_2/mol acetate (at a 0.85 V applied voltage), which achieved a 92 % current recovery as hydrogen [3]. Then, by minimizing the distance between the anode and the cathode electrodes, 2.1 mol H_2/mol acetate (at a 0.8 V applied voltage) could be generated. Of this, a volumetric hydrogen production rate of approximately 0.052 m^3 H_2/m^3 reactor liquid volume/day was generated [2], which is 2.6 times faster than the 0.02 m^3 H_2/m^3 reactor liquid volume/day reported by Rozendal et al. [4]. A compact cuboid two-chambered MEC reactor, which generated 1.10 $m^3/(m^3$ d) of hydrogen gas from acetate at a 0.6 V applied voltage, was subsequently developed using ammonia-treated graphite granules and an anion exchange membrane [23].

Based on the optimization of the reactor configuration to reduce internal resistance, a single-chambered MEC was introduced [8, 24–26]. The first single-chambered MECs were equipped with a disk-shaped gas diffusion electrode-membrane assembly (i.e., a cation exchange membrane or anion exchange membrane) and were tested by Rozendal et al. [8]. Call and Logan [24] reported the generation of 3.12 ± 0.02 m^3 $H_2/(m^3$ d) at a 0.8 V applied voltage using a single-chambered MEC lacking a membrane.

For practical application of MECs in polishing processes, continuous flow reactor architectures have been studied [27–33]. A serpentine-shaped flow channel was installed in an MEC to prevent stagnant areas from developing around the electrode [27]. Lee et al. [28] and Guo et al. [29] designed upflow single-chambered MECs by arranging the cathode on top of the MECs so as to efficiently collect hydrogen. Tartakovsky et al. [30] investigated a membraneless single-chambered MEC under continuous mode. They found that by removing the membrane and reducing the distance between electrodes (0.3 mm), a relatively high hydrogen production rate of 6.32 m^3 $H_2/(m^3$ d) could be achieved at a 1.0 V applied voltage.

To commercialize MEC systems, investigations into determining appropriate designs for scaling up MECs, and for combining other biological wastewater treatment processes, have been conducted [31–39]. For example, Kyazze et al. [31] used a two concentric tubular MEC reactor design having an anode chamber volume of 0.33 L and a cathode chamber volume of 1.29 L. This tubular design produced hydrogen at a rate of 0.53 m^3 $H_2/(m^3$ d) at a 0.9 V applied voltage. Rader and Logan [32] then demonstrated that the use of a multi-electrode design having eight separate graphite brush anode-stainless steel mesh cathode pairs could enhance the scalability of MECs. More recently, another tubular MEC design was introduced by Gil-Carrera et al. [33]. This tubular single-chambered MEC having a 2 L chamber volume achieved a moderate hydrogen production rate of 55.6 mL/d from wastewater at a 4 h hydraulic retention time and 1.0 V applied voltage. The sleeve-shaped MEC design proposed by Feng et al. [34] obtained a hydrogen production rate of 2.36 L/(L d), using seven graphite felt anodes and a Pt-coated mipor titanium tube cathode. In other studies, upflow anaerobic sludge blanket-type MEC, submersible MEC, and anaerobic-baffled reactor MEC designs were developed in attempts to easily combine with existing biological wastewater treatment systems [35–37]. To test pilot-scale MECs, a continuous flow MEC (1000 L) equipped with 144 pairs of a graphite fiber brush anodes and stainless steel-304 mesh cathodes was operated using winery wastewater [38]. A pilot-scale MEC (120 L) having six cassette style modules was constructed using carbon felt anodes, and stainless steel wire wool cathodes were operated using domestic wastewater [39]. These pilot-scale MEC tests did not reach the desired hydrogen production performance, but rather highlighted important challenges for scaling up the architecture of MECs.

9.4 Factors Affecting MEC Performance

Factors affecting the performance of MECs include the power supply, electrode, electrolyte, substrate, temperature, catalyst, and separator [2, 12, 40, 41]. To improve the hydrogen production yield of MECs and realize their practical application, the above performance factors must be investigated and optimized; in this section, factors affecting the performance of MECs are reviewed.

9.4.1 Anode Electrode Materials and Anodic Biocatalysts

The anode is a primary component that affects the hydrogen production capability of MECs; it is thus essential to select an appropriate anodic electrode material. The prerequisites for anode electrode materials include large surface area, noncorrosive, high electronic conductivity, good biocompatibility, high chemical stability, high durability, and reasonable cost [42, 43]. Table 9.2 describes the anode electrode materials employed in previous MEC research. Carbonaceous materials, such as carbon cloth [3], carbon felt, graphite plates, graphite granules, graphite felt, graphite rods, carbon brushes, and graphite brushes, have been the most common materials used as anode electrodes in MECs; carbonaceous materials satisfy most of the essential requirements for the anode electrodes of MECs. In particular, the use of carbon and graphite fiber brushes provides highly extended surfaces for the

Table 9.2 Summary of reported electrode materials applied to the anode of microbial electrolysis cells

Electrode material (size)	Hydrogen yield ($m^3/(m^3\ d)$) or current density (A/m^2)	Ref.
Carbon cloth[a]	0.88 A/m^2	[3]
Carbon cloth[b]	2.5 $m^3/(m^3\ d)$	[25]
Carbon fiber brush[b]	0.28 $m^3/(m^3\ d)$	[46]
Carbon fiber brush[a]	0.55 $m^3/(m^3\ d)$	[47]
Carbon felt[b] (10 × 5 × 0.5 cm)	5.4 $m^3/(m^3\ d)$	[54]
Carbon paper[a] (4.0 × 4.0 cm)	0.015 $m^3/(m^3\ d)$	[55]
Carbon nanotube-reticulated vitreous carbon (1.0 × 1.0 × 0.66 cm)	68 A/m^2	[50]
Graphite brush[b]	292 A/m^3	[24]
Graphite felt[a] (0.65 cm thickness)	6.5 A/m^2	[56]
Graphite granule[a]	1.1 $m^3/(m^3\ d)$	[23]
Graphite plate[b] (1.5 × 1.0 cm)	–	[57]
Graphite rod[a]	–	[58]
Oxidized stainless steel felt[a]	19.2 A/m^2	[51]
Stainless steel mesh + graphite granule[a]	2.8 A/m^2	[59]
Fe-NP-decorated[a] graphite disk (5.2 cm^2)	42.5 $\mu A/cm^2$	[53]

[a]Two chamber
[b]Single chamber

inoculation of microorganisms and a high current density [44–49]. The installation of graphite granules in the anode can also increase the surface area of anode electrodes [23]. Flexer et al. [50] developed a three-dimensional scaffold carbon nanotube-reticulated vitreous carbon electrode. This carbon-nanostructured electrode achieved a very high current density (68 A/m^2) by increasing the rate of bacterial extracellular electron transfer and its high electrode surface. Other than carbonaceous materials, stainless steel felt has been exploited as an anode electrode [51]. Stainless steel is attractive candidate for use as an electrode for large-scale MECs because of its high conductivity and low cost, but to date, its poor biocompatibility restricts its application as an anode electrode in MECs. In an attempt to overcome this problem, Guo et al. [51] enhanced the biocompatibility of a stainless steel electrode by adopting flame oxidation, with flame oxidized stainless steel felt displaying one of the highest current densities at 19.2 A/m^2.

Several MEC studies have employed a modified anode electrode material to increase the performance of MECs. For example, ammonia treatment improves the adhesive properties between electrode surfaces and bacteria by making the electrode surface more positively charged; bacteria are generally negatively charged [43]. Heat treatment promotes the adhesion of bacteria to the surface of anode electrode materials by increasing nitrogen functionalization [52]. Xu et al. [53] reported that MECs equipped with Fe nanoparticle-decorated graphite anodes achieved a 5.89-fold higher average current density than those equipped with plain graphite anodes; nanoparticle-decorated anodes remarkably upregulated gene-encoding biofilm formation.

Inoculum sources for anodic biocatalysts can also affect hydrogen production in MECs. The hydrogen production rate and recovery of MECs inoculated with pure or mixed cultures were compared by Call et al. [14]. Their results showed that at an applied voltage of 0.4 V, the MEC inoculated with a mixed culture showed a lower hydrogen recovery and hydrogen production rate because of an increase in the methane production due to methanogen in the mixed consortium.

9.4.2 Cathode Electrode Materials and Cathodic Catalysts

Hydrogen evolution occurs at the cathode in MECs; the efficiency and economic feasibility of MECs are very closely related to the cathode. Similar to the anode electrode materials, the MEC cathode electrode materials require a large surface area, high electronic conductivity, high chemical stability, high durability, and reasonable cost. In MECs, carbonaceous materials are also the most common materials used for cathode electrodes. Plain carbon cathode electrodes have a cathodic recovery efficiency of 25.4 ± 12.1 % due to their high overpotential for hydrogen production [60]. To reduce the high overpotential, catalysts can be deposited on the cathodes [11]. Platinum is widely used as a cathodic catalyst in MECs because of its superior catalytic performance for promoting hydrogen production. Call and Logan [24] reported that a membraneless single-chambered MEC equipped with a Pt-coated carbon cloth achieved a hydrogen production rate of

3.12 $m^3/(m^3$ d) and a cathodic recovery efficiency of 96 % at an applied voltage of 0.8 V. However, Pt is a costly precious metal, and this is a critical obstacle for the commercialization of MECs. Therefore, cost-effective noble metal-free catalysts and cathodic electrode materials have been investigated as substitutes for Pt.

Electrochemically active bacteria can serve as catalysts for hydrogen evolution reactions on the cathode electrode in MECs. For example, Rozendal et al. [27] first attempted to replace Pt with a mixed culture of electrochemically active bacteria in MECs, which is referred to as a biocathode. Jeremiasse et al. [61] also presented the proof of concept for a full biological MEC, in which both the anodic and cathodic catalysts were electrochemically active bacteria. In addition, an MEC having thermophilic microorganisms as cathodic catalysts, operated at 55 °C, achieved a comparable cathodic hydrogen recovery of 70 % [62].

It was identified that Ni and Ni-containing electrode materials are promising alternatives to Pt because of their stability and lower price [41]. Selembo et al. [63] tested different Ni-containing stainless steel alloys (SS A286, SS 304, SS 316, and SS 420) and Ni alloys (Ni 201, Ni 400, Ni 625, and Ni HX) as cathode electrodes in am MEC. According to their results, SS A286 showed the highest hydrogen production rate of 1.50 ± 0.04 $m^3/(m^3$ d) and a cathodic recovery rate of $61 \pm 3\%$. Furthermore, it was shown that the hydrogen production performance can be increased by electrodepositing NiO_x on the surface of stainless steel and Ni alloys. To this end, NiMo- and NiW-catalyzed carbon-fiber-weaved cloth electrodes were evaluated in single-chamber tubular MECs [25]. An MEC having an NiMo cathode exhibited comparable hydrogen production rate to an MEC having a Pt cathode at an applied voltage of 0.6 V. Nickel powders were then examined by Selemobo et al. [13] and Mamiel et al. [64] as cathode catalysts at different Ni loading rates and composition in MECs. To maximize the electrode surface area, a brush-type Ni containing a stainless steel electrode and an Ni foam electrode were used as a cathode [65, 66]. The Ni foam cathode generated high-purity hydrogen at a volumetric production rate of 50 $m^3/(m^3$ d) [66], which can be considered high, as compared to previous studies.

In addition to Ni, other alternative candidates to Pt catalysts in MECs have been explored. Notably, Huang et al. [60] developed a Pd nanoparticle-coated cathode. In their study, the Pd nanoparticle electrode achieved a better hydrogen production rate (2.6 ± 0.5 L/(m^2 d)) than a Pt-coated electrode (2.1 ± 0.3 L/(m^2 d)) [60]. Tokash et al. [67] used MoS_2, a well-known photocatalyst for hydrogen production, as the cathodic catalyst in an MEC. According to their research, the MoS_2-catalyzed cathode generated an analogous average current density of 10.7 ± 1.2 A/m^2 [67]. Xiao et al. [68] demonstrated that cost-effective nitrogen-containing core-shell-structured catalysts N-Fe/Fe_3C@C can be used as cathodic catalysts for hydrogen production in MECs. Chen et al. [69] developed a photocatalytic cathode using nanostructured TiO_2 nanorod arrays on conductive fluorine-doped tin oxide-coated glass, to produce hydrogen without using an applied voltage.

Carbon nanotubes have been employed as catalysts and catalyst supporters to enhance hydrogen production in several MEC studies because of their nanometer size, high conductivity, high surface area, and relatively cheap price [70–72]. The carbon nanotube-based cathodes tested in these studies had comparable performance to Pt-based cathodes in MECs (Table 9.3).

Table 9.3 Summary of cathodic electrode materials and catalysts used in MECs. H_2 production rates and cathodic recovery efficiencies (r_{cat}) were obtained from previous studies on microbial electrolysis cells using the cathodic electrode and catalysts

Catalyst	Electrode material	H_2 production rate ($m^3/(m^3\ d)$)	Maximum r_{cat} efficiency (%)	Ref.
–	Carbon paper	0.4 $L/(m^2\ d)$	25.4	[60]
Pt	Carbon cloth	3.12	96	[24]
Pt	Titanium mesh	0.3	–	[8]
Pt	Carbon paper	2.1	41.6	[60]
Pt	Carbon cloth	0.29	89.1	[73]
Biocatalyst	Graphite felt	0.04	21	[61]
Biocatalyst	Graphite felt	0.63	49	[27]
Biocatalyst[a]	Carbon cloth	0.376 $mol/(m^2\ d)$	70	[62]
–	Stainless steel A268	1.50	61	[63]
–	Stainless steel 304	0.59	53	[63]
–	Stainless steel 316	0.35	27	[63]
–	Stainless steel 420	0.58	43	[63]
–	Stainless steel brush	1.7	84	[65]
–	Ni mesh	0.28	75.7	[73]
–	Ti mesh	0.23	73.3	[73]
–	Ni 210	0.38	27	[63]
–	Ni 400	0.41	31	[63]
–	Ni 625	0.79	43	[63]
–	Ni HX	0.55	40	[63]
NiO_x	Stainless steel A286	0.76	52	[63]
NiO_x	Ni 625	0.76	52	[63]
NiMo	Carbon-fiber-weaved cloth	2.0	86	[25]
NiW	Carbon-fiber-weaved cloth	1.5	52	[25]
Ni210	Carbon cloth	1.3	79	[13]
Ni210/ Carbon block	Carbon cloth	1.2	94	[13]
Ni	Carbon paper	5.4	–	[54]
Ni	Carbon paper	4.14	103.3	[64]
Ni	Ni foam	50	–	[66]
Pd	Carbon paper	2.6 $L/(m^2\ d)$	46.4	[60]
MoS_2	Carbon cloth	10.7 A/m^2	–	[67]
N-Fe/Fe_3@C nanorods	Carbon cloth	0.0181	79.8	[68]
Nanostructured TiO_2 nanorod arrays on conductive fluorine-doped tin oxide-coated glass		4.2 $mL/(m^3\ d)$	–	[69]

(continued)

Table 9.3 (continued)

Catalyst	Electrode material	H_2 production rate ($m^3/(m^3$ d))	Maximum r_{cat} efficiency (%)	Ref.
Pt/multiwalled carbon nanotube	Carbon cloth	1.42	65	[70]
Multiwalled carbon nanotube	Carbon cloth	1.20	62	[70]
MoS_2/carbon nanotube	Carbon cloth	0.01	49.0	[71]
Polyaniline/multiwalled carbon nanotube	Carbon cloth	1.04	57	[72]

[a]MEC operated at 55 °C

9.4.3 *Chamber Volume, Electrode Size, and Electrode Position*

The H-type two-chambered reactors in early MECs had a low hydrogen production rate (2.9 mol H_2/mol acetate) and a low current density due to their high internal resistance [3]. Various factors can affect the internal resistance, but this can be reduced by optimizing the electrode surface area, electrode position, and chamber volume. As examples, Call and Logan [23] improved the hydrogen production rate by increasing the anodic surface area using graphite granules. Wang et al. [74] decreased the internal resistance in an MEC by shortening the distance between the anode and cathode electrodes from 14 to 4 cm. Cheng and Logan [75] reported that the hydrogen production rate in an MEC can be increased by reducing the electrode spacing; the maximum hydrogen production rate they achieved in an MEC had a 2 cm electrode spacing. Gil-Carrera et al. [76] investigated the optimum electrode size and arrangement in flat-plate MECs. They reported that the optimum electrode size and arrangement in flat-plate MECs was a two-layer carbon felt anode having a 10 mm thickness and a single gas diffusion cathode. Liang et al. [77] further showed that optimizing the anode arrangement effectively reduced the internal resistance. In their study, an MEC separately positioning two anode electrodes at either side of the cathode in parallel reached a higher current density of 621.3 ± 20.6 A/m^3 and hydrogen production rate of 5.56 m^3/m^3 than an MEC having two anodes at one side of the cathode (360 A/m^3 and 2.55 m^3/m^3).

9.4.4 *Separator*

A typical MEC design is a two-chambered reactor that consists of an anode, a cathode, and a separator. Separators play an important role in MECs. Separators physically divide the anode and cathode chambers and theoretically prevent mass

Table 9.4 Summary of separators reported in previous microbial electrolysis cell studies

Separator	Hydrogen yield ($m^3/(m^3$ d))	Current density (A/m^2)	Ref.
Cation exchange membrane[a] (256 cm^2)	0.33	2.25	[8]
Anion exchange membrane[a] (256 cm^2)	0.31	2.37	[8]
Anion exchange membrane[b] (30 mm diameter)	1.10	–	[23]
No separator[c]	3.12	292 A/m^3	[24]
Cation exchange membrane[b] (Nafion)	About 0.5 mL/h	–	[79]
Anion exchange membrane[b]	2.0 mL/h	–	[79]
No separator[c]	6.32	3.7	[30]
Cation exchange membrane[a]	1.22	1.8	[30]
Anion exchange membrane[b]	0.43	109 A/m^3	[80]
Cation exchange membrane[b]	0.36	92 A/m^3	[80]
Bipolar membrane[b] (8 cm^2)	0.018	–	[83]
Microporous membrane[c]	0.015	–	[39]
Cation exchange membranc[b] (Nafion; 25 cm^2)	12.9 mL	–	[84]
Cation exchange membrane[b] (sulfonated polyether ether ketone based; 25 cm^2)	14.4 mL	–	[84]

[a]Single chamber
[b]Two chamber
[c]Scale-up cassette style design

transport of substrate, hydrogen gas, methane gas, and microorganisms, from occurring between the anode and cathode chambers, except for protons. Cation exchange membranes are commonly used for separators in MECs [8, 78–80], but they have several crucial disadvantages: they require pH gradients between the anode and cathode chambers, and they have a high cost [42]. Several studies have compared MECs equipped with cation exchange and anion exchange membranes [8, 78–80]. The performance of MECs equipped with an anion exchange membrane is better than those having a cation exchange membrane because the anion exchange membrane-based MECs have a lower ion transport resistance [78]. However, anion exchange membranes generally show high substrate permeability and configuration deformation [81, 82]. Call and Logan [24] removed the membrane to thereby increase the hydrogen production rate. Their membraneless single-chambered MEC achieved a relatively high production rate of 3.12 ± 0.01 $m^3/(m^3$ d), though concerns about purity of hydrogen remained due to methane production by methanogen. A bipolar membrane was then employed as a separator in an MEC to overcome a thermodynamic barrier for hydrogen production [83]. This membrane produced hydrogen at a rate of 0.01 $m^3/(m^3$ d) with no applied voltage. In a pilot-scale MEC test, microporous membranes were used as the separator because of their relatively low price and high proton transfer ability [39] (Table 9.4).

Table 9.5 Summary of two previous studies on hydrogen production yield or efficiency as a function of the applied voltage level in microbial electrolysis cells

Applied voltage	Cell design	Substrate	Hydrogen yield ($m^3/(m^3\ d)$) or efficiency (%)	Ref.
0.4 V	Single chamber	Acetate	0.20 m^3/m^3 d	[26]
0.6 V	Anode: carbon cloth		0.53 m^3/m^3 d	
0.6 V	Cathode: carbon cloth with Pt		0.69 m^3/m^3 d	
0.1 V	Two chamber		Negligible amount	[2]
0.2 V	Anode: carbon felt		Negligible amount	
0.3 V	Cathode: stainless steel with Pt		About 30 %	
0.5 V			About 40 %	
0.6 V			About 40 %	
0.8 V			52.5 %	

9.4.5 *Power Supply*

Hydrogen production reactions in MECs are nonspontaneous, i.e., to overcome thermodynamic barriers for the hydrogen production reaction, an external power supply is required [3]. Indeed, the applied voltage is a crucial factor for determining the hydrogen production performance in MECs. Previous research has thus investigated the effect of applied voltage on MEC performance [2, 26]. Theoretically, if more than a 0.14 V external voltage is applied, the hydrogen production reaction can occur in MECs. However, in the actual operation of MECs, a minimum of 0.3 V applied voltage is required due to the overpotential [2]. By increasing the applied voltage to over 0.3 V, the corresponding hydrogen production performance in MECs improves [2, 26]. This increase implies that the applied voltage is directly related to the energy requirement of MECs for hydrogen production. For developing more cost-effective MECs, the hydrogen production needs to be improved without increasing the applied voltage (Table 9.5).

To achieve this goal, alternative renewable power sources have been proposed to apply a voltage to the MEC. Table 9.6 presents various alternative power sources that have been used with MECs. Notably, Ajayi et al. [85, 86] and Chae et al. [87] successfully developed solar-powered MECs by installing dye-sensitized solar cells. Quan et al. [88] modified an MEC by setting up a nanowire photocathode in a cathode chamber, which is now referred to as a microbial photoelectrochemical cell. This modified MEC successfully generated hydrogen under white light illumination. More recently, a photoelectrochemical cell consisting of a Pt cathode, TiO_2 nanowire-arrayed photoanode, and microbial fuel cell has been integrated to produce hydrogen with solar power assistance [89].

Microbial fuel cells have also been investigated as a power source for MECs [46, 55]. Sun et al. [55] developed an electricity-assisting microbial fuel cell-MEC-coupled system. In this case, an MEC comprised of three series-connected microbial fuel cells produced hydrogen at a rate of 14.54 ± 0.12 $m^3/(m^3\ d)$.

Table 9.6 Alternative power supply sources for microbial electrolysis cells

Power source	Hydrogen yield or cathode efficiency	Ref.
Solar powered (dye-sensitized solar cell)	0.07 $m^3/(m^3\ d)$	[87]
Solar powered (dye-sensitized solar cell)	78 ± 2.5 %	[85]
Solar powered (dye-sensitized solar cell)	78 ± 2.5 %	[86]
Solar assisted (nanowire photocathode)	20 mW/cm^2	[88]
Solar assisted (photochemical cell)	4 mL (AM 1.5G, 100 mW/cm^2)	[89]
Microbial fuel cell	14.54 ± 0.12 $m^3/(m^3\ d)$	[55]
Microbial fuel cell	0.28 $m^3/(m^3\ d)$	[46]
Reverse electrodialysis stack	0.8 to 1.6 $m^3/(m^3_{anolyte}\ d)$	[90]

Kim and Logan [90] used concentration difference in the energy conversion to sustainably produce hydrogen by inserting reverse electrodialysis stacks between the anode and cathode chambers of an MEC, which generate electricity from the salinity gradient between high and low salt concentration solutions.

9.4.6 Substrates

For hydrogen production using MECs, organic carbon compounds such as acetate, glucose, butyric acid, cellulose, protein, and wastewater have been used as substrates (Table 9.7). In contrast, simple-structure carbon compounds including volatile acid and glucose were used in early MECs studies. Acetate, however, remains the most commonly utilized substrate. Hydrogen has been produced from acetate at the highest rate of 50 $m^3/(m^3\ d)$ in a continuous-mode two-chambered MEC equipped with an Ni foam cathode electrode at an applied voltage of 1.0 V [24]. According to Cheng and Logan [23], comparable hydrogen production rates have been obtained in glucose-fed and lactic acid-fed MECs using acetate-fed MECs, with lower hydrogen production rates achieved in butyric acid-, propionic acid-, and valeric acid-fed MECs.

More complex-structure organic compounds, such as wastewater and waste biomass, have been employed as substrates in MECs. For example, glycerol—generated as a by-product from the biodiesel manufacturing process—has been tested as a substrate in MECs [91]. A two-chambered MEC fed with ultrapure glycerol obtained a comparable hydrogen production rate of 2.01 ± 0.41 $m^3/(m^3\ d)$ to that fed with glucose (1.87 ± 0.30 $m^3/(m^3\ d)$), though the MEC fed with the glycerol by-product of biodiesel fuel achieved a much lower hydrogen production rate (0.41 ± 0.13 $m^3/(m^3\ d)$, i.e., the glycerol generated from the biodiesel manufacturing process is impure. To achieve a higher hydrogen conversion efficiency from glycerol, Chookaew et al. [92] integrated an MEC with a dark fermentation process.

Table 9.7 Summary of reported substrate types used for microbial electrolysis cells and hydrogen yields obtained from previous microbial electrolysis cell studies with the substrate

Substrate	Hydrogen yield ($m^3/(m^3$ d))	Ref.
Acetate	50	[66]
Acetate	1.10	[23]
Butyric acid	0.45	[23]
Lactic acid	1.04	[23]
Propionic acid	0.72	[23]
Valeric acid	0.14	[23]
Formic acid	0.62	[94]
Glucose	1.23	[23]
Sucrose	–	[101]
Bovine serum albumin (protein)	0.54	[80]
Ultrapure glycerol	2.01	[91]
Glycerol by-product of biodiesel fuel	0.41	[91]
Crude glycerol	106 mL/g COD	[92]
Cellobiose	0.96	[94]
Lignocellulose	1.0	[94]
Potato wastewater	0.74	[96]
Winery wastewater	0.28	[95]
Domestic wastewater	0.17	[95]
Swine wastewater	1.0	[48]
Industrial wastewater	–	[102]
Sewage sludge	7.7 mL	[103]
Waste-activated sludge	0.91	[97]
Fermentation liquid of waste-activated sludge	1.2 mL/mg COD	[99]
Effluent from anaerobic-baffled reactor	1.31	[100]
Primary effluent	–	[104]

As another potential source, lignocellulose has attracted attention as a promising feedstock for hydrogen production in MECs because of its abundance and renewability [93]. However, the direct use of lignocellulose as a substrate for hydrogen production in MECs is difficult; lignocellulose is difficult to degrade using electrochemically active bacteria. About half of the hydrogen production rate was achieved from cellulose 0.11 $m^3/(m^3$ d) compared with that from acetate [23]. A two-stage process combining dark fermentation and an MEC was then adopted to produce hydrogen from lignocellulose [94]. Dark fermentation was used to convert lignocellulose into acetic, lactic, succinic, and formic acids and ethanol, which are significantly more easily degradable organic compounds by electrochemically active bacteria in MECs. The hydrogen production from the effluent of lignocellulose and cellobiose fermentation was 1.0 ± 0.19 $m^3/(m^3$ d) and 0.96 ± 0.16 $m^3/(m^3$ d) in a two-stage MEC [94].

Various types of wastewater have been examined as substrates for MECs, but lower hydrogen production rates were obtained compared with acetate-fed MECs.

For example, hydrogen was generated from swine wastewater at a rate of 0.9–1.0 $m^3/(m^3\ d)$ in a single-chambered MEC having a graphite fiber brush anode [48]. Cusick et al. [95] investigated hydrogen production performance and economics of MECs fed with domestic wastewater and winery wastewater; the respective hydrogen production rates were $0.28 \pm 0.04\ m^3/(m^3\ d)$ and $0.17 \pm 0.09\ m^3/(m^3\ d)$. The resulting hydrogen production costs were \$4.51/kg H_2 for winery wastewater and \$3.01/kg H_2 for domestic wastewater [95]. Based on these results, the first pilot-scale continuous flow MEC (1000 L) having 144 electrode pairs in 24 modules was tested using winery wastewater [38].

In other studies, an MEC supplied with potato processing wastewater generated hydrogen at a production rate of 0.74 $m^3/(m^3\ d)$ at an applied voltage of 0.9 V [96]. Waste-activated sludge, which contains a large amount of carbohydrates undegradable by electrochemically active bacteria, has been tested as a substrate in MECs [97, 98]. From the raw waste-activated sludge, very low hydrogen was produced ($0.056 \pm 0.008\ m^3/(m^3\ d)$ in a two-chambered MEC, though with alkaline-pretreated waste-activated sludge, a larger amount of hydrogen was generated ($0.91 \pm 0.10\ m^3/(m^3\ d)$) [97]. As with lignocellulose, a higher hydrogen production rate was achieved from waste-activated sludge in MECs by combining an MEC with a fermentation process [99]. Finally, an anaerobic-baffled reactor (ARB) was utilized to degrade complex organic compounds into volatile short-chain fatty acids and ethanol, with the ARB effluent then being supplied to MECs for hydrogen production [100].

9.4.7 Electrolyte

The conditions of electrolytes, such as pH and conductivity, are important factors that affect the performance of MECs; electrolyte pH can influence the activity of electrochemically active bacteria and be used to control the redox reaction potentials on the electrode [7]. However, the electrolyte conductivity can affect the internal resistance of MECs [105]. There have been several research works that explore the effect of electrolytes on the performance of MECs as described next.

Matthew et al. [105] reported that specific electrolytes can increase the performance of MECs by reducing the cathode overpotential or solution resistance according to its pH; phosphate and acetate electrolytes can improve the performance of MECs by decreasing the overpotential at pH 5. However, at a higher pH (pH 9), carbonate electrolytes increased the performance of MECs by reducing the solution resistance. Munoz et al. [106] showed that for MECs equipped with a stainless steel cathode electrode, the existence of phosphate species in an electrolyte improved the current density for hydrogen production because of the cathodic deprotonation reaction. Yossan et al. [44] tested five groups of catholytes in MECs, including a phosphate buffer, NaCl solution, deionized water, tap water, and acidified water. The MEC containing a 100 mM phosphate catholyte exhibited the highest hydrogen production rate because of its high buffer capacity, followed by NaCl solution due to its high conductivity, and then acidified water as it provided extra protons for hydrogen production. Liu et al. [47] investigated the optimal anolyte pH in MECs and found that the optimal pH was 9 for a maximum hydrogen production rate of 0.55 $m^3/(m^3\ d)$.

9.4.8 Other Operational Factors

Temperature can also affect the activity and selection of microorganisms in MECs. Omidi and Sathasivan [58] demonstrated, based on the COD removal rate and amount of biomass in MECs, that 31 °C was the optimum condition for MEC operation. In general, MEC tests are now conducted at a controlled temperature range of around 30 °C. However, hydrogen has been successfully generated in a single-chambered MEC enriched and operated at temperatures as low as 4 °C and 9 °C [107]. Under this psychrophilic condition, methane production by methanogen is effectively inhibited.

The effects of hydrodynamic force and dissolved oxygen on the MEC performance and anode biofilm have been evaluated as operational factors [108]. It was found that the hydrogen production in an MEC was markedly influenced by the hydrodynamic force, but not significantly affected by the anode biofilm exposure to dissolved oxygen.

As final considerations in this subsection, it has been shown that the organic loading rate and hydraulic retention time can be crucial operational factors under the continuous-mode operation of MECs [33, 109] (Table 9.8).

9.5 Hydrogen Yield of Organic Waste-Fed and Scaled-Up MECs

It is often difficult to directly compare the results of different studies because the MEC operations have been conducted under different experimental conditions and, in some cases, the key parameters or hydrogen production performance indicators were not provided. However, in this section, a comparative analysis of hydrogen production yields for organic waste-fed MECs and scaled-up MECs obtained from several studies is attempted.

9.5.1 Hydrogen Yield from Organic Waste in MECs

The ultimate goal of MEC technology is to sustainably produce commercially available hydrogen from organic waste. To achieve this goal, the current trend of

Table 9.8 Summary of previously reported factors affecting the performance of microbial electrolysis cells

Operational factor	Conditions	Performance evaluation parameter	Ref.
Temperature		COD removal rate	[58]
	25 °C	16.8 mg/(L h)	
	29 °C	24.7 mg/(L h)	
	30 °C	34.0 mg/(L h)	
	31 °C	36.2 mg/(L h)	
		Hydrogen yield	[107]
	4 °C	0.23 $m^3/(m^3\ d)$	
	9 °C	0.32 $m^3/(m^3\ d)$	
Hydrodynamic force		Hydrogen production	[108]
	Reynolds number 900	198 μmol	
	Reynolds number 4900	314 μmol	
Anode biofilm exposure time to dissolved oxygen		Hydrogen production	[108]
	0 h	About 500 μmol	
	48 h	About 550 μmol	
	120 h	About 700 μmol	
Hydraulic retention time		Cathodic conversion efficiency	[33]
	10 h	8.7 %	
	7 h	9.3 %	
	10 h	23.9 %	
		Hydrogen production	[109]
	6.5 h	2.03 mol/mol	
	3.1 h	1.88 mol/mol	
	1.6 h	1.81 mol/mol	

hydrogen production performance (Fig. 9.2) from organic waste needs to be realized. As mentioned in the previous section, using pure chemical substrates, such as acetate and glucose, MECs have already achieved an economical hydrogen production yield on a lab scale. However, using organic waste, lower hydrogen production yields have generally been observed, likely due to waste organics containing complex-structure carbon compounds that are not easily degradable by electrochemically active bacteria. Among the organic waste substrates, the effluent from an ethanol-hydrogen coproducing fermentation reactor feeding into a single-chambered MEC (working volume of 26 mL) having a carbon fiber brush anode and Pt-coated carbon cloth cathode could achieve a maximum hydrogen production rate of 2.11 $m^3/(m^3\ d)$ [110]. However, during the same experiment, without the assistance of fermentation prior to being fed into the MEC, a much lower hydrogen production rate of 1.41 $m^3/(m^3\ d)$ was obtained [110]. Overall, it was found that the utilization of organic waste that was pretreated by other processes gave relatively

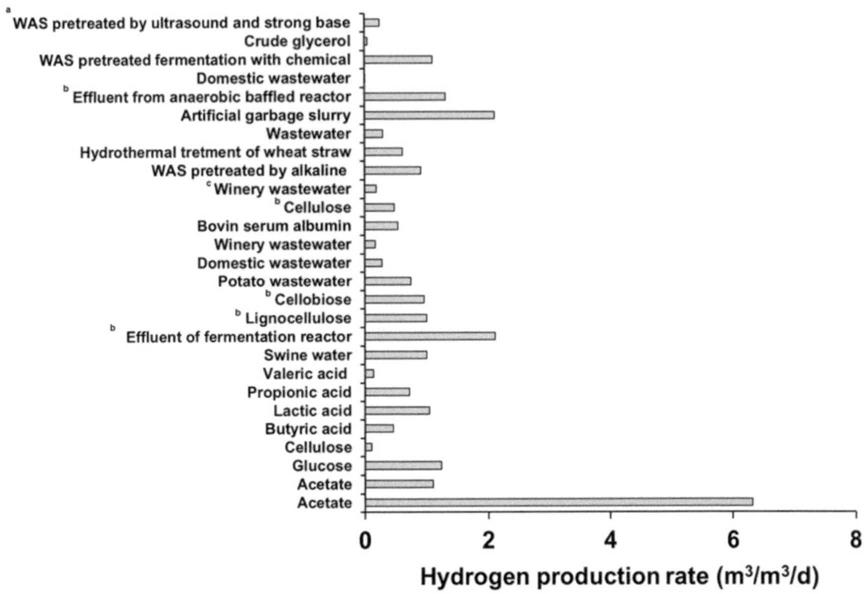

Fig. 9.2 Hydrogen production rates in organic waste-fed MECs; [a] *WAS* waste-activated sludge; [b] integration with other biological process; [c] pilot-scale MEC (1000 L)

higher hydrogen production rates [46, 94, 100, 110]; these pretreatments convert complex compounds into more readily available organic compounds for consumption by electrochemically active bacteria.

9.5.2 Hydrogen Yield in Scaled-Up MECs

Many MEC tests have been conducted using milliliter-scale reactors under relatively optimum operating conditions (Fig. 9.3a). In a lab-scale test, a maximum hydrogen production rate of 50 $m^3/(m^3\ d)$ was successfully achieved using a two-chambered MEC (effective anode chamber volume of about 20 mL) equipped with a graphite felt anode and an Ni foam cathode [66]. However, to develop MECs that can operate at a practical site, larger-scale MECs are required. Several investigations of large-scale MECs with 1–1000 L volume reactors have been performed (Fig. 9.3b). In general, the larger reactor volume exhibits a lower hydrogen production rate, likely due to the increase in overpotential and internal resistance. A two-chambered MEC (anode chamber volume of 3.3 L) fed with acetate showed a hydrogen production rate of about 0.02 $m^3/(m^3\ d)$ [4]. Single-chambered MECs (chamber volume of 3.3 L [8] and 3.2 L [59]) fed with acetate generated hydrogen at a rate of over 0.3 $m^3/(m^3\ d)$ [8] and 0.12 $m^3/(m^3\ d)$ [59], respectively. In the best

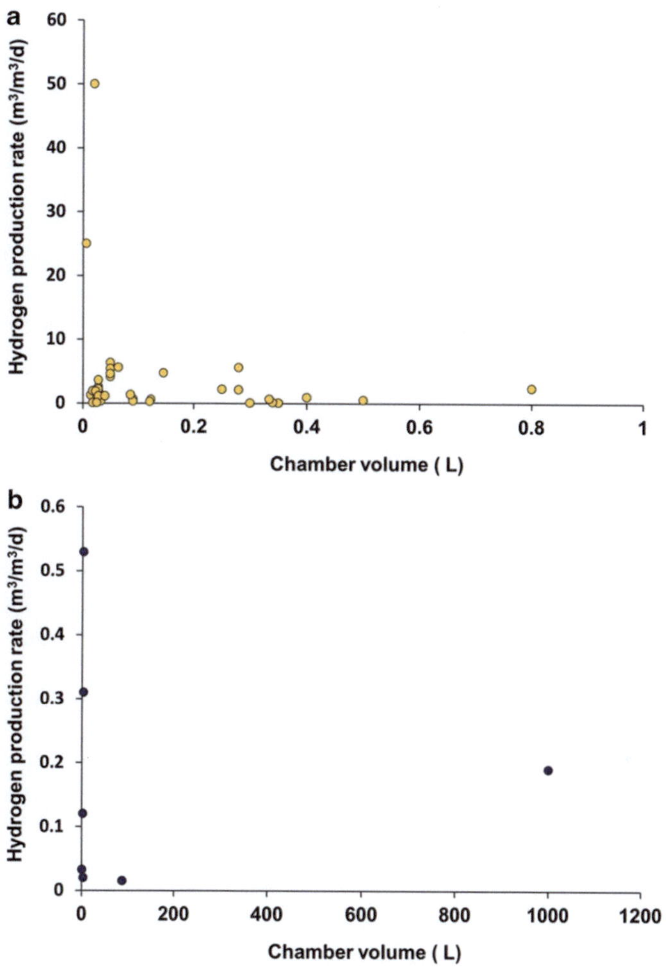

Fig. 9.3 Hydrogen production rate in (**a**) lab-scale MECs and (**b**) liter-scale and pilot-scale MECs. In the case of two-chambered MECs, chamber volumes were calculated based on the anode chamber volume

case, an acetate-fed MEC having eight pairs of electrodes (chamber volume of 2.5 L) achieved a hydrogen production rate of 0.53 $m^3/(m^3\ d)$, but this MEC was afflicted with hydrogen losses due to methanogens [32]. A pilot-scale MEC (chamber volume of 1000 L) obtained a hydrogen production rate of 0.19 $m^3/(m^3\ d)$, even as most of the produced gas was transformed into methane ($86 \pm 6\%$). Another pilot-scale MEC (working chamber volume of 88 L) fed with wastewater exhibited a lower hydrogen production rate of 0.015 $m^3/(m^3\ d)$ compared with the previous pilot-scale MEC (1000 L), but had a lower level of methane production [39].

9.6 Anodic Bacterial Community

Electrochemically active bacteria play a key role in hydrogen production in MECs; they oxidize organic matter and release electrons obtained from metabolic processes into anode electrodes. Anodic microbial communities in MECs inoculated with a mixed culture are generally composed of electrochemically active bacteria and other bacteria symbiotically interacting with the electrochemically active bacteria [111]. However, the diversity of bacterial species can change according to MEC operational conditions (e.g., inoculum, temperature, and substrate). Several studies have reported the species composition of anodic microbial community in MECs operated in different environments, based on 16S rDNA sequencing and pyrosequencing analyses (Table 9.9) [13, 33, 35, 37, 95, 96, 107, 111, 112].

In general, *δ-Proteobacteria*, *Bacteroidetes*, and *Firmicutes* have been the predominant microorganism classes used in all MEC research. Anodic bacterial communities of acetate-fed single-chambered MECs, inoculated with the effluent of acetate-fed microbial fuel cells, one having a Pt-catalyzed cathode and the other an Ni-catalyzed cathode, were compared by Selembo et al. [13]. *Pelobacter propionicus* and *Geobacter sulfurreducens* were the most dominant species in both MEC reactors; it was found that the cathodic catalysts did not affect the anodic bacterial communities [13]. Cusick et al. [95] examined the composition of anodic bacteria from a consortia of winery wastewater-fed and daily wastewater-fed MECs. The difference between the two MECs in terms of bacterial community present was observed. In the MEC fed with winery wastewater, the most dominant species was *Geobacter sulfurreducens* (44 %), but in the MEC fed with domestic wastewater, *Geobacter metallireducens* (23 %) was dominant. In addition, Kiely et al. [96] compared MECs fed with acetate, potato wastewater, and daily wastewater. Similar to the results reported by Selembo et al. [13], the predominance of *Pelobacter propionicus* was observed in the acetate-fed MECs, and the utilization of acetate as a substrate reduced the bacterial species diversity. Both potato wastewater-fed and daily wastewater-fed MECs showed more diverse bacterial species, but had different compositions. Potato wastewater-fed MECs were predominantly *Geobacter* species; daily wastewater-fed MECs were dominated by *Clostridium* (*Firmicutes*).

Lu et al. [107] investigated the microbial community structures of MECs operated at different temperatures. It was found that temperature considerably affected the anodic bacterial consortia. In MECs operated at temperatures of 4 °C and 9 °C, *Geobacter psychrophilus* was the predominant species, whereas *Geobacter chapelleii* were dominant in MECs operated at 25 °C. Moreover, hydrogenotrophic methanogens were detected in MECs operated at 25 °C, though methanogen growth was inhibited in MECs operated at 4 °C and 9 °C. Lu et al. [112] also explored the anodic bacterial communities of glucose-fed MECs at low temperature (4 °C), in which more diverse bacteria species were observed in MECs fed with glucose, compared to MECs fed with acetate. In the glucose-fed MECs at 4 °C, methanogens and homoacetogenesis were negligible.

Table 9.9 Summary of the dominant anodic bacterial class or species for microbial electrolysis cells reported in previous studies

Most dominant class or species	Inoculum	Substrate	Cell type	Ref.
Pelobacter propionicus (54.3 %)	Acetate-fed microbial fuel cell	Acetate	SC (with Ni catalyst)	[13]
Pelobacter propionicus (56.5 %)	Acetate-fed microbial fuel cell	Acetate	SC (with Pt catalyst)	
Geobacter sulfurreducens (44 %)	–	Winery wastewater	SC	[95]
Geobacter metallireducens (23 %)	–	Domestic wastewater	SC	
G. metallireducens G. lovleyi	–	Potato wastewater	SC	[96]
Clostridium	–	Daily wastewater	SC	
Pelobacter propionicus	–	Acetate	SC	
Geobacter psychrophilus (63 %) *Pseudomonas* spp. (14 %)	Domestic water	Acetate	SC (4 °C)	[107]
Geobacter psychrophilus (55 %) *Pseudomonas* spp. (7 %)	Domestic water	Acetate	SC (9 °C)	
Geobacter chapelleii (68 %)	Domestic water	Acetate	SC (25 °C)	
Unclassified genus Geobacter (12.23 %)	Waste-activated sludge	Waste-activated sludge	TC	[98]
Unclassified genus Paludibacter Geobacter (3.55 %)	Waste-activated sludge	Waste-activated sludge	SC	
Shewanella Geobacter	Raw sludge	Acetate	TC	[111]
Dysgonomonas	Glucose-fed microbial fuel cell	Glucose	SC (4 °C)	[112]
Unclassified genus	Glucose-fed microbial fuel cell	Glucose	SC (25 °C)	
Clostridium sp. (28 %) *Bacteroidetes bacterium* (11 %)	Domestic wastewater fed MEC	Domestic wastewater	TMEC	[33]
Chlorobium limicola (31 %) *Geobacter metallireducens* (15 %)	Domestic wastewater-fed MEC	Domestic wastewater	TMEC	
Bacteroidetes (55.5 %)	Wastewater	Acetate	SMEC	[37]
Proteobacteria, Firmicutes	–	Wastewater	BR	[35]

SC single chamber, *TC* two chamber, *SMEC* submersible microbial electrolysis cell, *BR* baffled reactor type, *TMEC* tubular-type microbial electrolysis cell

Waste-activated sludge-fed MECs were examined based on a pyrosequencing analysis [98], in which a diverse range of electrochemically active bacteria and acid-producing bacteria species were found. It was shown that there are symbiotic interactions between the acid-production bacteria and electrochemically active bacteria: acid-producing bacteria convert sugars into organic acids, and then electrochemically active bacteria utilize these organic acids or directly use the specific sugars. Further, acetoclastic methanogens were dominantly found. Gil-Carrera et al. [33] reported that complex organic matter was degraded to simpler organic matters and that electrochemically active bacteria then use this simpler organic matter in a semi-pilot tubular MEC.

9.7 Technological Challenges for Practical Implementation

Lab-scale MEC systems have achieved satisfactory levels of hydrogen production [12]. However, it is now imperative to overcome the diverse technological challenges to scale up these systems for use in field applications. In this section, core technical challenges related to each MEC component shown in Fig. 9.4 are addressed.

9.7.1 Challenges Associated with the Anode and Electrolyte

9.7.1.1 Metabolic Diversity

Electron and protons are generated from the oxidation of organic waste by electrochemically active bacteria in the anode. Pure simple-structure organic compounds can be readily used by electrochemically active bacteria. However, during

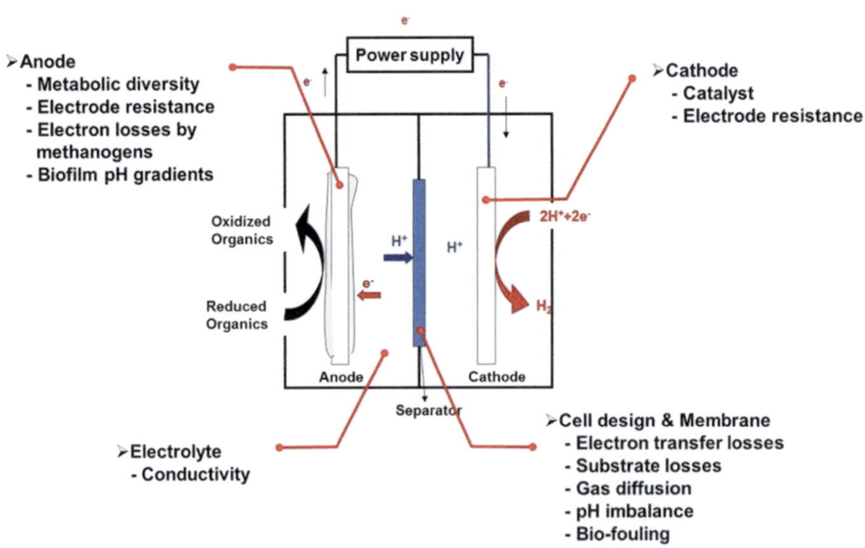

Fig. 9.4 Technological challenges for MECs

hydrogen production using MECs at practical sites, various types of organic waste are used as electron donors. This organic waste usually consists of a wide spectrum of organic matter; to degrade this complex organic waste, vast and protein microbial food webs need to be established [113]. However, previous research has shown that organic waste-fed MECs achieve lower hydrogen production performance compared to when a simpler pure substrate is used, even though more diverse bacterial species that consist of electrochemically active bacteria, and bacteria symbiotically related to electrochemically active bacteria, were found in organic waste-fed MECs [13, 33, 35, 37, 95, 96, 107, 111, 112]. This indicates that anodic bacterial communities cannot effectively degrade complex organic compounds, and as such that more investigation of metabolic diversity is required than for microbial species diversity [113]. The use of pretreatments to convert organic waste into easily degradable compounds has been recommended [46, 94, 100, 110].

9.7.1.2 Electron Losses by Methanogens

Substantial electron losses due to competition with methanogens for substrates reduce the hydrogen production performance in MECs [26, 107, 114]. In particular, there are notable electron losses due to competition with methanogens in MECs fed with complex organic compounds, such as glucose and wastewater, due to their fermentable characteristics [114, 115]. Moreover, hydrogen produced in the cathodes of MECs can be converted into methane by methanogens under circumstances that allow hydrogen to come into contact with methanogens [32, 38]. For this reason, it is necessary to effectively suppress methanogenic activities to improve hydrogen production from organic waste in MECs. Methods for operating MECs under adverse conditions for methanogens, such as intermittent air-exposure [24] and low temperature [107, 112], and chemical treatments, such as the inoculation of specific methanogen inhibitors and ultraviolet irradiation [116], might also be needed.

9.7.1.3 Electrode Resistance

One further challenge associated with the anode material is the electrical resistivity of the anodic electrode. Popular anode materials are carbonaceous materials because of their good biocompatibility, low overpotential, and reasonable cost [42, 43]. However, carbonaceous materials have high electrical resistivity. For example, graphite has a much higher electrical resistivity (1375 μΩcm) than iron (9.71 μΩcm) [113]. In lab-scale MECs, though ohmic losses due to the electrical resistivity of electrodes are not negligible, when scaling up MECs, these ohmic losses might reach considerable levels. To reduce ohmic losses from the electrodes,

electrodes can be modified using nanoparticles [53], and electrodes can be fabricated using new materials, such as graphene [117]; these can be considered as countermeasures to improve hydrogen production.

9.7.1.4 Electrolyte Buffer Capacity and Conductivity

The low buffer capacity and conductivity of the electrolyte is another challenge that must be overcome to make MEC use more practical [113]. Low numbers of protons, released by electrochemically active bacteria, transport out of the anodic biofilm at a low buffer strength, which leads to a local pH drop in the biofilm. According to Torres et al. [118], the viability of electrochemically active bacteria can be suppressed by a local pH drop in the biofilm, and this negatively affects the MEC performance; anolytes in MECs should have a high buffer capacity. However, wastewater having low protons, released by electrochemically active bacteria, transporting out of the anodic biofilm at a low buffer strength generally has a low buffer capacity. Indeed, wastewater commonly has a low conductivity, which will then significantly increase the ohmic resistance. In practical applications, this might restrict the performance of MECs. Adding buffers and salts to wastewater is one possible option for resolving this problem.

9.7.2 Challenges Associated with the Cathode

9.7.2.1 Expensive Catalysts and High Potential Losses

High potential losses can be observed at the cathode in MECs, although Pt has commonly been applied to the cathode. This loss is likely due to MECs operating at a neutral pH and under ambient temperature [113]. Sufficient protons need to be supplied to the cathode for the hydrogen production reaction. However, operation at a neutral pH cannot guarantee high concentration of available protons. Similarly, operation at ambient temperature does not assist hydrogen production reaction kinetics. If wastewater is being supplied to MECs, Pt catalysts might be damaged by irreversible poisoning by H_2S. Furthermore, Pt is a precious metal; the use of Pt as a catalyst can considerably reduce the economic feasibility of MECs. Therefore, alternative catalysts or cathode electrodes to Pt are required to improve MEC performance and economic efficiency. Excellent alternatives might include inexpensive metal catalysts such as Ni and Ti [63, 73], biocatalysts [27, 61, 62], and extremely high surface area cathodic electrodes [66, 70, 72].

9.7.3 Challenges Associated with Cell Design and Separator

9.7.3.1 pH Imbalance Between Anode and Cathode Chambers

In MECs, to both physically separate the anode and cathode chambers and to allow ion transport between the two chambers, ion exchange membranes are used. Protons, produced by electrochemically active bacteria in the anode chamber, theoretically move into the cathode through ion exchange membranes as electrons, generated by electrochemically active bacteria in the anode, and transfer to the cathode through an external circuit in order to maintain the charge balance. Other ionic species dominantly transfer through ion exchange membranes during actual MEC operation; these ionic species exist at much higher concentrations in the electrolyte. As a result, protons accumulate in the anode chamber and become depleted in the cathode chamber, leading to a pH decline in the anode chamber and incline in the cathode chamber. According to Sleutels et al. [78], this pH imbalance contributes to quite a large portion of the overall potential losses in MECs.

9.7.3.2 Biofouling on Surface of Membranes

A tremendous population of microorganisms exists in the anode chamber. These microorganisms can adhere to and grow on the surface of membranes, that is, biofouling. Biofouling possibly damages membranes and finally can reduce the performance of the MEC [119]. In practical applications, biofouling might increase capital cost of the MECs because severely fouled membranes might need to be replaced by new membranes.

9.7.3.3 Gas Crossover Through Membranes

High purity of hydrogen production is an advantage of MECs. This might reduce hydrogen production cost because expensive gas purification process will be not required [42]. However, membranes, such as ion exchange membranes and porous membranes, used in MECs are quite permeable to gases [84]. Hydrogen generated in the cathode can be contaminated by biogases, such as methane and carbon dioxide, produced in the anode chamber, and undesirable hydrogen losses can possibly occur.

9.7.3.4 Membraneless Single-Chambered Design

To resolve the challenges associated with separators, membraneless single-chambered designs have been introduced [24, 120]. These designs resolve the pH imbalance between the anode and cathode chambers and biofouling on the surface of membranes. Furthermore, these designs reduce ohmic losses derived from membranes and reduce the capital cost of MECs by eliminating membranes. However, these designs have also faced a number of critical challenges. Importantly, the drop in hydrogen purity due to biogases generated by microorganisms cannot be avoided. In addition, hydrogen can be utilized by hydrogenotrophic methanogens for methane production [32, 38].

9.8 Conclusions and Future Outlook

Despite the remaining technical challenges that have yet to be overcome before upscaling and commercialization, the outlook of MEC technology for hydrogen generation is still promising. It is expected that MEC technology will reduce the overall cost of wastewater treatment and be able to sustainably produce hydrogen. The challenges and suggestions for MECs discussed in this chapter can be expected to lead developments toward resolving energy losses and reducing the capital costs of MECs. Overall, the following conditions need to be developed to fulfill the potential of MECs in the future.

1. More research on operating MECs with actual wastewater and improving the ability of MECs to degrade complex organic matter needs to be conducted.
2. Designing an appropriate configuration for larger-scale MECs is required.
3. Low-cost and comparable or improved performance for current components for full-scale MECs needs to be studied.
4. Further research on renewable and alternative energy supply sources for MECs is required.
5. Cost-effective substitute materials for anode and cathode electrodes and separators need to be identified prior to the practical implementation of MECs.
6. More studies on MEC assisting systems or processes, such as anaerobic digestion [103] and dynamically adaptive control systems [121], are required to improve the performance of MECs.

References

1. Rabaey K, Rodriguez J, Blackall LL, Keller J, Gross P, Batstone D, Verstraete W, Nealson KH. Microbial ecology meets electrochemistry: electricity-driven and driving communities. ISME J. 2007;1(1):9–18.
2. Chae K-J, Choi M-J, Lee J, Ajayi F, Kim IS. Biohydrogen production via biocatalyzed electrolysis in acetate-fed bioelectrochemical cells and microbial community analysis. Int J Hydrogen Energy. 2008;33(19):5184–92.
3. Liu H, Grot S, Logan BE. Electrochemically assisted microbial production of hydrogen from acetate. Environ Sci Technol. 2005;39(11):4317–20.

4. Rozendal RA, Hamelers HV, Euverink GJ, Metz SJ, Buisman CJ. Principle and perspectives of hydrogen production through biocatalyzed electrolysis. Int J Hydrogen Energy. 2006;31(12):1632–40.
5. Rabaey K, Clauwaert P, Aelterman P, Verstraete W. Tubular microbial fuel cells for efficient electricity generation. Environ Sci Technol. 2005;39(20):8077–82.
6. Kim BH, Chang IS, Gil GC, Park HS, Kim HJ. Novel BOD (biological oxygen demand) sensor using mediator-less microbial fuel cell. Biotechnol Lett. 2003;25(7):541–5.
7. Gil GC, Chang IS, Kim BH, Kim M, Jang JK, Park HS, Kim HJ. Operational parameters affecting the performance of a mediator-less microbial fuel cell. Biosens Bioelectron. 2003;18(4):327–34.
8. Rozendal RA, Hamelers HV, Molenkamp RJ, Buisman CJ. Performance of single chamber biocatalyzed electrolysis with different types of ion exchange membranes. Water Res. 2007;41(9):1984–94.
9. Logan B, Grot S, Mallouk TE, Liu H. Bio-electrochemically assisted microbial reactor that generates hydrogen gas and methods of generating hydrogen gas. US Patent 7,709,113, 4 May 2010
10. Logan BE. Scaling up microbial fuel cells and other bioelectrochemical systems. Appl Microbiol Biotechnol. 2010;85(6):1665–71.
11. Logan BE, Call D, Cheng S, Hamelers HV, Sleutels TH, Jeremiasse AW, Rozendal RA. Microbial electrolysis cells for high yield hydrogen gas production from organic matter. Environ Sci Technol. 2008;42(23):8630–40.
12. Zhang Y, Angelidaki I. Microbial electrolysis cells turning to be versatile technology: recent advances and future challenges. Water Res. 2014;56:11–25.
13. Selembo PA, Merrill MD, Logan BE. Hydrogen production with nickel powder cathode catalysts in microbial electrolysis cells. Int J Hydrogen Energy. 2010;35(2):428–37.
14. Call DF, Wagner RC, Logan BE. Hydrogen production by Geobacter species and a mixed consortium in a microbial electrolysis cell. Appl Environ Microbiol. 2009;75(24):7579–87.
15. Rosenbaum M, Cotta MA, Angenent LT. Aerated Shewanella oneidensis in continuously fed bioelectrochemical systems for power and hydrogen production. Biotechnol Bioeng. 2010;105(5):880–8.
16. Bond DR, Lovley DR. Electricity production by Geobacter sulfurreducens attached to electrodes. Appl Environ Microbiol. 2003;69(3):1548–55.
17. Gorby YA, Yanina S, McLean JS, Rosso KM, Moyles D, Dohnalkova A, Beveridge TJ, Chang IS, Kim BH, Kim KS, Culley DE, Reed SB, Romine MF, Saffarini DA, Hill EA, Shi L, Elias DA, Kennedy DW, Pinchuk G, Watanabe K, Ishii S, Logan B, Nealson KH, Fredrickson JK. Electrically conductive bacterial nanowires produced by Shewanella oneidensis strain MR-1 and other microorganisms. Proc Natl Acad Sci U S A. 2006;103(30):11358–63.
18. Lovley DR. Bug juice: harvesting electricity with microorganisms. Nat Rev Microbiol. 2006;4(7):497–508.
19. Kim BH, Kim HJ, Hyun MS, Park DH. Direct electrode reaction of Fe(III)-reducing bacterium, Shewanella putrefaciens. J Microbiol Biotechnol. 1999;9(2):127–31.
20. Chang IS, Moon H, Bretschger O, Jang JK, Park HI, Nealson KH, Kim BH. Electrochemically active bacteria (EAB) and mediator-less microbial fuel cells. J Microbiol Biotechnol. 2006;16(2):163–77.
21. Hallenbeck PC, Ghosh D. Advances in fermentative biohydrogen production: the way forward? Trends Biotechnol. 2009;27(5):287–97.
22. Kim IS, Chae K-J, Choi M-J, Verstraete W. Microbial fuel cells: recent advances, bacterial communities and application beyond electricity generation. Env Eng Res. 2008;13(2):51–65.
23. Cheng S, Logan BE. Sustainable and efficient biohydrogen production via electrohydrogenesis. Proc Natl Acad Sci. 2007;104(47):18871–3.
24. Call D, Logan BE. Hydrogen production in a single chamber microbial electrolysis cell lacking a membrane. Environ Sci Technol. 2008;42(9):3401–6.

25. Hu H, Fan Y, Liu H. Hydrogen production in single-chamber tubular microbial electrolysis cells using non-precious-metal catalysts. Int J Hydrogen Energy. 2009;34(20):8535–42. doi:10.1016/j.ijhydene.2009.08.011.
26. Hu H, Fan Y, Liu H. Hydrogen production using single-chamber membrane-free microbial electrolysis cells. Water Res. 2008;42(15):4172–8.
27. Rozendal RA, Jeremiasse AW, Hamelers HV, Buisman CJ. Hydrogen production with a microbial biocathode. Environ Sci Technol. 2007;42(2):629–34.
28. Lee H-S, Torres CI, Parameswaran P, Rittmann BE. Fate of H2 in an upflow single-chamber microbial electrolysis cell using a metal-catalyst-free cathode. Environ Sci Technol. 2009;43 (20):7971–6.
29. Guo K, Tang X, Du Z, Li H. Hydrogen production from acetate in a cathode-on-top single-chamber microbial electrolysis cell with a mipor cathode. Biochem Eng J. 2010;51(1):48–52.
30. Tartakovsky B, Manuel M-F, Wang H, Guiot S. High rate membrane-less microbial electrolysis cell for continuous hydrogen production. Int J Hydrogen Energy. 2009;34(2):672–7.
31. Kyazze G, Popov A, Dinsdale R, Esteves S, Hawkes F, Premier G, Guwy A. Influence of catholyte pH and temperature on hydrogen production from acetate using a two chamber concentric tubular microbial electrolysis cell. Int J Hydrogen Energy. 2010;35(15):7716–22.
32. Rader GK, Logan BE. Multi-electrode continuous flow microbial electrolysis cell for biogas production from acetate. Int J Hydrogen Energy. 2010;35(17):8848–54.
33. Gil-Carrera L, Escapa A, Carracedo B, Morán A, Gómez X. Performance of a semi-pilot tubular microbial electrolysis cell (MEC) under several hydraulic retention times and applied voltages. Bioresour Technol. 2013;146:63–9.
34. Feng Y, Cheng Y, Du Y, Teng Q, Li H. Hydrogen production from acetate in a sleeve shape microbial electrolysis cell with a mipor cathode. Int J Electrochem Sci. 2014;9:6993–7002.
35. Ran Z, Gefu Z, Kumar JA, Chaoxiang L, Xu H, Lin L. Hydrogen and methane production in a bioelectrochemical system assisted anaerobic baffled reactor. Int J Hydrogen Energ. 2014;39 (25):13498–504.
36. Zhang J, Zhang Y, Liu B, Dai Y, Quan X, Chen S. A direct approach for enhancing the performance of a microbial electrolysis cell (MEC) combined anaerobic reactor by dosing ferric iron: enrichment and isolation of Fe (III) reducing bacteria. Chem Eng J. 2014;248:223–9.
37. Zhang Y, Angelidaki I. Innovative self-powered submersible microbial electrolysis cell (SMEC) for biohydrogen production from anaerobic reactors. Water Res. 2012;46 (8):2727–36.
38. Cusick RD, Bryan B, Parker DS, Merrill MD, Mehanna M, Kiely PD, Liu G, Logan BE. Performance of a pilot-scale continuous flow microbial electrolysis cell fed winery wastewater. Appl Microbiol Biotechnol. 2011;89(6):2053–63.
39. Heidrich E, Dolfing J, Scott K, Edwards S, Jones C, Curtis T. Production of hydrogen from domestic wastewater in a pilot-scale microbial electrolysis cell. Appl Microbiol Biotechnol. 2013;97(15):6979–89.
40. Wang A, Liu W, Ren N, Zhou J, Cheng S. Key factors affecting microbial anode potential in a microbial electrolysis cell for H_2 production. Int J Hydrogen Energy. 2010;35(24):13481–7.
41. Kundu A, Sahu JN, Redzwan G, Hashim M. An overview of cathode material and catalysts suitable for generating hydrogen in microbial electrolysis cell. Int J Hydrogen Energy. 2013;38(4):1745–57.
42. Kim IS, Chae K-J, Choi MJ, Verstraete W. Microbial fuel cells: recent advances, bacterial communities and application beyond electricity generation. Environ Eng Res (EER). 2008;13 (2):51–65.
43. Sarathi V, Nahm KS. Recent advances and challenges in the anode architecture and their modifications for the applications of microbial fuel cells. Biosens Bioelectron. 2013;43:461–75.
44. Yossan S, Xiao L, Prasertsan P, He Z. Hydrogen production in microbial electrolysis cells: choice of catholyte. Int J Hydrogen Energy. 2013;38(23):9619–24.

45. Zhang B, Wen Z, Ci S, Chen J, He Z. Nitrogen-doped activated carbon as a metal free catalyst for hydrogen production in microbial electrolysis cells. RSC Adv. 2014;4(90):49161–4.
46. Wang A, Sun D, Cao G, Wang H, Ren N, Wu W-M, Logan BE. Integrated hydrogen production process from cellulose by combining dark fermentation, microbial fuel cells, and a microbial electrolysis cell. Bioresour Technol. 2011;102(5):4137–43.
47. Liu Y-P, Wang Y-H, Wang B-S, Chen Q-Y. Effect of anolyte pH and cathode Pt loading on electricity and hydrogen co-production performance of the bio-electrochemical system. Int J Hydrogen Energy. 2014;39(26):14191–5.
48. Wagner RC, Regan JM, Oh S-E, Zuo Y, Logan BE. Hydrogen and methane production from swine wastewater using microbial electrolysis cells. Water Res. 2009;43(5):1480–8.
49. Linji X, Wenzong L, Yining W, Lee P, Aijie W, Shuai L. Trehalose enhancing microbial electrolysis cell for hydrogen generation in low temperature (0° C). Bioresour Technol. 2014;166:458–63.
50. Flexer V, Chen J, Donose BC, Sherrell P, Wallace GG, Keller J. The nanostructure of three-dimensional scaffolds enhances the current density of microbial bioelectrochemical systems. Energy Environ Sci. 2013;6(4):1291–8.
51. Guo K, Donose BC, Soeriyadi AH, Prévoteau A, Patil SA, Freguia S, Gooding JJ, Rabaey K. Flame oxidation of stainless steel felt enhances anodic biofilm formation and current output in bioelectrochemical systems. Environ Sci Technol. 2014;48(12):7151–6.
52. Saito T, Mehanna M, Wang X, Cusick RD, Feng Y, Hickner MA, Logan BE. Effect of nitrogen addition on the performance of microbial fuel cell anodes. Bioresour Technol. 2011;102(1):395–8. http://dx.doi.org/10.1016/j.biortech.2010.05.063.
53. Xu S, Liu H, Fan Y, Schaller R, Jiao J, Chaplen F. Enhanced performance and mechanism study of microbial electrolysis cells using Fe nanoparticle-decorated anodes. Appl Microbiol Biotechnol. 2012;93(2):871–80.
54. Hrapovic S, Manuel MF, Luong JHT, Guiot SR, Tartakovsky B. Electrodeposition of nickel particles on a gas diffusion cathode for hydrogen production in a microbial electrolysis cell. Int J Hydrogen Energy. 2010;35(14):7313–20.
55. Sun M, Sheng G-P, Mu Z-X, Liu X-W, Chen Y-Z, Wang H-L, Yu H-Q. Manipulating the hydrogen production from acetate in a microbial electrolysis cell–microbial fuel cell-coupled system. J Power Sources. 2009;191(2):338–43.
56. Sleutels TH, Hamelers HV, Buisman CJ. Effect of mass and charge transport speed and direction in porous anodes on microbial electrolysis cell performance. Bioresour Technol. 2011;102(1):399–403.
57. Ivanov I, Ren L, Siegert M, Logan BE. A quantitative method to evaluate microbial electrolysis cell effectiveness for energy recovery and wastewater treatment. Int J Hydrogen Energy. 2013;38(30):13135–42.
58. Omidi H, Sathasivan A. Optimal temperature for microbes in an acetate fed microbial electrolysis cell (MEC). Int Biodeter Biodegr. 2013;85:688–92.
59. Jia YH, Ryu JH, Kim CH, Lee WK, Tran TVT, Lee HL, Zhang RH, Ahn DH. Enhancing hydrogen production efficiency in microbial electrolysis cell with membrane electrode assembly cathode. J Ind Eng Chem. 2012;18(2):715–9.
60. Huang Y-X, Liu X-W, Sun X-F, Sheng G-P, Zhang Y-Y, Yan G-M, Wang S-G, Xu A-W, Yu H-Q. A new cathodic electrode deposit with palladium nanoparticles for cost-effective hydrogen production in a microbial electrolysis cell. Int J Hydrogen Energy. 2011;36 (4):2773–6.
61. Jeremiasse AW, Hamelers HV, Buisman CJ. Microbial electrolysis cell with a microbial biocathode. Bioelectrochemistry. 2010;78(1):39–43.
62. Fu Q, Kobayashi H, Kuramochi Y, Xu J, Wakayama T, Maeda H, Sato K. Bioelectrochemical analyses of a thermophilic biocathode catalyzing sustainable hydrogen production. Int J Hydrogen Energy. 2013;38(35):15638–45. http://dx.doi.org/10.1016/j.ijhydene.2013.04.116.
63. Selembo PA, Merrill MD, Logan BE. The use of stainless steel and nickel alloys as low-cost cathodes in microbial electrolysis cells. J Power Sources. 2009;190(2):271–8.

64. Manuel M-F, Neburchilov V, Wang H, Guiot S, Tartakovsky B. Hydrogen production in a microbial electrolysis cell with nickel-based gas diffusion cathodes. J Power Sources. 2010;195(17):5514–9.
65. Call DF, Merrill MD, Logan BE. High surface area stainless steel brushes as cathodes in microbial electrolysis cells. Environ Sci Technol. 2009;43(6):2179–83.
66. Jeremiasse AW, Hamelers HV, Saakes M, Buisman CJ. Ni foam cathode enables high volumetric H_2 production in a microbial electrolysis cell. Int J Hydrogen Energy. 2010;35 (23):12716–23.
67. Tokash JC, Logan BE. Electrochemical evaluation of molybdenum disulfide as a catalyst for hydrogen evolution in microbial electrolysis cells. Int J Hydrogen Energy. 2011;36 (16):9439–45.
68. Xiao L, Wen Z, Ci S, Chen J, He Z. Carbon/iron-based nanorod catalysts for hydrogen production in microbial electrolysis cells. Nano Energy. 2012;1(5):751–6.
69. Chen Q-Y, Liu J-S, Liu Y, Wang Y-H. Hydrogen production on TiO_2 nanorod arrays cathode coupling with bio-anode with additional electricity generation. J Power Sources. 2013;238:345–9.
70. Wang L, Chen Y, Huang Q, Feng Y, Zhu S, Shen S. Hydrogen production with carbon nanotubes based cathode catalysts in microbial electrolysis cells. J Chem Technol Biotechnol. 2012;87(8):1150–6.
71. Yuan H, Li J, Yuan C, He Z. Facile synthesis of MoS2@ CNT as an effective catalyst for hydrogen production in microbial electrolysis cells. ChemElectroChem. 2014;1 (11):1828–33.
72. Yang Q, Jiang Y, Xu Y, Qiu Y, Chen Y, Zhu S, Shen S. Hydrogen production with polyaniline/multi-walled carbon nanotube cathode catalysts in microbial electrolysis cells. J Chem Technol Biotechnol. 2014;70(7):1263–9.
73. Farhangi S, Ebrahimi S, Niasar MS. Commercial materials as cathode for hydrogen production in microbial electrolysis cell. Biotechnol Lett. 2014;36(10):1987–92.
74. Wang A, Liu W, Ren N, Cheng H, Lee D-J. Reduced internal resistance of microbial electrolysis cell (MEC) as factors of configuration and stuffing with granular activated carbon. Int J Hydrogen Energy. 2010;35(24):13488–92.
75. Cheng S, Logan BE. High hydrogen production rate of microbial electrolysis cell (MEC) with reduced electrode spacing. Bioresour Technol. 2011;102(3):3571–4.
76. Gil-Carrera L, Mehta P, Escapa A, Morán A, García V, Guiot SR, Tartakovsky B. Optimizing the electrode size and arrangement in a microbial electrolysis cell. Bioresour Technol. 2011;102(20):9593–8.
77. Liang D-W, Peng S-K, Lu S-F, Liu Y-Y, Lan F, Xiang Y. Enhancement of hydrogen production in a single chamber microbial electrolysis cell through anode arrangement optimization. Bioresour Technol. 2011;102(23):10881–5.
78. Sleutels THJA, Hamelers HVM, Rozendal RA, Buisman CJN. Ion transport resistance in microbial electrolysis cells with anion and cation exchange membranes. Int J Hydrogen Energy. 2009;34(9):3612–20.
79. Cheng S, Logan BE. Evaluation of catalysts and membranes for high yield biohydrogen production via electrohydrogenesis in microbial electrolysis cells (MECs). Water Sci Technol. 2008;58(4):853.
80. Lu L, Xing D, Xie T, Ren N, Logan BE. Hydrogen production from proteins via electrohydrogenesis in microbial electrolysis cells. Biosens Bioelectron. 2010;25(12):2690–5. http://dx.doi.org/10.1016/j.bios.2010.05.003.
81. Kim JR, Cheng S, Oh SE, Logan BE. Power generation using different cation, anion, and ultrafiltration membranes in microbial fuel cells. Environ Sci Technol. 2007;41(3):1004–9.
82. Zhang X, Cheng S, Huang X, Logan BE. Improved performance of single-chamber microbial fuel cells through control of membrane deformation. Biosens Bioelectron. 2010;25 (7):1825–8. doi:10.1016/j.bios.2009.11.018.

83. Wang Y-H, Wang B-S, Liu Y-P, Chen Q-Y. Electricity and hydrogen co-production from a bio-electrochemical cell with acetate substrate. Int J Hydrogen Energy. 2013;38(16):6600–6.
84. Chae K-J, Kim K-Y, Choi M-J, Yang E, Kim IS, Ren X, Lee M. Sulfonated polyether ether ketone (SPEEK)-based composite proton exchange membrane reinforced with nanofibers for microbial electrolysis cells. Chem Eng J. 2014;254:393–8.
85. Ajayi FF, Kim K-Y, Chae K-J, Choi M-J, Kim S-Y, Chang I-S, Kim IS. Study of hydrogen production in light assisted microbial electrolysis cell operated with dye sensitized solar cell. Int J Hydrogen Energy. 2009;34(23):9297–304.
86. Ajayi FF, Kim K-Y, Chae K-J, Choi M-J, Chang IS, Kim IS. Optimization studies of bio-hydrogen production in a coupled microbial electrolysis–dye sensitized solar cell system. Photochem Photobiol Sci. 2010;9(3):349–56.
87. Chae K-J, Choi M-J, Kim K-Y, Ajayi FF, Chang I-S, Kim IS. A solar-powered microbial electrolysis cell with a platinum catalyst-free cathode to produce hydrogen. Environ Sci Technol. 2009;43(24):9525–30.
88. Qian F, Wang G, Li Y. Solar-driven microbial photoelectrochemical cells with a nanowire photocathode. Nano Lett. 2010;10(11):4686–91.
89. Wang H, Qian F, Wang G, Jiao Y, He Z, Li Y. Self-biased solar-microbial device for sustainable hydrogen generation. ACS Nano. 2013;7(10):8728–35.
90. Kim Y, Logan BE. Hydrogen production from inexhaustible supplies of fresh and salt water using microbial reverse-electrodialysis electrolysis cells. Proc Natl Acad Sci. 2011;108 (39):16176–81.
91. Selembo PA, Perez JM, Lloyd WA, Logan BE. High hydrogen production from glycerol or glucose by electrohydrogenesis using microbial electrolysis cells. Int J Hydrogen Energy. 2009;34(13):5373–81. doi:10.1016/j.ijhydene.2009.05.002.
92. Chookaew T, Prasertsan P, Ren ZJ. Two-stage conversion of crude glycerol to energy using dark fermentation linked with microbial fuel cell or microbial electrolysis cell. N Biotechnol. 2014;31(2):179–84.
93. Kadier A, Simayi Y, Kalil MS, Abdeshahian P, Hamid AA. A review of the substrates used in microbial electrolysis cells (MECs) for producing sustainable and clean hydrogen gas. Renew Energy. 2014;71:466–72.
94. Lalaurette E, Thammannagowda S, Mohagheghi A, Maness P-C, Logan BE. Hydrogen production from cellulose in a two-stage process combining fermentation and electrohydrogenesis. Int J Hydrogen Energy. 2009;34(15):6201–10.
95. Cusick RD, Kiely PD, Logan BE. A monetary comparison of energy recovered from microbial fuel cells and microbial electrolysis cells fed winery or domestic wastewaters. Int J Hydrogen Energy. 2010;35(17):8855–61.
96. Kiely PD, Cusick R, Call DF, Selembo PA, Regan JM, Logan BE. Anode microbial communities produced by changing from microbial fuel cell to microbial electrolysis cell operation using two different wastewaters. Bioresour Technol. 2011;102(1):388–94.
97. Lu L, Xing D, Liu B, Ren N. Enhanced hydrogen production from waste activated sludge by cascade utilization of organic matter in microbial electrolysis cells. Water Res. 2012;46 (4):1015–26.
98. Lu L, Xing D, Ren N. Pyrosequencing reveals highly diverse microbial communities in microbial electrolysis cells involved in enhanced H2 production from waste activated sludge. Water Res. 2012;46(7):2425–34.
99. Liu W, Huang S, Zhou A, Zhou G, Ren N, Wang A, Zhuang G. Hydrogen generation in microbial electrolysis cell feeding with fermentation liquid of waste activated sludge. Int J Hydrogen Energy. 2012;37(18):13859–64.
100. Wu T, Zhu G, Jha AK, Zou R, Liu L, Huang X, Liu C. Hydrogen production with effluent from an anaerobic baffled reactor (ABR) using a single-chamber microbial electrolysis cell (MEC). Int J Hydrogen Energy. 2013;38(25):11117–23.
101. Zhang J, Zhang Y, Quan X, Chen S, Afzal S. Enhanced anaerobic digestion of organic contaminants containing diverse microbial population by combined microbial electrolysis

cell (MEC) and anaerobic reactor under Fe (III) reducing conditions. Bioresour Technol. 2013;136:273–80.
102. Ullery ML, Logan BE. Comparison of complex effluent treatability in different bench scale microbial electrolysis cells. Bioresour Technol. 2014;170:530–7.
103. Guo X, Liu J, Xiao B. Bioelectrochemical enhancement of hydrogen and methane production from the anaerobic digestion of sewage sludge in single-chamber membrane-free microbial electrolysis cells. Int J Hydrogen Energy. 2013;38(3):1342–7.
104. Brown RK, Harnisch F, Wirth S, Wahlandt H, Dockhorn T, Dichtl N, Schröder U. Evaluating the effects of scaling up on the performance of bioelectrochemical systems using a technical scale microbial electrolysis cell. Bioresour Technol. 2014;163:206–13.
105. Merrill MD, Logan BE. Electrolyte effects on hydrogen evolution and solution resistance in microbial electrolysis cells. J Power Sources. 2009;191(2):203–8.
106. Munoz LD, Erable B, Etcheverry L, Riess J, Basséguy R, Bergel A. Combining phosphate species and stainless steel cathode to enhance hydrogen evolution in microbial electrolysis cell (MEC). Electrochem Commun. 2010;12(2):183–6.
107. Lu L, Ren N, Zhao X, Wang H, Wu D, Xing D. Hydrogen production, methanogen inhibition and microbial community structures in psychrophilic single-chamber microbial electrolysis cells. Energy Environ Sci. 2011;4(4):1329–36.
108. Ajayi FF, Kim K-Y, Chae K-J, Choi M-J, Kim IS. Effect of hydrodynamic force and prolonged oxygen exposure on the performance of anodic biofilm in microbial electrolysis cells. Int J Hydrogen Energy. 2010;35(8):3206–13.
109. Lee H-S, Rittmann BE. Significance of biological hydrogen oxidation in a continuous single-chamber microbial electrolysis cell. Environ Sci Technol. 2009;44(3):948–54.
110. Lu L, Ren N, Xing D, Logan BE. Hydrogen production with effluent from an ethanol–H_2-coproducing fermentation reactor using a single-chamber microbial electrolysis cell. Biosens Bioelectron. 2009;24(10):3055–60.
111. Liu W, Wang A, Sun D, Ren N, Zhang Y, Zhou J. Characterization of microbial communities during anode biofilm reformation in a two-chambered microbial electrolysis cell (MEC). J Biotechnol. 2012;157(4):628–32.
112. Lu L, Xing D, Ren N, Logan BE. Syntrophic interactions drive the hydrogen production from glucose at low temperature in microbial electrolysis cells. Bioresour Technol. 2012;124:68–76.
113. Rozendal RA, Hamelers HVM, Rabaey K, Keller J, Buisman CJN. Towards practical implementation of bioelectrochemical wastewater treatment. Trends Biotechnol. 2008;26 (8):450–9.
114. Chae K-J, Choi M-J, Kim K-Y, Ajayi F, Chang I-S, Kim IS. Selective inhibition of methanogens for the improvement of biohydrogen production in microbial electrolysis cells. Int J Hydrogen Energy. 2010;35(24):13379–86.
115. Freguia S, Rabaey K, Yuan Z, Keller J. Electron and carbon balances in microbial fuel cells reveal temporary bacterial storage behavior during electricity generation. Environ Sci Technol. 2007;41(8):2915–21.
116. Hou Y, Luo H, Liu G, Zhang R, Li J, Fu S. Improved hydrogen production in the microbial electrolysis cell by inhibiting methanogenesis using ultraviolet irradiation. Environ Sci Technol. 2014;48(17):10482–8.
117. Wang H, Wang G, Ling Y, Qian F, Song Y, Lu X, Chen S, Tong Y, Li Y. High power density microbial fuel cell with flexible 3D graphene–nickel foam as anode. Nanoscale. 2013;5 (21):10283–90.
118. Torres CI, Kato Marcus A, Rittmann BE. Proton transport inside the biofilm limits electrical current generation by anode-respiring bacteria. Biotechnol Bioeng. 2008;100(5):872–81.
119. Choi M-J, Chae K-J, Ajayi FF, Kim K-Y, Yu H-W, Kim C-W, Kim IS. Effects of biofouling on ion transport through cation exchange membranes and microbial fuel cell performance. Bioresour Technol. 2011;102(1):298–303. doi:10.1016/j.biortech.2010.06.129.

120. Ye Y, Wang L, Chen Y, Zhu S, Shen S. High yield hydrogen production in a single-chamber membraneless microbial electrolysis cell. Water Sci Technol. 2010;61(3):721–7.
121. Andersen SJ, Pikaar I, Freguia S, Lovell BC, Rabaey K, Rozendal RA. Dynamically adaptive control system for bioanodes in serially stacked bioelectrochemical systems. Environ Sci Technol. 2013;47(10):5488–94.

Chapter 10
Solar Hydrogen Production

Athanasios G. Konstandopoulos, Chrysoula Pagkoura, Dimitrios A. Dimitrakis, Souzana Lorentzou, and George P. Karagiannakis

Abstract This chapter summarizes the current status of solar-aided hydrogen production technologies, with special emphasis on high temperature thermochemical concepts. The required high temperatures are achieved via concentrated solar irradiation through the respective systems, e.g., solar towers and solar dishes. Customized, efficient, and robust solar reactor concepts are important to ensure optimum coupling of the thermochemical phenomenon with the solar source. Of fundamental importance for such thermochemical processes is the development of active materials and key components. Some of the most studied and promising active materials are presented in this chapter along with their relevant advantages and challenges. Solar hydrogen (/fuels) production is found to constitute an in principle promising alternative and supplementary solution to currently employed renewables. Nevertheless, further development is required to increase solar-to-fuel efficiencies and to overcome long-term stability issues. Favorable solutions strongly depend on the identification of more active and robust materials as well as on the definition of solar reactor designs that will ensure optimum exploitation of solar irradiation.

Keywords Solar hydrogen • Hydrogen production • Solar fuels • Concentrated irradiation • Thermochemical processes • Water-splitting solar reactors

10.1 Introduction

This chapter provides an overall picture of the past, present, and future perspectives on the technologies applied in the field of solar hydrogen production with emphasis on high temperature thermochemical cycles. This work discusses technologies that

A.G. Konstandopoulos (✉) • D.A. Dimitrakis
Aerosol and Particle Technology Laboratory, Center for Research and Technology Hellas, 57001 Thessaloniki, Greece

Department of Chemical Engineering, Aristotle University, 54006 Thessaloniki, Greece
e-mail: agk@cperi.certh.gr

C. Pagkoura • S. Lorentzou • G.P. Karagiannakis
Aerosol and Particle Technology Laboratory, Center for Research and Technology Hellas, 57001 Thessaloniki, Greece

Z. Fang et al. (eds.), *Production of Hydrogen from Renewable Resources*, Biofuels and Biorefineries 5, DOI 10.1007/978-94-017-7330-0_10

have the potential to be coupled with solar facilities and are able to achieve high temperatures and eventually substitute energy requirements currently covered from fossil fuel combustion. The replacement of conventional energy sources with solar power is a major challenge for the future development of sustainable and renewable fuels.

The current status of energy consumption worldwide is presented, as well as possible viable alternatives able to meet the energy-intensive modern way of life in a sufficient and fossil fuel-independent manner. One of the most promising solutions to this end lies in the utilization of the vast solar potential. Current technologies used for collecting solar radiation and their capabilities are discussed in the course of this chapter as well.

The majority of existing industrial scale solar facilities are used for electricity production. Currently, such systems are coupled with fossil fuel combustion to produce electricity without interruption that is caused by the intermittent nature of solar irradiance. These back-up systems are necessary for the support of the electricity network during operation under insufficient solar radiation conditions or even during the night. An appealing alternative to solar electricity is that of solar hydrogen production, i.e., the use of solar energy for the production of hydrogen. Moreover, the transformation of solar energy into chemical energy addresses the issue of storage. There are numerous hydrogen sources and possible alternative routes leading to its production. In this chapter, a section summarizes the possible alternatives to solar hydrogen production.

Currently, materials for solar hydrogen production are being pursued by many groups around the world while several promising solar reactor concepts have been developed. Within this chapter, the production of solar hydrogen from fossil fuels is presented, followed by a more detailed description of solar hydrogen production routes from renewable sources that have the potential of providing a sustainable long-term solution to the energy problem. A particularly interesting approach combines solar hydrogen with the production of carbon monoxide to form synthesis gas which can be directly used for the synthesis of conventional fuels.

10.2 The Growing Energy Demand Challenge

Plato in his laws (360 BC) stated: "*...The land must be sufficient to maintain a certain number of inhabitants in a moderate way of life, more than this is not required....*" In its modern version, the above statement is described by the Nobel laureate in Chemistry 1996 R.E. Smalley, as the "Terawatt Challenge" [1]. The future energy demands should be calculated in a fair mode, i.e., aiming at the provision of adequate amounts of resources for the entire global population. The ideal solution would be to meet the high energy demands of Earth's population from carbon-free fuels.

Earth's population (~7 billion people) is expected to increase by 1.7 billion until 2035 (~25 % increase). This increase is estimated to be accompanied by an even higher increase in energy demand (almost as high as ~40 %), since significant growth is expected to occur in the non-OECD countries [2, 3]. These predictions highlight the imperative need for finding alternative ways, in order to cover the future energy demands in an efficient and environmental-friendly way. Currently, the economy is highly dependent on fossil fuels, and although significant progress has been recorded with respect to energy efficiency and thus to optimization of consumed fuels per produced work, further effort must be put on finding less polluting – i.e., carbon-free – energy sources. Renewable energy sources are currently developing very fast and have already become immersed in the energy mixture but still have a long way to go to allow "energy security" and prevention of "climate change." It is estimated that until the year 2035, the contribution of renewables to global energy requirements will not amount to more than 10 %. Until 2035, fossil fuels (i.e., mainly oil and natural gas) are still expected to be playing a key role as an energy source for more than half (~50 % [4]) of the future energy demands.

A very promising solution to meet future energy demands is by exploiting the Earth's solar potential. The amount of solar energy reaching Earth surpasses by far the existing and future energy requirements [5]. While the vital significance of solar energy has always been indisputable, intensive efforts to convert it into other forms of energy, besides thermal, have only begun during the last decades, with its conversion to electricity prevailing. If alternative technologies related to long-term storage of solar energy into fuels are further developed and if the technologies related with the collection of the solar potential become economically more affordable, solar energy will have higher potential of contributing significantly to Earth's energy balance.

10.3 Solar Technologies

This section provides the main principles of the technologies that deal with concentrated solar radiation, i.e., technologies that are capable of concentrating solar irradiance at a target and achieving concentration factors typically higher than ~30. The principles used in concentrated solar technologies (CST) may be exploited by several thermochemical technologies that require elevated temperatures for their operation, ranging from a few hundreds up to more than 1000°.

To depict what happens when solar irradiation "hits" a surface, a simplified scheme is presented in Fig. 10.1. By applying a thermal equilibrium calculation, a generic equation for the theoretical maximum temperature ($T_{stagnation}$, Eq. 10.4) at the focal point on the receiver can be established [6].

The following equations are applicable to the illustration of Fig. 10.1.

$$q_{useful} = q_{absorbed} - q_{reradiation} - q_{loss} \tag{10.1}$$

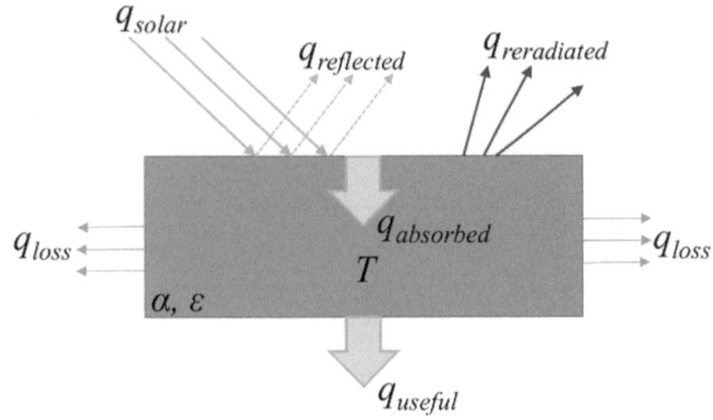

Fig. 10.1 Schematic representation of the heat transfer occurring when solar energy reaches a surface

$$q_{\text{useful}} = \alpha q_{\text{solar}} - \varepsilon \sigma T^4 - q_{\text{loss}} \tag{10.2}$$

$$q_{\text{useful}} = \alpha C I - \varepsilon \sigma T^4 - q_{\text{loss}} \tag{10.3}$$

The q_{useful} corresponds to the heat that is actually transferred to the working fluid employed; the factors α and ε refer to the absorptivity and emissivity, respectively, of the irradiated surface; C is the concentration ratio of solar irradiance; I the solar irradiation, typically taken as $I = 1000\ \text{W/m}^2$ (one sun); σ the Stefan Boltzmann constant (5.67 × 10–8 W/(m^{2}K^4); T the temperature developed on the irradiated surface; and q_{loss} the heat losses to the environment. In Eq. 10.3, when q_{useful} is set to 0, the system is perfectly insulated and has no heat losses, and the point of concentration is considered to be a flat area of a perfect receiver that has an absorptivity and emissivity $\alpha_{\text{eff}} \approx \varepsilon_{\text{eff}} \approx 1$. The stagnation temperature can be calculated as follows (Eq. 10.4).

$$T_{\text{stagnation}} = \left(\frac{C.I}{\sigma}\right)^{0.25} \tag{10.4}$$

The high temperatures required for the production of either steam or heat for electricity generation or for thermochemical hydrogen production are achieved by concentrating the solar energy upon a receiver. Currently, there are four different configurations that can achieve sufficient concentration of solar energy: the parabolic trough, the linear Fresnel reflector, the solar tower, and the parabolic dish. An additional configuration is that of a solar furnace which is a combination of a sun-tracking heliostat field (similar to the one employed in solar tower systems) and a nonmoving parabolic dish that concentrates the solar irradiance on a stable receiver located at its focal point. Since the latter system is mainly used for research purposes and has no practicality for scale-up/commercial purposes, it will not be further analyzed here.

The above optical configurations can be classified based on the concentration ratios they achieve and are divided into three categories. The first category – where parabolic troughs (Fig. 10.2a, e) and linear Fresnel systems (Fig. 10.2b, f) fall into – achieves relatively low concentration ratios, i.e., the parameter C ranges between

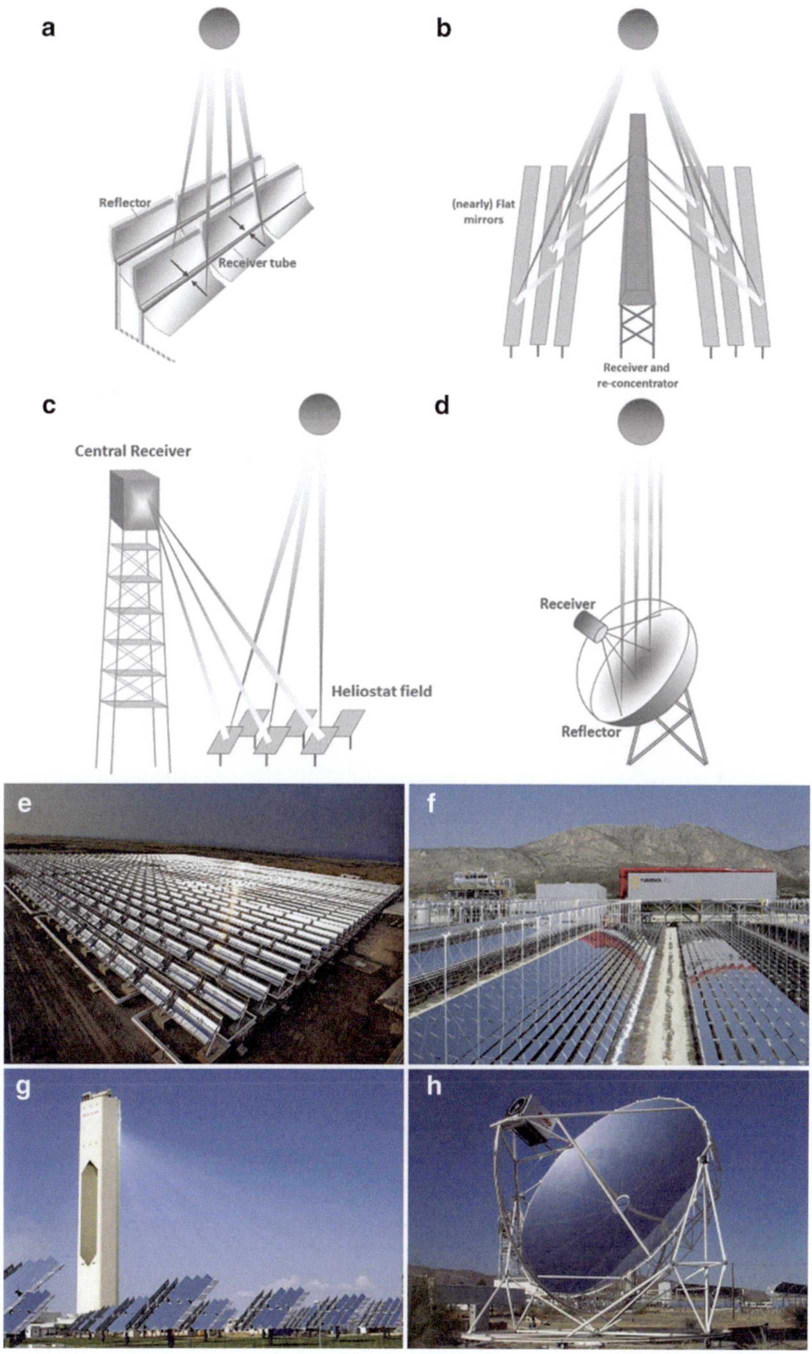

Fig. 10.2 Schematic representation of the available concentrated solar power (CSP) technologies: (**a**) solar trough, (**b**) linear Fresnel, (**c**) solar tower, (**d**) solar dish (Adapted from open access image by Greenpeace International [7]). Images of CSP plants around the world: (**e**) Holaniku at Keahole Point solar power plant with parabolic troughs (open access image by Xklaim) [8],

~30 and 100. The second category refers to systems that achieve moderate concentrations, such as the solar tower system (Fig. 10.2c, g). These systems may achieve concentrations ranging from ~200 and up to 1000. The parabolic dish (Fig. 10.2d, h) is capable of achieving the highest concentration ratios – starting from $C = 1000$ and going up to ~13,000.

The actual concentration that can be achieved by each system depends on various parameters like the inherent properties of the reflective area or the geometric configuration of the mirrors. The former parameter is characterized by a certain reflectivity value, which is actually the percentage of irradiance that the mirrors reflect, while the latter refers to, e.g., curved or flat mirrors. In the following paragraphs, more details about these systems are presented.

Solar troughs and linear Fresnel configurations both employ a single axis tracking system that allows them to move throughout the day, usually from North to South (for systems in the Northern Hemisphere), following the movement of the sun. These two systems, concentrate the solar irradiance on a linear receiver/tube, located at their focal line. At this stage, it should be clarified that the term "focus" refers to a "narrow area" with its shape being strongly dependent on the configuration of the mirrors. Thus, in the cases of the solar troughs and the linear Fresnel systems, it is a linear area, while in the case of the solar towers, it is the projection of a parallelepiped mirror. The focus actually resembles a point only in the case of the solar dishes (Fig. 10.2a) that achieve the highest concentration.

Solar troughs are long reflectors of parabolic shape, mainly used in power stations for electricity production. The concentrated solar irradiance usually heats a transfer fluid (e.g., molten salts, oil, or synthetic fluids) that constantly flows through the receiver tube. Although among the CSP systems compared within this work parabolic troughs achieve the lowest concentration ratios, they currently possess the largest share in the market of solar power production. According to data provided by CSP World [12], the largest solar power facility in operation based on parabolic troughs was inaugurated in 1984, in the Mojave Desert, USA, with a total capacity of 354 MW, derived from nine solar plants located at different sites within the region (SEGS I to SEGS IX). In general, North America currently has the largest individual installations of solar troughs. Nevertheless, Southern Europe – throughout which more installations of significantly smaller scale are located – has cumulatively almost twice the capacity of all North American facilities presently in operation.

On the other hand, the Fresnel systems (Fig. 10.2b) employ an array of thin flat mirrors that direct the solar radiation on tubular receivers located at a common focus line. The receivers – similarly to the solar troughs – are tubes containing a working fluid that absorbs the heat. Contrary to the wide adaptation of the solar

Fig. 10.2 (continued) (**f**) Linear-Fresnel solar power station: Porto Errado 2 (open access image by Novatec Solar) [9], (**g**) Solucar PS-10 solar tower (open access image by Afloresm) [10], and (**h**) parabolic dish at the Plataforma Solar de Almería (PSA) in Spain (open access image by Schlaich Bergermann und Partner) [11]

troughs in the commercial power market, the first Fresnel commercial plant was built in 2008 in California and can deliver 5 MW of power (Kimberlina Solar Thermal Energy Plant). Currently, the largest installation is located in Spain (Puerto Errado 2, 30 MW), while new, larger projects are planned or are under construction (e.g., two plants of 100 MW power each, one in India and one in South Africa).

The solar tower plants (Fig. 10.2c) are systems with a solar field comprising a large number of heliostats, appropriately distributed around a central tower. On the top of the solar tower where the receiver is placed lies the focal point of the field. This technology is also less mature compared to parabolic troughs, as the first commercial solar tower was completed as late as 2007 (the PS10 solar plant in Spain). Interestingly, these facilities emerged almost two and half decades earlier than that, almost in parallel with the solar troughs, with the solar facilities of Themis in France and Solar One in the USA probably being among the first ones ever to be built (in 1983 and 1982, respectively). This technology did not evolve as fast, with the total number of commercially available central receiver solar plants being confined only to five including plants currently in operation. The respective figure for solar-only parabolic trough technologies – i.e., excluding any hybrid parabolic trough systems – is at least decuple [12]. The solar tower systems can achieve high concentrations and thus are capable of developing high temperatures on the central receiver, ranging from few hundred degrees up to a couple of thousand degrees. In addition to electricity production, this capacity can be utilized as the driving force of (typically endothermic) thermochemical reactions. Efforts in the direction of verifying the feasibility of the concept started earlier than the establishment of the first commercial solar towers for electricity production with early thermochemical experiments performed with the aid of solar simulators dating back to 1992 [13]. Around that period, a directly heated tubular reformer was installed at the solar tower of Weizmann Institute of Science [14]. The solar receivers/reactors developed since then are numerous, and the main relevant concepts will be described in the corresponding section of this chapter.

The solar dish technology is the last CST to be described here. Although parabolic dish technology has the capacity to achieve the highest concentration ratios, it is the least commercially developed one. Currently, only one pilot plant based on parabolic dishes, located in China (the 1 MW E Cube Energy Dish Pilot Plant), is in operation. The solar dishes are most commonly related with Stirling engines for electricity production, but they are also in principle capable of heating up a working fluid within a receiver or an actual reactor/receiver at high temperatures. Their main advantage – as compared to solar tower technologies – is their capability to operate individually, while scaling up is achieved by a simple increase of their number thus forming larger groups. Solar dishes have also been tested for thermochemical applications, with one of the first examples being that of the direct catalytic absorption receiver (DCAR) employed for solar carbon dioxide reforming of methane (in 1989 in the frame of a research project entitled Caesar in a solar facility located in Germany) [15].

10.4 Solar Hydrogen Production

Solar hydrogen has been characterized as the fuel of the future [16]. Hydrogen is actually an energy carrier, which can be derived from various sources, either carbonaceous (such as fossil fuels and biomass) or non-carbonaceous (e.g., water and ammonia) [17, 18]. Conventionally, hydrogen is produced from fossil fuels through very energy-intensive processes (e.g., reforming of natural gas and hydrocarbon pyrolysis), and the majority of it is not used to cover energy demands but is rather used as a chemical in industrial processes such as ammonia synthesis, in refineries, and in the steel industry [19]. What makes hydrogen a promising future energy carrier is the potential for its production from totally renewable sources such as water and solar energy.

With the conventional mainstream technologies employed, hydrogen is produced at the expense of a significant amount of energy, since either elevated temperatures or significant power consumption is required. The required energy usually derives from fossil fuel combustion. The replacement of this energy with a renewable one is expected to make hydrogen production environmentally friendlier and thus contribute positively to its penetration into the energy market. Among the available options for renewable hydrogen production are technologies that are based on solar energy utilization, a line of research envisioned several decades ago [20]. At this point, it should be clarified that depending on the feedstock, the product can be either pure hydrogen (e.g., in the case that the feedstock is water) or a mixture of hydrogen and carbon monoxide (i.e., syngas) when carbonaceous sources are employed. In the latter case, hydrogen can be either separated or the mixture is used "as is" for the production of synthetic hydrocarbons/synthetic fuels.

The existing alternative pathways for the production of solar hydrogen (or solar syngas) are summarized in the general scheme presented in Fig. 10.3. The alternative pathways presented are grouped on the basis of whether their operation requires very high temperatures – and thus the processes can be performed only with the use of CST capable of providing such temperatures – or not. In the latter case, they can be powered by technologies that employ non-concentrated solar irradiation. The last "group" of available processes can be further divided, based on whether the solar potential is directly used by the process (e.g., in case of photochemical or photobiological production via direct use of photon energy) or it is first converted to electricity, which is subsequently used for the hydrogen (or hydrogen/carbon monoxide depending on the feedstock) production via electrolysis.

As it can be seen from Fig. 10.3, CST systems can be employed for the production of hydrogen either directly or indirectly. Direct production involves supplying the necessary heat for direct thermal dissociation of water or for the thermochemical production of hydrogen (or syngas). In the case of direct thermal dissociation of water, very high temperatures (>2000 °C) are required, and the separation of hydrogen and oxygen at such high temperatures to avoid recombination is also necessary. Direct thermal dissociation of water has major technological challenges and thus is not the preferred option when it comes to hydrogen

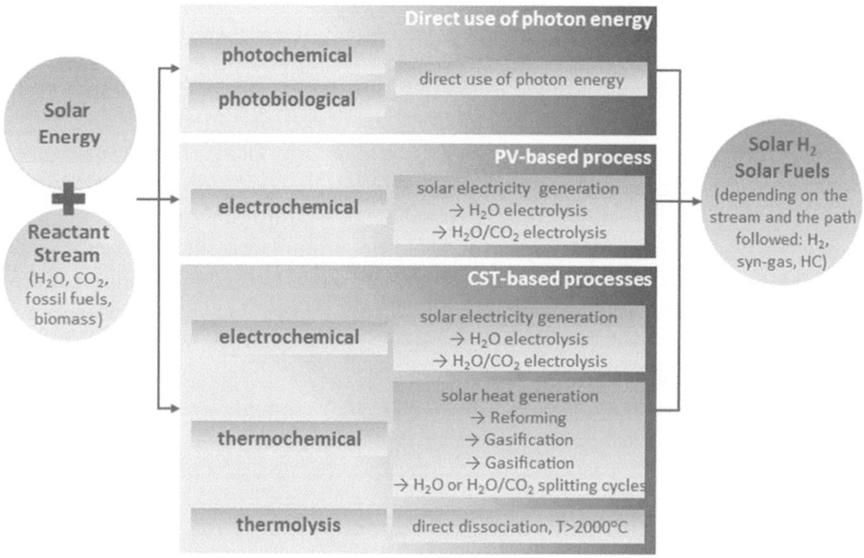

Fig. 10.3 Pathways for solar hydrogen/fuels production

production. Indirect hydrogen production via CST refers to generating electricity that is later used for electrolysis of water or water/carbon dioxide. Both these issues can be dealt with the use of thermochemical processes [21], e.g., reforming or via cyclic schemes employing two or more sequential steps. The thermochemical production of hydrogen (or syngas) has the advantage of using materials that achieve water or water/carbon dioxide splitting at significantly lower temperatures. In addition, due to the fact that such processes take place in more than one step issues related to product separation are inherently solved since the production of hydrogen (or syngas) is typically performed separately to oxygen generation. More details regarding the available thermochemical processes and the materials employed are provided in the respective sections of this chapter.

The coupling of hydrogen production with solar energy opens the potential of introducing new "players" to the energy sector, where importance is shifted from fossil fuel reserves to the available solar potential. The further development of promising thermochemical processes that are based on non-carbonaceous sources, e.g., water or water and carbon dioxide, is expected to further enhance the adaptation of solar technologies for the production of renewable energy, thereby offering a viable alternative to regions that are dependent on fossil fuels imports. Solar potential has the benefit of being distributed around the world more evenly than fossil fuel sources, and thus such technologies can be adapted by a large number of regions that could create a uniformly distributed hydrogen production network. On the other hand, the inherent challenge associated with solar power derives from its intermittent nature, and as a result solar-aided processes cannot be performed continuously.

10.5 Thermochemical Processes

In the previous section, pathways leading to solar hydrogen production were presented. This section intends to elaborate upon available thermochemical processes for hydrogen production. By definition, the term thermochemical refers to processes employing chemical reactions at elevated temperatures (typically above 500 °C). When the required energy input derives from solar power, the term expands to solar thermochemical processes. Regarding such processes employed for hydrogen production, there are numerous routes that can be followed and depend significantly on the reactant feedstock/hydrogen source used. In general, possible feedstock for hydrogen production can be divided in two categories: carbonaceous hydrogen sources and non-carbonaceous hydrogen sources. Typical examples of thermochemical processes employing carbonaceous fossil fuels for hydrogen production are steam reforming or cracking of natural gas and gasification of solid carbon sources (e.g., biomass or coal), while an abundant non-carbonaceous hydrogen source is water.

Currently, the majority of hydrogen produced is derived from the well-established technology of catalytic steam reforming of natural gas [18]. The overall reaction is depicted in Eq. 10.5 and is actually the combination of two reversible reactions proceeding in parallel: methane reforming (Eq. 10.6) and the water gas shift reaction (Eq. 10.7). Besides the two main reactions, and depending on the conditions employed, additional side reactions may occur, for example, methane combustion (Eq. 10.8) and the Boudouard reaction (Eq. 10.9), resulting to a mixture of hydrogen with carbon monoxide, carbon dioxide, and steam. Thus, the completion of the process requires an additional step where the main product (hydrogen) is separated. The ratio of CO to CO_2 produced in the final mixture depends on the operation conditions and the catalyst used. The methane reforming reaction usually requires temperatures in the range of 800–1000 °C. The reactors typically employed in such processes are tubular units with packed catalyst beds of transition metal-based or precious metal-based catalysts supported on ceramic or metallic substrates [22]. The combination of the conventional/industrial natural gas steam reforming processes with solar technologies is considered as an efficient way for the short-term achievement of high-throughput solar hydrogen production. Solar steam reforming has already been demonstrated by several research programs, among the first being that employing the cavity-type solar catalytic reactor SCR-3 which employed a closed-loop system [23] and the 170 kW reactor developed in the ASTERIX project and operated in Almeria, Spain [24]. Among the most recent examples of methane-reforming reactors are the 400 kW cavity reactor developed within the collaborative EU-funded SOLREF project [25] and the SolarGas reformer of the CSIRO organization [26].

$$CH_4 + 2H_2O \leftrightarrow CO_2 + 4H_2 \tag{10.5}$$

$$CH_4 + H_2O \leftrightarrow CO + 3H_2, \quad \Delta H^0 = -206 \text{ kJ/mol} \tag{10.6}$$

$$CO + H_2O \leftrightarrow CO_2 + H_2, \quad \Delta H^0 = 41 \text{ kJ/mol} \tag{10.7}$$

$$CH_4 \leftrightarrow C + 2H_2 \tag{10.8}$$

$$2CO \leftrightarrow C + CO_2 \tag{10.9}$$

A similar process to steam reforming is that of dry reforming (Eq. 10.10), a process with a product less enriched in hydrogen and thus more suitable for syngas production with relatively low H_2/CO ratio. Combinations between the two reforming approaches are also possible.

$$CH_4 + CO_2 \leftrightarrow 2CO + 2H_2 \tag{10.10}$$

A second route for solar hydrogen (or solar syngas) production that has also been extensively studied is that of solar gasification of, e.g., coal, heavy hydrocarbons, and biomass. Gasification involves the conversion of solid carbonaceous raw materials to higher-added-value products through a multistep process [27]. This can be achieved by various reactions such as pyrolysis, partial oxidation, and hydrogenation that all lead to a gaseous product containing, in addition to hydrogen and depending on the route followed, quantities of carbon monoxide and/or dioxide, water and gaseous hydrocarbons and amounts of char, and ash and condensable compounds (tars). For example, in the case of coal gasification, two basic steps take place [28], the decomposition of coal upon thermal treatment (pyrolysis, Eq. 10.11) followed by the steam or the carbon dioxide gasification (Boudouard reaction, reverse reaction of Eq. 10.9). Similar to the reaction steps involved in coal gasification are the reactions for biomass gasification. Currently, the term solar gasification is very closely related to biomass gasification, an approach that has gained a lot of interest during the last decades [28], mainly due to the broad availability of biomass and its renewable nature. As in the case of steam reforming, various reactor designs have been developed regarding solar gasification [29], employing either directly [30–32] or indirectly irradiated concepts [33] such as packed bed reactors [28], fluidized bed reactors [34], etc.

$$\text{Coal} \overset{\text{heat}}{\leftrightarrow} \text{Char(C)} + CO + CO_2 + H_2 + CH_4 + \textit{tars} \tag{10.11}$$

$$\text{Char(C)} + H_2O \leftrightarrow CO + H_2, \quad \Delta H = -131.4 \text{ kJ/mol} \tag{10.12}$$

Hydrogen can also be derived from direct solar thermal decomposition of methane (Eq. 10.13). This approach – as compared to the majority of the solar thermochemical routes employing fossil fuels for hydrogen production – has the advantage of not producing greenhouse gas emissions (e.g., CO_2) [35–37]. The final products are only hydrogen and elemental carbon (Carbon Black, CB) thus avoiding the requirement of gas separation steps. To be exact, the final gas product may contain in addition to hydrogen amounts of unreacted methane; however, these two gases can be easily separated by standard absorption or membrane separation techniques [38]. The prominent challenge of direct methane decomposition is the high temperature (>1200–1300 °C) required to achieve significant yields and reaction rates [37, 39, 40].

$$CH_4 + C \leftrightarrow C_s + 2H_2, \quad \Delta H^0 = 75.6 \text{ kJ/mol} \tag{10.13}$$

To achieve production of hydrogen at lower temperatures, catalytic methane decomposition is employed, but this option is accompanied by the requirement for regeneration of the deactivated catalyst [38]. In the literature, various concepts have been proposed, among which are those that employ solar reactors heated directly [41, 42] or indirectly [37, 43–45] and those that achieve direct methane decomposition [45] or catalytic methane decomposition by employing either carbon-based catalysts [46] or metal-based catalysts [38]. For such purposes, different reactor designs have been proposed and experimentally evaluated, e.g., fluidized bed reactors [47] and vortex flow reactors [42].

In the case of hydrogen production from the solar dissociation of water, the simplest approach would be through direct water splitting from concentrated solar power (Eq. 10.14). To achieve high conversion rates and avoid production of intermediate products [42, 48], very high temperatures (>2000 °C) should be employed, while relatively complex strategies to prevent recombination of the produced hydrogen and oxygen back to water are necessary.

$$2H_2O \overset{\text{energy}}{\leftrightarrow} 2H_2 + O_2 \tag{10.14}$$

One of the first attempts to achieve solar water splitting at low temperatures with the aid of a two-step thermochemical cycle was the use of the ZnO/Zn redox pair [20], involving the solar thermal dissociation of ZnO to Zn (Eq. 10.15). The reduced Zn would subsequently be employed for the dissociation of H_2O via its oxidation back to ZnO (Eq. 10.16).

$$2ZnO \rightarrow 2Zn + O_2 \tag{10.15}$$

$$2Zn + 2H_2O \rightarrow 2ZnO + 2H_2 \tag{10.16}$$

Until 1978, a list with more than 100 different thermochemical multistep routes (that in some cases required up to eight reaction steps) was already available in the literature for indirect thermochemical dissociation of water for hydrogen production at temperatures significantly lower than direct water thermal dissociation [49]. This list by 2010 [50] had surpassed the value of 200 possible routes and in its most updated version did not even include cerium oxide- and perovskite-based thermochemical cycles that have arose over the last years.

This vast number of routes springs from the extensive theoretical studies that have already been published. For example, studies regarding hydrogen production from two-step thermochemical processes based on metal oxide pairs are dated almost a decade earlier than the aforementioned experimental solar dissociation

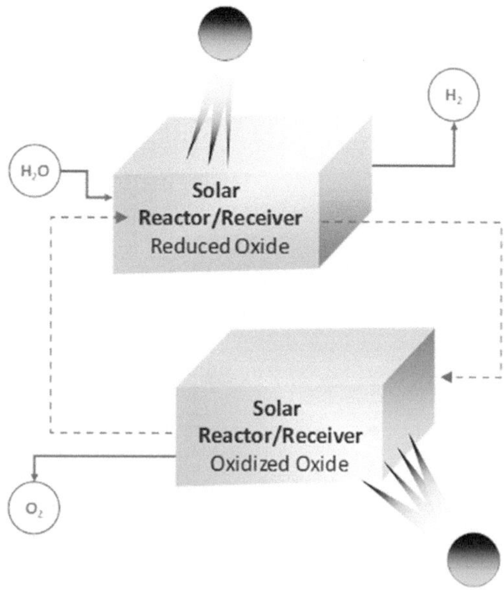

Fig. 10.4 A simplified schematic depicting a two-step thermochemical cycle for solar hydrogen production from water splitting

of ZnO [16], in a work by Funk et al. in 1966 [51]. In the latter comparative analysis, it was stated that low efficiency yields should be expected from any two-reaction step process performed at temperatures lower than 1100 °C. In accordance with this, in a comparative study based on thermodynamic calculations carried out few years later, it was stated that a minimum number of three reaction steps is required to achieve sufficient yields at temperatures between 1000 and 300 K [52]. This struggle to achieve adequately high efficiencies at low temperature ranges at the expense of process simplicity – i.e., by employing multi-reaction step processes – was imposed by the fact that at the time, hydrogen production had to be coupled with heat derived from nuclear energy [53]. The challenge of operating at higher temperatures could be, at least partially, surpassed by coupling the technologies with concentrated solar irradiation. To this end, almost in parallel with the work of Bilgen et al. [20], Nakamura [53] was investigating a second redox pair – Fe_3O_4/FeO, iron oxides pair – for solar hydrogen production. This research later on generated the investigation of a whole new class of materials for solar water splitting: the ferrites. Since then, both the ferrites [54–58] and the Zn/ZnO pair [59–62] have received great attention from numerous research teams.

As it has already been stated, thermochemical cycles with redox materials may proceed via a two-step process (Fig. 10.4). For the case of multivalent metal oxides, during the first step of the process (Eq. 10.17), the oxide obtains its reduced state (denoted as MO_{red}) and therefore is considered to be prone to oxidation. Thus, during the second step, the oxide is capable of absorbing oxygen from the steam in contact with it and converting back to its initial oxidized form (denoted as MO_{ox}). During this step, pure hydrogen is produced provided that reaction conditions favor 100% steam conversion. In the subsequent step of the process cycle, the oxidized form (MO_{ox}) is thermally reduced (typically under elevated temperatures and inert gas flow), thereby releasing oxygen and hence a cyclic operation is established. As long as the oxygen released is equal to the oxygen absorbed, the material could participate in a perpetual operation emitting only hydrogen and oxygen at two distinct steps in the course of the cycle. The whole procedure can be performed with the aid of solar energy under the prerequisite that a proper temperature swing is applied between the two reaction steps, also accompanied by the respective swing in the applied gas composition. Although the hydrogen production step (performed when water stream is applied) is an exothermic reaction, relatively high temperatures (e.g., >800 °C) are required for reaction kinetic purposes. The regeneration step is endothermic and requires even higher temperatures (on the order of 1400 °C) to achieve sufficient reaction extent and kinetics.

$$MO_{red} + H_2O(g) \rightarrow MO_{ox} + H_2(g)(\text{exothermic}) \tag{10.17}$$

$$MO_{ox} \rightarrow MO_{red} + 1/2O_2(g)(\text{endothermic}) \tag{10.18}$$

The process described above has been confirmed to be able to produce hydrogen in a semicontinuous way with the aid of concentrated solar radiation deriving from a solar tower facility, developed and validated within the research program HYDROSOL [63]. Additional details about the HYDROSOL concept are provided in Sect. 7 of this chapter, while a list of materials employed in hydrogen production from water splitting is stated in the following section.

10.6 Materials for Hydrogen Production

This section focuses on materials employed in two-step thermochemical cycles for the production of hydrogen from steam dissociation. In such cycles (Fig. 10.3), redox pairs are usually employed for the facilitation of the steam dissociation process. As described earlier, the materials used during thermochemical cycling are subjected to repeated changes between two states. Generally, two categories of materials/thermochemical cycles can be identified: volatile, if a phase change occurs in the material during the hydrogen production cycle, and nonvolatile. In volatile cycles, phase changes of the material occur during the thermal cycling, e.g., during the reduction step, the required temperature usually exceeds the vaporization temperature of the metal oxide [64]. This phase change favors the reaction kinetics. It also enhances the performance of the material since more oxygen can be released during the phase change per mass of material than in the solid state. The main disadvantage of volatile cycles is that once the material is gasified, it has to either be separated from the rest of the flow or rapidly quenched, procedures that both consume significant amounts of energy [65]. Typical materials used in volatile cycles are ZnO, CdO, and SnO_2. Materials employed in nonvolatile cycles are those that do not undergo a phase transition during the thermochemical reactions and include iron oxides, ceria, and a large number of spinel phases where usually at least one of the metals employed is iron, i.e., $Me_xFe_{3-x}O_4$ and lately perovskites.

Examples of these two categories of thermochemical cycles are given in Eqs. 10.19 [66] and 10.20 [53]. In the case of zinc oxide, the ZnO solid phase is reduced to metallic Zn gas, while in the case of the nonvolatile cycle of iron oxide, the solid phase of Fe_3O_4 transforms into the solid phase of FeO. In both cases, the reduction reactions lead to oxygen production.

$$Zn^{2+}O \rightarrow Zn^0(g) + 1/2O_2 \tag{10.19}$$

$$Fe_2^{3+}Fe^{2+}O_4 \rightarrow 3Fe^{2+}O + 1/2O_2 \tag{10.20}$$

An additional example of materials employed in nonvolatile cycles is mixed oxides, e.g., ferrites doped with a second metal (e.g., $Me_xFe_{3-x}O_4$, where Me is the second metal). The reduction step proceeds in a similar way with the example presented above and a solid solution of the two component oxides is formed (Eq. 10.21) [67]. Moreover, examples of mixtures comprising of more than one oxide have also been reported in the literature for hydrogen production via thermochemical cycles. In this case, a different mechanism is employed during the thermal cycling known as metal replacement reaction (Eq. 10.22) [67].

$$Me_xFe_{3-x}O_4 \rightarrow xMeO + (3-x)FeO + 1/2O_2 \tag{10.21}$$

$$CoFe_2O_4 + 3Al_2O_3 \rightarrow CoAl_2O_4 + 2FeAl_2O_4 + 1/2O_2 \tag{10.22}$$

All cycles described above proceed stoichiometrically. For the case of nonvolatile cycles, it is quite common for several such schemes to occur in a nonstoichiometric way. Typical such examples are those of water splitting with mixed ferrites ($Me_xFe_{3-x}O_4$, where usually Me: Ni, Co [55]), cerium oxide, and perovskites. In the nonstoichiometric mechanism, the oxide is not subjected to a complete structural transformation but involves a transition between a fully oxidized state and an oxygen-deficient one. In Eq. 10.23 and Eq. 10.24, δ represents the oxygen vacancy concentration of the oxygen-deficient state of the material [68]. Perovskites are represented by the chemical form of $ABO_{3-\delta}$, where both cation sites, A and B, can be substituted with metal ions or rare earths. The recent consideration of such materials for two-step redox thermochemical cycles further opened new possibilities in the field of solar-aided water splitting. The first publications reporting the water-splitting activity of perovskites (e.g., $La_{1-x}Sr_xO_{3-\delta}$) seem very promising [69, 70].

$$Me_xFe_{3-x}O_4 \rightarrow Me_xFe_{3-x}O_{4-\delta} + \delta/2O_2 \tag{10.23}$$

$$CeO_2 \rightarrow CeO_{2-\delta} + \delta/2O_2 \tag{10.24}$$

Each category of material/cycle described above possesses its own pros and cons. Stoichiometric cycles have the capacity of achieving high hydrogen production efficiencies per mass of redox material, as the maximum hydrogen production is proportionally related to the amount of oxygen subtracted from the oxide. On the other hand, in the case of nonstoichiometric cycles, only a rather small portion of oxygen participates in the redox reactions, thereby limiting hydrogen production potential. Based on the phase diagrams provided in the literature, the vacancy concentration (δ) that ceria can achieve at temperatures higher than 1000 °C is ~0.25 [71], a value that corresponds to a theoretical hydrogen yield of 2.9 mmol H_2/g. This value seems rather low in comparison with the theoretical yield that can be achieved by the Fe_3O_4/FeO couple (4.3 mmol H_2/g) or by the ZnO/Zn pair (12.3 mmol H_2/g). In practice, sufficient reduction of ceria is achieved only at significantly higher temperatures, while doping of ceria with other materials has also been reported in the literature as a potentially efficient means of achieving reduction at lower temperature [72] and therefore increases the oxygen storage capacity [73]. Main advantages of the nonstoichiometric mechanism are the faster kinetic rates achieved and the chemical and morphological stability [65] that leads to good cyclability. Common challenges with stoichiometric cycles are sintering and consequently deactivation of the materials since temperatures needed for reduction can often be higher than the melting point of at least one of the two oxides. This is the case for the reduction of Fe_3O_4 to FeO where the reduction temperature is >2200 °C [74], while Fe_3O_4 melting point is 1538 °C [75] and the melting point of FeO lies even lower. Such a reaction scheme is highly dependent on the available surface of the redox active material so that operation at these temperatures inevitably would lead to deactivation of the material due to loss of surface area caused by sintering.

10.7 Solar Reactor Concepts

A significant number of concepts for solar reactors can be found in the literature referring to hydrogen production from high temperature thermochemical processes. The different reactor approaches can be classified according to the following categories:

- Directly and indirectly irradiated
- Structured and nonstructured
- Moving active material or moving components and nonmoving/fixed reactors
- Volatile or nonvolatile material cycles

The distinction in direct or indirect irradiated reactors refers to the transfer mechanism employed to heat the reaction zone via concentrated solar radiation. In the directly irradiated type (Fig. 10.5a), the reactor and the working fluid are directly heated by the concentrated solar radiation and are thus directly heated by it. In such reactors, the front face is usually covered by a transparent window that allows the incoming radiation to reach the reactor. The window is made of high temperature-resistant materials (e.g., quartz), capable of withstanding large temperature gradients. The existence of a quartz window at the front is in many cases

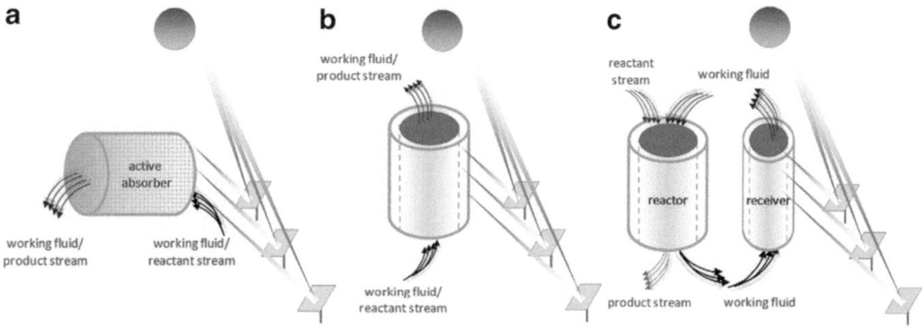

Fig. 10.5 General schematic representation of the transfer mechanism applied in reactors irradiated directly (**a**) and indirectly (**b**, **c**) by concentrated solar irradiation. In image (**a**) a representation of a volumetric receiver is shown, in (**b**) of a tubular reactor, and in (**c**) of an allothermal reactor

necessary to separate the reaction zone from the external environment. The radiation reaching the reactor directly heats the working fluid subsequently entering the reaction zone. For the case in which the reaction zone is in porous media, the so-called volumetric effect takes place, according to which concentrated radiation is first absorbed within the volume of the porous structure. Subsequently, a part of the absorbed heat is transferred to a fluid passing through the structure. The minimization of the heat transfer steps and the intermediate reactor components achieved in the case of directly heated reactors is a thermally more efficient reactor concept. The main challenge of this concept relates to the practical difficulties imposed by the use of the transparent (typically quartz) window.

For the case of indirect irradiated reactors, solar irradiance is concentrated upon an opaque surface – characterized by high absorptivity and thermal conductivity – which develops high temperatures and serves as a heat transfer medium to provide the required energy to the reactor. The simplest example of an indirect heated solar reactor for hydrogen production is that of a tubular receiver (Fig. 10.5b). In this case, the tube is the absorbing surface placed at the hot spot of the CSP installation. The heat that develops on the exterior of the tube is transferred to the interior surface of the opaque walls by conduction and from there to the working fluid of the reactor – or in the case that the reactor structure is enclosed in the tube to both the structure and the working fluid employed – by (forced) convection and conduction. From the latter example, the indirect irradiated solar reactors can be further distinguished into two subcategories depending on whether the reactor and the thermal receiver are decoupled (allothermal reactor, Fig. 10.5c) or not (Fig. 10.5b). In the first case, the heat absorbed by the working fluid has to be further transferred in a following step to the reactor.

The second classification refers to the reactive material used for the production of hydrogen that can be divided into two large categories depending on whether it is found in structured or nonstructured state (e.g., particles or powder formulations). In the case of structured reactors, the reactive material is "ordered" within the volume of the reactor to create a specific geometry, while in the case of

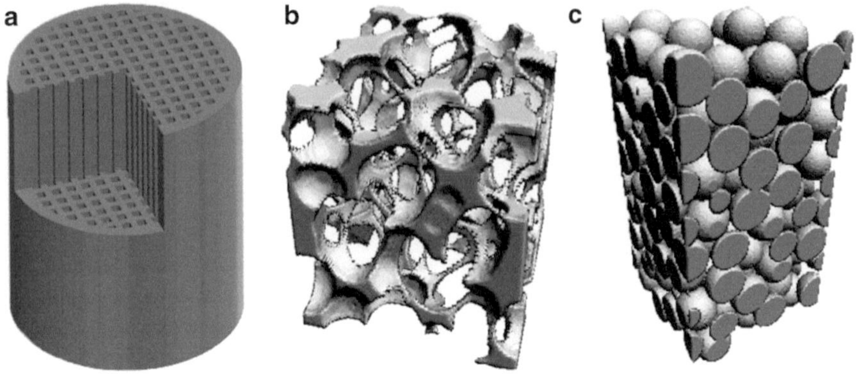

Fig. 10.6 Schematic representation of (**a**) a honeycomb structure, (**b**) a digital image of foam structure, and (**c**) a digital image of (nonstructured) packed bed

nonstructured reactors, the active material is randomly arranged. The structured reactors usually employ a geometry made from an inactive material with respect to reaction conditions (substrate). Prominent examples are porous medium such as monolithic honeycombs (Fig. 10.6a), foams (Fig. 10.6b), tubes, or membranes. In these cases, the active material responsible for the hydrogen production is coated upon the structure. Recently, efforts of shaping the active material into the desired final structures have been recorded in lab-scale experiments, in an attempt to increase the mass of active material per reactor volume [76–78], thereby achieving more compact formulations. Nonstructured reactors on the other hand usually employ particles of the reactive material that can be either randomly arranged in packed beds (Fig. 10.6c) or are forced to behave as fluids, e.g., by mechanical agitation, fluidized or spouted beds, and falling particle curtain. In this case, the reactor could even be filled with small pellets shaped from the active material [79].

Another classification categorizes the reactors depending on their operation principle. Thus, they could be discriminated into fixed and non-fixed/moving reactors. The first category refers to reactors that do not involve the movement of any of their components during operation, while the second refers to reactors where operation is based either on the movement of the active particles or actual movement of some of the reactor components. Two typical examples of moving reactors are those of the ROCA and the CR5 reactors. The CR5 reactor (Fig. 10.7a) consists of counterrotating rings/disks, at the perimeter of which the active phase is deposited upon fins of proper thickness that are appropriately allocated, in order to facilitate the solar flux penetration and the transportation of product and reactant gases [80]. Each ring rotates in the opposite direction to its neighboring one. Since part of the upper side of the rings is directly exposed to concentrated solar irradiance, there is a temperature gradient along the perimeter of the rings. Thus, as the rings rotate, two separate reaction zones are established that allow two different reaction steps to take place simultaneously but with two different products evolving in separate reactor compartments.

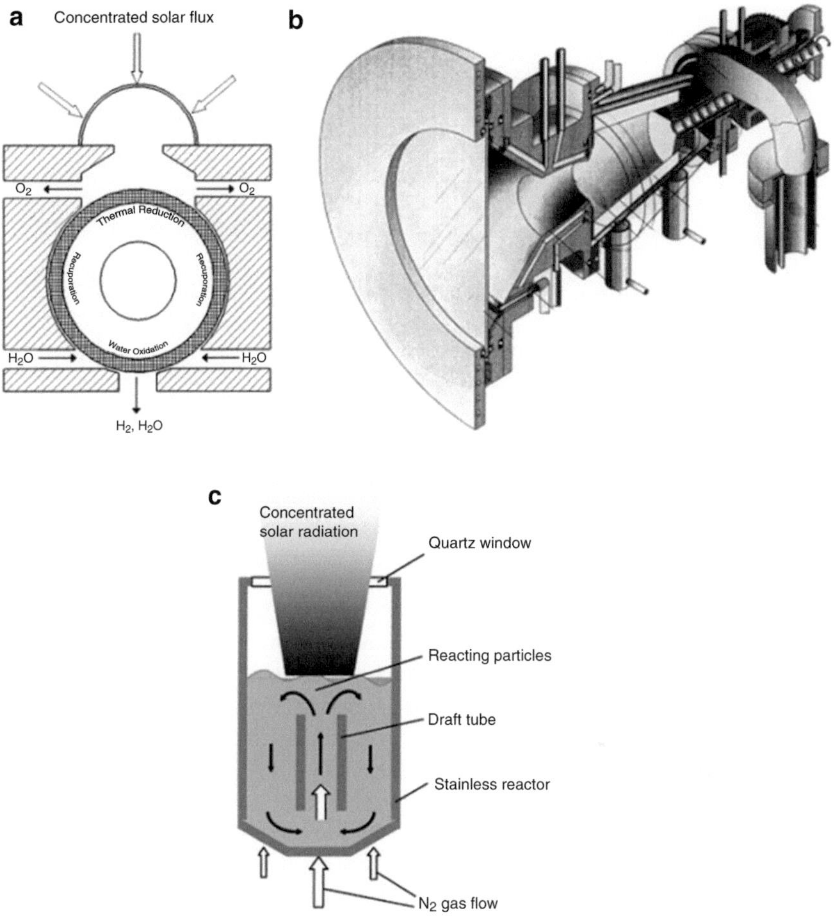

Fig. 10.7 Schematic representation of (**a**) reactor with counterrotating rings/disks, CR5 reactor (Reprinted with permission from Alonso and Romero [84]. Copyright © 2015, Elsevier), (**b**) a rotating kiln reactor, ROCA reactor (Reprinted with permission from Haueter et al. [82]. Copyright © 1999, Elsevier), and (**c**) a fluidized bed solar reactor (Reprinted with permission from Gokon et al. [83]. Copyright © 2008, Elsevier)

The ROCA reactor on the other hand is a rotating kiln (Fig. 10.7b), directly exposed to the concentrated solar irradiance (the reactor is a windowed cavity), where the active particles are fed into the reaction zone rotating cavity with the aid of a screw feeder. After their activation (i.e., solar decomposition of ZnO to Zn), the particles have to be transferred to a second (nonsolar) reactor where the least energy demanding reaction of Zn oxidation occurs and leads to the production of H_2 and/or CO [81, 82], depending on the feedstock employed. A different reactor concept also used for ZnO dissociation employs very fine particles of ZnO dispersed in an argon flow which is directed within the reaction zone [62]. This reactor, in addition to the

Fig. 10.8 Images of the solar reactors employed in the HYDROSOL projects [86]: (**a**) first HYDROSOL reactor (10 kW) used for the proof of concept, (**b**) first dual-chamber reactor for semicontinuous hydrogen production and (**c**) side view of the 100 kW_{th} dual-chamber HYDROSOL reactor scale tested at PSA in CIEMAT (Almeria, Spain)

constantly moving/flowing reaction zone (i.e., the aerosolized ZnO), contains also a vibration tray that controls the particle flow rate into a spinning wheel feeder, where shearing of the particles occurs with the aid of a rotating titanium wheel.

In other cases, the reaction zone is a fluidized bed of reactant particles (Fig. 10.7c). In such a formulation [83], the reactor zone has two coaxial tubes where the active particles are shifted upwards through the inner tube with the aid of gas streams and move back downwards through the space in-between the two tubes (annulus section). The concentration of solar radiation occurs at the upper side of the reactor, with the solar irradiation heating the particles directly through a quartz window. This constant particle circulation enables the energy transfer from the top region of the fluidized bed to the bottom and thus eliminates unfavorable heat gradients within the reactor. Another advantage of a fluidized bed reactor – as compared to a fixed bed configuration – is that effects such as sintering and agglomeration are significantly confined preventing thus the need of intermediate/periodic steps of pulverization/grinding of the sintered particles. As stated in the previous paragraph, the movement of the particles within the reactors can be realized via several ways, the most common of which are either by forcing the particles to move or by just relying on gravity.

There are reactors that have no moving parts such as the HYDROSOL reactors (Fig. 10.8) [63]. The particular reactor type is a volumetric receiver, where a honeycomb monolithic structure made of siliconized silicon carbide (Si-SiC) coated by active ferrites is directly irradiated. The front face of the reactor is separated from the external environment by a quartz window. Since the reactor does not contain any moving parts, switching between the separate thermochemical steps (i.e., thermal splitting step and material regeneration) is achieved by control of the solar field to reach the two different required temperatures (water splitting occurs at lower temperatures than material reduction). In the first reactors tested, the active material coated upon the Si-SiC monoliths was a Zn ferrite, while currently Ni ferrites [55] are employed. Recent work has also reported the employment of similar reactors, manufactured from ceria-based materials, to achieve co-splitting of steam and carbon dioxide for syngas production [85], thus allowing the synthesis of solar fuels.

The final classification of solar reactors, as stated in the beginning of this section, is based on the nature of the cycles that the reactors accommodate. While a range of cycles has been proposed in the literature for the production of solar hydrogen, for two step cycles, it is quite common to distinguish them into volatile and nonvolatile ones. A detailed presentation of the various cycles and advantages/challenges per case has already been provided in the section referring to materials.

All aforementioned classifications share a common objective, to accommodate as much of the active material as possible in the reactor with the aim of increasing product yield. It is also crucial to consider the kinetics of the reactions employed, especially for the thermal reduction step for which typically very high temperatures and low oxygen partial pressures are needed. To address this issue, temperature swing is employed between the two steps of the cycle. During the water splitting, it must be ensured that the conditions employed are those favoring the oxidation reaction kinetics and in parallel exclude the occurrence of the reverse step (i.e., the reduction of the material that would release oxygen). The reduction step usually occurs at a higher temperature, and the low oxygen partial pressure is achieved by employing an inert purge gas stream or (less commonly) sufficient vacuum. Depending on the nature of the thermochemical cycle, the water-splitting step usually takes place between 1000 and 1400 K, while thermal reduction requires up to ~400 K higher temperatures. Although in principle the temperature gradient between the two steps allows adequate control of the two separate steps, for such temperature swings, it is highly plausible to define heat recuperation strategies in order to substantially increase process efficiency.

The overall efficiency of a thermochemical reactor (Eq. 10.25) depends on two separate factors: chemical efficiency (n_{ch} in Eq. 10.26) and thermal efficiency (n_{th} in Eq. 10.27).

$$n_{\mathrm{total}} = n_{\mathrm{th}} + n_{\mathrm{ch}} \tag{10.25}$$

$$n_{\mathrm{ch}} = \frac{n_{\mathrm{R,in}} \cdot X_{\mathrm{R}} \cdot \Delta H_{\mathrm{reaction}} \left(T_{\mathrm{in}} \rightarrow T_{\mathrm{r}}\right)}{Q_{\mathrm{solar}} + Q_{\mathrm{aux}}} \tag{10.26}$$

$$n_{\mathrm{th}} = \frac{n_{\mathrm{R,in}} X_{\mathrm{R}} \Delta H_{r\ (T_{\mathrm{in}} \rightarrow T_{\mathrm{r}})} + n_{\mathrm{R,in}}(1 - X_{\mathrm{R}}) \int_{T_{\mathrm{in}}}^{T_{\mathrm{r}}} C_{p_{\mathrm{R}}} dT + n_i \int_{T_{\mathrm{in}}}^{T_{\mathrm{r}}} C p_i dT}{Q_{\mathrm{solar}} + Q_{\mathrm{aux}}} \tag{10.27}$$

The terms appearing in the above equations ar*e the molar flow of the reactant(s) at the inlet of the reactor* and *of the inert gas* ($n_{R,in}$ and n_i, respectively, in moles), the *conversion rate of the reactant(s)* (X_R, in moles/s), *the reaction(s) enthalpy(ies)* $\left(\Delta H_{r\ (T_{\mathrm{in}} \rightarrow T_{\mathrm{r}})}, \text{ in J/moles}\right)$ *from the inlet temperature to the temperature of the reactor* ($T_{\mathrm{in}} \rightarrow T_{\mathrm{r}}$, in K), the *heat capacities of reactant(s) and inert gas* (C_{PR} and C_{Pi} respectively in J/moles K), the *incoming solar radiation to the reactor*, and the *auxiliary power needed to drive the reactions* (Q_{solar} and Q_{aux} respectively in W).

10.8 Solar Fuels

Naturally, the major challenge in proper exploitation of solar energy is to identify efficient ways of converting and storing it. An appealing approach towards this is its conversion to chemical energy via thermochemical processes towards the production of "solar fuels" [87]. This term usually includes any energy carrier that can be stored and transported and is produced via solar energy exploitation process. It can refer to solar hydrogen, solar synthesis gas (syngas), and any other hydrocarbon that can occur by further treatment of these two streams (e.g., synthetic fuels like methanol, gasoline, diesel, and kerosene) but also to metal powders deriving from reduction of the respective metal oxides [88].

In the section referring to solar thermochemical processes, the routes presented included pathways not only for hydrogen production but to carbon monoxide and hydrogen/carbon monoxide mixtures (solar syngas) as well. Syngas is a product of great value since it can be further exploited via transformation into synthetic fuels via commercially established technologies, such as the Fischer-Tropsch process. By using either an organic feedstock (e.g., in the cases of steam reforming of natural gas and gasification of solid hydrocarbons) or by combining water splitting with carbon dioxide splitting, syngas can be directly produced. The latter route is a research field that has gained a lot of interest during the last years. Given the similarities of the mechanisms of solar thermochemical water and carbon dioxide splitting, there are numerous examples in the literature employing the same materials for both CO_2 and H_2O splitting, via either co-feeding or H_2O and CO_2 splitting occurring separately. Literature references include thermochemical cycles based on ZnO [89, 90], ferrites [73], CeO_2 [78, 83, 91, 92], or even combinations of ceria with iron oxide, e.g., CeO_2-modified Fe_2O_3 [88, 93] and very recently also perovskites [69].

The most profound advantage of producing renewable fuels – in addition to reducing dependence on fossil-derived ones – is that a new sustainable energy source in principle capable of being directly incorporated into existing infrastructures and industrial processes is becoming available. Technological advances achieved over decades in the evolution of internal combustion engines maintain their value, since solar fuels can be the source for the production of liquid hydrocarbons. Additionally, the advantages of hydrocarbon fuels in terms of transportation, storage, and energy content have placed them in a unique position in the energy mix. However, the environmental, financial, and political impacts of fossil hydrocarbons have urged societies to look for alternatives. Solar fuels offer indeed a promising future alternative regarding the replacement of fossil fuels and overcoming problems and drawbacks associated with their use in terms of economy, energy dependence, geopolitical tensions, and environmental problems.

A conceptually viable option for solar fuel production is the coupling of a plant for their production to a conventional power plant burning carbonaceous fuels. The emitted CO_2 can be collected directly at the source, subsequently fed to the solar plant, and the produced solar fuels can then be returned to the power plant. Until this point though, multiple challenges have to be addressed. As pointed out in the

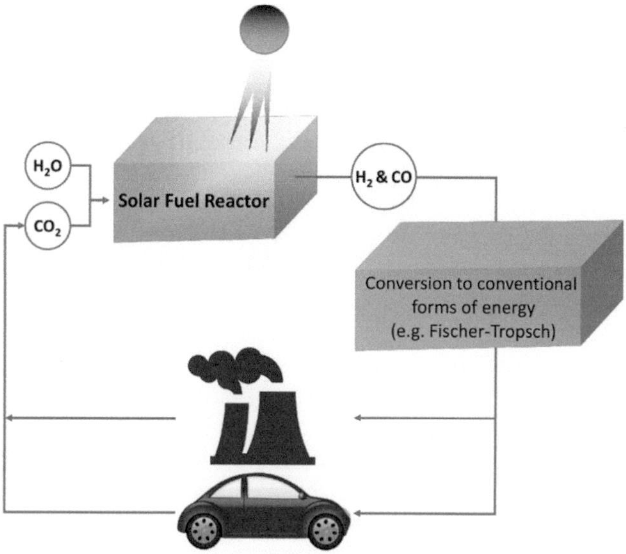

Fig. 10.9 Vision of a sustainable energy future based on renewable solar fuels

previous sections, challenges related to the production of solar hydrogen and carbon monoxide require advances in materials and solar reactor designs that will improve the sun-to-fuel efficiency of the technology and eventually make it competitive to conventional fuels. Additional development of separation technologies [94] is also needed for the development of solar fuel technologies. The advancement of solar fuel technology and an opportunity for commercial applications will be influenced by the fluctuation of fossil fuel costs and imposed emission taxations (CO_2 credits).

The vision of a transition from the current fossil fuel-dependent society to a carbon neutral one (Fig. 10.9) could take place over an intermediate technological step. During such a step, solar upgrading of existing carbonaceous feedstock (e.g., fossil fuels, biomass, C-containing wastes) can be used for the transformation into higher-added-value products, e.g., via solar steam reforming, solar gasification, and solar decomposition processes.

10.9 Conclusions and Future Outlook

From the previous sections, it becomes clear that existing options for solar hydrogen production are numerous, while the field of research is vast as scientists continue to seek and evaluate more materials and reactor concepts in an effort to increase hydrogen production efficiency. Solar hydrogen has only been tested at the pilot-scale level so far with all commercial solar plants being dedicated to electricity generation [12]. The implementation of the available hydrogen production technologies in CST facilities still has key challenges to overcome. The reasons are primarily based on market economy. Based on the rule of supply and demand, until the hydrogen penetration as fuel/primary energy carrier occurs, its demand is expected to remain low and no alternative production paths are likely to be sought. Currently, the conventional techniques that utilize fossil fuels for hydrogen production are competitively priced, since both the cost of fossil resources and carbon dioxide taxation are still at relatively low levels. Therefore, there is no need yet to seek alternative renewable routes that would imply investment of large initial capitals for the establishment of solar hydrogen facilities. This trend could be reversed if market requirements gradually started to change and energy intensive sectors, such as the transportation sector, started to rely more on hydrogen. This would probably force the development of solar hydrogen technology, since in this case the technology would attract industrial funding and scaling up thereby accelerating commercialization of most promising technologies.

From the technical point of view, and based on the current market trends in the hydrogen production sector that currently relies on dissociation of carbonaceous raw materials, it seems that large-scale hydrogen production via thermochemical processes has potential. For solar thermochemical hydrogen production, it seems like from the available CST systems, only two could be coupled with thermochemical routes: the solar tower plants and the solar dish technology, as only these two can achieve the required high temperatures. The parabolic troughs and the Fresnel systems would only be employed for indirect production of hydrogen through electrolysis.

The adaptation of solar dishes to commercial hydrogen production is a very promising option. Due to their limited size – as compared to solar tower plants – solar dishes can easily be installed on versatile geographic morphologies and also have the advantage of being able to be scaled up progressively, depending both on land and capital availability. Such systems have also the benefit of being standalone systems and offer small-scale decentralized hydrogen production stations. Although the latter characteristics sound very promising, solar dish implementation in high temperature thermochemical processes requires addressing first several technical challenges during their operation. To achieve the required temperatures, the solar reactor has to be placed at the focal point of the solar dish, which is constantly aligned with the sun motion. This means that novel flexible materials have to be used for the piping employed for the removal of the very hot reactor gases. To this aspect, solar towers probably have the least challenges to face, as their focal point is controlled by the heliostat field which is decoupled from the solar reactor. The disadvantages of solar tower systems mainly lie on the initial capital cost of the installation and their high requirements in terms of land area. Development of alternative solar tower technologies with lower towers and higher land coverage with respect to the heliostat field [95, 96] could improve the conditions and make solar tower systems more modular.

Regarding solar reactors, commercialization has still a long way to go. Certain trends can be observed, arising from the challenges that current solar reactor designs are facing. Commercialization should target designs that would provide the maximum efficiency, so directly irradiated solar reactors seem to have an advantage over indirect irradiated reactors. The expected lifetime of the reactor is also a matter of interest – e.g., solar reactors with moving parts in very high temperatures are very challenging and so is the performance and the long-term stability of their key components. With respect to the raw reactant stream employed as a source of hydrogen, the choice should be based on various parameters, including its hydrogen content, its market value, and the reaction energy requirements. If the target is the use of completely renewable sources of hydrogen only, then water would be the choice, and if the process would be further combined with carbon dioxide capture and reutilization, then a long-term option for renewable synthetic fuels via using syngas as raw material can be considered.

To sum up, in this chapter, the importance of clean solar hydrogen has been presented along with the pathways and technologies currently available and/or under investigation for its production. Simpler concepts in terms of technological challenges appear more attractive. Two main aspects have to be discussed: increase the overall efficiency in terms of heat demands and recuperation and develop materials providing desired yields at reasonable costs. If this research manages to reach pre-commercialization levels, a viable option for a CO_2 neutral (or even negative) and sustainable global energy system will emerge. Despite the current challenges in key aspects of solar hydrogen and solar fuel production, breakthrough research in the field is being conducted, and there are clear signs that new frontiers in the areas of energy production and storage will be generated. Under such a positive development, young researchers and innovative companies and industries will collaborate in the framework of a fruitful ground for novel achievements towards a more sustainable energy future.

Acknowledgments We thank the European Research Council (ERC) and the General Secretariat of Research and Technology (GSRT) for supporting this work through the ERC Advanced Grant Project ARMOS (ERC-2010-AdG 268049-ARMOS).

References

1. Smalley RE. Future global energy prosperity: the terawatt challenge. Mater Matters Bull. 2005;30:412–7.
2. BP plc. Sustainability review 2013. 2014. (http://www.bp.com/content/dam/bp/pdf/sustainability/groupreports/ BP_Sustainability_Review_2013.pdf). (Last accessed on Sept 2015).
3. International Energy Agency. Medium-term renewable energy market report 2014: market analysis and forecasts to 2020. 2014. ISBN 978-92-64-21821-5. OECD/IEA, Paris.
4. OECD/International Energy Agency. World Energy Investment Outlook. 2014. International Energy Agency, Paris. (https://www.iea.org/publications/freepublications/publication/WEIO2014.pdf. Last accessed Sept 2015).
5. OECD/International Energy Agency. Technology roadmap solar thermal electricity. 2014. ISBN 978-92-64-21821-5. OECD/IEA, Paris.
6. Hirsch D, Epstein M, Steinfeld A. The solar thermal decarbonization of natural gas. Int J Hydrog Energy. 2001;26:1023–33.
7. Richter C, Teske S, Short R. Concentrating solar power global outlook 09. 2009. Greenpeace International. (https://energypedia.info/wiki/Greenpeace_International_Concentrating_Solar_

Power_Global_Outlook. Open access material: https://energypedia.info/wiki/File:Greenpeace_International_Technologies_CSP.png).
8. Image by Xklaim (Own work) [CC BY-SA 3.0 (http://creativecommons.org/licenses/by-sa/3.0) or GFDL (http://www.gnu.org/copyleft/fdl.html)], via Wikimedia Commons (http://commons.wikimedia.org/wiki/File:Holaniku.png).
9. Image by Novatec Solar (Own work) [CC BY-SA 3.0 (http://creativecommons.org/licenses/by-sa/3.0)], via Wikimedia Commons (http://commons.wikimedia.org/wiki/File:Novatec_Solar_Puerto_Errado_2_BoP_PI.jpg).
10. Image by afloresm. BY 2.0 (http://creativecommons.org/licenses/by/2.0), via Wikimedia Commons (http://commons.wikimedia.org/wiki/File:PS10_solar_power_tower.jpg).
11. Image by Schlaich Bergermann und Partner. By Lumos3 at en.wikipedia [Public domain], from Wikimedia Commons (http://commons.wikimedia.org/wiki/File:EuroDishSBP_front.jpg).
12. CSP World Map Website: http://www.cspworld.org. (Last accessed on Sept 2015).
13. Diver RB, Fish JD, Levitan R, Levy M, Meirovitch E, Rosin H, Paripatyadar SA, Rich-ardson JT. Solar test of an integrated sodium reflux heat pipe receiver/reactor for thermochemical energy transport. Sol Energy. 1992;48(1):21–30.
14. Epstein M. Reforming technology for syngas production. IAEA-TECDOC-923, Ja-karta; 1995. p. 165–78.
15. Buck R, Muir JF, Hogan RE. Carbon dioxide reforming of methane in a solar volumetric receiver/reactor: the CAESAR project. Sol Energy Mater. 1991;24(1–4):449–63.
16. Pagliaro M, Kostandopoulos AG. Solar hydrogen: fuel of the future. Cambridge, UK: Royal Society of Chemistry Publishing; 2012. ISBN 978-1-84973-195-9.
17. Kodama T, Gokon N. Thermochemical cycles for high-temperature solar hydrogen production. Chem Rev. 2007;107(10):4048–77.
18. Dincer I, Joshi AS. Solar based hydrogen production systems. Springer Briefs in Energy. 2014. ISBN 978-1-4614-7430-2. Springer-Verlag New York.
19. Pregger T, Graf D, Krewitt W, Sattler C, Roeb M, Möller S. Prospects of solar thermal hydrogen production processes. Int J Hydrog Energy. 2009;34(10):4256–67.
20. Bilgen E, Ducarroir M, Foex M, Sebieude F, Trombe F. Use of solar energy for direct and two-step water decomposition cycles. Int J Hydrog Energy. 1977;2(3):251–7.
21. Zini G, Tartarini P. Solar hydrogen energy systems science and technology for the hydrogen economy. Italy: Springer-Verlag Italia; 2012. ISBN 978-88-470-1997-3.
22. Epstein M. Solar thermal reforming of methane, 2nd SFERA Winter School: solar fuels & materials. Zürich; 2011. p. 125–168.
23. Anikeev VI, Parmon VN, Kirillov VA, Zamaraev KI. Theoretical and experimental studies of solar catalytic power plants based on reversible reactions with participation of methane and synthesis gas. Int J Hydrog Energy. 1990;15(4):275–86.
24. Böhmer M, Langnickel U, Sanchez M. Solar steam reforming of methane. Sol Energy Mater. 1991;24:441–8.
25. Möller S. SOLREF – solar steam reforming, solar-SMR. Technology platform operation review days, Brussels; 8–9 Dec 2005.
26. Hinkley J. Solar Fuels Research at CSIRO. Presentation. December 20, 2013. http://www.iitj.ac.in/CSP/material/20dec/fuels.pdf. Last accessed Sept 2015.
27. Radwan AM. An overview on gasification of biomass for production of hydrogen rich gas. Der Chem Sin. 2012;3(2):323–35.
28. Gregg DW, Aiman WR, Otsuki HH, Thorsness CB. Solar coal gasification. Sol Energy. 1980;24(3):313–21.
29. Puig-Arnavat M, Tora EA, Bruno JC, Coronas A. State of the art on reactor designs for solar gasification of carbonaceous feedstock. Sol Energy. 2013;97:67–84.
30. Gregg DW, Taylor RW, Campbell JH, Taylor JR, Cotton A. Solar gasification of coal, activated carbon, coke and coal and biomass mixtures. Sol Energy. 1980;25:353–64.
31. Taylor RW, Berjoan R, Coutures JP. Solar gasification of carbonaceous materials. Sol Energy. 1983;30:513–25.

32. Gokon N, Izawa T, Abe T, Kodama T. Steam gasification of coal cokes in an internally circulating fluidized bed of thermal storage material for solar thermochemical processes. Int J Hydrog Energy. 2014;39(21):11082–93.
33. Melchior T, Perkins C, Lichty P, Weimer AW, Steinfeld A. Solar-driven biochar gasification in a particle-flow reactor. Chem Eng Process Process Intensif. 2009;48:1279–87.
34. Kodama T, Gokon N, Enomoto S, Itoh S, Hatamachi T. Coal coke gasification in a windowed solar chemical reactor for beam-down optics. J Dol Energy Eng. 2010;132(4):041004.
35. Zedtwitz P, Petrasch J, Trommer D, Steinfeld A. Hydrogen production via the solar thermal decarbonization of fossil fuels. Sol Energy. 2006;80:1333–7.
36. Abanades A, Rubbia C, Salmieri D. Thermal cracking of methane into hydrogen for a CO_2-free utilization of natural gas. Int J Hydrog Energy. 2013;38(20):8491–6.
37. Abanades S, Kimura H, Otsuka H. Hydrogen production from CO_2-free thermal decomposition of methane: design and on-sun testing of a tube-type solar thermochemical reactor. Fuel Process Technol. 2014;122:153–62.
38. Amin AM, Croiset E, Epling W. Review of methane catalytic cracking for hydrogen production. Int J Hydrog Energy. 2011;36(4):2904–35.
39. Abbas HF, Wan Daud WMA. Hydrogen production by methane decomposition: a review. Int J Hydrog Energy. 2010;35(3):1160–90.
40. Abbas HF, Wan Daud WMA. Hydrogen production by thermocatalytic decomposition of methane using a fixed bed activated carbon in a pilot scale unit: apparent kinetic, deactivation and diffusional limitation studies. Int J Hydrog Energy. 2010;35(22):12268–76.
41. Kogan M, Kogan A. Production of hydrogen and carbon by solar thermal methane splitting I The unseeded reactor. Int J Hydrog Energy. 2003;28(11):1187–98.
42. Hirsch D, Steinfeld A. Solar hydrogen production by thermal decomposition of natural gas using a vortex-flow reactor. Int J Hydrog Energy. 2004;29(1):47–55.
43. Rodat S, Abanades S, Flamant G. Experimental evaluation of indirect heating tubular reactors for solar methane pyrolysis. Int J Chem React Eng. 2010;8: Art No. 25.
44. Rodat S, Abanades S, Sans JL, Flamant G. A pilot-scale solar reactor for the production of hydrogen and carbon black from methane splitting. Int J Hydrog Energy. 2010;35 (15):7748–58.
45. Rodat S, Abanades S, Flamant G. Co-production of hydrogen and carbon black from solar thermal methane splitting in a tubular reactor prototype. Sol Energy. 2011;85(4):645–52.
46. Muradov N, Smith F, Bockerman G, Scammon K. Thermocatalytic decomposition of natural gas over plasma-generated carbon aerosols for sustainable production of hydrogen and carbon. Appl Catal Gen. 2009;365(2):292–300.
47. Steinfeld A, Kirillov V, Kuvshinov G, Mogilnykh Y, Reller A. Production of filamentous carbon and hydrogen by solarthermal catalytic cracking of methane. Chem Eng Sci. 1997;52 (20):3599–603.
48. Lapicque F, Lédé J, Villermaux J. Design and optimization of a reactor for high temperature dissociation of water and carbon dioxide using solar energy. Chem Eng Sci. 1986;41 (4):677–84.
49. Bamberger CE. Hydrogen production from water by thermochemical cycles; a 1977 update. Cryogenics. 1978;18(3):170–83.
50. McQuillan BW, Brown LC, Besenbruch GE, Tolman R, Cramer T, Russ BE, Vermillion BA, Earl B, Hsieh HT, Chen Y, Kwan K, Diver R, Siegal N, Weimer A, Perkins C, Lewandowski A. High efficiency generation of hydrogen fuels using solar thermochemical splitting of water. Annual report for the Period 10/01/2003 through 09/30/2004. No GA-A24972. General Atomics. 2010.
51. Funk JE, Reinstorm RM. Industrial and engineering chemistry process design and development. Ind Eng Chem Process Des Dev. 1966;5(3):336–42.
52. Abraham BM, Shhreiner F. General principles underlying chemical cycles which thermally decompose water into the elements. Ind Eng Chem Fundam. 1974;13(4):305–10.
53. Nakamura T. Hydrogen production from water utilizing solar heat at high temperatures. Sol Energy. 1977;19(5):467–75.

54. Agrafiotis CC, Pagkoura C, Zygogianni A, Karagiannakis G, Kostoglou M, Konstandopoulos AG. Hydrogen production via solar-aided water splitting thermochemical cycles: combustion synthesis and preliminary evaluation of spinel redox-pair materials. Int J Hydrog Energy. 2012;37(11):8964–80.
55. Agrafiotis C, Zygogianni A, Pagkoura C, Kostoglou M, Konstandopoulos AG. Hydrogen production via solar-aided water splitting thermochemical cycles with nickel ferrite: experiments and modeling. AIChE J. 2013;59(4):1213–25.
56. Fresno F, Fernández-Saavedra R, Gómez-Mancebo MB, Vidal A, Sánchez M, Rucandio MI, Quejido AJ, Romero M. Solar hydrogen production by Two-step thermochemical cycles: evaluation of the activity of commercial ferrites. Int J Hydrog Energy. 2009;34(7):2918–24.
57. Charvin P, Abanades S, Flamant G, Lemort F. Two-step water splitting thermochemical cycle based on iron oxide redox pair for solar hydrogen production. Energy. 2007;32:1124–33.
58. Tamaura Y, Kaneko H. Oxygen-releasing step of $ZnFe_2O_4$/(ZnO + Fe_3O_4)-system in air using concentrated solar energy for solar hydrogen production. Sol Energy. 2005;78(5):616–22.
59. Schunk LO, Haeberling P, Wepf S, Wuillemin D, Meier A, Steinfeld A. A receiver-reactor for the solar thermal dissociation of zinc oxide. J Dol Energy Eng. 2008;130(2):021009-1—6.
60. Charvin P, Abanades S, Neveu P, Lemont F, Flamant G. Dynamic modeling of a volumetric solar reactor for volatile metal oxide reduction. Chem Eng Res Des. 2008;86(11):1216–22.
61. Wieckert C, Frommherz U, Kräupl S, Guillot E, Olalde G, Epstein M, Santén S, Osinga T, Steinfeld A. A 300kW solar chemical pilot plant for the carbothermic production of zinc. J Dol Energy Eng. 2006;129(2):190–6.
62. Perkins C, Lichty PR, Weimer AW. Thermal ZnO dissociation in a rapid aerosol reactor as part of a solar hydrogen production cycle. Int J Hydrog Energy. 2008;33(2):499–510.
63. Roeb M, Säck JP, Rietbrock P, Prahl C, Schreiber H, Neises M, de Oliveira L, Graf D, Ebert M, Reinalter W, Meyer-Grünefeldt M, Sattler C, Lopez A, Vidal A, Elsberg A, Stobbe P, Jones D, Steele A, Lorentzou S, Pagkoura C, Zygogianni A, Agrafiotis C, Konstandopoulos AG. Test operation of a 100 kW pilot plant for solar hydrogen production from water on a solar tower. Sol Energy. 2011;85(4):634–44.
64. Perret R. Solar thermochemical hydrogen production research (STCH): thermochemical cycle selection and investment priority. Sandia National Laboratories Report: 1–117. SAND2011-3622. 2011. Sandia National Laboratories, Albuquerque (http://www1.eere.energy.gov/hydrogenandfuelcells/pdfs/solar_thermo_h2.pdf. Last accessed Sept 2015).
65. Scheffe JR, Steinfeld A. Oxygen exchange materials for solar thermochemical splitting of H_2O and CO_2: a review. Mater Today. 2014;17(7):341–8.
66. Palumbo RD, Fletcher EA. High temperature solar electro-thermal processing III zinc from zinc oxide at 1200-using a non-consumable anode. Energy. 1988;13(4):319–32.
67. Scheffe JR, Li J, Weimer AW. A spinel ferrite/hercynite water-splitting redox cycle. Int J Hydrog Energy. 2010;35(8):3333–40.
68. Chueh WC, Haile SM. A thermochemical study of ceria: exploiting an Old material for new modes of energy conversion and CO_2 mitigation. Philos Trans A. 2010;368:3269–94.
69. McDaniel AH, Ambrosini A, Coker EN, Miller JE, Chueh WC, OHayre R, Tong J. Nonstoichiometric perovskite oxides for solar thermochemical H_2 and CO production. Energy Procedia. 2014;49:2009–18.
70. Muhich CL, Evanko BW, Weston KC, Lichty P, Liang X, Martinek J, Musgrave CB, Weimer AW. Efficient generation of H_2 by splitting water with an isothermal redox cycle. Science. 2013;341(6145):540–2.
71. Eyring L. In: Meyer G, Morss LR, editors. The binary lanthanide oxides: synthesis and identification, synthesis of lanthanide and actinide compounds. Dordrecht: Kluwer Academic Publishers; 1991. p. 187–224.
72. Abanades S, Legal A, Cordier A, Peraudeau G, Flamant G, Julbe A. Investigation of reactive cerium-based oxides for H_2 production by thermochemical two-step water-splitting. J Mater Sci. 2010;45(15):4163–73.
73. Kuhn M, Bishop SR, Rupp JLM, Tuller HL. Structural characterization and oxygen nonstoichiometry of ceria-zirconia ($Ce_{1-x}Zr_xO_{2-\delta}$) solid solutions. Acta Mater. 2013;61:4277–88.

74. Sibieude F, Ducarroir M, Tofighiv A, Ambriz J. High temperature experiments with a solar furnace: the decomposition of Fe_3O_4, Mn_3O_4, CdO. Int J Hydrog Energy. 1982;7(1):79–88.
75. Green DW, Perry RH, editors. Perry's chemical engineers handbook. 8th ed. New York: McGraw-Hill; 2008. ISBN 9780071422949.
76. Miller JE, Allendorf MD, Diver RB, Evans LR, Siegel NP, Stuecker JN. Metal oxide composites and structures for ultra-high temperature solar thermochemical cycles. J Mater Sci. 2008;43(4):4714–28.
77. Lorentzou S, Bakatselou E, Pagkoura C, Karagiannakis G, Konstandopoulos AG. H_2 and CO production via two-step thermochemical splitting of H_2O and CO_2 over redox powders and redox porous structures. International Congress on Particle Technology (PARTEC 2013), V12 Applications, Paper no 426, Nuremberg, 23–25 Apr 2013.
78. Lorentzou S, Karagiannakis G, Pagkoura C, Zygogianni A, Konstandopoulos AG. Thermochemical CO_2 and CO_2/H_2O splitting over $NiFe_2O_4$ for solar fuels synthesis. Energy Procedia. 2014;49:1999–2008.
79. Chambon M, Abanades S, Flamant G. Thermal dissociation of compressed ZnO and SnO_2 powders in a moving-front solar thermochemical reactor. AIChE J. 2011;57(8):2264–73.
80. Diver RB, Miller JE, Allendorf MD, Siegel NP, Hogan RE. Solar thermochemical water-splitting ferrite-cycle heat engines. J Dol Energy Eng. 2008;130(4):041001-1–8.
81. Loutzenhiser PG, Meier A, Steinfeld A. Review of the two-step H_2O/CO_2-splitting solar thermochemical cycle based on Zn/ZnO redox reactions. Materials. 2010;3:4922–38.
82. Haueter P, Moeller S, Palumbo R, Steinfeld A. The production of zinc by thermal dissociation of zinc oxide-solar chemical reactor design. Sol Energy. 1999;67(1-3):161–7.
83. Gokon N, Takahashi S, Yamamoto H, Kodama T. Thermochemical two-step water-splitting reactor with internally circulating fluidized Bed for thermal reduction of ferrite particles. Int J Hydrog Energy. 2008;33(9):2189–99.
84. Alonso E, Romero M. Review of experimental investigation on directly irradiated particles solar reactors. Renew Sustain Energy Rev. 2015;41:53–67.
85. Lorentzou S, Karagiannakis G, Dimitrakis D, Pagkoura C, Zygogianni A, Konstandopoulos AG. Thermochemical redox cycles over Ce-based oxides. Energy Procedia. 2015; 69:1800–1809; Also in Proceedings of SolarPACES 2014 international conference, Beijing; 16–19 Sept 2015.
86. Official Hydrosol Projects Website: http://160.40.15.244/hydrosol/images.html (Last accessed on Sept 2015).
87. Kodama T. High-temperature solar chemistry for converting solar heat to chemical fuels. Prog Energy Combust Sci. 2003;29(6):567–97.
88. COMETNANO project website on CORDIS. http://cordis.europa.eu/project/rcn/91283_de.html.
89. Loutzenhiser PG, Meier A, Gstoehl D, Steinfeld A. CO_2 splitting via the solar thermochemical cycle based on Zn/ZnO redox reactions. In Yun Hang Hu (ed.): Advances in CO_2 conversion and utilization, vol. 1056. 2010; p. 25–30. American Chemical Society: Washington, DC.
90. Stamatiou A, Steinfeld A, Jovanovic ZR. On the effect of the presence of solid diluents during Zn oxidation by CO_2. Ind Eng Chem Res. 2013;52(5):1859–69.
91. Furler P, Scheffe J, Gorbar M, Moes L, Vogt U, Steinfeld A. Solar thermochemical CO_2 splitting utilizing a reticulated porous ceria redox system. Energy Fuels. 2012;26(11):7051–9.
92. Chueh WC, Falter C, Abbott M, Scipio D, Furler P, Haile SM, Steinfeld A. High-flux solar-driven thermochemical dissociation of CO_2 and H_2O using nonstoichiometric ceria. Science. 2010;330(6012):1797–801.
93. Galvita VV, Poelman H, Bliznuk V, Detavernier C, Marin GB. CeO_2-modified Fe_2O_3 for CO_2 utilization via chemical looping. Ind Eng Chem Res. 2013;52(25):8416–26.
94. Miller JE, Ambrosini A, Coker EN, Allendorf MD, McDaniel AH. Advancing oxide materials for thermochemical production of solar fuels. Energy Procedia. 2014;49:2019–26.
95. Schramek P, Mills DR. Multi-tower solar array. Sol Energy. 2003;75:249–60.
96. Schramek P, Mills DR. Heliostats for maximum ground coverage. Energy. 2004;29:701–13.

Part IV
Separations and Applications with Fuel Cells

Chapter 11
Separation and Purification of Hydrogen Using CO_2-Selective Facilitated Transport Membranes

Varun Vakharia and W.S. Winston Ho

Abstract A crucial step for the use of hydrogen as a clean, renewable source of energy towards a "hydrogen economy" is the purification of hydrogen from other gaseous compounds, mainly carbon dioxide and hydrogen sulfide. This chapter reviews the carbon dioxide- and hydrogen sulfide-selective facilitated transport membranes for low-pressure and high-pressure applications. Hydrophilic polymeric materials have been investigated that are blended with amino acid salts and polyamines as mobile and fixed-site CO_2 carriers, respectively.

For low-pressure applications (1–2 atm), novel facilitated transport membranes have been synthesized in the lab scale containing sterically hindered amines as the CO_2 carries in the cross-linked polyvinyl alcohol networks. The membranes have demonstrated high CO_2 permeability and high CO_2/H_2 selectivities. For high-pressure applications (15–30 atm), an improved stability of the membranes was demonstrated by incorporating fumed silica and multiwalled carbon nanotubes (MWNTs) to reinforce the mechanical strength of the polymer matrix.

The facilitated transport membranes highlighted in this work are first of a kind demonstrating the steric hindrance effect of the amine carriers and the presence of carbon nanotubes in enhancing the CO_2 transport capacity and long-term stability. These membranes have shown exceptional potential for industrial applications with enhanced H_2 recovery, including low-pressure H_2 purification for fuel cells and high-pressure syngas purification in an IGCC power plant or steam reforming of hydrocarbons.

V. Vakharia
William G. Lowrie Department of Chemical and Biomolecular Engineering, The Ohio State University, 151 West Woodruff Avenue, Columbus, OH 43210-1350, USA

W.S.W. Ho (✉)
William G. Lowrie Department of Chemical and Biomolecular Engineering, The Ohio State University, 151 West Woodruff Avenue, Columbus, OH 43210-1350, USA

Department of Materials Science and Engineering, The Ohio State University, 151 West Woodruff Avenue, Columbus, OH 43210-1350, USA
e-mail: ho.192@osu.edu

Z. Fang et al. (eds.), *Production of Hydrogen from Renewable Resources*, Biofuels and Biorefineries 5, DOI 10.1007/978-94-017-7330-0_11

Keywords Amine-containing facilitated transport membrane • H_2 purification • Mixed matrix membrane • Fuel cell • IGCC • Precombustion carbon capture

11.1 Introduction

The global energy demand is expected to double by 2050 due to growing industrialization in the developing nations [1, 2]. The need for an alternative source of energy is inevitable, due to the escalating energy demand and restricted supply of conventional energy resources. One of the potential long-term solutions to the global energy crisis is the adoption of "hydrogen economy" by the use of hydrogen as an energy carrier [3].

Hydrogen is industrially produced using the steam methane reforming (SMR) process, followed by the water-gas-shift (WGS) reaction (Fig. 11.1). The hydrogen content in the final product stream is enriched via the WGS reaction by the reaction of carbon monoxide and steam to generate hydrogen and carbon dioxide (Eq. 11.1).

$$CO + H_2O \rightleftharpoons CO_2 + H_2 \tag{11.1}$$

Hydrogen can be recovered from the H_2-rich gas streams in the refineries or from synthesis gas (syngas). Syngas is a mixture of CO and H_2O along with H_2, CO_2, CH_4, N_2, and H_2S. The syngas is typically sent to the WGS reactor to enrich the H_2 content in the resultant product stream. The hydrogen rich product stream can be used for different applications like power generation or synthesis of value-added chemicals [4, 5]. The hydrogen rich gas stream can be used as an energy carrier for fuel cells or it can be directed to the gas turbine (GT) in an integrated gasification combined cycle (IGCC) power plant for the large-scale power generation [6–8]. The extent of H_2 purification depends upon the purity requirements for different applications. The successful operation of a fuel cell is critically dependent on the supply of a high-purity hydrogen stream. The use of syngas as an energy carrier for the proton-exchange membrane fuel cell (PEMFC) could deteriorate the fuel cell performance if the CO content in the syngas exceeds 10 ppm [7]. Thus, for an efficient fuel cell operation, the CO levels in the syngas are required to be below 10 ppm along with efficient removal of H_2S and other impurities. On the other hand, the syngas in an IGCC power plant is sent to a CO_2 and H_2S separation unit prior to GT to limit the H_2S, SO_2, and CO_2 emission from the GT exhaust [6, 8].

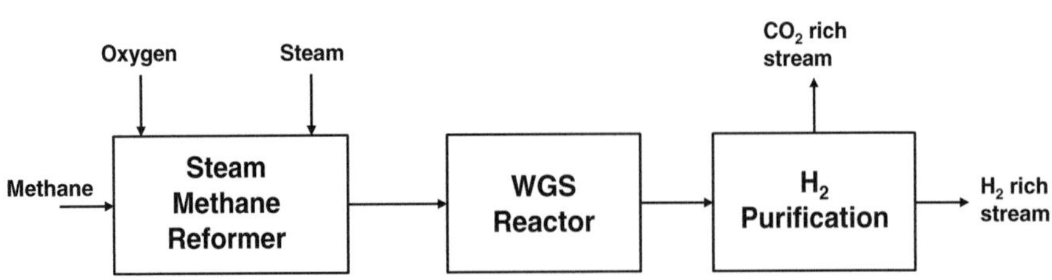

Fig. 11.1 Block diagram of the H_2 production using the steam methane reforming (SMR) process

The most commonly well-known and industrially used processes for the H_2 purification are the adsorption, absorption, and membrane processes. Pressure swing adsorption (PSA) is a well-developed technology, industrially used to purify H_2 up to 99.99 %. PSA is typically associated with high-pressure operating condition using adsorbents like activated carbon or zeolite [9]. The Selexol® process (physical absorption and solvent regeneration technology) is a well-established absorption process for high-pressure CO_2 and H_2S removal from syngas [6, 10]. However, PSA and Selexol® processes are energy intensive with significant parasitic power consumption along with high capital costs. On the other hand, membrane technology holds great potential for the H_2 purification application by virtue of high energy efficiency, compact design, maintenance simplicity, and hence low capital and operating costs. In some cases, membrane technology could be coupled with other technologies like in the case of the WGS membrane reactor to achieve a higher H_2 recovery and purity (<10 ppm of CO content).

11.2 Membranes for H_2 Purification

The separation performance of a membrane can be characterized by two parameters, namely, selectivity (or separation factor) and permeability. The selectivity of a two component mixture is defined as

$$\alpha_{ij} = \frac{y_i/y_j}{x_i/x_j} \tag{11.2}$$

The permeability, P_i, is defined as

$$P_i = \frac{J_i}{\Delta p_i/\ell} \tag{11.3}$$

The common unit of P_i is Barrer, which is equivalent to 10^{-10} cm^3 (STP) • cm/(cm^2 • s • cmHg). The permeance of a membrane for "ith" component is defined as P_i/ℓ. The common unit of permeance is the gas permeation unit (GPU), which is equivalent to 10^{-6} cm^3 (STP)/(cm^2 • s • cmHg). The selectivity (separation factor) is a process-dependent parameter. Thus, a more intrinsic parameter for the separation performance of the membrane is the ideal selectivity, which is defined as the ratio of permeability of the two components P_i and P_j:

$$\alpha_{i/j} = \frac{P_i}{P_j} \tag{11.4}$$

Inorganic membranes, polymeric membranes, and mixed matrix membranes have been reported in literature for H_2 purification. There are various types of inorganic membranes for H_2 purification: metallic, ceramic, silica, and others [11–16]. The

Table 11.1 The characteristics of membrane materials with merits and demerits for H_2 purification

Membrane type	Transport mechanism	Merits	Demerits
Inorganic			
Metallic	Modified solution–diffusion (Sieverts' law)	High-temperature stability	Sulfur contamination
Ceramic	Molecular sieving	Superior H_2 selective transport performance	Dimensional stability
Silica	Molecular sieving		Scale up
Organic			
Polymer	Solution–diffusion	Low-cost scale up	High-temperature operation
	Facilitated transport	Sulfur tolerance	

separation mechanisms for the inorganic membranes differ based on the membrane material and structure. Dense metallic membranes like palladium and platinum membranes are highly selective for H_2 via the modified solution–diffusion mechanism [12]. On the contrary, the gas separation in the ceramic and silica membranes are typically associated with molecular sieving and site-hopping diffusion mechanism, respectively [13–16]. Dense metallic membranes exhibit superior thermal stability and gas transport performance with theoretically infinite H_2 selectivity. However, the expensive scale-up fabrication, surface contamination (on exposure to sulfur-containing species), and H_2 embrittlement are major challenges towards commercialization [12]. Silica-based porous membranes are relatively cheap and easy to be fabricated as compared to the dense metallic membranes. However, they exhibit instability in the form of disruptive stresses, densification, and structural changes in the porous layers during operation [11]. Furthermore, inorganic membranes like palladium and platinum alloys, ceramics, carbons, and zeolites are difficult to roll and convert into a large-scale module for practical applications. Compared to the inorganic membrane materials, polymeric membranes offer low permeance and selectivity but the low cost of fabrication of a scale-up robust module and tolerance towards sulfur contaminants makes the polymeric membranes attractive for industrial H_2 purification [8, 11]. These characteristics of membrane materials are summarized in Table 11.1.

Dense polymeric membranes are CO_2 selective or H_2 selective based on the mechanism of mass transport. The H_2-selective polymeric membranes separate gas molecules based on their molecule size (kinetic diameter) difference via the solution–diffusion mechanism. Glassy polymers are typically used to synthesize H_2-selective polymeric membranes [11, 17]. The rigid structure in the glassy polymer matrix results in the desired H_2/CO_2 diffusional selectivity [11, 18–21]. The smaller-size H_2 molecule (2.89 Å) would experience a lower diffusional resistance, leading to an enhanced flux for H_2 permeation.

CO_2-selective polymeric membranes separate the gaseous components by the solution–diffusion mechanism or facilitated transport mechanism. The CO_2-selective solution–diffusion membranes are usually derived from rubbery polymers. The kinetic size of H_2 molecule (2.89 Å) is considerably smaller than the CO_2 molecule (3.3 Å). Thus, the diffusional selectivity (D_{CO2}/D_{H2}) always favors H_2 permeation. The performance of CO_2-selective polymeric solution–diffusion membranes can be improved by enhancing the solubility selectivity (S_{CO2}/S_{H2}), keeping the diffusional selectivity (D_{CO2}/D_{H2}) as close to 1 as possible. The higher CO_2 solubility in the polymer matrix relative to the H_2 molecule leads to selective transport of CO_2 molecules through the membranes [11, 21–23]. However, such an increase in the overall CO_2/H_2 selectivity has been observed at low operating temperatures for the polymeric CO_2-selective solution–diffusion membranes. At low temperatures, the diffusion of the gaseous components is restricted, and the sorption of CO_2 molecules relative to H_2 molecules becomes pronounced. But, at high temperatures, the solubility selectivity reduces significantly, resulting in low CO_2/H_2 selectivity. Therefore, these polymeric CO_2-selective solution–diffusion membranes cannot be operated at high temperatures due to the low CO_2/H_2 selectivity.

In view of high-temperature performance, CO_2-selective facilitated transport membranes have shown desirable CO_2/H_2 selectivity under such conditions. Facilitated transport membranes consist of reactive carriers dispersed in the polymer matrix on a polymeric support. The carriers react reversibly with CO_2 or H_2S in the feed gas mixture. The reaction–diffusion mechanism by virtue of gas-carrier reaction and diffusion of the gas-carrier reaction product results in a remarkably high CO_2/H_2 and H_2S/H_2 selectivity [24–34].

Membranes for the fuel cell (low pressure) or IGCC power plant (high pressure) require high operating temperatures (>100 °C) for economic feasibility [34, 35]. Thus, high CO_2 and H_2S permeances along with remarkably high CO_2/H_2 and H_2S/H_2 selectivities exhibited by the polymeric facilitated transport membranes at high operating temperatures offer great potential for industrial H_2 purification applications. In view of an economically attractive membrane technology for H_2 purification, this work reviews the CO_2- and H_2S-selective polymeric facilitated transport membranes for low-pressure and high-pressure conditions.

11.3 Polymeric Facilitated Transport Membranes for H_2 Purification

The gas molecules (CO_2 or H_2S) react reversibly with the carrier molecules that are dispersed in the polymeric facilitated transport membranes. Thus, the CO_2 or H_2S solubility is greatly enhanced in the selective polymer layer. These reactive carrier molecules are classified as mobile carriers or fixed-site carries. In the case of mobile carriers, the gas-carrier reaction product can diffuse through the membrane, whereas the transport of the reaction product for the fixed-site carriers occurs in the form of "hopping" mechanism [36, 37]. The CO_2 or H_2S partial pressure is relatively low on the permeate side of the membrane. Thus, the gas-carrier reaction product can release the gas on the permeate side through the reversible chemical

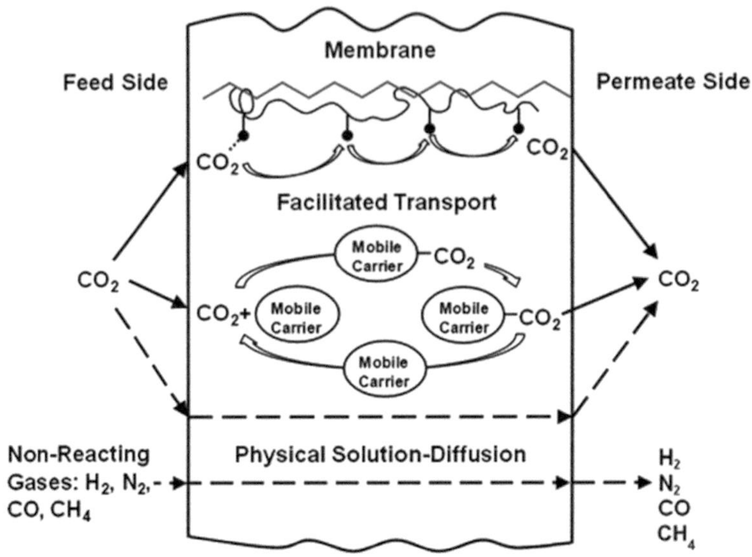

Fig. 11.2 Schematic representation of gas permeation through a CO_2-selective facilitated transport membrane (Reprinted with permission from Zhao and Ho [38], Copyright © 2013 American Chemical Society)

reaction. Figure 11.2 shows a schematic representation of the transport mechanism in facilitated transport membranes.

Early facilitated transport membranes like the supported liquid membranes or the ion-exchange membranes exhibited desirable CO_2/H_2 and CO_2/CH_4 separation performances [34, 39, 40]. Despite the high selectivity, it is challenging to obtain the long-term stability of these membranes. Issues like carrier leakage, low resistance to transmembrane pressure gradients, slow reaction kinetics, and solvent evaporation have deterred the commercialization of these membranes. However, there has been significant improvement recently by the emergence of new, nonvolatile carriers in form of dendrimers and ionic liquids [41–43].

Facilitated transport membranes based on water-swollen polyelectrolytes have demonstrated attractive transport results and stability at 23–35 °C [44]. Polyelectrolytes are hydrophilic polymers with a high ionic content. Polyelectrolyte membranes containing fluoride (F^-) and acetate (CH_3COO^-) ions are considered for H_2 purification by the virtue of the CO_2-selective facilitated transport mechanism. The strong interaction of the fluoride and acetate ions with the hydrogen atom of the water molecules (via the hydrogen bonding) is expected to increase the electron density on the oxygen atom of the water molecule. This enhances the reaction kinetics for the CO_2–H_2O reaction. The reaction product diffuses from the feed side to the permeate side via the concentration gradient and eventually releasing the CO_2 molecules on the permeate side (by the reversible reaction), thereby facilitating the CO_2 transport via the reaction–diffusion transport mechanism. The ions (fluoride and acetate ions) facilitate the CO_2 transport and remain well dispersed and strongly bonded in the polymer matrix unlike the ion-exchange membranes. It is essentially the diffusion of the CO_2–H_2O reaction product, i.e., $H_2CO_3 = H^+ + HCO_3^-$, from the feed side to the sweep side that enhances the CO_2 transport. Membranes containing poly(vinylbenzyltrimethylammonium fluoride) (PVBTAF) are reported in the literature with a CO_2/H_2 selectivity higher than 80 [44–48]. Quinn et al. investigated the transport performance of two stacked polyelectrolyte layers. These dual layered membranes exhibited >200 CO_2/H_2 selectivity [44]. Furthermore, an improved transport performance was demonstrated by incorporating various fluoride- and acetate-containing salts into the polyelectrolyte membranes.

Among the hydrophilic polymers, those based on carriers blended with a cross-linked polymer have shown promising high-temperature performances at both high-pressure and low-pressure conditions. Ho and coworkers have studied the transport performances of various membranes with amines as CO_2 carriers that are blended in a cross-linked poly(vinyl alcohol) matrix. Different mobile carriers (like 2-aminoisobutyric acid-potassium salt (AIBA-K), glycine-Li, and dimethylglycine-Li) and fixed carriers (e.g., commercial poly(vinyl amine) (PVAm), polyethylenimine (PEI) and polyallylamine (PAA)) have been investigated at temperatures greater than 100 °C [24–33, 38, 49, 50]. These membranes simultaneously separate H_2S and CO_2 from the feed gas based on the reversible reaction and diffusion mechanism. The CO_2- and H_2S-selective natures of the membranes not only can be used for carbon capture applications but also can be employed for hydrogen sulfide cleanup, required for the protection of fuel cell catalysts and low-temperature WGS reaction catalysts.

The transport performance and the long-term stability of membranes containing various amines as mobile and fixed-site carriers at different pressures are described in the following sections. Moreover, the potential industrial applications of the aforementioned amine-containing membranes at different operating conditions are discussed.

11.4 Membranes for Low-Pressure H_2 Purification

11.4.1 CO_2 Transport Properties

Ho and coworkers have studied various amines as mobile and fixed-site carriers for the CO_2-selective facilitated transport membranes. Zou et al. investigated the transport performance of membranes containing AIBA-K and KOH as mobile carriers and PAA as a fixed-site carrier [26, 29]. Various temperatures between 100 and 180 °C were investigated for the transport measurements, and CO_2 permeability of >5,000 Barrers and CO_2/H_2 selectivity of >300 were reported at 2 atm feed pressure. Moreover, these membranes demonstrated remarkably high CO_2/CO selectivity at >100 °C operating temperature with 1 % CO content in the feed gas [26, 29, 34].

Zhao and Ho developed the next generation of CO_2-selective facilitated transport membranes by exploring the steric hindrance effect of the amine-containing fixed-site carriers [38, 50]. The reversible reaction between CO_2 and a conventional amine consumes 2 mol of amines to convert 1 mol of CO_2 into carbamate ion. However, in case of a sterically hindered amine, the formed carbamate ion is unstable (due to the attached bulky group) and undergoes hydrolysis to yield a bicarbonate ion [38, 50–53]. Hence, it results in an enhanced CO_2 loading of 1 mol of hindered amine consumed for each mole of CO_2 molecule reacted (as shown in Eq. 11.5) [38].

$$CO_2 + R_1 - NH - R_2 + H_2O \rightleftharpoons R_1R_2 - NH_2{}^+ + HCO_3{}^- \qquad (11.5)$$

The steric hindrance effect in the membrane was first demonstrated by Zhao and Ho, using different sets of membranes containing 70 % polyamines and 30 % cross-linked PVA at 110 °C [50]. The membranes with sterically hindered PAA exhibited an improved CO_2 permeability (average 297 Barrers; although not high, but greater than a factor of 5 relative to sterically unhindered PAA) and an improved CO_2/H_2 selectivity (by a factor >2) compared to membranes with sterically unhindered PAA [50]. The CO_2 permeability was further enhanced by incorporating mobile carriers along with polyamines in the cross-linked PVA matrix. The membranes containing sterically hindered poly-N-isopropylallylamine (PAA-C_3H_7) as a fixed-site carrier along with 2-aminoisobutyric acid-potassium salt (AIBA-K) and potassium hydroxide (KOH) as the mobile carriers were synthesized and investigated for transport performance [38]. PAA-C_3H_7 is a sterically hindered polyamine derived from PAA by the modification reaction [54]. The mobile and fixed-site carriers were dispersed into the polymeric matrix and cast into a thin film on a microporous polysulfone support. Poly(vinyl alcohol) (PVA) was used as the hydrophilic polymeric matrix because of the good film-forming capacity, good compatibility with carriers, and strong reactivity of the hydroxyl groups for cross-linking. PVA was polycondensed with (3-aminopropyl)triethoxysilane (APTEOS) and cross-linked by glutaraldehyde (GA) to form a strong polymer (PVA-POS) network. The transport performances of the membranes were compared with the membranes of similar composition but containing sterically unhindered PAA as a fixed-site carrier. Figure 11.3 shows the transport performances of membranes containing PAA and PAA-C_3H_7 as fixed carriers, tested at 110 °C, and 2 atm feed pressure. The 25-μm selective layer comprised of 20 wt.% PVA-POS (100 mol% degree of cross-linking), 17.2 wt.% PAA or PAA-C_3H_7 as fixed-site carriers, 25.7 wt.% AIBA-K, 16.6 wt.% KOH, 12 wt.% SiO_2, and 8 wt.% APTEOS [38].

As seen from Fig. 11.3, significant enhancements in both CO_2 permeability and CO_2/H_2 selectivity were achieved by incorporating sterically hindered polyamine as the fixed-site carrier. The total solid content in the membrane contained more moles of amines with PAA as a fixed-site carrier as compared to the membranes with PAA-C_3H_7 as a fixed-site carrier (at the same weight content). Moreover, the reaction rate constant of the CO_2-PAA reaction would be larger than the rate constant for CO_2-(PAA-C_3H_7) reaction (due to the steric hindrance effect).

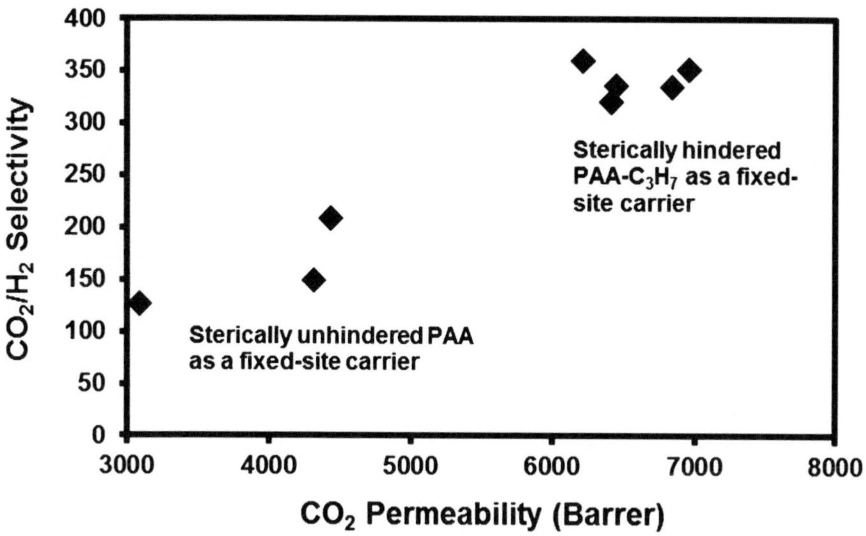

Fig. 11.3 Gas separation results of cross-linked PVA/poly(siloxane)-based facilitated transport membranes containing amines (25-μm selective layer thickness), tested at 110 °C, and 2 atm feed pressure using a feed gas composition of 20 % CO_2, 40 % H_2, and 40 % N_2 (on dry basis) with 40 % water vapor content on the feed side (The figure has been plotted from the transport results published in Zhao and Ho [38])

However, the enhancement in CO_2 loading for CO_2-(PAA-C_3H_7) reaction due to the steric hindrance effect resulted in more moles of free amines available for the reaction with CO_2. Thus, improved transport performances were demonstrated by membranes containing (PAA-C_3H_7) as a fixed-site carrier.

Ho and coworkers have investigated the performances of membranes containing commercial polyvinylamine (PVAm) as a fixed-site carrier. The synthesized membrane contained AIBA-K and KOH as mobile carriers. The commercial PVAm composed of more than 60 % of salts with the rest containing the polymer. The salt content was believed to aid in water retention, thereby enhancing the transport performance of the commercial PVAm-containing membranes. The mobile and fixed-site carriers were dispersed in a cross-linked PVA matrix using GA as the cross-linking agent. The transport performances of the membranes containing various compositions of AIBA-K and commercial PVAm as CO_2 carriers are shown in Fig. 11.4. The thickness of the amine-containing selective layer was varied from 10 to 20 μm. These membranes were tested at 106 °C and 2 atm feed pressure with 40 % steam content on the feed side using 20 % CO_2, 40 % H_2, and 40 % N_2 as the dry feed gas composition. In spite of the performance variation shown in this figure due to various membrane compositions and thicknesses, an average CO_2 permeability greater than 3,000 Barrers and an average CO_2/H_2 selectivity greater than 200 were exhibited by the membranes. The AIBA-K and KOH contents (mobile carriers) for these membranes were varied from 20–27 % to 13–18 %, respectively. The commercial PVAm (fixed-site carrier) content was

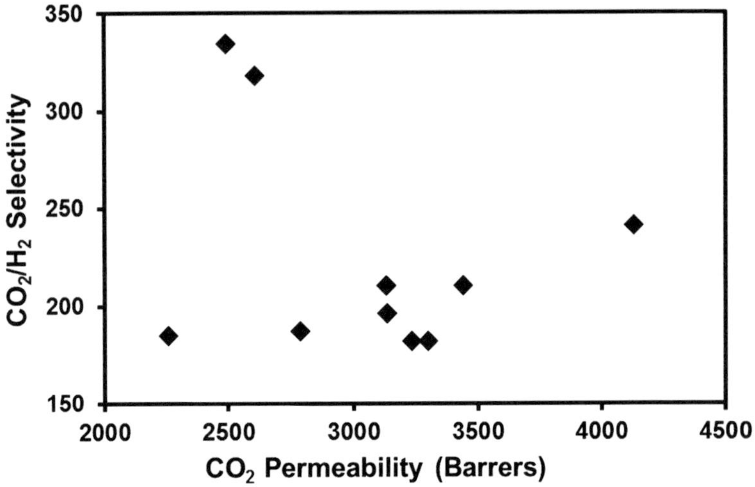

Fig. 11.4 Gas separation results of cross-linked PVA-based facilitated transport membranes containing various compositions of AIBA-K and commercial PVAm as CO_2 carriers, tested at 110 °C, and 2 atm feed pressure using a feed gas composition of 20 % CO_2, 40 % H_2, and 40 % N_2 (on dry basis) with 40 % and 57 % water vapor contents on the feed and sweep sides, respectively (feed flow rate = 60 cm^3/min (dry basis) and sweep flow rate = 30 cm^3/min (dry basis))

varied from 17 % to 24 %, and the cross-linked PVA content was varied from 25 % to 34 %. The mobile carrier and fixed-site carrier contents are required to be optimized for a desirable transport performance. These membranes demonstrated highly attractive CO_2/H_2, CO_2/CO, H_2S/CO_2, and H_2S/H_2 selectivities [55].

Considerable research activities are currently focused on using novel mobile carries such as triethylenetetramine (TETA), glycine-K, and N,N-dimethylglycine-K along with the use of different fixed carriers such as PEI and high molecular-weight PVAm for low-pressure H_2 purification application. Ho's membrane research group has also made significant progress towards the synthesis of thin membranes (<1 μm) for the CO_2 removal from the simulated flue gas [55]. Although the CO_2 removal from flue gas was not related to H_2 purification application, the potential of fabricating a very thin amine-containing polymeric layer for CO_2 removal is emphasized. A thinner selective layer would increase the CO_2 and H_2S permeance because of a lower diffusional resistance. The transport results of H_2S for the amine-containing membranes are discussed in the next section.

11.4.2 H_2S Transport Properties

Amine-containing facilitated transport membranes (discussed in the previous section) are selective for CO_2 and H_2S transport. Amine carriers (mobile and fixed-site carriers) dispersed in a cross-linked PVA matrix have a higher reaction rate with

H_2S as compared to CO_2 by the virtue of the proton transfer reaction (as shown in Eq. 11.6). Thus, H_2S molecules are able to permeate through the amine-containing hydrophilic polymeric matrix much faster than CO_2 [29, 55].

$$H_2S + RNH_2 \rightleftharpoons HS^- + RNH_3^+ \tag{11.6}$$

The initial H_2S transport results reported in the literature are with membranes containing AIBA-K and KOH as mobile carriers and PAA as a fixed-site carrier [56]. The PVA was cross-linked using formaldehyde as the cross-linking agent. The transport measurements were carried out at 2 atm and 1 atm pressures on the feed and sweep sides, respectively, using a dry feed gas composition containing 50 ppm of H_2S, 1 % CO, 17 % CO_2, 37 % N_2, and 45 % H_2 (40 % water vapor content on the feed side). The H_2S permeability was approximately three times the CO_2 permeability. A H_2S permeability of greater than 20,000 Barrers and a H_2S/H_2 selectivity of higher than 600 were demonstrated [34, 55].

The H_2S transport properties were also explored using the membranes containing AIBA-K and KOH as the mobile carriers and commercial PVAm as a fixed-site carrier [55]. Membranes with a selective layer thickness varying from 10 to 15 μm exhibited high H_2S and CO_2 permeability. The transport measurements were carried out at 2 atm feed pressure with a 40 % water vapor content on the feed side. The membranes demonstrated greater than 3,000 Barrers of CO_2 permeability with the H_2S/CO_2 selectivity in the range of 2.5–3.5.

11.4.3 Membrane Stability

Membrane stability is important for commercial viability of a membrane technology. The long-term operational membrane stability is a challenging task for the polymeric facilitated transport membranes. Early facilitated transport membranes experienced several instabilities due to carrier leakage and solvent/carrier evaporation. But the new generation of facilitated transport membranes has shown promisingly stable performances. A membrane stability of over 25 days was demonstrated by fluoride-containing cross-linked poly(vinyl amine) membranes (membranes containing PVBTAF/CsF) [43, 46, 47], poly(vinyl alcohol)/polyethylene glycol/polyethylenimine blend membranes [57], and diethanolamine/PVA-based membranes [58]. The transport stability was obtained at a low-temperature range of 23–35 °C. On the other hand, the membrane technology is economically more attractive for H_2 purification application when operated at high temperatures to reduce the energy penalty. In the context of high-temperature membrane stability, amine-containing membranes have shown promising performance. The membranes containing PAA-C_3H_7 and AIBA-K as CO_2 carriers have demonstrated 430 h (>18 days) of stable transport performance at 110 °C and 2 atm feed pressure [50].

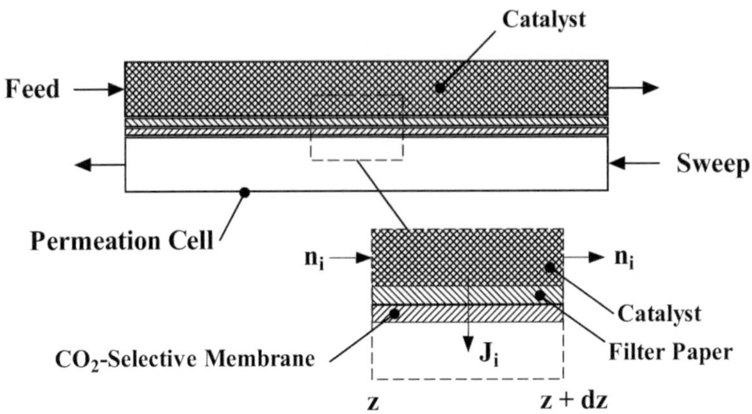

Fig. 11.5 Schematic of the water-gas-shift membrane reactor (Reprinted with permission from Zou et al. [29], Copyright © 2007 American Chemical Society)

The membrane process with attractive transport performance can be integrated with other processes to improve the final product quality and/or process intensification. One of the examples of such integration is the use of a membrane reactor. The CO_2- and H_2S-selective natures of the membrane can be combined with the water-gas-shift (WGS) reaction kinetics in an integrated system of WGS membrane reactor. The following section illustrates some of the transport performances using the WGS membrane reactor.

11.4.4 Water-Gas-Shift (WGS) Membrane Reactor

The CO_2- and H_2S-selective facilitated transport membranes can be applied for CO cleanup using the WGS membrane reactor, in which H_2S is cleaned up before the WGS where a H_2S-sensitive catalyst is used. Figure 11.5 shows the schematic representation of the membrane reactor [55, 56]. As seen from the figure, a membrane reactor is the use of a semipermeable membrane in a chemical reactor. The reversible WGS reaction was shifted in the forward direction by continuous removal of the reaction product CO_2 using the CO_2-selective membrane.

Ho and coworkers have demonstrated the use of amine-containing CO_2- and H_2S-selective polymeric facilitated transport membranes in the WGS membrane reactor for CO cleanup [29]. The CO content was shown to be reduced to less than 10 ppm in the H_2-rich product stream using the feed gas containing 50 ppm H_2S, 1 % CO, 17 % CO_2, 45 % H_2, and 37 % N_2. These experimental results were demonstrated using an effective membrane area of 343 cm^2 at 2 atm feed pressure and 150 °C. Commercial Cu/ZnO/Al2O3 was used as a low-temperature WGS catalyst. Less than 10 ppm of CO content in the H_2-rich product stream (equivalent to 100 % conversion) was demonstrated for various feed flow rates ranging from

Fig. 11.6 Pilot-scale membrane coating machine consisting of the coating section, rinse tank, and thin-film-casting (TFC) assembly

20 to 70 cc/min. The experimental results agreed reasonably well with those predicted by the WGS membrane reactor modeling using the WGS reaction kinetics and membrane mass transport [56].

The CO_2 and H_2S transport results using the amine-containing facilitated transport membranes have demonstrated optimum transport performance in the temperature range of 100–120 °C. On the other hand, a higher operating temperature (140–150 °C) would aid the CO cleanup (higher CO conversion) via the WGS reaction kinetics in the WGS reactor. Hence, the two advantages were coupled together in a two-stage configuration. The membrane separation process was used as a stand-alone unit for CO_2 removal before the conventional low-temperature WGS reactor. The first-stage membrane unit demonstrated >99 % CO_2 removal at 120 °C and the subsequent low-temperature WGS reactor (operated at 150 °C) reduced the CO content from the membrane unit retentate to less than 10 ppm [59].

11.4.5 Pilot-Scale Membrane Fabrication

The CO_2- and H_2S-selective membranes discussed above were synthesized in the laboratory using the casting knife (doctor's blade) with a controlled gap setting. Recently, the pilot-scale continuous membrane fabrication machine was successfully installed by Ho's membrane research group (Figs. 11.6 and 11.7). The

Fig. 11.7 Pilot-scale membrane casting machine for fabricating nanoporous membranes using the phase inversion process

machine consists of two sections: the coating machine and the casting machine. The casting machine can be used to fabricate nanoporous substrates using the phase inversion process. The coating machine comprises the coating assembly, rinse tank, and thin-film-casting (TFC) assembly. As a whole, the machine can be used to fabricate state-of-the-art and new gas separation and desalination membranes. However, the TFC assembly could be operated independently as stand-alone equipment for the fabrication of multilayer composite membranes by the knife-casting technique. The TFC unit can be considered as a scale-up unit of the laboratory membrane synthesis technique using the casting knife. The schematic representation of the TFC assembly for the fabrication of polymer membranes is shown in Fig. 11.8.

Ho and coworkers have fabricated pilot-scale membranes containing AIBA-K and commercial PVAm using the TFC assembly. More than 300 ft long and 14 in. wide membranes have been fabricated via a semicontinuous process for pilot-scale membrane fabrication. The transport results of small flat-sheet membrane samples (3.4 cm^2) taken from the pilot-scale long fabricated membranes appeared to be in agreement with those exhibited by laboratory-synthesized membranes [55]. An important step towards commercialization was demonstrated by polymer membrane fabrication using the pilot-scale continuous membrane fabrication machine.

All the above sections have summarized the performance of CO_2- and H_2S-selective facilitated transport membranes operated at close to 2 atm pressure. The following section reviews the transport results of the CO_2-selective facilitated transport membranes operated at high pressures.

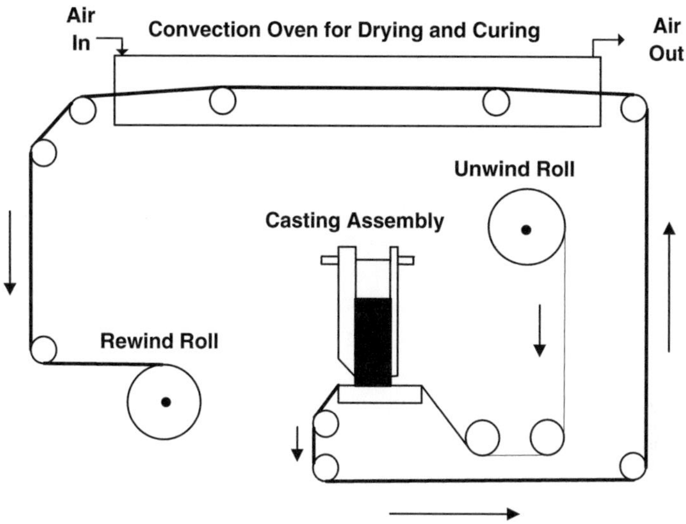

Fig. 11.8 Schematic representation of the pilot-scale thin-film-casting (TFC) machine for scale-up fabrication of thin-film composite membranes

11.5 Membranes for High-Pressure H_2 Purification

The development of polymeric facilitated transport membranes for high-pressure H_2 purification applications is more challenging. In an IGCC power plant, an operating pressure greater than 15 atm is preferred for the membrane technology to be economically more competitive with the well-known absorption process. However, the CO_2 permeability is typically lowered at high pressure due to the characteristic "carrier saturation" phenomenon of facilitated transport membranes. The increase in the feed pressure would increase the available driving force for both the gas components (H_2 and CO_2). The H_2 molecules are transported through the membrane by the solution–diffusion mechanism, and the flux of the H_2 molecules would then proportionately increase on the increasing the feed pressure. However, in case of CO_2, the flux does not increase proportionally and eventually does not change appreciably as the feed pressure increases. This can be attributed to the reaction–diffusion transport mechanism and the finite number of CO_2-reactive carriers available in the membrane matrix. Thus, the CO_2 permeance and CO_2/H_2 selectivity are typically lowered on increasing the feed pressure. Moreover, it is quite challenging to develop a polymer matrix that is durable against the trans-membrane pressure difference and compressive forces (during >10 atm operating pressure). Despite these challenges, CO_2-selective facilitated transport membranes still hold a competitive edge towards the development of economically feasible membrane technology for high-pressure and high-temperature H_2 purification.

Attractive transport performances have been reported for membranes containing PVAm and NH_4F at a relatively high CO_2 partial pressure of 2 atm [43, 60]. PVAm was incorporated as a fixed-site carrier to facilitate the CO_2 transport. Moreover, NH_4F provided the ionic character to the membrane matrix that enhanced the CO_2/H_2 selectivity by reducing the H_2 solubility. Some published papers also highlighted the role of fluoride ions in facilitating the CO_2 transport [61]. Due to the hydrogen bond interaction between the hydrogen atom of the water molecule and the fluoride ion, the water molecule was believed to be more basic for the reaction with CO_2. Although the hydrogen transport data were not reported, the PVAm- and NH_4F-containing membranes demonstrated 1,143 CO_2/CH_4 selectivity at a high CO_2 partial pressure of 2 atm.

Significant research efforts are focused on improving the transport performance and durability of facilitated transport membranes under harsh industrial operating conditions. New carriers with a higher reaction rates are desired to improve the performance. Furthermore, a stronger polymer matrix is imperative for the durability of the membrane at harsh conditions. Mixed matrix membranes have recently gained significant attention for coupling the two advantages: (1) transport performance offered by carriers and (2) the durability offered by incorporating inorganic particles within the polymer matrix. Various inorganic fillers have been reported in the literature, including fumed silica [28], sol–gel silica [62–66], carbon nanotubes [67, 68], and carbon-silica microspheres [69]. The addition of nanofillers resulted in an increased free volume and a higher CO_2 sorption due to disruptions in the polymer chain packing. The dimension of the increase in the size of the free-volume element would vary with different types and dimensions of nanofillers. Optimizing the carrier and the inorganic additive contents in the polymer matrix is vital for a stable membrane performance.

11.5.1 Mixed Matrix Membranes

Polymer/silica mixed matrix membranes have gained importance in recent years especially for high-pressure applications. Ho's research group has done several studies using different amines as efficient CO_2 carriers for high-pressure transport measurements. Membranes containing amine carriers like AIBA-K and PAA, along with silica particles as inorganic nanofillers, have demonstrated attractive transport results. The mechanical and thermal properties of the membrane were improved by incorporating silica as nanofillers. Moreover, silica particles were believed to offer swelling stability to the membrane against excessive compressive forces [28].

Amine-containing facilitated transport membranes without consisting of fillers demonstrated attractive transport results at high pressures initially. However, the CO_2 permeability and CO_2/H_2 selectivity decreased with time, up to 16 % and 19 %, respectively, during 170 h of testing at 15 atm feed pressure and 107 °C [70]. Membrane compaction was believed to be one of the governing reasons for the transport result reduction at high pressures. However, Xing and Ho improved in the gas transport performance and stability of the amine-containing facilitated transport membranes by incorporating sol–gel silica and fumed silica nanoparticles in a hydrophilic cross-linked PVA matrix [28]. Mixed matrix membranes containing AIBA-K and KOH as mobile carriers and PAA as fixed carriers were investigated for high-pressure syngas purification. CO_2 permeability greater than 1,200 Barrers and CO_2/H_2 selectivity as large as 87 were reported at 15 atm feed gas pressure and 107 °C. The temperature sensitivity studies were also carried out and 107 °C was concluded to be an optimum temperature for operation. The fumed silica content was varied from 4 % to 31 %, and a fumed silica loading of 23 % demonstrated optimum transport result for a given amine (CO_2 carriers) content in the membrane [28].

Numerous studies have demonstrated improved mechanical properties (especially, compression effect) of PVA-carbon nanotubes composites [71–75]. Zhao et al. developed advanced CO_2-selective facilitated transport membranes for high-pressure testing by incorporating multiwalled carbon nanotubes (MWNTs) in the polymer matrix. MWNTs were used as mechanical reinforcing fillers within the amine-containing cross-linked PVA-POS matrix. The transport results and the stability of the amine-containing membranes with untreated MWNTs and acid-treated MWNTs as nanofillers were investigated. Amine-containing membranes with 2 wt.% of untreated MWNTs loading demonstrated a significant improvement in the membrane stability during the 792 h of testing. The membrane showed a stable CO_2 permeability of 836 Barrers and a stable CO2/H2 and CO2/N2 selectivities of 43 and 407, respectively, for initial 444 h of testing. However, a dropping trend in the transport results was observed after the 444th hour. The membrane has shown significant improvement in stability as compared to the membranes with no fillers. The membranes containing 3 wt.%, 4 wt.% ,and 8 % untreated MWNTs were also tested for transport stability. It was concluded that 2 wt.% of untreated MWNTs was the optimal loading for the high-pressure stability tests.

The gas transport results and membrane stability were also explored for the membranes containing acid-treated MWNTs. A series of membranes containing 2–8 wt.% acid-treated MWNTs were tested. The membrane containing 4 wt.% acid-treated MWNTs demonstrated an impressive stability for 250 h in a gas permeation experiment carried out at 15 atm feed pressure and 107 °C. The membranes containing 4–6 wt.% acid-treated MWNTs showed higher CO2 permeability than the rest. Moreover, the membranes containing acid-treated MWNT showed higher selectivity as compared to the membranes with untreated MWNT [70].

Zhao et al. showed that the mixed matrix membranes prepared by randomly dispersing MWNTs (acting as mechanical reinforcing fillers) in the polymeric facilitated transport membranes can potentially achieve both improved membrane stability and the separation performance at high pressures and high temperatures. Moreover, the previously discussed amine-containing facilitated transport membranes operated at low pressures (2 atm) have demonstrated attractive separation performances and transport stability. The well-developed membranes with impressive CO_2 and H_2S permeances and exceptional CO_2/H_2 and H_2S/H_2 selectivities can potentially be used for various industrial applications.

11.6 Potential Industrial Applications

The membrane separation performances at different pressures ranging from 2 to 15 atm and at >100 °C have been reviewed in the previous sections. Those membranes serve as an attractive platform for the development of an economically feasible separation technology for hydrogen purification applications. Different potential applications at low pressures as well as high pressures are discussed in the following subsections.

11.6.1 Low-Pressure H_2 Purification for Fuel Cells

In view of low-pressure H_2 purification, the transport results of the membranes with amine as CO_2 carriers were investigated at 2 atm feed pressure and temperatures ranging from 100 to 120 °C. The separation performances were investigated at different water vapor contents on the feed and sweep sides. These membranes offer great potential for the fuel cells by the virtue of exceptional selectivity at >100 °C. The gas stream containing H_2, CO_2, H_2S, CO, and other impurities can be sent to a stand-alone membrane separation unit for the removal of CO_2 and H_2S. The CO_2 removal is essentially targeted at enriching the H_2 (fuel) content in the resultant stream, whereas the H_2S removal is aimed at avoiding the downstream sulfur poisoning of the WGS catalyst or the fuel cell catalyst. The retentate stream from the stand-alone membrane separation unit can be sent to the WGS reactor for an overall CO cleanup to less than 10 ppm. In terms of a system design, the combination of membrane separation unit and a WGS reactor can result in efficient CO_2, H_2S, and CO cleanup to generate a H_2-rich product stream (Fig. 11.9). Moreover, a single unit of WGS membrane reactor can potentially be incorporated as stand-alone equipment for the WGS reaction along with separation of CO_2 and H_2S to enhance the H_2 content.

The membrane technology used for both configurations could significantly increase the fuel cell efficiency and reduce the overall costs by generating H_2 (fuel)-enriched product stream via an economic H_2 purification and recycle process.

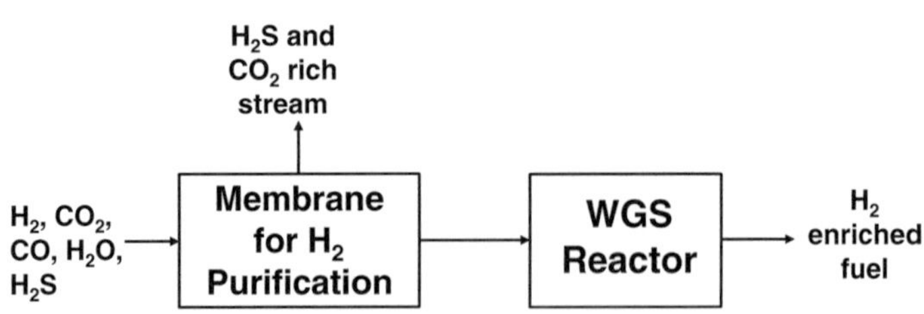

Fig. 11.9 Schematic representation of a two-stage H_2 purification process to enhance the fuel (H_2) recovery

11.6.2 High-Pressure H_2 Purification

Syngas purification in an IGCC power plant using CO_2-selective membranes has attracted considerable attention in recent years. The CO_2 removal from syngas would enhance the H_2 content in the resultant gaseous stream. The H_2-rich gas stream can either be used for large-scale power generation or the resultant stream can be directed towards the manufacture of valuable chemicals. Various techno-economic feasibility studies have been published in literature for use of membrane processes at an IGCC power plant in conjunction with CO_2 capture. Merkel et al. have published a techno-economic study for three different process configurations in an IGCC power plant with carbon capture: (1) CO_2-selective membrane, (2) H_2-selective membrane, and (3) a hybrid process (using CO_2-selective and H_2-selective membranes) [8]. The hybrid process was evaluated to be economically more feasible, resulting in an 18 % increase in the cost of electricity (COE). The process was designed for CO_2 capture with co-sequestration of H_2S from syngas in an IGCC power plant. An additional 5 % increase in the COE was estimated for integrating the Selexol® process for H_2S removal [6, 8].

In a process modeling study, Ho and coworkers estimated around 17 % increase in the COE for integrating CO_2-selective membrane technology to an IGCC power plant. The experimental separation performances exhibited by the amine-containing facilitated transport membranes of high CO_2 and H_2S selectivities were used for the techno-economic study of syngas purification in an IGCC power plant. The use of 2-stage membrane process for CO_2 removal resulted in >99.9 % H_2 recovery for 90 % CO_2 capture from the syngas in an IGCC power plant. Moreover, the cost of separating H_2S from the syngas using the Selexol® process was incorporated in the estimate of 17 % increase in the COE.

The cost of power generation from the IGCC power plant can further be reduced by developing a membrane technology that can potentially replace the energy-intensive Selexol® process for H_2S removal. The development of membranes with higher H_2S/CO_2 selectivity could potentially result in significant cost reduction by eliminating the use of expensive absorption process for H_2S removal. Thus, syngas purification in an IGCC power plant has the potential of incorporating cost-effective membrane processes for CO_2 and H_2S removal.

Hydrogen is industrially produced by the SMR process. Steam reforming is a process of generating a syngas stream containing H_2, CO_2, H_2O, and CO from a fuel, like methane, in a reformer. The fuel value (H_2 content) of the gaseous product stream is enriched by the high-pressure and high-temperature WGS reactor. Similar to the IGCC power plants, a membrane technology can be incorporated after the WGS reactor as a cost-efficient H_2 purification process. The development of advanced WGS membrane reactors for high-pressure applications can result in significant cost reductions for industrial H_2 production or large-scale power generation.

11.7 Conclusions and Future Outlook

In summary, the CO_2- and H_2S-selective facilitated transport membranes reviewed in this work have shown promising transport performance and stability at various industrially relevant operating conditions. The potential of incorporating a cost-effective membrane technology for different applications using such membranes has been emphasized. Considerable future directions for research have been available and discussed to develop a promising membrane technology for H_2 purification.

Some potentially important industrial applications of the membrane technology for H_2 purification have been discussed in this chapter. The development of membrane materials should be aimed at the economic feasibility of the overall system design. The CO_2-selective polymer facilitated transport membranes for low-pressure and high-pressure H_2 purification have shown promising CO_2/H_2 selectivity and impressive transport stability. However, the development of membranes with higher CO_2 permeance is essential. Membranes with greater CO_2 permeance (at a desirable CO_2/H_2 selectivity) will be cost effective as it will reduce the membrane area requirement for a given degree of separation. The CO_2 permeability of such membranes can be improved further by the development of novel mobile and fixed-site carriers. The new generation of mobile carriers with higher reaction rate constants with CO_2 (to reduce the reaction resistance) and a smaller size of the molecule (to reduce the diffusion resistance) is desired. For the high-pressure H_2 purification applications, the development of novel fixed-site carriers with a stronger polymer backbone (for stability against the compression forces) and higher interchain and intermolecular interactions (for stability against carrier leakage) is essential. For both low- and high-pressure H_2 purification, the development of advanced WGS membrane reactors can result in significant cost reductions for industrial H_2 production or large-scale power generation.

Acknowledgments We would like to gratefully acknowledge the Department of Energy/National Energy Technology Laboratory (DE-FE0007632), the Office of Naval Research/DJW Technology (N00014-14-C-098), the National Science Foundation (CBET 1033131 and IIP 1127812), and the Ohio Development Services Agency (OOE-CDO-D-13-05) for their financial support of the results published in this work. This work was partly supported by the Department of Energy under Award Number DE-FE0007632 with substantial involvement of the National Energy Technology Laboratory, Pittsburgh, PA, USA.

Nomenclature

J	Steady-state permeation flux (cm^3 (STP)/(cm^2 s))
ℓ	Membrane thickness (cm)
P	Permeability (Barrer)
p	Pressure (cmHg)
pf	Feed side pressure (cmHg)
ps	Sweep (permeate) side pressure (cmHg)
Δp	Pressure difference between the feed and permeate sides (cmHg)

x Retentate molar fraction
y Permeate molar fraction
D Diffusion coefficient
S Solubility

Greek Letter

α Selectivity

Subscripts

i, j Species

Abbreviations

COE	Cost of electricity
IGCC	Integrated gasification combined cycle
MWNT	Multiwalled carbon nanotubes
PEMFC	Proton-exchange membrane fuel cell
PSA	Pressure swing adsorption (PSA)
SMR	Steam methane reforming
WGS	Water-gas shift

References

1. Crabtree GW, Dresselhaus MS, Buchanan MV. The hydrogen economy. Phys Today. December 2004;39–44.
2. Turner JA. A realizable renewable energy future. Science. 1999;285:687–9.
3. Edwards PP, Kuznetsov VL, David WIF. Hydrogen energy. Phil Trans R Soc A. 2007;365:1043–56.
4. U.S. Department Of Energy. Robust polymer composite membranes for hydrogen separation. Industrial Technologies Program DOE/EE-0466; May 2011. EERE Information Center. Available online at http://energy.gov/sites/prod/files/2013/11/f4/polymer_composite_membranes.pdf
5. Zornoza B, Casado C. Advances in hydrogen separation and purification with membrane technology. Renewable hydrogen technologies: production, purification, storage, applications and safety. Elsevier; 2013. p. 245–266. Available at http://books.google.com/books?id=6pMuGXwPi7IC&dq=-hydrogen+economy+membrane+h2+purification&source=gbs_navlinks_s.
6. Black J. Cost and performance baseline for fossil energy plants. U.S. Department of Energy DOE/NETL-2010/1397; 2010. National Energy Technology Laboratory. Available online at

http://www.netl.doe.gov/File%20Library/Research/Energy%20Analysis/OE/BitBase_FinRep_Rev2a-3_20130919_1.pdf
7. Song C. Fuel processing for low-temperature and high-temperature fuel cells: challenges, and opportunities for sustainable development in 21st century. Catal Today. 2002;77:17–49.
8. Merkel TC, Zhou M, Baker R. Carbon dioxide capture with membranes at an IGCC power plant. J Membr Sci. 2012;389:441–50.
9. Ribeiro AM, Grande CA, Lopes FVS, Loureiro JM. Four beds pressure swing adsorption for hydrogen purification: case of humid feed and activated carbon beds. AIChE J. 2009;55:2292–302.
10. Sweny JW. Synthetic fuel gas purification by the Selexol® process, Allied Chemical Corporation. Available online on the web page https://web.anl.gov/PCS/acsfuel/preprint%20archive/Files/18_2_DALLAS_04-73_0142.pdf.
11. Shao L, Low BT, Chung T, Greenberg AR. Polymeric membranes for the hydrogen economy: contemporary approaches and prospects for the future. J Membr Sci. 2009;327:18–31.
12. Ockwig NW, Nenoff TM. Membrane for hydrogen separation. Chem Rev. 2007;107:4078–110.
13. Kusakabe K, Shibao F, Zhao G, Sotowa KI, Watanabe K, Saito T. Surface modification of silica membranes in a tubular-type module. J Membr Sci. 2003;215:321–6.
14. Iwamoto Y, Sato K, Kato T, Inada T, Kubo Y. A hydrogen-permselective amorphous silica membrane derived from polysilazane. J Eur Ceram Soc. 2005;25:257–64.
15. Diniz da Costa JC, Lua GQ, Rudolph V, Lin YS. Novel molecular sieve silica (MSS) membranes: characterization and permeation of single-step and two-step sol-gel membranes. J Membr Sci. 2002;198:9–21.
16. Gu YF, Oyama ST. High molecular permeance in a poreless ceramic membrane. Adv Mater. 2007;19:1636–40.
17. Adhikari S, Fernando S. Hydrogen membrane separation techniques. Ind Eng Chem Res. 2006;45:875–81.
18. Xu ZK, Dannenberg C, Springer J, Banerjee S, Maier G. Gas separation properties of polymers containing fluorene moieties. Chem Mater. 2002;14:3271–6.
19. Camacho-Zuniga C, Ruiz-Trevino FA, Zolotukhin MG, del Castillo LF, Guzman J, Chavez J. Gas transport properties of new aromatic cardo poly(aryl ether ketone)s. J Membr Sci. 2006;283:393–8.
20. Ghosal K, Freeman BD. Gas separation using polymer membranes: an overview. Polym Adv Tech. 2003;5:673–97.
21. Lin H, Van Wagner E, Freeman BD, Toy LG, Gupta RP. Plasticization-enhanced hydrogen purification using polymeric membranes. Science. 2006;311:639–42.
22. Lin H, Freeman BD. Gas solubility, diffusivity and permeability in poly(ethylene oxide). J Membr Sci. 2004;239:105–17.
23. Okamoto K, Fujii M, Okamyo S, Suzuki H, Tanaka K, Kita H. Gas permeation properties of poly(ether imide) segmented copolymers. Macromolecules. 1995;28:6950–6.
24. Ho WSW. Membranes comprising salts of aminoacids in hydrophilic polymers. 1997. U.S. Patent 5,611,843.
25. Ho WSW. Membrane comprising aminoacid salts in polyamine polymers and blends. 2000. U.S. Patent 6,099,621.
26. Zou J, Ho WSW. CO_2-selective polymeric membranes containing amines in crosslinked poly (vinyl alcohol). J Membr Sci. 2006;286:310–21.
27. Bai H, Ho WSW. Carbon dioxide-selective membranes for high-pressure synthesis gas purification. Ind Eng Chem Res. 2011;50:12152–61.
28. Xing R, Ho WSW. Crosslinked polyvinylalcohol-polysiloxane/fumed silica mixed matrix membranes containing amines for CO_2/H_2 separation. J Membr Sci. 2011;367:91–102.
29. Zou J, Huang J, Ho WSW. CO_2-selective water gas shift membrane reactor for fuel cell hydrogen processing. Ind Eng Chem Res. 2007;46:2272–9.
30. Huang J, El-Azzami L, Ho WSW. Modeling of CO_2-selective water gas shift membrane reactor for fuel cell. J Membr Sci. 2005;261:67–75.

31. Zou J, Ho WSW. Hydrogen purification for fuel cells by carbon dioxide removal membrane followed by water gas shift reaction. J Chem Eng Jpn. 2007;40:1011–20.
32. Hussain A, Hägg M. A feasibility study of CO_2 capture from flue gas by facilitated transport membrane. J Membr Sci. 2010;359:140–8.
33. Omidkhah M, Pedram MZ, Amooghin AE. Facilitated transport of CO2 through DEA-nediated poly(vinyl alcohol) membrane cross linked by formaldehyde. J Membra Sci Technol. 2013;3:119. doi:10.4172/2155-9589.1000119.
34. Ramasubramanian K, Zhao Y, Ho WSW. CO_2 capture and H_2 purification: prospects for CO_2-selective membrane process. AIChE J. 2013;59:1033–45.
35. Franz J, Scherer V. An evaluation of CO_2 and H_2 selective polymeric membranes for CO_2 separation in IGCC power plant. J Membr Sci. 2010;359:173–83.
36. Cussler EL, Aris R, Bhown A. On the limits of facilitated diffusion. J Membr Sci. 1989;43:149–64.
37. Cussler EL. Diffusion mass transfer in fluid systems. 3rd ed. Cambridge: Cambridge University Press; Cambridge, New York, NY 2009.
38. Zhao Y, Ho WSW. CO_2-selective membranes containing sterically hindered amines for CO_2/H_2 separation. Ind Eng Chem Res. 2013;52:8774–82.
39. Ho WSW, Sirkar KK. Membrane handbook. Reprint ed. Norwell: Kluwer Academic Publishers; 2001.
40. Hanioka S, Maruyama T, Sotani T, Teramoto M, Matsuyama H, Nakashima K, et al. CO_2 separation facilitated by task-specific ionic liquids using a supported liquid membrane. J Membr Sci. 2008;314:1–4.
41. Bara JE, Gabriel CJ, Carlisle TK, Camper DE, Finotello A, Gin DL, Noble RD. Gas separations in fluoroalkyl-functionalized room-temperature ionic liquids using supported liquid membranes. J Chem Eng. 2009;147:43–50.
42. Chen HZ, Li P, Chung T. PVDF/ionic liquid polymer blends with superior separation performance for removing CO_2 from hydrogen and flue gas. Int J Hydrogen Energy. 2012;37:11796–804.
43. Kovvali AS, Chen H, Sirkar KK. Dendrimer membranes: a CO_2-selective molecular gate. J Am Chem Soc. 2000;122:7594–5.
44. Hägg M, Quinn R. Polymeric facilitated transport membranes for hydrogen purification. MRS Bull. 2006;31:750–5.
45. Quinn R, Laciak DV, Pez GP. Process for separating acid gaseous mixtures utilizing composite membranes formed from salt-polymer blends. 2001. U.S. Patent 6,315,968.
46. Quinn R, Laciak DV, Appleby JB, Pez GP. Polyelectrolyte membranes for separation of acd gases. 1994. U.S. Patent 5,336,298.
47. Quinn R, Laciak DV. Polyelectrolyte membranes for acid gas separations. J Membr Sci. 1997;131:49–60.
48. Quinn R, Laciak DV, Pez GP. Polyelectrolyte-salt blend membranes for acid gas separations. J Membr Sci. 1997;131:61–9.
49. Bai H, Ho WSW. Recent developments in fuel-processing and proton-exchange membranes for fuel cells. Polym Int. 2011;60:26–41.
50. Zhao Y, Ho WSW. Steric hindrance effect on amine demonstrated in solid polymer membranes for CO_2 transport. J Membr Sci. 2012;415–416:132–8.
51. Sartori G, Ho WSW, Savage DW, Chludzinski GR, Wiechert S. Sterically-hindered amines for acid-gas absorption. Sep Purif Methods. 1987;16:171–200.
52. Chakraborty AK, Astarita G, Bischoff KB. CO_2 absorption in aqueous solutions of hindered amines. Chem Eng Sci. 1986;41:997–1003.
53. Park JY, Yoon SJ, Lee H. Effect of steric hindrance on carbon dioxide absorption into new amine solutions: thermodynamic and spectroscopic verification through solubility and NMR analysis. Environ Sci Technol. 2003;37:1670–5.
54. Yamaguchi T, Boetje LM, Koval CA, Noble RD, Bowman CN. Transport properties of carbon dioxide through amine functionalized carrier membranes. Ind Eng Chem Res. 1995;34:4071–7.

55. Ramasubramanian K. CO_2 (H_2S)-selective membranes for fuel cell hydrogen purification and flue gas carbon capture: an experimental and process modeling study. Ph.D. dissertation, The Ohio State University; 2013.
56. Huang J. CO_2 (H_2S) membrane separations and WGS membrane reactor modeling for fuel cells. Ph.D. dissertation, The Ohio State University; 2007.
57. Hamouda SB, Nguyen QT, Langevin D, Roudesli S. Poly(vinylalcohol)/poly(ethyleneglycol)/poly(ethyleneimine) blend membranes-structure and CO_2 facilitated transport. C R Chim. 2010;13:372–9.
58. Francisco GJ, Chakma A, Feng X. Separation of carbon dioxide from nitrogen using diethanolamine-impregnated poly(vinyl alcohol) membranes. Sep Purf Technol. 2010;71:205–13.
59. Zou J. Carbon dioxide selective membranes and their applications in hydrogen processing, Ph. D. dissertation, The Ohio State University; 2007.
60. Kim TJ, Li B, Hägg MB. Novel fixed-site-carrier polyvinylamine membrane for carbon dioxide capture. J Polym Sci B. 2004;42:4326.
61. Zhang L, Wang R. Salting-out effect on facilitated transport membranes for CO_2 separation: from fluorine salt to polyoxometalates. RSC Adv. 2012;2:9551–4.
62. Patel NP, Miller AC, Spontak RJ. Highly CO_2-permeable and selective polymer nanocomposite membranes. Adv Mat. 2003;15:729–33.
63. Chen HZ, Chung TS. CO_2-selective membranes for hydrogen purification and the effect of carbon monoxide (CO) on its gas separation performance. Int J Hydrogen Energy. 2012;37:6001–11.
64. Lau CH, Liu S, Paul DR, Xia J, Jean YC, Chen H, et al. Silica nanohybrid membranes with high CO_2 affinity for green hydrogen purification. Adv Energy Mater. 2011;1:634–42.
65. Xia J, Liu S, Lau CH, Chung TS. Liquidlike poly(ethylene glycol) supported in the organic-inorganic matrix for CO_2 removal. Macromolecules. 2011;44:5268–80.
66. Sforca ML, Yoshida IVP, Nunes SP. Organic-inorganic membranes prepared from polyether diamine and epoxy silane. J Membr Sci. 1999;159:197–207.
67. Murali RS, Sridhar S, Sankarshana T, Ravikumar YVL. Gas permeation behavior of Pebax-1657 nanocomposite membrane incorporated with multiwalled carbon nanotubes. Ind Eng Chem Res. 2010;49:6530–8.
68. Yu B, Cong H, Li Z, Tang J, Zhao SX. Pebax-1657 nanocomposite membranes incorporated with nanoparticles/colloids/carbon nanotubes for CO_2/N_2 and CO_2/H_2 separation. J Appl Polym Sci. 2013. doi:10.1002/app.39500.
69. De Clippel F, Khan AL, Cano-Odena A, Dusselier M, Vanherck K, Peng L, et al. CO_2 reverse selective mixed matrix membranes for H_2 purification by incorporation of carbon–silica fillers. J Mater Chem A. 2013;1:945–53.
70. Zhao Y, Jung BT, Ansaloni L, Ho WSW. Multiwalled carbon nanotube mixed matrix membranes containing amines for high pressure CO_2/H_2 separation. J Membr Sci. 2014;459:233–43.
71. Schadler LS, Giannaris SC, Ajayan PM. Load transfer in carbon nanotube epoxy composites. Appl Phys Lett. 1998;73:3842–4.
72. Shaffer MSP, Windle AH. Fabrication and characterization of carbon nanotube/poly(vinyl alcohol) composites. Adv Mat. 1999;11:937–41.
73. Cadek M, Coleman JN, Barron V, Hedicke K, Blau WJ. Morphological and mechanical properties of carbon-nanotube-reinforced semicrystalline and amorphous polymer composites. Appl Phys Lett. 2002;81:5123–5.
74. Coleman JN, Cadek M, Blake R, Nicolosi V, Ryan KP, Belton C, et al. High-performance nanotube-reinforced plastics: understanding the mechanism of strength increase. Adv Funct Mater. 2004;14:791–8.
75. Chen W, Tao X, Xue P, Cheng X. Enhanced mechanical properties and morphological characterizations of poly(vinyl alcohol)-carbon nanotube composite films. Appl Surf Sci. 2005;252:1404–9.

Chapter 12
Hydrogen Production for PEM Fuel Cells

Angelo Basile, Adolfo Iulianelli, Giuseppe Bagnato, and Francesco Dalena

Abstract Today, hydrogen is seen as the most convenient energy carrier for a number of applications and, particularly, for proton exchange membrane fuel cells (PEMFCs). In the specialized literature, many studies address the production of hydrogen derived from renewables. Therefore, the scope of this chapter is to review the recent findings about hydrogen generation from reforming processes of bio-sources combined with membrane reactor technology. A deep discussion is presented about the general classification of the membranes, with special attention being paid toward palladium-based membranes. Furthermore, an overview on the representative results on the reforming of ethanol and methanol as renewable sources performed in membrane reactors is given.

Keywords PEM fuel cell • Hydrogen • Membrane reactor • Steam reforming reaction

12.1 Introduction

The growing attention toward climate change and air pollution related to the emissions caused by the exploitation of fossil fuels, associated to their depletion, has pushed both academia and industry to seek alternative technologies and renewable energy sources to mitigate the effects of harmful emissions. Therefore, proton exchange membrane fuel cells (PEMFCs) can be considered, for example, since they are electrochemical devices capable of producing electricity directly from hydrogen and oxygen, without combustion, making the process clean and nonpolluting (Fig. 12.1). PEMFCs present several advantages such as

A. Basile (✉) • A. Iulianelli • G. Bagnato
Institute on Membrane Technology of the Italian National Research Council (ITM-CNR), University of Calabria, Via P. Bucci Cubo 17/C, Rende, CS 87036, Italy
e-mail: a.basile@itm.cnr.it

F. Dalena
Chemistry and Chemical Technologies Department, University of Calabria, Cubo 15/D, Via P. Bucci, Rende, CS 87036, Italy

Z. Fang et al. (eds.), *Production of Hydrogen from Renewable Resources*, Biofuels and Biorefineries 5, DOI 10.1007/978-94-017-7330-0_12

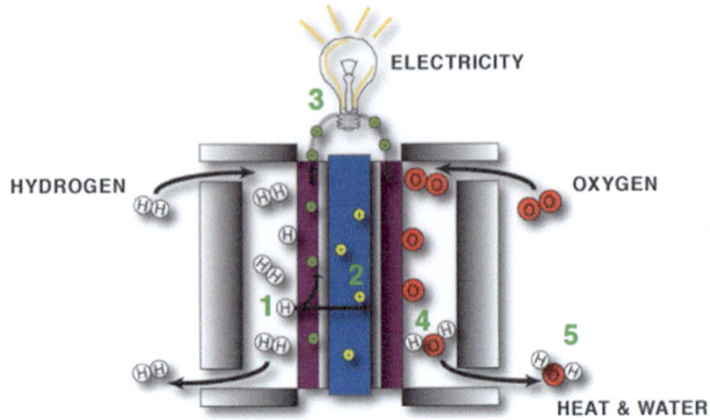

Fig. 12.1 General schematic of a proton exchange membrane fuel cell (From GreenSpec, Fuel Cells: Heat and Electricity (http://www.greenspec.co.uk/building-design/fuel-cells/))

low operating temperatures (60–100 °C), sustained operation at a high current density, compactness, fast start-up, and suitability for intermittent operation [1–5]. In the last decades, PEMFCs have been particularly attractive for power production in both automotive and stationary applications due to their ability to convert the chemical energy of a fuel (hydrogen) directly into electrical energy with relatively high efficiency, their adaptable size, and low operating temperature [6–8]. Furthermore, PEMFCs represent a viable energy solution, since they have zero greenhouse gas emission. Nevertheless, the main drawbacks affecting negatively their full commercialization are represented by the high cost of the membrane, the fuel crossover, and the anodic catalyst poisoning mainly caused by CO_x [9–11]. As a consequence, PEMFC supply imposes the purification of hydrogen, which currently takes place in second-stage processes, namely, water gas shift (WGS) reaction performed in two reactors operating in series at high and low temperatures, partial oxidation (PROX), and pressure swing adsorption (PSA) [12].

The aforementioned stages of hydrogen purification negatively affect the overall process in terms of costs and efficiency. Thus, at the scientific level, much attention has devoted to the development of alternative technologies to generate high purity hydrogen or, at least, COx-free for PEMFC supply. Among them, membrane reactor (MR) technology plays an important role as an alternative solution to the conventional reactors (CRs), besides further stage of hydrogen purification devices, in terms of combination in a single stage of the reforming reaction for generating hydrogen and its purification without the need for any further processing [13, 14]. As a characteristic of the aforementioned technology, inorganic MRs possess several benefits over CRs [15–17]. This is reflected by the extensive literature on hydrogen production through the use of these devices based on both dense and supported Pd-based membranes [18–23], particularly because they possess high permselectivity to hydrogen with respect to all other gases. The scope of this chapter is to illustrate the current state of the art of inorganic MR utilization for producing PEMFC-grade hydrogen.

12.2 Membrane Reactors

Since the 1950s, the concept of MRs has been associated with the use of new inorganic materials and the development of high-temperature membrane processes. Growing scientific interest toward the research and application of MR technology is, today, testified by extensive literature regarding different applications and scientific fields. As a general comment, an MR strictly depends on the membrane typology housed inside. As a general subdivision, MRs can be summarized as follows [14, 24–30]:

(a) Dense and porous inorganic membrane reactors
(b) Electrochemical membrane reactors (fuel cells, electrolytic cells, etc.)
(c) Zeolite membrane reactors
(d) Photocatalytic membrane reactors
(e) Polymeric membrane reactors
(f) Biomedical membrane reactors or membrane bio-reactors

The combination of membranes using chemical and biochemical reactions makes it possible to intensify the whole process. As a consequence, membrane subdivision plays an important role as discussed in the following sections.

12.2.1 Membrane Categories

Generally, membranes are categorized by their material or structure, and they are defined as a layer of material, acting as a selective barrier between two phases, remaining impermeable to specific particles, molecules, or substances under a specific driving force.

Fundamentals for performance evaluation of a generic membrane are (1) the permeating flux through the membrane and (2) its selectivity.

The permeating flux through a membrane can be expressed as *flux or permeation rate* (J), and it is defined as the volume flowing through the membrane per unit area and time. Since the transport through the membrane takes place as a result of a driving force acting on the components in the feed, the permeation rate through the membrane is proportional to this driving force as in Eq. 12.1:

$$J = -A\frac{dY}{dx} \tag{12.1}$$

In Eq. 12.1, A represents the phenomenological coefficient and dY/dx the driving force, expressed as the gradient of Y as a variable indicating temperature, concentration, and pressure along with a coordinate x perpendicular to the transport barrier.

Equation 12.1 is not limited to mass transport description, but is also useful for describing heat, volume, momentum, and electrical flux, as expression of the

Table 12.1 Phenomenological equations relating flux and force

Driving force	Relationship	Theory
Mass flux	$J_m = -D \quad dc/dx$	Fick's law
Heat flux	$J_h = -\lambda \quad dT/dx$	Fourier's law
Volume flux	$J_v = -L_p \quad dp/dx$	Darcy's law
Momentum flux	$J_n = -\upsilon \quad dv/dx$	Newton's law
Electrical flux	$J_i = -1/R \quad dE/dx$	Ohm's law

phenomenological equations (relating flux and force) shown in Table 12.1. The membrane *selectivity* can be expressed as the retention (*R*) or the separation factor (α). For dilute aqueous mixtures, consisting of a solvent and a solute, the selectivity can be expressed in terms of the retention toward the solute. The latter is partially or completely retained, whereas the solvent molecules pass freely through the membrane. The retention can be defined as reported below:

$$R = \frac{c_f - c_p}{c_f} = 1 - \frac{c_p}{c_f} \tag{12.2}$$

in which c_f is the solute concentration in the feed and c_p is the solute concentration in the permeate. R can vary from 0 (solute and solvent pass through the membrane freely) to 100 % (complete retention of the solute).

The separation factor (α), usually used for gas mixtures and mixtures of organic liquids, can be defined for a mixture consisting of two components as shown in Eq. 12.3:

$$\alpha_{A/B} = \frac{y_A / y_B}{x_A / x_B} \tag{12.3}$$

where y_A and y_B are the concentrations of the mixture components *A* and *B* in the permeate and x_A and x_B their concentrations in the feed. If the permeation rate of component *A* through the membrane is larger than that of component *B*, the separation factor is represented by the ratio $\alpha_{A/B}$; if component *B* permeates preferentially, then the separation factor is represented by $\alpha_{B/A}$. If $\alpha_{A/B} = \alpha_{B/A} = 1$, no separation occurs. However, another kind of membrane classification distinguishes these on the basis of their nature, geometry, and separation regime [31]. Concerning the classification by nature, membranes can be subdivided into biological and synthetic. Then, they are different for functionality and structure. Biological membranes are easy to manufacture, but they present different drawbacks such as limited operating temperatures (below 100 °C), limited pH range, not easy to clean up, and susceptibility to microbial attack due to their natural origin.

Regarding the synthetic membranes, a further classification distinguishes them into organic (polymeric) and inorganic (ceramic, metal), and the main difference is related to the operating temperature limit. Indeed, polymeric membranes normally operate at temperature below 200 °C, while inorganic ones operate at temperatures above 250 °C. Furthermore, the latter membranes show greater chemical, mechanical, and thermal stability than that of organic materials, even though they can become brittle in such conditions and they are more expensive.

Regarding the morphology and/or membrane structure, the inorganic membranes can be further classified into porous and dense. In fact, IUPAC defines porous membranes according to their pore diameter in microporous ($d_p < 2$ nm), mesoporous (2 nm $< d_p <$ 50 nm), and macroporous ($d_p >$ 50 nm) [31]. Dense metallic membranes can be classified into supported and unsupported ones [14].

12.2.2 Palladium-Based Membranes

Several types of inorganic materials can be used for preparing membranes such as ceramic, carbon, silica, zeolite, oxides (alumina, titania, zirconia), as well as palladium, silver, etc. and their alloys. Generally, inorganic membranes are used in a temperature range between 300 and 800 °C and, sometimes, up to 1,000 °C (ceramic membranes). Robustness is favorable among the characteristics of inorganic membranes, meaning that they can be operated in harsh environments, showing quite high resistance to large pressure drops. Furthermore, they are inert toward microbiological degradation, they are easy to clean after fouling, and they can be easily reactivated. On the other hand, inorganic membranes possess some drawbacks. Regarding the metallic membranes, they have high cost and can suffer from embrittlement phenomenon, particularly, in the case of palladium membranes. Furthermore, they possess low permeability when showing high hydrogen permselectivity at moderate temperatures.

Concerning the ceramic membranes, the difficulty of achieving high hydrogen permselectivity is the most significative limitation, while for both inorganic typologies, the low membrane surface per module volume and not easy membrane-to-module sealing at high temperatures represent other significant limitations. However, it is worth of noting that much literature exists on palladium-based membranes, particularly in the field of high-grade hydrogen generation due to their high hydrogen solubility and permselectivity [18].

Palladium absorbs about 600 times its volume of hydrogen at room temperature, even though some palladium alloys show higher hydrogen permeability than pure palladium. Therefore, in the specialized literature, several groups have studied the influence of metals alloyed with palladium, but particularly significant is the study of Hwang and Kammermeyer [32], in which the hydrogen permeating flux through a dense palladium membrane is depicted over the content of the alloyed material (Fig. 12.2).

Furthermore, palladium alloys present some benefits regarding the issue of hydrogen embrittlement. In particular, when pure palladium membranes are exposed to hydrogen flux, the amount of hydrogen absorbed into the membrane lattice induces a phase transition from α to β palladium hydride [15]. This

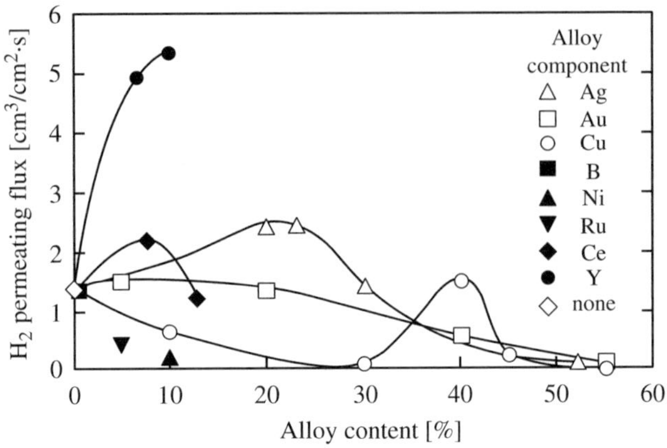

Fig. 12.2 Hydrogen flux through palladium alloy membranes against metal content (Adapted with permission from [32], Copyright © 1975 Wiley Interscience)

phenomenon is called "hydrogen embrittlement," and it is due to the dissolved hydrogen that makes several elongations of the metallic film (involving in the α-β hydride transformations), causing fractures after repeated thermal cycles. As an example, by considering the palladium-silver alloy, the membrane lattice is expanded by the silver atoms, making the alloy less influenced by the hydrogen permeation and, thus, less brittle than the pure palladium [33]. In dense self-supported Pd-based membranes, the molecular transport takes place via solution-diffusion mechanism, which involves the following activated steps [15]:

1. Dissociation of molecular hydrogen at the gas/metal interface
2. Adsorption of the atomic hydrogen on the membrane surface
3. Dissolution of atomic hydrogen into the palladium matrix
4. Diffusion of atomic hydrogen toward the opposite side
5. Recombination from atomic to molecular hydrogen at the gas/metal interface
6. Molecular hydrogen desorption

Each one of the aforementioned steps can be responsible for hydrogen permeation through the dense palladium membrane, all of which also depend on variables such as temperature, pressure, gas mixture composition, and thickness of the membrane.

The hydrogen transport through a generic membrane can be expressed as the hydrogen permeating flux, as follows:

$$J_{H2} = Pe_{H2}\left(p^{n}{}_{H2,retentate} - p^{n}{}_{H2,permeate}\right)/\delta \qquad (12.4)$$

where J_{H2} represents the hydrogen flux permeating through the membrane, Pe_{H2} the hydrogen permeability, δ the membrane thickness, $p_{H2\text{-}retentate}$ and $p_{H2\text{-}permeate}$ the hydrogen partial pressures in the retentate (reaction side) and permeate (side in which hydrogen permeating through the membrane is collected) zones, respectively, and n (variable in the range 0.5–1) the dependence factor of the hydrogen flux on the hydrogen partial pressure. For membranes having a thickness higher than 5 μm, Eq. 12.4 becomes the *Sieverts-Fick* law (12.5):

$$J_{H2,Sieverts-Fick} = Pe_{H2} \cdot \left(p^{0.5}{}_{H2,retentate} - p^{0.5}{}_{H2,permeate}\right)/\delta \tag{12.5}$$

At high pressures, the hydrogen-hydrogen interactions in the palladium bulk are not negligible, so that n becomes equal to 1:

$$J_{H2} = Pe_{H2} \cdot \left(p_{H2,retentate} - p_{H2,permeate}\right)/\delta \tag{12.6}$$

Furthermore, if the hydrogen permeability is expressed as an *Arrhenius*-like equation, the *Sieverts-Fick* law becomes *Richardson's* Eq. 12.7:

$$J_{H2} = Pe^{0}{}_{H2}[exp(-E_a/RT)] \cdot \left(p^{0.5}{}_{H2,retentate} - p^{0.5}{}_{H2,permeate}\right)/\delta \tag{12.7}$$

In the presence of contaminants such as hydrogen sulfide, SO_2, Hg vapor, thiophene, arsenic, unsaturated hydrocarbons, or chlorine carbon, dense Pd-based membranes are irreversibly poisoned [34]. The presence of CO strongly affects the hydrogen permeation performances of the membrane. Particularly, at lower temperatures (below 150 °C) or at high CO feed concentrations, this effect is due to the adsorbed CO that displaces the adsorbed hydrogen, covering the hydrogen adsorption sites [35]. A negative effect on the hydrogen permeation through dense Pd membranes can be also due to steam, which affects the water vapor dissociation/recombinative desorption, contaminating the palladium surface with adsorbed oxygen [36].

During the last years, special attention has been paid to composite Pd-based membranes with the intent of reducing the amount of palladium and, consequently, lowering the cost. In detail, composite membranes can be constituted by a thin dense layer of palladium or its alloy deposited onto a porous support among porous Vycor glass (silica gel), SiO_2, Al_2O_3, and B_2O_3 or porous stainless steel (PSS) and pencil [20, 37]. In particular, a thermal expansion coefficient of the aforementioned supports close to palladium makes the composite membrane durable and resistant to cracks due to the thermal cycles, simplifying the gas sealing. Sometimes, at relatively high temperatures, PSS supports alloy the palladium, lowering the hydrogen permeability. Then, as a consequence, a PSS-supported Pd-based membrane can offer an optimal compromise among moderate cost, high H_2 permeability and permselectivity, and good mechanical resistance. Nevertheless, thin selective Pd layers can be fulfilled directly onto the specific support, and depending on the membrane preparation process, a threshold about the minimum thickness obtainable by the deposition method can be encountered. This limits the performance of a composite membrane and it is more emphasized at higher support pore sizes. However, the reduction of the palladium layer involves two opposing effects. On the one hand, thin Pd layers can induce an atomic H diffusion time reduction with a consequent enhancement of the membrane permeance, meanwhile decreasing the cost due to a lower Pd content. On the other hand, the hydrogen diffusion resistance can be increased owing to the need of reducing the pore size of the support. Another important issue associated with the deposition of palladium on a support (porous or dense) is the entire membrane stability. For example, owing to different thermal expansion coefficients within the Pd layer and the selective layer volume variation due to the hydrogen diffusion in the palladium lattice, the membrane support could show an instability at the interface Pd layer/support in terms of adherence loss, flaking off, and cracking.

Furthermore, the interaction between the selective Pd layer with the support and the gaseous environment can introduce other forms of instability due to the Pd layer microstructure. Therefore, thin Pd layers, small grains, and high density of grain borders are related to grain size growth, impurity dissolution, grain border diffusion, and alloy segregation. All of the aforementioned issues on instability can result in a reduction of the membrane performance, particularly evident at higher temperatures.

The general difference between dense self-supported and composite Pd-based membranes is related to the hydrogen permselectivity with respect to all other gases. Indeed, for dense Pd membranes, it is full, while for the composite Pd membranes, it can vary depending on many parameters such as Pd layer, preparation technique, support, and support preparation [38]. The composite Pd-based membranes are commonly not fully hydrogen permselective and the hydrogen permeating flux can be determined by using Eq. 12.4, as given earlier in this chapter.

In this case, the n value, variable from 0.5 (used for indicating *Sieverts-Fick* law and applied for dense Pd membranes) to 1 (used when, at high pressures, the hydrogen-hydrogen interactions in the palladium bulk are not negligible), is calculated for a supported Pd-based membrane. A linear regression equation can be used with experimental points obtained by considering the experimental hydrogen permeating flux against transmembrane pressure at different "n," with the associated response factor R^2 (Fig. 12.3). The most coherent "n" factor is, then, the one associated with the linear regression and the maximum R^2 factor.

In the field of hydrogen separation and purification from CO_x for fuel cell supply, both dense self-supported and composite Pd-based membranes have the peculiarity of being hydrogen permselective with respect to all the other gases. Thus, both kinds of membranes, when housed in MRs, make it possible to overcome the thermodynamic restrictions of equilibrium-limited reactions due to the removal of hydrogen from the reaction side for the selective permeation through the membrane ("shift effect"). Indeed, due to *Le Chatelier's* principle, the reaction can be shifted toward the reaction products, with a consequent enhancement of the conversion and with the further benefit of collecting high-grade hydrogen on the permeate side of the MR. Therefore, dense self-supported Pd-based MRs seem to be more adequate to generate PEMFC-grade hydrogen due to the full hydrogen permselectivity of the membrane, while – depending on the finite value of hydrogen permselectivity of the composite membrane – the purified hydrogen can be supplied to other kinds of fuel cells or to high-temperature PEMFCs, whose CO content can be up to 20,000 ppm [40].

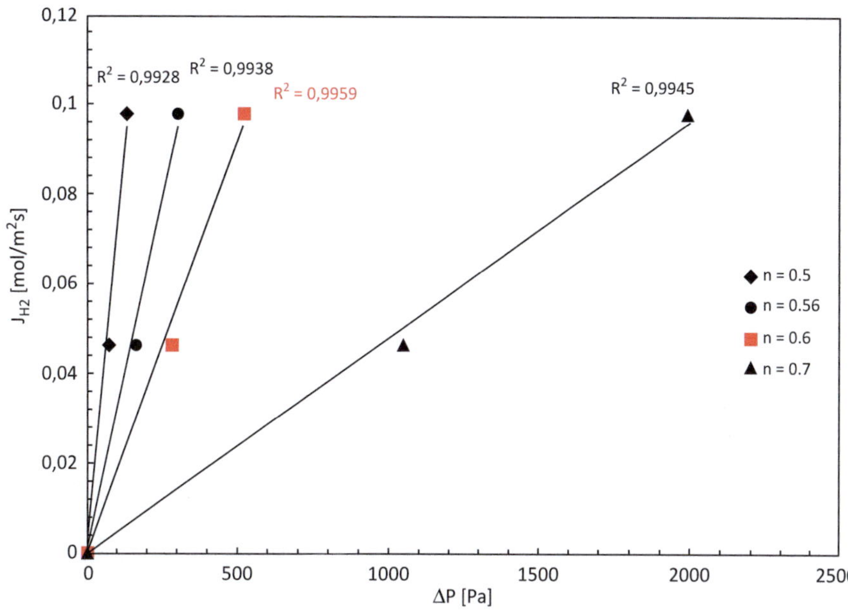

Fig. 12.3 Calculation of "n" factor for a composite Pd-based membrane based on the experimental hydrogen permeating flux versus transmembrane pressure (Reprinted with permission from [39], Copyright © 2015 Elsevier)

12.3 High-Grade Hydrogen Generation for Fuel Cells from Reforming of Renewables in MRs

In the last decade, MRs have been particularly studied in the field of reforming reactions of renewable sources to generate high-grade hydrogen. Among the renewables, ethanol seems to be the most promising because it can be produced renewably from biomass, even though other renewable bio-feedstocks such as methanol, glycerol, acetic acid, and diethylether have been also studied via MR technology. In the following sections, representative results regarding the hydrogen generation in MRs from renewables are reported.

12.3.1 Ethanol Steam Reforming in MRs

Ethanol steam reforming (ESR) reaction has been studied in the specialized literature with conventional fixed-bed reactors for producing hydrogen. It is worth of noting that, besides hydrogen, other undesirable by-products are produced during the reaction owing to a complex reaction system. In particular, ESR reaction has been studied in CRs by several groups, who have paid attention to the role of the

Table 12.2 Literature studies on ethanol steam reforming reaction with membrane reactors

Membrane material	Pd/Pd alloy layer [μm]	T [°C]	p [bar]	Conversion [%]	H_2 recovery [%]	H_2 yield [%]	H_2 purity [%]	Reference
Dense Pd-Ag	50	400	1.5	95	30	~20[a]	≈100	Iulianelli and Basile [42]
Dense Pd-Ag	50	400	3.0	~100	90	53[a]	≈100	Iulianelli et al. [43]
Dense Pd-Ag	50	400	1.0	50	<10	16	≈100	Iulianelli et al. [44]
Dense Pd-Ag	50	400	1.3	~100	15	~56	≈100	Basile et al. [45]
Dense Pd-Ag	50	500	3.5	99	~27	–	≈100	Basile et al. [46]
Dense Pd-Ag	150	450	5.0	–	93	70	≈100	Borgognoni et al. [47]
Dense Pd-Ru	50	450	1.0	–	–	~50	≈100	Mironova et al. [69]
Composite Pd-Ag	30	700	~7.0	–	–	75	–	Papadias et al. [48]
Composite Ni-Pd-Ag	<8	450	3.0	81	–	70	>90	Lin et al. [49]
Composite Pd on Al_2O_3	8	400	3.0	98	67	–	≈97	Iulianelli et al. [50]
Composite Pd on PSS	25	400	8.0	100	55	–	≈95	Basile et al. [51]
Composite Pd on PSS	25	400	12.0	87	12	17	≈95	Seelam et al. [52]
Composite Pd-Ag on PSS	30	650	4.0	100	–	~35	–	Hedayati et al. [70]
Composite Pd on Al_2O_3	4–5	480	10	100	–	80	≈100	Murmura et al. [71]

[a]Calculated

catalysts on the reaction performance in terms of conversion, hydrogen yield, and selectivity, strictly depending on the catalyst used [41, 42].

Some representative literature studies on ESR reaction performed in different MRs are summarized in Table 12.2, which provides the most important and up-to-date results in terms of high-grade hydrogen generation.

As addressed in a number of studies in the literature, Pd-based MRs seem to be dominant over other kinds of MRs, and Table 12.2 is split in two parts: (i) dense self-supported and (ii) composite supported Pd-based membranes in MRs. In the last several years, the need for decreasing the Pd content in a Pd-based membranes has strongly emerged for reducing the membrane cost, mainly affected by the Pd content, and for potential scaling up of this technology in industry. As shown in Table 12.2, the purity of hydrogen produced during the reaction is ~100 % in the case of dense Pd membranes and between 90 % and 100 % in the case of composite Pd membranes in MRs.

Our group at CNR-ITM has been involved in the study of the ESR reaction, paying special attention to both dense and composite Pd-based MRs [42–46, 50–52]. By including the various MRs studied previously, we obtained a variety of results, depending on the operating conditions adopted as well as the catalysts used.

However, most of the research of Basile's group was realized using dense Pd-Ag membranes that have a thickness of 50 μm which are produced at ENEA Frascati and that have full hydrogen permselectivity. More recently, Borgognoni et al. [47] used a 150 μm thick dense Pd-Ag membrane in an MR at 450 °C and 5.0 bar to obtain a 70 % hydrogen yield and more than 90 % hydrogen recovery with a hydrogen purity of 100 %. On the contrary, Mironova et al. [69] used a Pd-Ru dense membrane 50 μm thick in an MR, reaching around 50 % hydrogen yield with the hydrogen recovered being 100 % pure at 450 °C and 1.0 bar.

In the last few years, much attention has been devoted to developing composite Pd-based MRs. For example, Papadias et al. [48] performed the ESR reaction in an MR allocating a composite Pd-Ag-based membrane having a Pd-Ag layer of around 30 μm, reaching 75 % hydrogen yield at 700 °C and around 7.0 bar. Lin et al. [49] developed an MR housing a supported Ni-Pd-Ag membrane with an active layer <8 μm, allowing a hydrogen recovery of about 70 % with a purity >90 %, while reaching an ethanol conversion of 80 % at 450 °C and 3.0 bar. More recently, Hedayati et al. [70] prepared a composite Pd membrane with a dense layer of around 30 μm deposited via electroless plating technique onto a porous stainless steel support, globally obtaining at 650 °C and 4.0 bar complete ethanol conversion and a hydrogen yield around 35 %. Murmura et al. [71] developed an MR housing a really thin Pd layer supported on alumina, produced by ECN. In this case, 100 % ethanol conversion, 80 % hydrogen yield, and a 100 % hydrogen purity were obtained with this technological solution.

12.3.2 Methanol Steam Reforming in MRs

Not only ethanol can be produced renewably but also methanol, which can react with steam in a reforming process to generate hydrogen for fuel cell applications [12]. Methanol steam reforming (MSR) reaction is performed in conventional reformers at relatively low temperatures in the range between 240 and 260 °C.

CO formation as a by-product of hydrogen represents the main drawback of this reaction because it acts as a poison. Therefore, MR technology has been studied combined to MSR reaction for producing PEMFC-grade hydrogen [40]. Table 12.3 summarizes representative results in the field, particularly reporting the literature data about both dense and composite Pd-based MRs.

Table 12.3 Recent literature studies on methanol steam reforming reactions with membrane reactors

Membrane material	Pd/Pd alloy layer [μm]	T [°C]	p [bar]	Conversion [%]	H_2 recovery [%]	H_2 yield [%]	H_2 purity [%]	Reference
Dense Pd-Ag	50	300	3	–	80	70	≈100	Iulianelli et al. [54]
Dense Pd-Ru-In	200	200	7	≈90	≈24	–	≈100	Itoh et al. [61]
Dense Pd-Cu	25	300	10	>90	≈38	–	≈100	Wieland et al. [16]
Composite Pd-Ag/Al_2O_3	7	330	3.0	85	40	80	≈100	Liguori et al. [53]
Composite Pd-Ag/α-Al_2O_3	3.9	250	10	100	95	–	≈100	Israni and Harold [62]
Composite Pd-Ag/PSS	20/25	240	10	<40	18	–	–	Rei et al. [63]

As in the case of the ESR reaction, our group at CNR-ITM has reported many results on Pd-based MR technology [40, 53–60], paying much attention to different kinds of MRs. In these studies, the MSR performance in MRs was compared with those in CRs, by analyzing the influence of the parameters of reaction temperature, pressure, residence time, feed molar ratio, sweep gas flow rate, and oxygen addition. In all investigated cases, it was found that Pd-based MRs had clear superiority in terms of methanol conversion, selectivity, and productivity over the CRs, operating at the same experimental conditions. In Table 12.3, the most representative results regarding a dense self-supported Pd-Ag MR are reported in which MSR reaction was carried out reaching 80 % of hydrogen recovered with a purity of about 100 % and a hydrogen yield of 70 % at 300 °C and 3.0 bar [54]. Furthermore, a composite Pd-based MR was also studied to achieve 85 % of methanol conversion recovering 40 % of hydrogen with around 100 % of purity and a CO content lower than 10 ppm, as required by low-temperature PEMFCs.

However, regarding dense membranes, other researchers developed MRs (housing dense and unsupported Pd membranes at different thickness) to carry out MSR reaction by optimizing the experimental conditions, achieving complete methanol conversion and high hydrogen recovery with around 100 % of hydrogen purity [16, 61].

Table 12.3 shows that, by reducing the palladium thickness in the composite Pd-based membranes, the high H_2/other gas permselectivity is maintained sometimes at the same level of the dense ones showing high performance during MSR reaction in terms of methanol conversion and hydrogen recovery. For example, Israni and Harold [62] obtained 100 % of methanol conversion with a correspondent hydrogen recovery of 95 % at a purity of around 100 %, at 250 °C and 10 bar.

On the contrary, in the work of Rei et al. [63], low performance was achieved (less than 40 % of methanol conversion and 20 % of hydrogen recovery at 200 °C and 10 bar), confirming that the solution of thin Pd layers deposited onto porous supports is not really effective.

However, as a general consideration, the generation of hydrogen from renewables in MRs seems to be very attractive, and this chapter has only described ethanol and methanol steam reforming, even though other kinds of reactions could be applied to produce PEMFC-grade hydrogen, for example, from glycerol, acetic acid, etc., as reviewed in the existing literature [63–68].

12.4 Conclusions and Future Outlook

This chapter has described the generation of proton exchange membrane fuel cell-grade hydrogen from reforming processes of renewables such as ethanol and methanol by utilizing membrane reactor technology. This option instead of the conventional system could make it possible to enhance hydrogen-generating devices, allowing the integration of membrane reactors with the proton exchange membrane fuel cells. In fact, an important issue, not commonly addressed in the specialized literature, takes into account that natural gas, or other derived of fossil fuels, is essentially used for stationary applications, generating high-impact products for the ambient. Otherwise, it would be expected that renewable sources such as methanol and ethanol could drive the production of hydrogen by taking care the ambient and harmful emissions. Nevertheless, the development of membrane reactor technology demonstrates that efforts are still needed to solve some deficiencies related to its utilization on a larger industrial scale.

Acronyms

CR	Conventional reactor
ESR	Ethanol steam reforming
ID	Internal diameter
MR	Membrane reactor
MSR	Methanol steam reforming
PEMFC	Proton exchange membrane fuel cell
PPS	Porous stainless steel
PrOX	Preferential oxidation
WGS	Water gas shift

Symbols

A	Phenomenological coefficient
c_f	Solute concentration in the feed
c_p	Solute concentration in the permeate
d_P	Pore diameter
E_a	Apparent activation energy
H	Permeation rate
J_{H2}	Hydrogen flux permeating through the membrane
k	Rate coefficient of reactions
K	Adsorption constant
K_{eq}	Equilibrium constant of reactions
n	Dependence factor of the hydrogen flux to the hydrogen partial pressure
p	Partial pressure
Pe_0	Pre-exponential factor
Pe_{H2}	Hydrogen permeability
$p_{H2\text{-}permeate}$	Hydrogen partial pressures in the permeate side
$p_{H2\text{-}retentate}$	Hydrogen partial pressures in the retentate side
r	Reaction rate
R	Universal gas constant
T	Absolute temperature
X	Coordinate
x_A,x_B	Concentrations of the mixture components A and B in the feed
Y	Variable indicating temperature, concentration, pressure, etc.

y_A,y_B	Concentrations of the mixture components A and B in the permeate
α	Separation factor
δ	Membrane thickness

References

1. Chalk G, Miller JF, Wagner FW. Challenges for fuel cells in transport applications. J Power Sources. 2000;86:40–51.
2. Chu D, Jiang R, Gardner K, Jacobs R, Schmidt J, Quakenbush T, Stephens J. Polymer electrolyte membrane fuel cells for communication applications. J Power Sources. 2001;96:174–8.
3. Costamagna P, Srinivasan S. Quantum jumps in the PEMFC science and technology from the 1960s to the year 2000 part II. Engineering, technology, development and application aspects. J Power Sources. 2001;102:253–69.
4. Gamburzev S, Appleby AJ. Recent progress in performance improvement of the proton exchange membrane fuel cell (PEMFC). J Power Sources. 2002;107:5–12.
5. Mehta V, Cooper JS. Review and analysis of PEM fuel cell design and manufacturing. J Power Sources. 2003;114:32–53.
6. Barbir F. PEM fuel cells-theory and practice. Burlington: Elsevier Academic Press; 2005. p. 8.
7. EG&G Services P., Inc. Fuel cell handbook, contract no. DEAM26-99FT40575. Morgantown: US Department of Energy; 2000. p. 312.
8. Wee J. Applications of proton exchange membrane fuel cell systems. Renew Sustain Energy Rev. 2007;11:1720–38.
9. Barbir F, Gomez T. Efficiency and economic of proton exchange membrane (PEMFC) fuel cell. Int J Hydrogen Energy. 1996;21(10):891–901.
10. Schultz T, Zhou S, Sundmacher K. Current status of and recent developments in the direct methanol fuel cell. Chem Eng Technol. 2001;24:1223–33.
11. Pettersson LJ, Westerholm R. State of the art of multi-fuel reformers for fuel cell vehicles: problem identification and research needs. Int J Hydrogen Energy. 2001;26:243–64.
12. Chein RY, Chen YC, Lin YS, Chung JN. Hydrogen production using integrated methanol-steam reforming reactor with various reformer designs for PEM fuel cells. Int J Hydrogen Energy. 2012;36:466–76.
13. Lu GQ, Diniz de Costa JC, Duke M, Giessler S, Socolow R, Williams RH, Kreutz T. Inorganic membranes for hydrogen production and purification: a critical review and perspective. J Colloid Interface Sci. 2007;314:589–603.
14. Iulianelli A, Liguori S, Longo T, Basile A. Inorganic membrane and membrane reactor technologies for hydrogen production. In: Honery DR, Moriarty P, editors. Hydrogen production: prospects and processes, Energy Science, Engineering and Technology. Victoria: Nova Science Publishers; 2012. p. 377–98. Ch. 12. ISBN 978-1-62100-246-8.
15. Basile A, Iulianelli A, Longo T, Liguori S, De Falco M. Pd-based selective membrane state-of-the-art. In: Marrelli L, De Falco M, Iaquaniello G, editors. Membrane reactors for hydrogen production processes. London: Springer; 2011. p. 21–55. doi:10.1007/978-0-85729-151-6. Ch. 2. ISBN 978-0-85729-150-9.
16. Wieland S, Melin T, Lamm A. Membrane reactors for hydrogen production. Chem Eng Sci. 2002;57:1571–6.
17. Damle AS. Hydrogen production by reforming of liquid hydrocarbons in a membrane reactor for portable power generation – experimental studies. J Power Sources. 2009;186:167–77.
18. Basile A. Hydrogen production using Pd-based membrane reactors for fuel cells. Top Catal. 2008;51:107–22.

19. Lin YM, Rei MH. Process development for generating high purity hydrogen by using supported palladium membrane reactor as steam reformer. Int J Hydrogen Energy. 2000;25:211–9.
20. Mallada R, Menéndez M. Inorganic membranes: synthesis, characterization and applications, technology & engineering. Amsterdam: Elsevier Ltd; 2008. p. 460. ISBN 978-0-444-53070-7.
21. Tong J, Shirai R, Kashima Y, Matsumura Y. Preparation of a pinhole-free Pd-Ag membrane on a porous metal support for pure hydrogen separation. J Membr Sci. 2005;260:84–9.
22. Adhikari S, Fernand S. Hydrogen membrane separation techniques. Ind Eng Chem Res. 2006;45:875–81.
23. Arstad B, Venvik H, Klette H, Walmsley JC, Tucho WM, Holmestad R, Holmen A, Bredesen R. Studies of self-supported 1.6 μm Pd/23 wt.% Ag membranes during and after hydrogen production in a catalytic membrane reactor. Catal Today. 2006;118:63–72.
24. Westermann T, Melin T. Review – flow-through catalytic membrane reactors—principles and applications. Chem Eng Process. 2009;48:17–28.
25. Piemonte V, Di Paola D, De Falco M, Iulianelli A, Basile A. Hydrogen production using inorganic membrane reactors. In: Basile A, Iulianelli A, editors. Advances in hydrogen production, storage and distribution. Cambridge, UK: Woodhead Publishing; 2014. p. 283–316. Ch. 11, ISBN: 978-0-85709-768-2 (print), ISBN: 978-0-85709-773-6 (online).
26. Fong YY, Abdullah AZ, Ahmad AL, Bhatia S. Review – development of functionalized zeolite membrane and its potential role as reactor combined separator for para-xylene production from xylene isomers. Chem Eng J. 2008;139:172–93.
27. Andrić P, Meyer AS, Jensen PA, Johansen KD. Reactor design for minimizing product inhibition during enzymatic lignocellulose hydrolysis II. Quantification of inhibition and suitability of membrane reactors. Biotechnol Adv. 2010;28:407–25.
28. Chatenet M, Dubau L, Job N, Maillard F. The (electro)catalyst membrane interface in the proton exchange membrane fuel cell: similarities and differences with non-electrochemical catalytic membrane reactors. Catal Today. 2010;156:76–86.
29. Mozia S. Photocatalytic membrane reactors (PMRs) in water and wastewater treatment. A review. Sep Purif Technol. 2010;73:71–91.
30. Reij MW, Keurentjes JTF, Hartmans S. Review – membrane bioreactors for waste gas treatment. J Biotechnol. 1998;59:155–67.
31. Koros WJ, Ma YH, Shimidzu T. Terminology for membranes and membrane processes. J Membr Sci. 1996;120:149–59.
32. Hwang ST, Kammermeyer K. Techniques in chemistry: membranes in separation. New York: Wiley Interscience; 1975.
33. Howard BH, Killmeyer RP, Rothenberger KS, Cugini AV, Morreale BD, Enick RM, Bustamante F. Hydrogen permeance of palladium-copper alloy membranes over a wide range of temperatures and pressures. J Membr Sci. 2004;241:207–18.
34. Edlund DJ, Pledger WA. Thermolysis of hydrogen sulfide in a metal-membrane reactor. J Membr Sci. 1993;77:255–64.
35. Li A, Liang W, Hughes R. The effect of carbon monoxide and steam on the hydrogen permeability of a Pd/stainless steel membrane. J Membr Sci. 2000;165:135–41.
36. McCool BA, Lin YS. Nanostructured thin palladium-silver membranes: effects of grain size on gas permeation properties. J Mater Sci. 2001;36:3221–7.
37. Uemiya S. State-of-art-the-art of supported metal membranes for gas separation. Sep Purif Methods. 1999;28:51–85.
38. Liguori S, Iulianelli A, Dalena F, Pinacci P, Drago F, Broglia M, Huang Y, Basile A. Performance and long-term stability of Pd/PSS and Pd/Al_2O_3 for hydrogen separation. Membranes. 2014;4:143–62.
39. Iulianelli A, Liguori S, Huang Y, Basile A. Model biogas steam reforming in a thin Pd-supported membrane reactor to generate clean hydrogen for fuel cells. J Power Sources. 2015;273:25–32.

40. Iulianelli A, Ribeirinha P, Mendes A, Basile A. Methanol steam reforming for hydrogen generation via membrane reactors: a review. Renew Sustain Energy Rev. 2014;29:355–68.
41. Mas V, Kipreos R, Amadeo N, Laborde M. Thermodynamic analysis of ethanol/water system with the stoichiometric method. Int J Hydrogen Energy. 2006;31:21–8.
42. Iulianelli A, Basile A. Hydrogen production from ethanol via inorganic membrane reactors technology: a review. Catal Sci Technol. 2011;1:366–79.
43. Iulianelli A, Basile A. An experimental study on bio-ethanol steam reforming in a catalytic membrane reactor. Part I: temperature and sweep-gas flow configuration. Int J Hydrogen Energy. 2010;35:3170–7.
44. Iulianelli A, Liguori S, Longo T, Tosti S, Pinacci P, Basile A. An experimental study on bio-ethanol steam reforming in a catalytic membrane reactor. Part II: reaction pressure, sweep factor and WHSV effects. Int J Hydrogen Energy. 2010;35:3159–64.
45. Basile A, Gallucci F, Iulianelli A, Tosti S. CO-free hydrogen production by ethanol steam reforming in a Pd-Ag membrane reactor. Fuel Cells. 2008;1:62–8.
46. Basile A, Gallucci F, Iulianelli A, De Falco M, Liguori S. Hydrogen production by ethanol steam reforming: experimental study of a Pd-Ag membrane reactor and traditional reactor behavior. Int J Chem React Eng. 2008;6:A30.
47. Borgognoni F, Tosti S, Vadrucci M, Santucci A. Combined methane and ethanol reforming for pure hydrogen production through Pd-based membranes. Int J Hydrogen Energy. 2013;38:1430–8.
48. Papadias DD, Lee SHD, Ferrandon M, Ahmed S. An analytical and experimental investigation of high-pressure catalytic steam reforming of ethanol in a hydrogen selective membrane reactor. Int J Hydrogen Energy. 2010;35:2004–17.
49. Lin W-H, Liu Y-C, Chang H-F. Autothermal reforming of ethanol in a Pd-Ag/Ni composite membrane reactor. Int J Hydrogen Energy. 2010;35:12961–9.
50. Iulianelli A, Liguori S, Vita A, Italiano C, Fabiano C, Huang Y, Basile A. The oncoming energy vector: hydrogen produced in Pd-composite membrane reactor via bioethanol reforming over Ni/CeO_2 catalyst. Catalysis Today. (in press). doi:10.1016/j.cattod.2015.04.046.
51. Basile A, Pinacci P, Iulianelli A, Broglia M, Drago F, Liguori S, Longo T, Calabrò V. Ethanol steam reforming reaction in a porous stainless steel supported palladium membrane reactor. Int J Hydrogen Energy. 2011;36:2029–37.
52. Seelam PK, Liguori S, Iulianelli A, Pinacci P, Drago F, Calabrò V, Huuhtanen M, Keiski R, Piemonte V, Tosti S, De Falco M, Basile. A hydrogen production from bio-ethanol steam reforming reaction in a Pd/PSS membrane reactor. Catal Today. 2012;193:42–8.
53. Mateos-Pedrero C, Silva H, Tanaka DA, Liguori S, Iulianelli A, Basile A, Mendes A. CuO/ZnO catalysts for methanol steam reforming: the role of the support polarity ratio and surface area. Appl Catal B Environ. 2015;174:67–76.
54. Liguori S, Iulianelli A, Dalena F, Piemonte V, Huang Y, Basile A. Methanol steam reforming in an Al_2O_3 supported thin Pd-layer membrane reactor over Cu/ZnO/Al_2O_3 catalyst. Int J Hydrogen Energy. 2014;39:18702–10.
55. Iulianelli A, Longo T, Basile A. Methanol steam reforming in a dense Pd-Ag membrane reactor: the pressure and WHSV effects on CO-free hydrogen production. J Membr Sci. 2008;323:235–40.
56. Iulianelli A, Longo T, Basile A. Methanol steam reforming reaction in a Pd-Ag membrane reactor for CO-free hydrogen production. Int J Hydrogen Energy. 2008;33:5583–8.
57. Ghasemzadeh K, Liguori S, Morrone P, Iulianelli A, Piemonte V, Babaluo AA, Basile A. H2 production by low pressure methanol steam reforming in a dense Pd-Ag membrane reactor in co-current flow configuration: experimental and modeling analysis. Int J Hydrogen Energy. 2013;36:16685–97.
58. Basile A, Parmaliana A, Tosti S, Iulianelli A, Gallucci F, Espro C, Spooren. J Hydrogen production by methanol steam reforming carried out in membrane reactor on Cu/Zn/Mg-based catalyst. Catal Today. 2008;137:17–22.

59. Basile A, Tereschenko GF, Orekhova NV, Ermilova MM, Gallucci F, Iulianelli A. An experimental investigation on methanol steam reforming with oxygen addition in a flat Pd/Ag membrane reactor. Int J Hydrogen Energy. 2006;31:1615–22.
60. Basile A, Tosti S, Capannelli G, Vitulli G, Iulianelli A, Gallucci F, Drioli E. Co-current and counter-current modes for methanol steam reforming membrane reactor: experimental study. Catal Today. 2006;118:237–45.
61. Itoh N, Kaneko Y, Igarashi A. Efficient hydrogen production via methanol steam reforming by preventing back-permeation of hydrogen in a palladium membrane reactor. Ind Eng Chem Res. 2002;41:4702–6.
62. Israni S, Harold MP. Methanol steam reforming in single-fiber bed Pd-Ag membrane reactor: experiments and modeling. J Membr Sci. 2011;369:375–87.
63. Rei MH, Yeh GT, Tsai YH, Kao YL, Shiau LD. Catalysis-spillover-membrane. III: the effect of hydrogen spillover on the palladium membrane reactor in the steam reforming reactions. J Membr Sci. 2011;369:299–307.
64. Iulianelli A, Seelam PK, Liguori S, Longo T, Keiski R, Calabrò V, Basile A. Hydrogen production for PEM fuel cell by gas phase reforming of glycerol as byproduct of bio-diesel. The use of a Pd-Ag membrane reactor at middle reaction temperature. Int J Hydrogen Energy. 2011;36:3827–34.
65. Iulianelli A, Longo T, Liguori S, Basile A. Production of hydrogen via glycerol steam reforming in a Pd-Ag membrane reactor over Co-Al2O3 catalyst. Asia Pac J Chem Eng. 2010;5:138–45.
66. Basile A, Gallucci F, Iulianelli A, Borgognoni F, Tosti S. Acetic acid steam reforming in a Pd-Ag membrane reactor: the effect of the catalytic bed pattern. J Membr Sci. 2008;311:46–52.
67. Iulianelli A, Longo T, Basile A. CO-free hydrogen production by steam reforming of acetic acid carried out in a Pd-Ag membrane reactor: the effect of co-current and counter-current mode. Int J Hydrogen Energy. 2008;33:4091–6.
68. Iulianelli A, Dalena F, Liguori S, Calabrò V, Basile A. Membrane reactors for steam reforming of glycerol and acetic acid, Ch. 9, In: Basile A, Hai F, Di Paola L, Piemonte V, editors. Membrane reactors for energy applications and basic chemical production. Woodhead; Amsterdam. 2015. pp. 249–266.
69. Mironova EY, Ermilova MM, Orekhova NV, Muraviev DN, Yaroslavtsev AB. Production of high purity hydrogen by ethanol steam reforming in membrane reactor. Catal Today. 2014;236:64–9.
70. Hedayati A, Le Corre O, Lacarriere B, Llorca J. Exergetic study of catalytic steam reforming of bio-ethanol over Pd-Rh/CeO_2 with hydrogen purification in a membrane reactor. Int J Hydrogen Energy. 2015;40:3574–81.
71. Murmura MA, Patrascu M, Annesini MC, Palma V, Ruocco C, Sheintuch M. Directing selectivity of ethanol steam reforming in membrane reactors. Int J Hydrogen Energy. 2015;40:5837–48.

Index

A

Ablative pyrolysis, 126
Absorption, 8, 111, 151, 293, 317, 329, 333
Acetate, 5, 17, 25, 27, 30, 55, 61, 251, 252, 258, 260–263, 265, 266, 268, 269, 320, 321
Acetic acid, 13, 14, 16–18, 25, 27, 62, 127, 135, 347, 351
Acidified water, 263, 264
Acidogenesis, 61
Acid-thermal, 46, 47
Activated carbon, 53, 110, 111, 124, 155–157, 159, 186, 206, 208, 317
Active metal, 126, 127, 137, 170, 235
Activity, 7, 14, 22, 51, 55, 58, 59, 88– 90, 126–128, 131, 132, 136, 137, 139, 150, 155–157, 165, 201–208, 210, 227, 230, 233–235, 239–241, 263, 264, 271, 297, 324
Adsorption, 110–111, 205, 241, 317, 340, 344, 345
Adsorption enhanced reforming (AER), 110–111
AEM. *See* Anion exchange membrane (AEM)
AFEX. *See* Ammonia fiber explosion (AFEX)
Agitated granular sludge bed reactor (AGSBR), 54
Agricultural residues, 9, 11, 12, 16, 17, 19, 22, 43
Agriculture, 42, 55
Alkali catalyst, 201–205, 216
Alkaline electrolyzer, 225–227, 231
Alkaline-pretreated, 19, 24, 263
Alkaline pretreatment, 18, 19, 22, 26, 47, 48
Alkali-thermal, 47
Allothermal, 99, 101, 103, 107, 108, 299
Alternative, 26, 30, 51, 78, 88–91, 127, 129, 134, 136–141, 149–151, 159, 165, 167, 169, 180, 232–235, 256, 260, 261, 272, 274, 284, 285, 290, 291, 304, 306, 316, 339, 340
Ammonia, 48, 167, 180, 253, 290
  production, 152, 168, 169
  treatment, 255
Ammonia fiber explosion (AFEX), 49, 50
Ammonia recycled percolation (ARP), 50
Anaerobes, 5, 6, 52
Anaerobic, 5, 6, 12, 24–26, 42, 52–62, 90, 180, 274
  baffled reactor, 252, 253, 262, 263
Anaerobic sequencing batch reactor (ASBR), 57
Anion exchange membrane (AEM), 252, 253, 259
Anode, 225–230, 234, 250–255, 258–261, 263–268, 270–274
Anodic, 235–237, 239–241, 251, 254–256, 258, 268–272, 340
  bacterial community, 268–270
Anolyte, 264, 272
Application, 11, 19, 23, 26, 27, 29, 30, 48, 51, 54, 63, 83, 88, 91, 97, 98, 105, 141, 150, 152, 154, 156, 167–169, 174, 180, 182, 226, 229–233, 253–255, 270, 272, 273, 289, 305, 316–319, 321, 324, 325, 329–334, 340, 341, 349, 352
Applied voltage, 251–253, 255, 256, 259–261, 263
Aqueous phase, 86, 123, 140
Aqueous phase reforming, 140
Arabinose, 16, 20

Z. Fang et al. (eds.), *Production of Hydrogen from Renewable Resources*, Biofuels and Biorefineries 5, DOI 10.1007/978-94-017-7330-0

Architecture, 251–253
Assessment
  economic, 113, 231
  technical, 229–231
Auger reactors, 126
Autothermal, 99–102, 104, 140
Autothermic regime, 139

B

Bacteria, 5–8, 18, 23, 24, 27, 28, 30, 42, 52, 53, 55, 58, 59, 180, 249–251, 255, 256, 262, 263, 265, 266, 268–273
Bacterial
  community, 268–270
  growth, 42, 55, 58
Bagasse, 7, 9, 19, 20, 22, 24, 25, 43, 46, 47, 128, 130, 131
Barley straw, 9, 13, 15–18, 26, 29
Barrer, 317, 321–323, 325, 331, 334
Batch, 8, 20, 21, 24, 28, 30, 44, 46, 47, 55, 57, 59–61, 128, 130, 132, 133, 211–215
Biocatalytic conversion, 249
Biocatalyzed, 250
Biocatalyzed electrolysis, 250
Biocathode, 256
Biochar, 121, 124
Biochemical, 27, 53, 61, 79, 115, 341
Biocompatibility, 254, 255, 271
Bioconversion, 90
Biodegradability, 45
Biodegradable, 43, 187
Biodiesel, 57, 152, 169, 170, 261, 262
Bioelectrochemical cell (BEC), 249, 250
Bioelectrochemical hydrogen production, 249–274
Bioethanol, 19, 22
Biofilm, 8, 54, 255, 264, 265, 272
Biofouling, 273
Biofuels, 4, 12, 13, 19, 27, 29, 42, 167–169, 210, 227
Biogas, 26, 29, 55, 61–63, 150, 151, 167, 273, 274
BioHPR, 107
Biohydrogen, 5, 10–12, 17, 22, 23, 27, 29, 30, 41–63, 77–91, 250, 251
Biological, 4, 6, 8, 18, 24, 25, 29, 42, 45, 50–51, 53, 58, 60, 151, 180, 210, 253, 256, 266, 342
Biomass, 3–30, 43, 45, 46, 48, 49, 51, 53–55, 58, 59, 62, 63, 83–85, 97–115, 119–141, 150, 152, 165, 166, 168, 179–216, 223, 230, 261, 264, 290, 292, 293, 305, 347
Biomethane, 45, 55, 58, 63
Biomimetic cofactor, 88, 89
Bio-oil, 120, 121, 123–125, 127–129, 134–141
Bioreactor, 23, 30, 54–56, 59, 62
Biorefinery, 22
Bioremediation, 250
Bipolar membrane (BPM), 259
Bovine serum albumin (BSA), 262
Breaking, 45, 51, 183
Bubble effect, 226
Buffer capacity, 263, 272
Butyric acid, 261, 262

C

Cage effect, 192
Calcination, 131
Caldicellulosiruptor saccharolyticus, 6, 8, 12, 17, 28, 46
Carbohydrate(s), 5, 7, 8, 10, 26, 29, 42–45, 51, 52, 58, 62, 77–91, 124, 135, 180, 263
Carbon
  brush, 254
  cloth, 254, 255, 257, 258, 260, 265
  compound, 261, 265
Carbonaceous
  catalyst, 155, 157, 159–161, 170
  materials, 155, 157, 254, 255, 271
Carbon capture and sequestration (CCS), 150, 151, 153, 154, 166, 167, 232
Carbon dioxide (CO_2), 42, 84, 101, 150–152, 168, 169, 182, 185, 186, 196, 199–206, 208, 209, 273, 289, 291–293, 302, 304, 306, 307, 316
Carbon dioxide electrolysis, 232
Carbon felt, 253, 254, 258, 260
Carbon monoxide, 79, 98, 102, 103, 105, 110, 111, 124, 129, 131–133, 185, 192, 196, 198, 199, 202–210, 227, 232, 284, 290, 292, 293, 301, 304, 305, 316, 317, 321, 326, 327, 332, 333, 345, 346, 349, 351
Carbon nanotubes, 156, 157, 171, 254–256, 258, 330, 331
Carrier
  leakage, 320, 325, 334
  saturation, 329
Carrier-induced granular sludge bed reactor (CIGSB), 54
Carrot pulp, 9, 21, 24
Cassette style, 252, 253, 259
Catalyst
  deactivation, 129, 136, 137, 139, 141, 160, 171
  weight/biomass flow rate, 131

Catalytic
  cracking, 131, 134, 138–140
  cracking of bio-oil, 139–140
  pyrolysis, 119–141
  reforming of bio-oil, 120, 127, 129, 134–140
  SWG, 192, 193, 195, 197
Cathode, 225–230, 232, 241, 250–253, 255–261, 263, 265, 266, 268, 271–274
Cathode-on-top, 252
Cathodic, 239, 240, 255–258, 263, 265, 268, 272
  recovery efficiency, 255, 256
Catholyte, 263
Cation exchange membrane (CEM), 250, 252, 253, 259
CBM. *See* Cellulose-binding module (CBM)
CCS. *See* Carbon capture and sequestration (CCS)
Cellulase, 10–12, 22, 48, 85
Cellulose, 4, 7, 10–12, 14, 21–23, 26, 43, 45, 48–51, 83–85, 121, 122, 124, 184, 188, 190, 198, 202, 205–207, 213, 261, 262
Cellulose-binding module (CBM), 86, 87
CEM. *See* Cation exchange membrane (CEM)
Ceramic, 105, 106, 292, 317, 318, 342, 343
Ceria, 233, 234, 297, 298, 302, 304
Challenge, 4, 7, 8, 19, 22, 27–29, 42, 45, 78, 79, 104, 105, 139, 151, 152, 155, 160, 171, 187, 188, 205, 210, 216, 230, 235, 242, 253, 270–274, 284–285, 290, 291, 294, 295, 298, 299, 303–307, 318, 329
Chamber, 99, 107, 251–254, 258–260, 266, 267, 269, 273, 274, 302
Char, 121–126, 133, 140, 181, 186, 187, 196, 197, 199–201, 205, 207, 216
Charcoal, 101, 124
Cheese whey, 44, 188
Chemical
  looping, 138
  stability, 173, 254, 255
Chemical oxygen demand (COD), 21, 23, 44, 55–57, 199, 200, 210, 262, 264, 265
Clostridium butyricum, 6, 7, 20, 46, 52
Clostridium thermocellum, 6, 21, 23, 24, 46
Coal gasification, 114, 150, 152, 153, 165, 293
COD. *See* Chemical oxygen demand (COD)
Co-electrolysis, 232–233, 242
Cofactor, 79, 80, 85, 88–91
  recycling system, 88
Coffee drink, 44, 54
Coke
  deactivation, 126
  formation, 126, 138, 155
Collection, 8, 45, 238, 285
Combustion, 78, 98, 99, 107, 113, 122, 125, 126, 139, 150, 151, 180, 182, 284, 290, 292, 304, 339
Commercialization, 11, 30, 210, 216, 227, 230, 256, 274, 306, 307, 318, 320, 328, 340
Complex, 10, 28, 43, 50, 51, 88–90, 100, 121, 123, 124, 135, 171, 188, 192, 194, 261, 263, 265, 266, 270, 271, 274, 294, 347
Concentrated acid, 10, 13, 46
Concentrating solar technologies, 285
Condensed molasses fermentation soluble (CMS), 43
Conductivity, 172, 181, 230, 233, 234, 238, 254–256, 263, 264, 272, 299
Contamination, 6, 318
Continuous, 8, 20, 21, 30, 44, 46, 53–62, 130, 131, 139, 160, 162–164, 171–174, 184, 191, 252, 253, 261, 263, 264, 326–328
Continuously stirred anaerobic bioreactor (CSABR), 54, 56, 62
Continuous stirred tank reactors (CSTR), 8, 53–57, 59
Conventional, 54, 104, 112, 130, 133, 134, 151, 180, 181, 186, 224, 284, 290, 292, 304–306, 316, 322, 327, 340, 347, 349, 351
Conversion, 4, 8, 11, 18, 23, 27, 43, 51, 61, 78, 85, 91, 98, 104, 107, 111, 115, 120, 129, 136, 137, 155, 156, 158–161, 173, 181, 182, 186, 188, 189, 192, 194, 197, 206, 224, 227, 249, 251, 261, 265, 285, 293–295, 303, 304, 326, 327, 346, 348–350
Conversion efficiency, 78, 91, 161, 251, 261, 265
Coprecipitation, 131
Coproducts, 4, 5, 9, 26–27
Corncob, 43
Corn stalks, 15, 17
Corn stover, 9, 11, 17, 21, 24, 46, 47
Cost, 5, 6, 8, 11, 22, 28–30, 42, 43, 45, 50, 51, 79, 83–91, 105, 107, 109, 111, 113, 115, 127, 129, 139, 140, 150–152, 158, 159, 165–167, 170, 187, 205, 206, 226, 227, 230, 231, 238, 242, 254–256, 259, 260, 271, 273, 274, 306, 318, 333, 334, 340, 343, 345, 346, 349
Cost-effective, 6, 86, 152, 159, 256, 260, 274, 334

Cost of electricity (COE), 333
Cracking, 139–140
Critical point, 181–184
Crude glycerol, 56, 262
Cryogenic, 48, 109
Crystalline cellulose, 43
Current, 4, 22, 49, 76, 86, 88, 89, 150, 152, 166, 172, 187, 226, 227, 229–233, 235–238, 240–242, 251, 252, 254–256, 258, 259, 263, 264, 284, 305–307, 340
  recovery, 252

D

Dairy wastewater, 44
Dark fermentation, 4–7, 10, 28, 30, 42, 79, 83, 180, 251, 261, 262
Dark fermentative, 3–30, 53, 55, 58, 62
Dark fermentative hydrogen production (DFHP), 3–30, 53, 55, 58, 62
Dark hydrogen fermentation, 4–23, 29
Decarbonization, 149–174
Decarboxylation, 53
Degradable, 262, 265, 271
Degradation, 13, 14, 16, 18, 29, 43, 48, 51, 181, 189, 194, 200, 230, 232, 235–242, 343
Delamination, 235–237, 239–241
Delignification, 9, 12, 19, 22, 48, 51
Densifying, 45
Density, 18, 25, 48, 78, 80, 123, 124, 152, 154, 173, 183, 184, 191, 223, 224, 231, 235, 237, 242, 254–256, 258, 259, 263, 320, 340, 346
Design, 28, 62, 80–82, 84, 88, 101, 105, 107, 109, 127, 160, 162, 164, 167, 172–174, 187, 193, 210, 240, 252, 253, 258–260, 273–274, 317, 332, 334, 340
Destruction, 45
Detoxification, 50
Devolatilization, 123
DFHP. *See* Dark fermentative hydrogen production (DFHP)
Dielectric constant, 181, 183, 184, 191
Digester, 60, 61
Digestion, 45, 58, 90, 274
Dilute acid, 11, 13, 16–18, 29, 48–50
Dilute-acid pretreatment, 13, 16–18, 29, 48
Dish, 173, 286–289, 306
Disruption, 11, 19, 49, 330
Dissolved oxygen (DO), 264, 265
Diversity, 268, 270–271
Domestic wastewater, 253, 262, 263, 268, 269
Drying, 98, 104, 109, 181, 186, 210
Dry reforming of bio-oil, 138
Durability, 106, 127, 230, 235, 239–240, 242, 254, 255, 330
Dye sensitized solar cell, 260, 261

E

Economic, 4, 6, 13, 26, 29, 62, 63, 113–115, 120, 139, 150–152, 164–167, 171, 172, 181, 201, 231, 255, 263, 272, 319, 332, 334
  efficiency, 272
  feasibility, 13, 62, 255, 272, 319, 333, 334
Efficiency, 4, 8, 23, 26, 28–30, 43, 54, 55, 59, 78, 79, 86, 91, 106, 115, 139, 141, 150, 153, 161, 168, 180, 181, 185–187, 199–201, 203, 205, 210, 224, 226–230, 232, 233, 235, 251, 255–258, 260, 261, 265, 272, 285, 295, 297, 303, 305–307, 317, 332, 340
Effluent, 28, 30, 43, 48, 53, 56, 61, 186, 190, 192, 193, 197, 198, 201–203, 206, 262, 263, 265, 268
Electrical energy, 224, 228, 229, 249, 340
Electricity, 78, 79, 91, 150, 160, 166, 180, 224, 225, 227–232, 242, 260, 261, 284–286, 288–291, 305, 333, 339, 340
Electrochemically active bacteria (EAB), 249, 250, 256, 262, 263, 265, 266, 268–273
Electrode
  arrangement, 258
  resistance, 236, 238, 271–272
  size, 258
  spacing, 258
Electrolysis, 79, 114, 150, 152, 156, 166, 167, 169, 223–242, 250–252, 254, 257, 259–262, 269, 290, 291, 306
Electrolyte, 225–227, 229, 230, 233–242, 251, 263–264, 270–273
Electron, 5, 27, 28, 45, 48, 52, 53, 58, 59, 88, 105, 158, 159, 164, 226, 249–251, 255, 268, 270, 271, 273, 320
  acceptor, 5, 250
Electronic conductivity, 254, 255
Energy
  consumption, 26, 78, 180, 181, 187, 224–226, 284
  conversion efficiency, 78, 91, 161
  crops, 4, 8, 9, 22, 42
  efficiency, 29, 79, 139, 150, 153, 161, 168, 187, 210, 226, 229, 230, 285, 317
Engineering, 62, 88, 90, 160, 250

Entrained bed, 99, 100
Environmental sensors, 250
Enzymatic
  hydrolysis, 10–12, 23, 45, 47, 48, 51
Enzyme
  immobilization, 86
  purification, 86
  recycling, 87, 88
  treated, 47
ESR. *See* Ethanol steam reforming (ESR)
Ethanol, 5, 11–14, 19, 27, 29, 42, 61, 185, 188, 190, 193, 195, 197, 198, 206, 207, 211, 213, 262, 263, 347–349, 351, 352
Ethanol steam reforming (ESR), 347–349, 351
External
  circuit, 251, 273
  power, 251, 260
Extraction, 27, 55
Extractives, 121

F
Facilitated transport membrane, 315–335
Failure model, 235
Fast pyrolysis, 120, 122–125
Fats, 27, 42, 43, 169, 180
Feedstock, 12, 42–51, 55, 62, 100, 113, 123, 125, 128, 132, 135, 136, 139, 168, 181, 186–189, 194, 196–200, 202, 204–206, 210–214, 216, 223, 232, 242, 262, 290, 292, 301, 304, 305
Fermentability, 4, 8, 13–15, 17–19, 22, 24, 26, 30
Fermentable, 4, 8, 10, 17, 23, 29, 271
  sugars, 18, 22, 26, 45
Fermentation, 4–10, 12–19, 21–30, 41–63, 79, 80, 83, 88, 90, 150, 152, 180, 251, 261–263, 265
Fermenter, 51, 53–55, 58, 60, 62, 63
Ferrites, 295, 297, 302, 304
Ferrous, 58
Fick law, 345, 346
Field application, 270
Fixed bed, 99, 100, 127, 130, 135–139, 160, 302, 347
Fixed bed reactor, 135–139, 160, 347
Fixed-site carrier, 319, 321–324, 330, 334
Flame oxidation, 255
Fluidized bed, 54, 99, 100, 110, 111, 124–126, 128, 130–133, 136–138, 160, 171, 173, 191, 293, 294, 301, 302
Fluidized-bed reactor, 160, 171
Food processing, 43, 180
Food waste, 7, 43, 44, 55, 57, 61, 62
Forestry residues, 4, 10, 45
Formate, 5, 53
Formic acid, 14, 16, 185, 262
Fossil fuels, 78, 104, 113, 150, 151, 167, 168, 180, 210, 223, 230–232, 284, 285, 290–293, 304–306, 339, 352
Fractionation, 10, 11, 22, 123, 136, 137
Fresnel, 286–289, 306
Fructose, 20, 80–82, 84
Fuel cell, 63, 78, 88, 91, 97, 98, 105, 107, 141, 151, 180, 224, 227, 228, 239, 249–252, 260, 261, 268, 269, 316, 319, 321, 332, 339–352
Full-scale, 274
Fungi, 50
Furfural, 13–17, 49

G
Gas crossover, 273
Gas diffusion, 227, 252, 253, 258
Gasification, 6, 78, 97–115, 122, 129, 131, 137, 140, 150, 152, 153, 155, 165–167, 179–216, 223, 292, 293, 304, 305, 316
Gas yield, 62, 122, 131, 132, 136, 190, 194, 196, 198–200, 202, 204, 207–209
Gibbs free energy, 228
Glassy polymer, 318
Gluconeogenesis, 80
Glucose, 5, 7, 11, 12, 16, 20, 21, 23–25, 27, 43, 48, 59, 62, 79–85, 188, 190, 193, 196–199, 202–206, 208, 211, 213, 214, 251, 261, 262, 265, 268, 271
Glycerol, 56, 87, 184, 188, 190, 198, 203, 204, 211, 214, 261, 262, 347, 351
Granule, 54, 253–255, 258
Graphite
  felt, 253, 254, 257, 266
  fiber brush, 253, 254, 263
  granule, 253, 254, 258
  plate, 254
  rod, 254
Grass, 9, 43, 194
Grinding, 11, 302

H
Heat treatment, 255
Heliostat, 165, 167, 286, 289, 306
Hemicellulose, 4, 11–14, 16, 19, 22, 26, 45, 48–51, 84, 121, 122, 124, 188, 189

Heterogeneous reactions, 101, 102
Hexose, 5, 7, 13, 17, 19, 23, 24, 27, 28, 44, 46, 47, 54–57, 83, 84
High pressure, 11, 49, 104, 110, 111, 113, 184, 185, 191–193, 317, 319, 328–334, 345, 346
High temperature electrolysis, 229–231
HMF. *See* 5-hydroxymethylfurfural (HMF)
Homogenous reactions, 102, 183
HPR. *See* Hydrogen production rate (HPR)
HRT. *See* Hydraulic retention time (HRT)
H-type, 252, 256, 258
Hydraulic retention time (HRT), 6, 8, 51, 53, 54, 56, 57, 59, 61–62, 253, 264, 265
Hydrodynamic, 53, 55, 58, 264, 265
  force, 264, 265
Hydrogen
  bond, 183, 198, 320, 330
  cost, 139, 167, 321
  economy, 91, 150, 152, 316
  electrode, 227, 233–235, 241–242
  evolution reaction, 256
  fuel cell vehicles, 78, 88
  production, 3–30, 42–46, 51–62, 78–80, 83–90, 97–115, 119–141, 149–174, 179–216, 223–242, 249–274, 283–307, 339–351
  productivity, 23, 29
  purification, 134, 251, 332, 340
  recovery, 255, 256, 349, 351
  yield, 4–8, 10, 12, 13, 17, 19, 20, 23–30, 45, 54, 59, 62, 106, 109, 120, 127, 132, 134, 137, 188, 190–196, 198–205, 207–209, 252, 254, 259–262, 264–267, 298, 348, 349, 351
Hydrogenase, 53, 58, 59, 80, 82, 83, 89
Hydrogen peroxide, 27, 199, 200
Hydrogen-producing microflora, 59
Hydrogen production granule (HPG), 54
Hydrogen production rate (HPR), 6, 12, 24, 53–55, 58–62, 78, 80, 83, 85, 86, 88, 226, 252, 253, 255, 256, 258, 259, 261–267
Hydrogen production yield (HY), 25, 54, 55, 60, 79, 254, 260, 264, 265
Hydrogen-selective membranes, 134
Hydrolysis, 10–14, 16, 17, 23, 24, 26, 45–51, 83, 184, 185, 191, 192, 322
Hydrothermal, 49, 185, 187, 190, 191, 193, 194, 200, 202, 203, 205, 210, 211
  gasification, 185, 190, 191, 193, 194, 200, 202, 203, 210, 211
5-hydroxymethylfurfural (HMF), 13–17

I

IEM. *See* Ion exchange membrane (IEM)
IGCC. *See* Integrated gasification combined cycle (IGCC)
Immobilization, 53, 54, 86, 88
Immobilized cells, 54, 62
Infiltration, 234
Inhibitors, 12–14, 16, 17, 19, 22, 29, 30, 45, 48–51, 87, 271
Inoculation, 255, 271
Inoculum, 61, 255, 268, 269
Inorganic constituents, 121
Inorganic membrane, 317, 318, 341, 343
Integrated gasification combined cycle (IGCC), 316, 319, 329, 333
Internal combustion engine, 78, 98, 180, 304
Internal rate of return (IRR), 63, 167
In vitro synthetic biosystem, 79, 81, 82, 85, 90
In vitro synthetic (enzymatic) pathway, 77–91
Ion exchange membrane (IEM), 250–253, 259, 273, 320, 321
Ionic liquids (IL), 11, 22, 49, 50, 320
Ionic Product, 184, 185, 191, 192
Irradiation, 45, 48, 160, 161, 165, 271, 285, 286, 290, 295, 299, 302

L

Lab-scale, 55, 137, 266, 267, 270, 271, 300
Lactic acid, 24, 185, 261, 262
$La_{0.6}Sr_{0.4}Co_{0.2}Fe_{0.8}O_{3-\delta}$ (LSCF), 234, 235, 237–239
Levoglucosan, 124
Levulinic acid, 14, 16
Lignin, 10–14, 19, 22, 26, 27, 29, 43, 45, 48, 50, 51, 121, 122, 124, 135, 188, 199, 202, 205–208, 211, 214, 215
Lignocellulose, 14, 41–63, 262, 263
Lignocellulosic materials, 13, 16, 43, 45, 48, 49
Lime, 19, 24, 48, 50
Liquid fuels, 97, 150, 168, 232, 242
Long term stability, 127, 128, 307, 320, 321
Low-carbon, 150–152, 169
Low pressure, 105, 109, 184, 193, 319, 321–329, 331, 332, 334

M

Maize, 46
Mass transfer, 53, 80, 104, 181, 185, 186, 210
Material, 4, 7–13, 16–19, 23, 24, 26, 29, 30, 43, 45, 48–50, 58, 62, 83, 84, 86, 105, 110, 113, 120, 121, 125, 127, 138, 150, 152,

154, 155, 157, 168, 169, 174, 184, 185, 187, 188, 205, 207, 210, 216, 224, 227, 229, 230, 233, 234, 238, 241, 242, 254–258, 271, 272, 274, 284, 291, 293, 295–300, 302–307, 318, 334, 341, 343, 348, 350
MEC. *See* Microbial electrolysis cell (MEC)
Mechanical, 11, 46, 48, 107, 126, 136–138, 170, 173, 300, 330, 331, 343, 345
  strength, 138, 170
Mediator, 250, 251
Membrane, 8, 55, 80, 98, 127, 227, 250, 293, 317, 339
Membrane bioreactor (MBR), 55, 59
Membrane electrode assembly (MEA), 252
Membrane reactor, 53, 107, 127, 317, 326–327, 332–334, 340–348, 350–352
Mesophilic, 5, 7, 21, 27, 44, 46, 52, 60–62, 86
Metabolic
  compounds, 52
  pathway, 7, 58, 59
  processes, 268
Metabolism, 5, 42, 53, 79
Metal catalyst, 126, 204, 205, 207, 208, 235, 272
Metallic membrane, 105, 106, 318, 343
Metal oxides, 131, 228, 295, 304
Metals, 58–59, 78, 107, 126, 127, 131, 137, 173, 174, 182, 186, 201, 204–210, 227, 228, 232, 233, 235, 256, 272, 292, 294–297, 304, 342–344
Methanation, 111, 188, 189, 194, 196, 198, 205–207
Methane, 8, 27, 52, 59–63, 97, 102, 107, 113, 114, 133, 139, 149–174, 185, 186, 189, 190, 192, 196, 198, 203, 205–210, 230, 255, 259, 264, 267, 271, 273, 274, 289, 292–294, 316, 333
Methane production rate (MPR), 55
Methane production yield (MY), 55
Methanogen, 5, 8, 52, 59, 61, 255, 259, 264, 267, 268, 270, 271, 274
Methanol, 129, 140, 167, 169, 180, 184, 188–190, 192, 193, 195, 197, 198, 211, 215, 304, 347, 349–352
Methanol steam reforming (MSR), 349–351
MFC. *See* Microbial fuel cell (MFC)
Microbes, 53, 86, 90
Microbial electrolysis cell (MEC), 79, 250–274
Microbial fuel cell (MFC), 249–252, 260, 261, 268, 269
Microorganism, 4–6, 8, 12, 14, 18, 20, 22, 23, 27, 29, 30, 43, 44, 46, 47, 50–53, 56, 57, 60, 62, 79, 180, 249, 255, 256, 259, 264, 268, 273, 274
Microporous membrane, 259
Microturbine, 97, 98
Microwave plasma, 120, 132, 133
Microwaves, 45, 48, 49, 120, 128, 132, 133, 141, 164
MIEC. *See* Mixed ionic-electronic conductor (MIEC)
Milling, 11, 27, 45, 48
Miscanthus, 9, 19, 20, 24, 26
Mixed culture, 5, 6, 16, 20, 24, 44, 50, 56, 57, 59, 60, 255, 256, 268
Mixed ionic-electronic conductor (MIEC), 234, 237–238
Mixed matrix membrane, 317, 330–331
Mixing pattern, 53
Mobile carrier, 319, 321–325, 331, 334
Model compound, 127, 128, 135, 137
Modified, 18, 84, 131, 255, 260, 272, 304, 317
Molten alkali reactor, 132
MSR. *See* Methanol steam reforming (MSR)
Multi-electrode, 252, 253
Multi-step processes, 120, 121, 124, 129, 134–141
Multiwalled carbon nanotube (MWNT), 156, 157, 258, 331
Municipal solid waste, 43, 62, 121
MWNT. *See* Multiwalled carbon nanotube (MWNT)

N

NADH, 90
NADPH, 80–83
Nanoparticle, 59, 86, 87, 234–237, 240, 241, 255, 256, 272, 331
Nanostructure, 234, 240, 255–257
Nanowire photocathode, 260, 261
Natural gas, 4, 63, 78, 79, 96, 98, 113, 114, 135, 150–152, 160, 163, 165–169, 180, 181, 223, 285, 290, 292, 304, 352
Net energy gain, 43
Neutral pH, 272
Nickel, 58, 126, 155, 160, 186, 204–206, 208, 213, 215, 216, 226, 241, 256
  catalyst, 205, 206
  cermet, 234, 241
  foam, 256, 257, 261, 266
Nickel-based catalyst, 127, 128, 130, 131, 136, 138, 155, 172, 205, 306

Nitrogen, 43, 58, 88, 124, 125, 128, 129, 168, 182, 186, 187, 205, 255
Noble metals, 126, 127, 232
Noncorrosive, 205, 210, 254
Non-volatile, 296–298, 303, 320
Nuclear power, 229
Nutrients, 28, 58

O

OFMSW. *See* Organic fractions of municipal solid wastes (OFMSW)
Oil, 42–45, 56, 136, 137, 141, 150, 152, 167–169, 171, 199, 285, 288
Olive pulp, 44, 45
One-step pyrolysis, 133, 134
On-site, 63
Optimization, 22, 41, 77, 83, 85, 188, 253, 285
Organic fractions of municipal solid wastes (OFMSW), 43, 62
Organic solid wastes, 43
Organic wastes, 42, 43, 62, 63, 187
Organosolv, 12, 19, 26, 50, 51
Oxidative steam reforming, 138
Oxides, 127, 131, 170, 179, 228, 233, 234, 295, 297, 298, 304, 343
Oxidizing agents, 48, 122, 164, 170
Oxygenated compounds, 123
Ozonolysis, 48, 50

P

Palladium, 105, 107, 111, 186, 204, 208, 209, 318, 343–347, 351
Palm oil mill effluent (POME), 43, 56
Parabolic trough, 286–289, 306
Partial oxidation (PROX), 99, 102, 114, 138, 140, 199, 200, 230, 293, 340
Pentose phosphate pathway (PPP), 80
Performance, 51, 54–56, 61, 87, 104, 120, 126, 127, 131, 151, 168, 171, 172, 229–233, 252–254, 252–265, 271–274, 296, 307, 316–334, 341, 345, 346, 348, 351
  degradation, 230, 235–242
Permeability, 259, 317, 321–323, 325, 329–331, 334, 343–345
Permeance, 317–319, 324, 329, 331, 334, 346
Perovskites, 297, 304
pH, 5–8, 14, 22, 48, 50, 51, 55, 57, 59–60, 79, 87, 124, 187, 201, 251, 259, 263, 264, 272, 273, 342
  imbalance, 273
Phenolic compounds, 13, 14, 16, 17, 29, 124, 135
Phosphate buffer, 263
Phosphorous (P), 58
Phosphorylation, 80, 82
Photo-fermentation, 79
Physical, 24, 45, 48, 49, 125, 140, 173, 182–184, 191, 317
Physicochemical, 45, 49
Pilot-scale, 61, 62, 253, 259, 263, 266, 267, 305, 327–329
Platinum, 126, 186, 204, 208, 255, 318
Poisoning, 126, 206, 238, 239, 272, 332, 340
Polarization, 226, 230, 235–241
Polyelectrolyte, 320, 321
Polymer, 45, 105, 182, 227, 318–323, 329–331, 334
Polymerization, 11, 19, 45, 139, 186, 196
Polymer membrane, 328, 330
POME. *See* Palm oil mill effluent (POME)
Potassium hydroxide, 186, 201–203, 213, 215, 226, 321–323, 325, 331
Potato
  processing wastewater, 263
  waste, 45
Potential, 4, 5, 10–12, 14, 17, 19, 22, 26, 43, 53, 58, 61–63, 86–88, 91, 115, 121, 126, 127, 159, 166, 180, 181, 210, 226, 227, 235–237, 239, 241, 251, 262, 263, 274, 284, 285, 290, 291, 298, 306, 316, 317, 319, 321, 324, 332–333, 349
  loss, 272, 273
Power supply, 254, 260–261
Practical application, 29, 30, 83, 154, 253, 254, 272, 273, 318
Precious metal, 256, 272, 292
Pressure, 6, 8, 11, 22, 49, 101, 104, 105, 108–111, 113, 123, 127, 130, 133, 134, 137, 140, 160, 182–197, 199–201, 204, 207, 209, 211–214, 233, 236, 241, 303, 317, 319–334, 340, 341, 343–347, 351
Pressure swing absorption (PSA), 109–110, 134, 302, 317, 340
Pretreatment, 4, 9–24, 26, 29, 30, 43, 45–51, 61, 62, 186, 210, 266, 271
Primary effluent, 262
Product gas, 7, 97–99, 101–104, 107, 110, 111, 129, 131, 133, 134, 136, 138, 140, 181, 185, 186, 207, 208
Production rate, 6, 20, 24, 44, 46, 47, 53–62, 78–80, 83–86, 88, 181, 226, 227, 252, 253, 255–259, 261–267

Promoter, 205, 206
Propionate, 5, 61
Propionic acid, 24, 28, 60, 261, 262
Protein, 5, 26, 43, 86, 87, 261, 262, 271
Proton, 53, 58, 225, 227, 233, 251, 259, 264, 270, 272, 273, 316, 325
Proton exchange membrane (PEM) electrolyzer, 225, 227
PROX. *See* Partial oxidation (PROX)
PSA. *See* Pressure swing absorption (PSA)
Pulsed-electric-field (PEF), 50
Pulverizing, 45
Purity, 12, 19, 26, 78, 79, 105, 107, 109, 111, 126, 134, 181, 224, 226, 227, 251, 256, 259, 273, 274, 316, 317, 340, 348–351
Pyrolysis, 98, 99, 114, 119–141, 153–161, 166, 167, 169, 181, 186, 290, 293
Pyrolytic lignin/organic phase, 124
Pyruvate, 53

R

Reaction, 6, 51, 79, 98, 121, 150, 180, 224, 251, 289, 316, 340
Reaction-diffusion, 319, 320, 329
(Pyrolysis) reaction pathways, 122–123
Reactor
  configuration, 30, 134, 136–138, 253
  design, 98, 160, 210, 252, 253, 293, 294, 305, 306
Redox, 58, 88, 89, 263, 294–297
  materials, 295, 297, 298
Reduction, 4, 11, 12, 16, 30, 45, 49, 58, 105, 111, 122, 127, 131, 138, 149, 151, 167, 168, 198, 208, 210, 233, 241, 251, 296–298, 302–304, 330, 333, 334, 345, 346
Reformer, 106–108, 111, 136, 138, 139, 168, 289, 292, 333, 349
Reforming, 29, 78, 98, 102–104, 107, 109–111, 113, 114, 120, 126–129, 134–141, 150, 152–154, 165–169, 181, 186–190, 192, 193, 195–198, 206, 207, 223, 230, 289–293, 304, 316, 333, 340, 347–351
Regeneration, 80, 82, 88, 125, 128, 139, 160, 171–173, 294, 296, 302, 317
Renewable, 4, 29, 43, 84, 85, 90, 120, 141, 150, 151, 165, 180–182, 223, 224, 226, 230–232, 242, 260, 274, 284, 285, 290, 291, 293, 304–307, 339, 347–352
  energy, 151, 181, 223, 224, 231, 232, 242, 285, 291, 339
  technologies, 181, 230, 232, 290, 291, 339
Residence time, 101, 120, 123, 124, 126, 128, 161, 188, 189, 193–196, 198–200, 203, 204, 210, 351
Resistance, 59, 126, 131, 136, 137, 181, 226, 229, 230, 232, 234–236, 238, 251, 253, 258, 259, 263, 266, 271–272, 318, 320, 324, 334, 343, 345, 346
Resistivity, 271
Reticulated vitreous carbon, 254, 255
Reverse electrodialysis stack, 261
Reversible operation, 239–240
Rhodium, 186, 204, 208
Rice
  husk, 47, 128, 130–132, 137, 188
  straw, 9, 13, 17, 20, 21, 46, 56, 62, 130
  winery, 44
Rotating cone, 126
Rubbery polymer, 318
Ruthenium, 186, 204, 206–209

S

Scale up, 126, 137, 172, 173, 259, 270, 286, 318, 328
Screw reactors, 126
Selectivity, 51, 126, 127, 136, 140, 159, 181, 184, 186, 194, 200, 202, 206, 208–210, 317–323, 325, 329–334, 341, 342, 348, 351
Selexol® process, 317, 333
Self-flocculation, 54
Separation, 8, 19, 23, 43, 48, 55, 60, 85, 90, 104–112, 125, 135, 140, 173, 174, 204, 290, 291, 293, 305, 315–335, 342, 346
Separator, 107, 140, 254, 258–260, 273–274
Sewage sludge, 44, 54, 59, 121, 188, 190, 191, 198, 200, 202, 212, 262
Shredding, 45
Sieverts' law, 318
Simple-structure, 261, 270
Single-chamber, 253, 255, 263, 264, 266, 268, 273–274
Single culture, 5
Size reduction, 45, 48
Sleeve shape, 252, 253
Slow pyrolysis, 122
Sludge, 5, 21, 24, 44, 48, 53, 54, 57, 59–61, 121, 187, 188, 190, 191, 198, 200, 202, 212, 252, 253, 262, 263, 266, 269
SMR. *See* Steam methane reforming (SMR)
Socioeconomic, 151–154

Sodium carbonate (Na_2CO_3), 126, 130, 186, 201–203, 213, 214
Sodium hydroxide, 9, 19, 24
SOEC. *See* Solid oxide electrolysis cells (SOEC)
SOFC. *See* Solid oxide fuel cells (SOFC)
Softwoods, 11, 12, 49, 121
Solar
  assisted, 261
  fuels, 302, 304–305, 307
  hydrogen, 283–307
  powered, 260, 261
  reactors, 161, 284, 294, 298–303, 305–307
Solid oxide electrolysis cells (SOEC), 224, 225, 227–242
Solid oxide electrolyzer (SOE), 224
Solid oxide fuel cells (SOFC), 227–230, 235, 239, 241
Solubility, 19, 184–187, 201, 216, 319, 330, 343
Soluble microbial products (SMP), 61
Solution-diffusion, 318, 319, 329
Spouted bed reactor, 137
Sr-doped LaMnO3 (LSM), 233–237, 239–241
Stability, 22, 50, 85–90, 106, 126–128, 173, 174, 205, 206, 227, 230, 233–235, 237, 238, 240–242, 254–256, 298, 307, 318, 320, 321, 325–326, 330, 331, 334, 343, 346
Stainless steel mesh, 253, 254
Starch, 7, 10, 28, 43, 44, 60, 80, 83–84, 215
Steam
  electrolysis, 232–233
  explosion, 11, 24, 46, 49, 50
  gasification, 102, 103, 129, 137
  pretreatment, 11, 12, 29
  reforming, 29, 104, 109, 120, 127, 128, 140, 150, 152–154, 166, 167, 169, 181, 186, 189, 190, 198, 292, 293, 304, 305, 333, 347–351
  reforming of bio-oil, 129, 135–139
Steam methane reforming (SMR), 113, 114, 152, 167, 316
Sterically hindered, 322
Sterilization, 5, 50
Submersible, 252, 253, 269
Substrate
  channeling, 88, 89
  concentration, 24, 51, 55–58, 83, 88
Sucrose, 7, 20, 25, 43, 54, 55, 59, 60, 80, 83–85, 262
Sugar, 4, 5, 7, 9–12, 14, 17–30, 43–47, 49, 56, 57, 80, 83–85, 88, 91, 124, 184, 269, 270
  beet, 9, 10, 26, 27, 84
  beet pulp, 10, 26, 27
  mobilization efficiency, 26
  yield, 11, 12, 30
Sugarcane
  bagasse, 7, 9, 20, 22, 43, 46
  distillery effluent, 61
Sugary wastewater, 43, 62, 63
Sulfuric acid, 13, 17, 24, 48, 49, 51
Supercritical
  fluid, 182–187
  water, 104, 111, 181–187, 189, 192, 194, 200, 210, 216
Supercritical water gasification (SWG), 6, 104, 114, 140, 179–216
Surface, 11, 45, 54, 101, 106, 107, 126, 132, 136, 138, 155, 157, 159, 170, 171, 173, 187, 205, 208, 216, 226, 236–241, 255, 256, 273, 285, 286, 298, 299, 318, 343–345
  area, 19, 28, 48, 50, 126, 155–157, 254–256, 258, 259, 272, 298
Suspended, 53
Sustainable, 7, 8, 29, 63, 78, 149, 151, 152, 164, 165, 180, 181, 216, 223, 224, 242, 284, 304, 305, 307
Sweet sorghum
  bagasse, 9, 19, 20, 22, 24, 25
  residues, 46
Swine
  manure, 57
  wastewater, 262, 263
Switchgrass, 22, 207
Syngas, 98, 105, 109–111, 115, 138, 230, 232, 242, 290, 291, 293, 302, 304, 307, 316, 317, 331, 333
Synthesis gas, 98, 110, 129, 163, 284, 304, 316
Synthetic wastewater, 54

T

Tars, 99, 100, 110, 111, 123, 181, 185, 186, 210, 216, 293
Tar yield, 131
TEC. *See* Thermal expansion coefficient (TEC)
Temperature, 5, 43, 78, 99, 120, 151, 181, 224, 254, 283, 319, 340
Textile wastewater, 44, 57, 188, 212
Thauer limit, 79

Thermal, 6, 11, 17, 46, 47, 50, 78, 79, 87, 97, 98, 106, 120, 121, 126, 129, 134, 138, 150, 151, 154, 156, 159–161, 163–165, 172–174, 181, 189, 223, 228–230, 233–235, 238, 241, 285, 289, 290, 293, 294, 296, 297, 299, 302, 303, 318, 330, 343–346
  decomposition (of biomass), 78, 120, 121, 223, 293
Thermal expansion coefficient (TEC), 234, 345, 346
Thermochemical
  conversion, 98
  processes, 152, 180, 291–296, 298, 304, 306
Thermodynamic
  barrier, 7, 251, 259, 260
  properties, 229
Thermophiles, 5, 6, 24, 27
Thermophilic, 5–8, 12, 19, 21, 23, 24, 28, 44, 47, 52, 53, 60–62, 86, 256
  microbe, 86
Thermostable enzyme, 85, 86
Thermotoga neapolitana, 6
Titanium, 253, 257, 302
Tolerance, 8, 52, 90, 105, 318
Tower, 166, 173, 286–289, 296, 306
Transfer, 53, 58, 80, 104, 105, 121, 125, 126, 138, 160, 181, 185, 186, 191, 210, 250, 251, 255, 259, 273, 286, 288, 298, 299, 302, 325
Transition metals, 126, 228, 292
Transmission electron microscopy (TEM), 59
Transport, 8, 104, 105, 183, 184, 237, 238, 241, 251, 259, 272, 273, 315–335, 341, 344
Transportation, 45, 91, 139, 151, 180, 242, 300, 304, 306
Treatment, 11, 19, 26, 27, 43, 48, 51, 54, 55, 58, 98, 111, 124, 170, 187, 190, 200, 215, 236, 237, 253, 255, 271, 274, 293, 304
TTN. *See* Turnonumber (TTN)
Tubular, 137–140, 160, 162, 163, 172, 229, 252, 253, 256, 269, 270, 288, 289, 292, 299
Turnover number (TTN), 86, 91
Two-chamber, 250, 252, 258, 261, 263, 266, 267
Two-stage, 58, 60, 110, 137, 138, 262, 327, 332

U
UASB. *See* Upflow anaerobic sludge blanket (UASB)
Ultrasonic, 48, 140
Ultrasound, 48
Ultrastructure, 48
Upflow, 53, 54, 140, 252, 253
Upflow anaerobic sludge blanket (UASB), 54, 56

V
Valeric acid, 261, 262
Vapor pressure, 49, 183, 241
Vapor residence time, 123, 124, 126
VFA. *See* Volatile fatty acids (VFA)
Viability, 139, 150, 152, 159, 167, 171, 172, 231, 272, 325
Volatile, 8, 42, 57, 60, 62, 125, 241, 261, 263, 296–298, 303, 320
Volatile fatty acids (VFA), 8, 42, 60, 61
Volumetric, 23, 62, 78, 252, 256, 299, 302

W
Washout, 8, 53, 54, 62
Waste activated sludge, 48, 262, 263, 266, 269
Wastewater, 5, 23, 42–44, 51, 54–57, 61, 63, 187, 188, 190, 191, 193, 194, 200, 201, 212, 250, 253, 261–263, 267–269, 271, 272, 274
  treatment, 43, 51, 55, 250, 253, 274
Water
  biophotolysis, 251
  content, 26, 121, 123, 186, 210
  electrolysis, 79, 150, 224–232, 251
  retention, 323
  splitting, 78–82, 84, 224, 234, 294–297, 302–304
Water gas shift (WGS) reaction, 98, 104, 108–111, 126, 129, 134, 139, 140, 180, 185, 188–192, 196–198, 201–205, 207, 208, 232, 292, 316, 317, 326–327, 332–334, 340
Wet oxidation, 22, 23, 48
WGS reaction. *See* Water gas shift (WGS) reaction
Wheat
  bran, 10, 26, 27
  straw, 9, 15, 17, 19, 20, 22, 23, 26, 43, 46, 128
Winery wastewater, 44, 253, 262, 263, 268, 269

X

Xylose, 12, 16, 20, 21, 25, 49, 51, 80, 82–85, 189, 190, 195, 212

Y

Yield, 4–14, 16–20, 23–28, 30, 44–47, 49, 51, 54–62, 77–91, 106, 109, 112, 120, 122–125, 127, 129, 131–134, 136, 137, 156, 184, 186–196, 198–209, 252, 254, 259–262, 264–267, 294, 295, 298, 303, 307, 322, 348–351

Y_2O_3-stabilized ZrO_2 (YSZ), 227, 233, 234, 236–241

Yttria stabilized zirconia, 227

Z

Zeolite, 317, 318, 341, 343

MIX
Papier aus verantwortungsvollen Quellen
Paper from responsible sources
FSC® C105338

If you have any concerns about our products,
you can contact us on
ProductSafety@springernature.com

In case Publisher is established outside the EU,
the EU authorized representative is:
Springer Nature Customer Service Center GmbH
Europaplatz 3, 69115 Heidelberg, Germany

Printed by Libri Plureos GmbH
in Hamburg, Germany